HARCOURT BRACE JOVANOVICH COLLEGE OUTLINE SERIES

LINEAR ALGEBRA

Hwei P. Hsu

Fairleigh Dickinson University

Books for Professionals
Harcourt Brace Jovanovich, Publishers
San Diego New York London

Library of Congress Cataloging-in-Publication Data

Hsu, Hwei P. (Hwei Piao), 1930–
Linear algebra/Hwei P. Hsu.
p. cm. — (Harcourt Brace Jovanovich college outline series)
ISBN 0-15-601526-9
1. Algebras, Linear. I. Title. II. Series.
QA184.H78 1990
512′.5 — dc20 90-32663
CIP

Printed in the United States of America

First edition

A B C D E

PREFACE

Linear algebra is a subject of central importance in both basic and applied mathematics. It is now studied by students in a wide variety of disciplines. In this Outline we provide an introduction to the fundamental concepts of linear algebra and matrix theory.

This Outline is designed for students in mathematics and various disciplines of science, engineering, and business. It may be used as a text and/or a supplement for a formal course in linear algebra or matrix theory, or it may be used for independent study. This Outline combines the advantages of both the textbook and the so-called review book: It provides the textual explanations and illustrative examples of a textbook; and, in the direct way characteristic of the review book, it gives hundreds of completely solved problems that use essential theory and techniques. Moreover, the solved problems constitute an integral part of the text, illustrating and amplifying the fundamental concepts and developing the techniques of linear algebra.

The only formal prerequisite for using this Outline is a good knowledge of high school algebra. Examples and sections involving topics from calculus (including Section 9-3) are clearly indicated, and they are optional.

The approach used in this text is gradual. Chapter 1 introduces systems of linear equations, along with basic material on matrices and their arithmetic properties. Chapter 2 introduces determinants through their basic properties rather than by the classical permutation approach. It provides a base for the study of more advanced topics in linear algebra and matrix theory to follow. Chapter 3 slowly develops the concepts and the basic results about real vector spaces or linear spaces. Since the concepts of vector spaces are very abstract, this subject is approached through the introduction of vectors in a two-dimensional space and a three-dimensional space; then the general concept of a vector (with ample examples) is slowly developed. Chapter 4 focuses on the vector spaces for which an inner product can be defined. Chapter 5 deals with linear transformations from one vector space into another vector space and their representations by matrices. Chapter 6 gives a brief introduction to complex numbers and then proceeds to the development of complex vector spaces. Chapter 7 deals with the important eigenvalue problems and diagonalization of a matrix. Chapter 8 gives applications of linear algebra to problems of approximation and Fourier series. Chapter 9 deals with more applications of linear algebra to the Markov processes, systems of difference equations, and differential equations.

* * * * * * * *

The author acknowledges the useful comments and superb editorial efforts of Executive Editor Emily Thompson. The author is grateful to his wife Daisy, whose patient understanding and constant support (including the typing!) were necessary factors to the completion of this book.

HWEI P. HSU

CONTENTS

1 SYSTEMS OF LINEAR EQUATIONS AND MATRICES

THIS CHAPTER IS ABOUT

- ☑ Systems of Linear Equations
- ☑ Matrix Notation and Matrix Representation
- ☑ Basic Matrix Operations
- ☑ Inverse of a Matrix
- ☑ Transpose of a Matrix
- ☑ Special Matrices

1-1. Systems of Linear Equations

A. A linear equation

DEFINITION A **linear equation in n unknowns** is an equation of the form

$$a_1x_1 + a_2x_2 + \cdots + a_nx_n = b \tag{1.1}$$

where $a_1, a_2, \ldots, a_n$ and b are constants (usually real numbers) and $x_1, x_2, \ldots, x_n$ are unknowns.

The x_i's are called **variables**, and a_k is called the **coefficient** of x_k in the equation. A **solution of the linear equation** (1.1) is an n-tuple of numbers $(s_1, s_2, \ldots, s_n)$ such that

$$a_1s_1 + a_2s_2 + \cdots + a_ns_n = b \tag{1.2}$$

EXAMPLE 1-1: Find the solution of the following linear equation in two unknowns:

$$2x_1 - x_2 = 3$$

Solution: This equation can be rewritten as

$$x_2 = 2x_1 - 3$$

Let $x_1 = t$ where t is an arbitrary number; then we get $x_2 = 2t - 3$. Thus

$$x_1 = t, \qquad x_2 = 2t - 3 \qquad \text{or} \qquad (t, 2t - 3)$$

is the solution of the equation in terms of an arbitrary parameter t.

An infinite number of numerical solutions can be obtained by substituting specific values for t. For example, $t = 1$ yields the solution $(1, -1)$ and $t = 2$ yields the solution $(2, 1)$.

B. Systems of linear equations

DEFINITION A **system of m linear equations in n unknowns** is a set of the form

$$\begin{aligned}
a_{11}x_1 + a_{12}x_2 + \cdots + a_{1n}x_n &= b_1 \\
a_{21}x_1 + a_{22}x_2 + \cdots + a_{2n}x_n &= b_2 \\
\vdots \qquad \vdots \qquad\qquad \vdots \quad &\quad \vdots \\
a_{m1}x_1 + a_{m2}x_2 + \cdots + a_{mn}x_n &= b_m
\end{aligned} \tag{1.3}$$

where the a_{ij}'s and b_i's are all real numbers and $x_1, x_2, \ldots, x_n$ are the n unknowns.

An n-tuple of numbers $(s_1, s_2, \ldots, s_n)$ is called a **solution of the system** (1.3) if it is a solution of every equation in the system. The system is called a **consistent system** if it has at least one solution, and it is called an **inconsistent system** if it has no solution.

EXAMPLE 1-2: Solve the following system of two equations in two unknowns:

$$3x_1 + x_2 = 10$$
$$2x_1 - x_2 = 5$$

Solution: From the second equation we get

$$x_2 = 2x_1 - 5$$

Substituting this value into the first equation, we can find x_1:

$$\begin{aligned} 3x_1 + 2x_1 - 5 &= 10 \\ 5x_1 &= 15 \\ x_1 &= 3 \end{aligned}$$

And now we can find x_2:

$$x_2 = 2x_1 - 5 = 6 - 5 = 1$$

Thus the solution of the system is (3, 1).

EXAMPLE 1-3: Solve the following system of two equations in three unknowns:

$$x_1 + 2x_2 + x_3 = 0$$
$$2x_1 + 4x_2 - x_3 = 0$$

Solution

(1) Multiplying the first equation by 2 and then subtracting it from the second equation, we get

$$\begin{array}{r} 2x_1 + 4x_2 - x_3 = 0 \\ 2x_1 + 4x_2 + 2x_3 = 0 \\ \hline -\,3x_3 = 0 \end{array} \quad (-$$

Thus, $x_3 = 0$.

(2) Substituting this value into the given equations, we get

$$x_1 + 2x_2 + 0 = 0$$
$$2x_1 + 4x_2 - 0 = 0$$

Thus, $x_1 = -2x_2$.

(3) Let $x_2 = t$ where t is an arbitrary number; then $x_1 = -2t$. Now we know that $x_1 = -2t$, $x_2 = t$, and $x_3 = 0$, so $(-2t, t, 0)$ is the solution of this system in terms of an arbitrary parameter t. And there are infinitely many numerical solutions of this system. For example, setting $t = 1$ yields the solution $(-2, 1, 0)$.

1-2. Matrix Notation and Matrix Representation

A. Matrix notation

In the system of m linear equations in n unknowns (1.3) we notice that there are three different types of quantities: There are n unknowns, $x_1, x_2, \ldots, x_n$; there is a column of numbers on the right side, $b_1, b_2, \ldots, b_m$; and there is a set of $m \times n$ numerical coefficients a_{ij} on the left-hand side. We can now introduce matrix notation to describe the system of equations (1.3).

- We can write the n unknowns in (1.3) as

$$\mathbf{x} = \begin{bmatrix} x_1 \\ x_2 \\ \vdots \\ x_n \end{bmatrix} \tag{1.4}$$

This is called an ***n*-vector**, which is an n-tuple of numbers written vertically.

- We can also write the column of numbers on the right side in (1.3) as

$$\mathbf{b} = \begin{bmatrix} b_1 \\ b_2 \\ \vdots \\ b_m \end{bmatrix} \tag{1.5}$$

This is called an ***m*-vector**, which is an m-tuple of numbers written vertically.

- We can write the coefficients as a rectangular array with m rows and n columns:

$$A = \begin{bmatrix} a_{11} & a_{12} & \cdots & a_{1n} \\ a_{21} & a_{22} & \cdots & a_{2n} \\ \vdots & \vdots & & \vdots \\ a_{m1} & a_{m2} & \cdots & a_{mn} \end{bmatrix} \tag{1.6}$$

Here A is called the **coefficient matrix**.

In writing (1.4), (1.5), and (1.6), we have rewritten the three types of quantities in a system of linear equations as a set of *matrices*. A **matrix** is simply a rectangular array of numbers, and the numbers in the array are called the **entries** or **elements** of the given matrix. Note that in the coefficient matrix a_{ij} denotes the entry of the matrix A that is in the ith row and the jth column. (Thus, for example, a_{21} is in the 2nd row and the 1st column.) Since A has m rows and n columns, we call it an $m \times n$ ("m by n") matrix, and $m \times n$ is sometimes referred to as the **size** or **dimension** of the matrix A.

note: We frequently abbreviate the notation in (1.6) as

$$A = [a_{ij}]_{m \times n}$$

Notice also that an n-vector (1.4) can be considered an $n \times 1$ matrix and an m-vector (1.5) can be considered an $m \times 1$ matrix. A **vector**, then, is a single-column (or -row) matrix, whose entries are called components.

- The matrix A is called a **square matrix** when $m = n$; we can also say that, when $m = n$, A is a square matrix of order n.

B. Matrix representation

1. **Matrix equality**

Two matrices $A = [a_{ij}]_{m \times n}$ and $B = [b_{ij}]_{m \times n}$ are equal if and only if they are of the same size (i.e., have the same number of rows and columns) and if $a_{ij} = b_{ij}$ (i.e., each entry in A is equal to the corresponding entry in B); that is,

EQUALITY OF MATRICES

$$A = B \Leftrightarrow a_{ij} = b_{ij} \tag{1.7}$$

2. **Matrix multiplication ($m \times n$ by $n \times 1$)**

Now we put matrix notation to use. We can write the system (1.3) in the simplified matrix form

$$A\mathbf{x} = \mathbf{b} \tag{1.8}$$

Written out in full, this form is

$$\overset{A}{\begin{bmatrix} a_{11} & a_{12} & \cdots & a_{1n} \\ a_{21} & a_{22} & \cdots & a_{2n} \\ \vdots & \vdots & & \vdots \\ a_{m1} & a_{m2} & \cdots & a_{mn} \end{bmatrix}} \overset{\mathbf{x}}{\begin{bmatrix} x_1 \\ x_2 \\ \vdots \\ x_n \end{bmatrix}} = \overset{\mathbf{b}}{\begin{bmatrix} b_1 \\ b_2 \\ \vdots \\ b_m \end{bmatrix}} \tag{1.9}$$

This matrix multiplication will be defined exactly so that it reproduces the original system (1.3):

$$\overset{A}{\begin{bmatrix} a_{11} & a_{12} & \cdots & a_{1n} \\ a_{21} & a_{22} & \cdots & a_{2n} \\ \vdots & \vdots & & \vdots \\ a_{m1} & a_{m2} & \cdots & a_{mn} \end{bmatrix}} \overset{\mathbf{x}}{\begin{bmatrix} x_1 \\ x_2 \\ \vdots \\ x_n \end{bmatrix}} = \overset{A\mathbf{x}}{\begin{bmatrix} a_{11}x_1 + a_{12}x_2 + \cdots + a_{1n}x_n \\ a_{21}x_1 + a_{22}x_2 + \cdots + a_{2n}x_n \\ \vdots \qquad \vdots \qquad \vdots \\ a_{m1}x_1 + a_{m2}x_2 + \cdots + a_{mn}x_n \end{bmatrix}} = \overset{\mathbf{b}}{\begin{bmatrix} b_1 \\ b_2 \\ \vdots \\ b_m \end{bmatrix}} \tag{1.10}$$

note: Here we have multiplied the entries in the first column of A by the entry in the first row of $\mathbf{x}$, the entries in the second column of A by the entry in the second row of $\mathbf{x}$, and so on. Notice that the product of a $1 \times n$ matrix (sometimes called a **row vector**) and an $n \times 1$ matrix (sometimes called a **column vector**) is a 1×1 matrix, which is a single number; i.e.,

PRODUCT RULE

$$[a_{i1} \quad a_{i2} \quad \cdots \quad a_{in}] \begin{bmatrix} x_1 \\ x_2 \\ \vdots \\ x_n \end{bmatrix} = [a_{i1}x_1 + a_{i2}x_2 + \cdots + a_{in}x_n], \quad i = 1, 2, \ldots, m \tag{1.11}$$

As eq. (1.10) indicates,

- The product of an $m \times n$ matrix and an $n \times 1$ matrix results in an $m \times 1$ matrix.

EXAMPLE 1-4: (**a**) Write the system $\begin{matrix} 3x_1 + x_2 = 10 \\ 2x_1 - x_2 = \ \ 5 \end{matrix}$ of Example 1-2 in matrix form; check your answer by reproducing the original system of equations from its matrix form. (**b**) Using the matrix form, check the solution (3, 1) of this system.

Solution

(**a**)

$$A = \begin{bmatrix} 3 & 1 \\ 2 & -1 \end{bmatrix}, \quad \mathbf{x} = \begin{bmatrix} x_1 \\ x_2 \end{bmatrix}, \quad \mathbf{b} = \begin{bmatrix} 10 \\ 5 \end{bmatrix}$$

Thus,

$$\overset{A}{\begin{bmatrix} 3 & 1 \\ 2 & -1 \end{bmatrix}} \overset{\mathbf{x}}{\begin{bmatrix} x_1 \\ x_2 \end{bmatrix}} = \overset{\mathbf{b}}{\begin{bmatrix} 10 \\ 5 \end{bmatrix}}$$

Check:

$$\overset{A}{\begin{bmatrix} 3 & 1 \\ 2 & -1 \end{bmatrix}} \overset{\mathbf{x}}{\begin{bmatrix} x_1 \\ x_2 \end{bmatrix}} = \overset{A\mathbf{x}}{\begin{bmatrix} (3)(x_1) + (1)(x_2) \\ (2)(x_1) + (-1)(x_2) \end{bmatrix}} = \begin{bmatrix} 3x_1 + x_2 \\ 2x_1 - x_2 \end{bmatrix} = \overset{\mathbf{b}}{\begin{bmatrix} 10 \\ 5 \end{bmatrix}}$$

and by the definition of matrix equality (1.7), if $\begin{bmatrix} 3x_1 + x_2 \\ 2x_1 - x_2 \end{bmatrix} = \begin{bmatrix} 10 \\ 5 \end{bmatrix}$, then

$$3x_1 + x_2 = 10$$
$$2x_1 - x_2 = 5$$

(b) If the solution to this system is (3, 1), we can let

$$\begin{bmatrix} x_1 \\ x_2 \end{bmatrix} = \begin{bmatrix} 3 \\ 1 \end{bmatrix}$$

so we get

$$\begin{bmatrix} 3 & 1 \\ 2 & -1 \end{bmatrix}\begin{bmatrix} 3 \\ 1 \end{bmatrix} = \begin{bmatrix} (3)(3) + (1)(1) \\ (2)(3) + (-1)(1) \end{bmatrix} = \begin{bmatrix} 9 + 1 \\ 6 - 1 \end{bmatrix} = \begin{bmatrix} 10 \\ 5 \end{bmatrix}$$

1-3. Basic Matrix Operations

A. Matrix addition

The sum C of two $m \times n$ matrices A and B is an $m \times n$ matrix whose ij-components are the sums of the corresponding ij components in A and B.

DEFINITION If $A = [a_{ij}]_{m \times n}$, $B = [b_{ij}]_{m \times n}$, and $C = [c_{ij}]_{m \times n}$, then

MATRIX ADDITION

$$C = A + B \Leftrightarrow c_{ij} = a_{ij} + b_{ij} \tag{1.12}$$

That is, matrix addition is performed by adding the corresponding entries in each position.

note: The sum of two matrices that are *not* same size is not defined.

EXAMPLE 1-5: Compute $A + B$, $A + D$, and $B + D$, where

$$A = \begin{bmatrix} 1 & 2 & 3 \\ 4 & 5 & 6 \end{bmatrix}, \qquad B = \begin{bmatrix} 1 & 1 & 2 \\ 3 & 2 & 1 \end{bmatrix}, \qquad D = \begin{bmatrix} 1 & 0 \\ 0 & 3 \end{bmatrix}$$

Solution: Since both A and B are same size, their sum is defined and

$$A + B = \begin{bmatrix} 1 & 2 & 3 \\ 4 & 5 & 6 \end{bmatrix} + \begin{bmatrix} 1 & 1 & 2 \\ 3 & 2 & 1 \end{bmatrix} = \begin{bmatrix} 1+1 & 2+1 & 3+2 \\ 4+3 & 5+2 & 6+1 \end{bmatrix} = \begin{bmatrix} 2 & 3 & 5 \\ 7 & 7 & 7 \end{bmatrix}$$

$A + D$, however, is not defined since the sizes of A and D are not the same. Similarly, $B + D$ is not defined.

B. Multiplication of a matrix by a scalar

When discussing matrices, it is common to refer to numerical quantities as **scalars**.

DEFINITION If $A = [a_{ij}]_{m \times n}$, α is any scalar, and $B = [b_{ij}]_{m \times n}$, then

MULTIPLICATION OF A MATRIX BY A SCALAR

$$B = \alpha A \Leftrightarrow b_{ij} = \alpha a_{ij} \tag{1.13}$$

That is, αA, the product of the matrix A and the scalar α, is the matrix whose entries are α times each of the corresponding entries of A. Thus, for example, the product of the scalar y by the matrix A, where $A = \begin{bmatrix} 2 & 4 \\ 1 & 3 \end{bmatrix}$, is $\begin{bmatrix} 2y & 4y \\ y & 3y \end{bmatrix}$.

EXAMPLE 1-6: Compute $3A$ and πA, where

$$A = \begin{bmatrix} 3 & 4 \\ 1 & 2 \end{bmatrix}$$

Solution: By definition (1.13),

$$3A = 3\begin{bmatrix} 3 & 4 \\ 1 & 2 \end{bmatrix} = \begin{bmatrix} 3(3) & 3(4) \\ 3(1) & 3(2) \end{bmatrix} = \begin{bmatrix} 9 & 12 \\ 3 & 6 \end{bmatrix}$$

and

$$\pi A = \pi \begin{bmatrix} 3 & 4 \\ 1 & 2 \end{bmatrix} = \begin{bmatrix} 3\pi & 4\pi \\ \pi & 2\pi \end{bmatrix}$$

C. Zero matrix

DEFINITION If all of the entries of an $m \times n$ matrix are zero, that matrix is called a **zero matrix**, which is denoted by O, or by $O_{m \times n}$ if it is important to indicate its size.

If $n = 1$ in a zero matrix, so that $O_{m \times 1}$ is an m-vector, we use the notation $\mathbf{0}$. It is important to note that, by the definition of matrix (1.12), if A is any $m \times n$ matrix, then

$$A + O = A = O + A \qquad \text{where } O = O_{m \times n} \tag{1.14}$$

D. Negative of a matrix

DEFINITION Let $A = [a_{ij}]_{m \times n}$ and $B = [b_{ij}]_{m \times n}$. Then we define the **negative of a matrix** A to be

NEGATIVE OF A MATRIX

$$B = -A = (-1)A \Leftrightarrow b_{ij} = -a_{ij} \tag{1.15}$$

That is, $-A$, the negative of the matrix A, is the matrix whose entries are the negatives of the corresponding entries of A. Note that, if A is any $m \times n$ matrix, then

$$A - A = O \qquad \text{where } O = O_{m \times n} \tag{1.16}$$

If A and B are any two $m \times n$ matrices, then we denote $A + (-B)$ by $A - B$.

EXAMPLE 1-7: Compute $A - B$ and $B - A$, where

$$A = \begin{bmatrix} 1 & 2 & 3 \\ 4 & 5 & 6 \end{bmatrix}, \qquad B = \begin{bmatrix} 1 & 1 & 2 \\ 3 & 2 & 1 \end{bmatrix}$$

Solution

$$A - B = A + (-1)B = \begin{bmatrix} 1 & 2 & 3 \\ 4 & 5 & 6 \end{bmatrix} + \begin{bmatrix} -1 & -1 & -2 \\ -3 & -2 & -1 \end{bmatrix} = \begin{bmatrix} 0 & 1 & 1 \\ 1 & 3 & 5 \end{bmatrix}$$

$$B - A = B + (-1)A = \begin{bmatrix} 1 & 1 & 2 \\ 3 & 2 & 1 \end{bmatrix} + \begin{bmatrix} -1 & -2 & -3 \\ -4 & -5 & -6 \end{bmatrix} = \begin{bmatrix} 0 & -1 & -1 \\ -1 & -3 & -5 \end{bmatrix}$$

E. Matrix multiplication ($m \times n$ by $n \times p$)

In Section 1-2B we gave the matrix product of an $m \times n$ matrix A and an $n \times 1$ matrix $\mathbf{x}$. The general definition of matrix multiplication is as follows:

DEFINITION If $A = [a_{ij}]_{m \times n}$ is an $m \times n$ matrix and $B = [b_{ij}]_{n \times p}$ is an $n \times p$ matrix, then the product of A and B is C, which is the $m \times p$ matrix defined by

PRODUCT OF TWO MATRICES

$$AB = C = [c_{ij}]_{m \times p} \tag{1.17}$$

where

$$c_{ij} = a_{i1}b_{1j} + a_{i2}b_{2j} + \cdots + a_{in}b_{nj} = \sum_{k=1}^{n} a_{ik}b_{kj} \qquad \text{for all } i, j \tag{1.18}$$

To show matrix multiplication more clearly, we can write the matrices as

$$\begin{bmatrix} a_{11} & a_{12} & \cdots & a_{1k} & \cdots & a_{1n} \\ \vdots & \vdots & & \vdots & & \vdots \\ a_{i1} & a_{i2} & \cdots & a_{ik} & \cdots & a_{in} \\ \vdots & \vdots & & \vdots & & \vdots \\ a_{m1} & a_{m2} & \cdots & a_{mk} & \cdots & a_{mn} \end{bmatrix} \begin{bmatrix} b_{11} & \cdots & b_{1j} & \cdots & b_{1p} \\ b_{21} & \cdots & b_{2j} & \cdots & b_{2p} \\ \vdots & & \vdots & & \vdots \\ b_{k1} & \cdots & b_{kj} & \cdots & b_{kp} \\ \vdots & & \vdots & & \vdots \\ b_{n1} & \cdots & b_{nj} & \cdots & b_{np} \end{bmatrix} = \begin{bmatrix} c_{11} & \cdots & c_{1j} & \cdots & c_{1p} \\ \vdots & & \vdots & & \vdots \\ c_{i1} & \cdots & c_{ij} & \cdots & c_{ip} \\ \vdots & & \vdots & & \vdots \\ c_{m1} & \cdots & c_{mj} & \cdots & c_{mp} \end{bmatrix} \tag{1.19}$$

The ijth entry of C is found by multiplying the corresponding entries in the ith row of A and the jth column of B and adding. For example, we can multiply the two matrices $\begin{bmatrix} 1 & 3 \\ 2 & 2 \\ 3 & 1 \end{bmatrix}$ and $\begin{bmatrix} 1 & 2 & 3 \\ 3 & 2 & 1 \end{bmatrix}$ like this:

$$\begin{bmatrix} 1 & 3 \\ 2 & 2 \\ 3 & 1 \end{bmatrix} \begin{bmatrix} 1 & 2 & 3 \\ 3 & 2 & 1 \end{bmatrix} = \begin{bmatrix} 1(1)+3(3) & 1(2)+3(2) & 1(3)+3(1) \\ 2(1)+2(3) & 2(2)+2(2) & 2(3)+2(1) \\ 3(1)+1(3) & 3(2)+1(2) & 3(3)+1(1) \end{bmatrix}$$

$$= \begin{bmatrix} 1+9 & 2+6 & 3+3 \\ 2+6 & 4+4 & 6+2 \\ 3+3 & 6+2 & 9+1 \end{bmatrix}$$

$$= \begin{bmatrix} 10 & 8 & 6 \\ 8 & 8 & 8 \\ 6 & 8 & 10 \end{bmatrix}$$

note: The matrix product AB is defined *only* when the number of columns of A is equal to the number of rows of B. In this case, A and B are said to be **conformable**.

Now notice that the number of rows of C is the same as the number of rows of A and the number of columns of C is the same as the number of columns of B. The relations are illustrated as

$$A_{m \times n} B_{n \times p} = C_{m \times p}$$

Thus, if A and B are both $n \times n$ matrices, then AB and BA will also be $n \times n$ matrices.

But matrix multiplication has a couple of complications we should be aware of. For one thing, in general,

$$AB \neq BA$$

that is, the product of matrices is *not commutative*. Another complication of the matrix product is that it is possible to have $AB = O$ without having $A = O$ or $B = O$ (see Example 1-9). Thus, the equation $AB = AC$ does not imply that $B = C$, even if A, B, and C are not zero matrices (see Example 1-10).

Finally, notice that

$$AO = O, \qquad OA = O \tag{1.20}$$

if the indicated operations can be performed.

EXAMPLE 1-8: Compute **(a)** AC, **(b)** CA, and **(c)** AB, where

$$A = \begin{bmatrix} 2 & 1 \\ 3 & 0 \\ 0 & 5 \end{bmatrix}, \qquad B = \begin{bmatrix} 1 & 1 \\ 4 & 2 \\ -1 & 0 \end{bmatrix}, \qquad C = \begin{bmatrix} 1 & 2 & 1 \\ 3 & 1 & 2 \end{bmatrix}$$

Solution

(a) $$AC = \begin{bmatrix} 2 & 1 \\ 3 & 0 \\ 0 & 5 \end{bmatrix} \begin{bmatrix} 1 & 2 & 1 \\ 3 & 1 & 2 \end{bmatrix} = \begin{bmatrix} 2(1)+1(3) & 2(2)+1(1) & 2(1)+1(2) \\ 3(1)+0(3) & 3(2)+0(1) & 3(1)+0(2) \\ 0(1)+5(3) & 0(2)+5(1) & 0(1)+5(2) \end{bmatrix} = \begin{bmatrix} 5 & 5 & 4 \\ 3 & 6 & 3 \\ 15 & 5 & 10 \end{bmatrix}$$

(b) $$CA = \begin{bmatrix} 1 & 2 & 1 \\ 3 & 1 & 2 \end{bmatrix} \begin{bmatrix} 2 & 1 \\ 3 & 0 \\ 0 & 5 \end{bmatrix} = \begin{bmatrix} 1(2)+2(3)+1(0) & 1(1)+2(0)+1(5) \\ 3(2)+1(3)+2(0) & 3(1)+1(0)+2(5) \end{bmatrix} = \begin{bmatrix} 8 & 6 \\ 9 & 13 \end{bmatrix}$$

Note that $AC \neq CA$.

(c) AB is not defined since the number of columns of A is 2 and the number of rows of B is 3; that is, A and B are not conformable.

EXAMPLE 1-9: Compute AB and BA, where

$$A = \begin{bmatrix} 1 & 0 \\ 0 & 0 \end{bmatrix}, \qquad B = \begin{bmatrix} 0 & 0 \\ 0 & 1 \end{bmatrix}$$

Solution

$$AB = \begin{bmatrix} 1 & 0 \\ 0 & 0 \end{bmatrix} \begin{bmatrix} 0 & 0 \\ 0 & 1 \end{bmatrix} = \begin{bmatrix} 1(0)+0(0) & 1(0)+0(1) \\ 0(0)+0(0) & 0(0)+0(1) \end{bmatrix} = \begin{bmatrix} 0 & 0 \\ 0 & 0 \end{bmatrix} = O$$

$$BA = \begin{bmatrix} 0 & 0 \\ 0 & 1 \end{bmatrix} \begin{bmatrix} 1 & 0 \\ 0 & 0 \end{bmatrix} = \begin{bmatrix} 0(1)+0(0) & 0(0)+0(0) \\ 0(1)+1(0) & 0(0)+1(0) \end{bmatrix} = \begin{bmatrix} 0 & 0 \\ 0 & 0 \end{bmatrix} = O$$

Note that even though $A \neq O$ and $B \neq O$, $AB = BA = O$.

EXAMPLE 1-10: Let

$$A = \begin{bmatrix} 0 & 2 \\ 0 & 1 \end{bmatrix}, \qquad B = \begin{bmatrix} 2 & 3 \\ 1 & 4 \end{bmatrix}, \qquad C = \begin{bmatrix} -1 & 1 \\ 1 & 4 \end{bmatrix}$$

Compute AB and AC and compare the results.

Solution

$$AB = \begin{bmatrix} 0 & 2 \\ 0 & 1 \end{bmatrix} \begin{bmatrix} 2 & 3 \\ 1 & 4 \end{bmatrix} = \begin{bmatrix} 0(2)+2(1) & 0(3)+2(4) \\ 0(2)+1(1) & 0(3)+1(4) \end{bmatrix} = \begin{bmatrix} 2 & 8 \\ 1 & 4 \end{bmatrix}$$

$$AC = \begin{bmatrix} 0 & 2 \\ 0 & 1 \end{bmatrix} \begin{bmatrix} -1 & 1 \\ 1 & 4 \end{bmatrix} = \begin{bmatrix} 0(-1)+2(1) & 0(1)+2(4) \\ 0(-1)+1(1) & 0(1)+1(4) \end{bmatrix} = \begin{bmatrix} 2 & 8 \\ 1 & 4 \end{bmatrix}$$

So we obtained $AB = AC$, even though $B \neq C$.

F. Properties of basic matrix operations

The following properties of basic matrix operations provide some useful rules for doing matrix arithmetic.

THEOREM 1-3.1 Each of the following statements is valid for any matrices A, B, and C for which the indicated operations are defined and for any scalars α and β:

$$A + B = B + A \qquad \textbf{Additive commutativity} \qquad \textbf{(1.21)}$$

$$A + (B + C) = (A + B) + C \qquad \textbf{Additive associativity} \qquad \textbf{(1.22)}$$

$$(AB)C = A(BC) \qquad \textbf{Multiplicative associativity} \qquad \textbf{(1.23)}$$

$$A(B + C) = AB + AC \tag{1.24}$$

$$(A + B)C = AC + BC \tag{1.25}$$

Distributivity

$$\alpha(A + B) = \alpha A + \alpha B \tag{1.26}$$

$$(\alpha + \beta)A = \alpha A + \beta A \tag{1.27}$$

$$(\alpha\beta)A = \alpha(\beta A) \tag{1.28}$$

Multiplicative associativity

$$\alpha(AB) = (\alpha A)B = A(\alpha B) \tag{1.29}$$

note: In view of the associative property (1.23), that is, $A(BC) = (AB)C$, we may simply drop the parentheses and write

$$A(BC) = (AB)C = ABC \tag{1.30}$$

EXAMPLE 1-11: Verify **(a)** the commutative property (1.21) and **(b)** the distributive property (1.24), that is, $A + B = B + A$ and $A(B + C) = AB + AC$, respectively.

Solution

(a) Let $A = [a_{ij}]_{m\times n}$ and $B = [b_{ij}]_{m\times n}$. Then by the definition of matrix addition (1.12), it follows that

$$A + B = [a_{ij} + b_{ij}]_{m\times n} = [b_{ij} + a_{ij}]_{m\times n} = B + A$$

(b) Let $A = [a_{ij}]_{m\times n}$, $B = [b_{ij}]_{n\times p}$, and $C = [c_{ij}]_{n\times p}$. Then by the definitions of matrix addition (1.12) and multiplication (1.18), it follows that

$$\begin{aligned} A(B + C) &= \sum_{k=1}^{n} a_{ik}(b_{kj} + c_{kj}) \\ &= \sum_{k=1}^{n} (a_{ik}b_{kj} + a_{ik}c_{kj}) \\ &= \sum_{k=1}^{n} a_{ik}b_{kj} + \sum_{k=1}^{n} a_{ik}c_{kj} \\ &= AB + AC \end{aligned}$$

EXAMPLE 1-12: Verify that property (1.23), $A(BC) = (AB)C$, and property (1.24), $A(B + C) = AB + AC$, are valid for

$$A = \begin{bmatrix} 1 & 2 \\ 3 & 4 \end{bmatrix}, \qquad B = \begin{bmatrix} 1 & -2 \\ 2 & 3 \end{bmatrix}, \qquad C = \begin{bmatrix} 1 & 1 \\ 2 & 0 \end{bmatrix}$$

Solution

(a)

$$BC = \begin{bmatrix} 1 & -2 \\ 2 & 3 \end{bmatrix}\begin{bmatrix} 1 & 1 \\ 2 & 0 \end{bmatrix} = \begin{bmatrix} 1(1) + (-2)(2) & 1(1) + (-2)(0) \\ 2(1) + \quad 3(2) & 2(1) + \quad 3(0) \end{bmatrix} = \begin{bmatrix} -3 & 1 \\ 8 & 2 \end{bmatrix}$$

$$AB = \begin{bmatrix} 1 & 2 \\ 3 & 4 \end{bmatrix}\begin{bmatrix} 1 & -2 \\ 2 & 3 \end{bmatrix} = \begin{bmatrix} 1(1) + 2(2) & 1(-2) + 2(3) \\ 3(1) + 4(2) & 3(-2) + 4(3) \end{bmatrix} = \begin{bmatrix} 5 & 4 \\ 11 & 6 \end{bmatrix}$$

$$AC = \begin{bmatrix} 1 & 2 \\ 3 & 4 \end{bmatrix}\begin{bmatrix} 1 & 1 \\ 2 & 0 \end{bmatrix} = \begin{bmatrix} 1(1) + 2(2) & 1(1) + 2(0) \\ 3(1) + 4(2) & 3(1) + 4(0) \end{bmatrix} = \begin{bmatrix} 5 & 1 \\ 11 & 3 \end{bmatrix}$$

$$A(BC) = \begin{bmatrix} 1 & 2 \\ 3 & 4 \end{bmatrix}\begin{bmatrix} -3 & 1 \\ 8 & 2 \end{bmatrix} = \begin{bmatrix} 1(-3) + 2(8) & 1(1) + 2(2) \\ 3(-3) + 4(8) & 3(1) + 4(2) \end{bmatrix} = \begin{bmatrix} 13 & 5 \\ 23 & 11 \end{bmatrix}$$

$$(AB)C = \begin{bmatrix} 5 & 4 \\ 11 & 6 \end{bmatrix}\begin{bmatrix} 1 & 1 \\ 2 & 0 \end{bmatrix} = \begin{bmatrix} 5(1)+4(2) & 5(1)+4(0) \\ 11(1)+6(2) & 11(1)+6(0) \end{bmatrix} = \begin{bmatrix} 13 & 5 \\ 23 & 11 \end{bmatrix}$$

Thus, $A(BC) = (AB)C$.

(b)

$$B + C = \begin{bmatrix} 1 & -2 \\ 2 & 3 \end{bmatrix} + \begin{bmatrix} 1 & 1 \\ 2 & 0 \end{bmatrix} = \begin{bmatrix} 1+1 & (-2)+1 \\ 2+2 & 3+0 \end{bmatrix} = \begin{bmatrix} 2 & -1 \\ 4 & 3 \end{bmatrix}$$

$$A(B + C) = \begin{bmatrix} 1 & 2 \\ 3 & 4 \end{bmatrix}\begin{bmatrix} 2 & -1 \\ 4 & 3 \end{bmatrix} = \begin{bmatrix} 1(2)+2(4) & 1(-1)+2(3) \\ 3(2)+4(4) & 3(-1)+4(3) \end{bmatrix} = \begin{bmatrix} 10 & 5 \\ 22 & 9 \end{bmatrix}$$

$$AB + AC = \begin{bmatrix} 5 & 4 \\ 11 & 6 \end{bmatrix} + \begin{bmatrix} 5 & 1 \\ 11 & 3 \end{bmatrix} = \begin{bmatrix} 5+5 & 4+1 \\ 11+11 & 6+3 \end{bmatrix} = \begin{bmatrix} 10 & 5 \\ 22 & 9 \end{bmatrix}$$

Hence, $A(B + C) = AB + AC$.

1-4. Inverse of a Matrix

A. The identity matrix

DEFINITION If $A = [a_{ij}]$ is an $n \times n$ matrix $\begin{bmatrix} a_{11} & a_{12} & \cdots & a_{1n} \\ a_{21} & a_{22} & \cdots & a_{2n} \\ \vdots & \vdots & & \vdots \\ a_{n1} & a_{n2} & \cdots & a_{nn} \end{bmatrix}$, we call $a_{11}, a_{22}, \ldots, a_{nn}$ its **diagonal entries** (or **elements**). Then the $n \times n$ matrix

IDENTITY MATRIX

$$I_n = \begin{bmatrix} 1 & 0 & 0 & \cdots & 0 \\ 0 & 1 & 0 & \cdots & 0 \\ 0 & 0 & 1 & \cdots & 0 \\ \vdots & \vdots & \vdots & \ddots & \vdots \\ 0 & 0 & 0 & \cdots & 1 \end{bmatrix} \qquad \textbf{(1.31)}$$

whose diagonal entries are all 1 and whose other entries are all 0, is called the **$n \times n$ identity matrix** (or **identity matrix of nth order**).

For example $I_2 = \begin{bmatrix} 1 & 0 \\ 0 & 1 \end{bmatrix}$ and $I_3 = \begin{bmatrix} 1 & 0 & 0 \\ 0 & 1 & 0 \\ 0 & 0 & 1 \end{bmatrix}$ are the identity matrices of second order and third order, respectively.

It will be convenient to introduce the symbol δ_{ij}, known as the **Kronecker delta**, which is defined by

$$\delta_{ij} = \begin{cases} 1 & \text{if } i = j \\ 0 & \text{if } i \neq j \end{cases} \qquad \textbf{(1.32)}$$

Then the identity matrix I_n can be written as

IDENTITY MATRIX (KRONECKER DELTA FORM)

$$I_n = [\delta_{ij}]_{n \times n} \qquad \textbf{(1.33)}$$

THEOREM 1-4.1 Let A be an $m \times n$ matrix and $\mathbf{x}$ be an n-vector (i.e., an $n \times 1$ matrix). Then

$$AI_n = I_m A = A \qquad \textbf{(1.34)}$$

$$I_n \mathbf{x} = \mathbf{x} \qquad \textbf{(1.35)}$$

That is, when an $m \times n$ matrix A or an n-vector $\mathbf{x}$ is multiplied by its identity matrix, the product is the matrix A or the vector $\mathbf{x}$. Thus, an identity matrix is sometimes called a **unit matrix**.

EXAMPLE 1-13: Compute AI_2 and I_3A, where

$$A = \begin{bmatrix} 1 & 2 \\ -1 & 0 \\ 3 & 2 \end{bmatrix}$$

Solution

$$AI_2 = \begin{bmatrix} 1 & 2 \\ -1 & 0 \\ 3 & 2 \end{bmatrix} \begin{bmatrix} 1 & 0 \\ 0 & 1 \end{bmatrix} = \begin{bmatrix} 1(1) + 2(0) & 1(0) + 2(1) \\ -1(1) + 0(0) & -1(0) + 0(1) \\ 3(1) + 2(0) & 3(0) + 2(1) \end{bmatrix} = \begin{bmatrix} 1 & 2 \\ -1 & 0 \\ 3 & 2 \end{bmatrix}$$

$$I_3A = \begin{bmatrix} 1 & 0 & 0 \\ 0 & 1 & 0 \\ 0 & 0 & 1 \end{bmatrix} \begin{bmatrix} 1 & 2 \\ -1 & 0 \\ 3 & 2 \end{bmatrix} = \begin{bmatrix} 1(1) + 0(-1) + 0(3) & 1(2) + 0(0) + 0(2) \\ 0(1) + 1(-1) + 0(3) & 0(2) + 1(0) + 0(2) \\ 0(1) + 0(-1) + 1(3) & 0(2) + 0(0) + 0(2) \end{bmatrix} = \begin{bmatrix} 1 & 2 \\ -1 & 0 \\ 3 & 2 \end{bmatrix}$$

EXAMPLE 1-14: Verify the identity property (1.35); that is, $I_n\mathbf{x} = \mathbf{x}$.

Solution: Let $\mathbf{x}$ be the n-vector

$$\mathbf{x} = \begin{bmatrix} x_1 \\ x_2 \\ \vdots \\ x_n \end{bmatrix}$$

Then since $I_n\mathbf{x}$ is also an n-vector ($n \times 1$ matrix), we can write

$$I_n\mathbf{x} = [\delta_{ij}]_{n \times n} \begin{bmatrix} x_1 \\ x_2 \\ \vdots \\ x_n \end{bmatrix} = \begin{bmatrix} a_1 \\ a_2 \\ \vdots \\ a_n \end{bmatrix}$$

Then by (1.18) and (1.33),

$$\delta_{i1}x_1 + \delta_{i2}x_2 + \cdots + \delta_{in}x_n = a_i$$

And since $\delta_{ij} = 0$ except when $i = j$, in which case $\delta_{ii} = 1$, we have

$$\delta_{ii}x_i = x_i = a_i, \qquad \text{where } i = 1, 2, \ldots, n$$

Thus, by the definition of the equality of matrices (1.7), it follows that

$$I_n\mathbf{x} = \mathbf{x}$$

B. Inverse of a matrix

DEFINITION If A is an $n \times n$ matrix and if there exists an $n \times n$ matrix B such that

$$AB = BA = I_n \qquad \textbf{(1.36)}$$

then A is called **nonsingular** (or **invertible**) and B is called an **inverse** of A. The inverse of A is denoted by A^{-1}. Thus

INVERSE OF A $$AA^{-1} = A^{-1}A = I_n \tag{1.37}$$

If A^{-1} exists, then it is uniquely determined by A. But if A^{-1} does not exist, then A is said to be **singular**.

EXAMPLE 1-15: Show that if B and C are both inverses of the matrix A, then $B = C$.

Solution: If $AB = BA = I_n$ and $AC = CA = I_n$, then by properties (1.34), (1.36), and (1.23),

$$B = BI_n = B(AC) = (BA)C = I_nC = C$$

EXAMPLE 1-16: Show that A and B are inverse to each other, where

$$A = \begin{bmatrix} 1 & 1 \\ 2 & 3 \end{bmatrix}, \qquad B = \begin{bmatrix} 3 & -1 \\ -2 & 1 \end{bmatrix}$$

Solution

$$AB = \begin{bmatrix} 1 & 1 \\ 2 & 3 \end{bmatrix}\begin{bmatrix} 3 & -1 \\ -2 & 1 \end{bmatrix} = \begin{bmatrix} 1(3) + 1(-2) & 1(-1) + 1(1) \\ 2(3) + 3(-2) & 2(-1) + 3(1) \end{bmatrix} = \begin{bmatrix} 1 & 0 \\ 0 & 1 \end{bmatrix} = I_2$$

$$BA = \begin{bmatrix} 3 & -1 \\ -2 & 1 \end{bmatrix}\begin{bmatrix} 1 & 1 \\ 2 & 3 \end{bmatrix} = \begin{bmatrix} 3(1) + (-1)(2) & 3(1) + (-1)(3) \\ -2(1) + \quad 1(2) & -2(1) + \quad 1(3) \end{bmatrix} = \begin{bmatrix} 1 & 0 \\ 0 & 1 \end{bmatrix} = I_2$$

Thus $B = A^{-1}$. Conversely, we also see that $A = B^{-1}$.

Example 1-16 illustrates a general principle that

- If B is the inverse of A, then A is the inverse of B.

EXAMPLE 1-17: Show that the matrix $A = \begin{bmatrix} 1 & 0 \\ 0 & 0 \end{bmatrix}$ has no inverse.

Solution: Suppose that there is a matrix

$$B = \begin{bmatrix} b_{11} & b_{12} \\ b_{21} & b_{22} \end{bmatrix}$$

such that $AB = I_2$; then

$$\begin{bmatrix} 1 & 0 \\ 0 & 0 \end{bmatrix}\begin{bmatrix} b_{11} & b_{12} \\ b_{21} & b_{22} \end{bmatrix} = \begin{bmatrix} b_{11} & b_{12} \\ 0 & 0 \end{bmatrix} = \begin{bmatrix} 1 & 0 \\ 0 & 1 \end{bmatrix}$$

But this is impossible for any choice of $b_{11}, b_{12}, b_{21}, b_{22}$; therefore, A has no inverse.

---

note: A procedure for deciding whether a matrix has an inverse and for finding the inverse is discussed in Chapter 2.

THEOREM 1-4.2 If A and B are nonsingular $n \times n$ matrices, then AB is also nonsingular and

$$(AB)^{-1} = B^{-1}A^{-1} \tag{1.38}$$

EXAMPLE 1-18: Verify Theorem 1-4.2.

Solution: If $(AB)C = A(BC)$ (property 1.23), then

$$(AB)(B^{-1}A^{-1}) = A(BB^{-1})A^{-1} = AI_nA^{-1} = AA^{-1} = I_n$$

$$(B^{-1}A^{-1})(AB) = B^{-1}(A^{-1}A)B = B^{-1}I_nB = B^{-1}B = I_n$$

Thus, AB is nonsingular and its inverse is given by $B^{-1}A^{-1}$.

Theorem 1-4.2 can be extended to the product of any finite number of nonsingular matrices (see Problems 1-22 and 1-23).

C. Systems of linear equations revisited

- A system of n linear equations in n unknowns is *homogeneous* when all the right-hand-side constants b vanish, such that the system always has at least one solution—the n-tuple of zeros, which is called the *trivial solution*.

We now return to considering the system of n linear equations in n unknowns written in matrix notation as $A\mathbf{x} = \mathbf{b}$, where A is an $n \times n$ coefficient matrix and $\mathbf{x}$ and $\mathbf{b}$ are n-vectors.

When $\mathbf{b} = \mathbf{0}$ in $A\mathbf{x} = \mathbf{b}$, then

$$A\mathbf{x} = \mathbf{0} \tag{1.39}$$

is called **homogeneous**. Note that $\mathbf{x} = \mathbf{0}$ automatically satisfies (1.39) and hence $\mathbf{x} = \mathbf{0}$ is called the **trivial solution**. Any other solution is called **nontrivial**. If only one nontrivial solution exists, that solution is said to be **unique**.

THEOREM 1-4.3 If an $n \times n$ matrix A is nonsingular, then the system $A\mathbf{x} = \mathbf{b}$ has a unique solution, and that solution is $A^{-1}\mathbf{b}$.

EXAMPLE 1-19: Verify Theorem 1-4.3.

Solution: If A is nonsingular, then A^{-1} exists by definition. Then, if we let

$$\mathbf{x} = A^{-1}\mathbf{b}$$

we get

$$A\mathbf{x} = A(A^{-1}\mathbf{b}) = (AA^{-1})\mathbf{b} = I_n\mathbf{b} = \mathbf{b}$$

Thus, $A^{-1}\mathbf{b}$ is the solution.

To see that the solution is unique, suppose that $\mathbf{x}_1$ and $\mathbf{x}_2$ are two solutions to the system; that is,

$$A\mathbf{x}_1 = \mathbf{b} \qquad \text{and} \qquad A\mathbf{x}_2 = \mathbf{b}$$

Then

$$A\mathbf{x}_1 = A\mathbf{x}_2$$

Multiplying both sides by A^{-1} from the left yields

$$A^{-1}(A\mathbf{x}_1) = A^{-1}(A\mathbf{x}_2)$$

or

$$(A^{-1}A)\mathbf{x}_1 = (A^{-1}A)\mathbf{x}_2$$

so that

$$I_n\mathbf{x}_1 = I_n\mathbf{x}_2$$

or

$$\mathbf{x}_1 = \mathbf{x}_2$$

in view of the identity property (1.35).

THEOREM 1-4.4 The following are equivalent:

(a) A is nonsingular
(b) $A\mathbf{x} = \mathbf{0}$ has only the trivial solution $\mathbf{0}$.

EXAMPLE 1-20: Verify that if A is nonsingular, $A\mathbf{x} = \mathbf{0}$ must have only the trivial solution $\mathbf{0}$. (Theorem 1-4.4).

Solution: If A is nonsingular, then A^{-1} exists. Let $\hat{\mathbf{x}}$ be a solution to $A\mathbf{x} = \mathbf{0}$; that is, let

$$A\hat{\mathbf{x}} = \mathbf{0}$$

Then, by (1.35), (1.37), and (1.20),

$$\hat{\mathbf{x}} = I_n\hat{\mathbf{x}} = (A^{-1}A)\hat{\mathbf{x}} = A^{-1}(A\hat{\mathbf{x}}) = A^{-1}\mathbf{0} = \mathbf{0}$$

Thus, $A\mathbf{x} = \mathbf{0}$ has only the trivial solution.

COROLLARY 1-4.5 $A\mathbf{x} = \mathbf{0}$ has a nonzero solution if and only if A is singular.

This follows from Theorem 1-4.4.

1-5. Transpose of a Matrix

A. Transpose operation

DEFINITION The **transpose** of an $m \times n$ matrix $A = [a_{ij}]_{m \times n}$, denoted by A^T, is the $n \times m$ matrix $B = [b_{ij}]_{n \times m}$ defined by

TRANSPOSE OF A MATRIX

$$b_{ij} = a_{ji} \tag{1.40}$$

It follows from (1.40) that A^T is the matrix obtained by interchanging the rows and columns of A. For example, the transpose of $\begin{bmatrix} 1 & 2 \\ 3 & 4 \end{bmatrix}$ is $\begin{bmatrix} 1 & 3 \\ 2 & 4 \end{bmatrix}$.

EXAMPLE 1-21: Find the transpose of the following matrices:

$$A = \begin{bmatrix} 1 & 2 & 3 \\ 0 & 4 & -1 \end{bmatrix}, \qquad B = [1 \quad 3 \quad 5], \qquad C = \begin{bmatrix} 1 & -1 \\ 2 & 3 \end{bmatrix}, \qquad D = \begin{bmatrix} 1 \\ 2 \\ 3 \end{bmatrix}$$

Solution

$$A^T = \begin{bmatrix} 1 & 0 \\ 2 & 4 \\ 3 & -1 \end{bmatrix}, \qquad B^T = \begin{bmatrix} 1 \\ 3 \\ 5 \end{bmatrix}, \qquad C^T = \begin{bmatrix} 1 & 2 \\ -1 & 3 \end{bmatrix}, \qquad D^T = [1 \quad 2 \quad 3]$$

B. Properties of transpose operation

There are five basic properties involving transpose operation:

$$(A+B)^T = A^T + B^T \tag{1.41}$$

$$(\alpha A)^T = \alpha A^T \tag{1.42}$$

$$(AB)^T = B^T A^T \tag{1.43}$$

$$(A^T)^T = A \tag{1.44}$$

$$(A^T)^{-1} = (A^{-1})^T \qquad \text{when } A \text{ is nonsingular} \tag{1.45}$$

EXAMPLE 1-22: Show that $(A+B)^T = A^T + B^T$ and $(AB)^T = B^T A^T$, where

$$A = \begin{bmatrix} 1 & 1 \\ 3 & 4 \end{bmatrix}, \qquad B = \begin{bmatrix} 3 & 1 \\ -2 & -1 \end{bmatrix}$$

Solution: We know that

$$A + B = \begin{bmatrix} 1 & 1 \\ 3 & 4 \end{bmatrix} + \begin{bmatrix} 3 & 1 \\ -2 & -1 \end{bmatrix} = \begin{bmatrix} 4 & 2 \\ 1 & 3 \end{bmatrix}, \qquad AB = \begin{bmatrix} 1 & 1 \\ 3 & 4 \end{bmatrix}\begin{bmatrix} 3 & 1 \\ -2 & -1 \end{bmatrix} = \begin{bmatrix} 1 & 0 \\ 1 & -1 \end{bmatrix}$$

and

$$A^T = \begin{bmatrix} 1 & 3 \\ 1 & 4 \end{bmatrix}, \qquad B^T = \begin{bmatrix} 3 & -2 \\ 1 & -1 \end{bmatrix}$$

Now, we see that

$$(A+B)^T = \begin{bmatrix} 4 & 1 \\ 2 & 3 \end{bmatrix}$$

and

$$A^T + B^T = \begin{bmatrix} 1 & 3 \\ 1 & 4 \end{bmatrix} + \begin{bmatrix} 3 & -2 \\ 1 & -1 \end{bmatrix} = \begin{bmatrix} 4 & 1 \\ 2 & 3 \end{bmatrix}$$

Thus, $(A+B)^T = A^T + B^T$.

Next, we see that

$$(AB)^T = \begin{bmatrix} 1 & 1 \\ 0 & -1 \end{bmatrix} \quad \text{and} \quad B^T A^T = \begin{bmatrix} 3 & -2 \\ 1 & -1 \end{bmatrix}\begin{bmatrix} 1 & 3 \\ 1 & 4 \end{bmatrix} = \begin{bmatrix} 1 & 1 \\ 0 & -1 \end{bmatrix}$$

Thus, $(AB)^T = B^T A^T$.

EXAMPLE 1-23: Verify property (1.43); that is, $(AB)^T = B^T A^T$.

Solution: Let $C = AB$ and denote the ijth entries of A^T, B^T, and C^T by a_{ij}^T, b_{ij}^T, and c_{ij}^T, respectively. Then, by definition (1.40),

$$c_{ij}^T = c_{ji}, \qquad a_{ij}^T = a_{ji}, \qquad b_{ij}^T = b_{ji}$$

Using matrix multiplication (1.18), we find the ijth entry of $B^T A^T$ by

$$\sum_{k=1}^{n} b_{ik}^T a_{kj}^T = \sum_{k=1}^{n} b_{ki} a_{jk} = \sum_{k=1}^{n} a_{jk} b_{ki}$$

and the ijth entry of $C^T = (AB)^T$ is given by

$$c_{ij}^T = c_{ji} = \sum_{k=1}^{n} a_{jk} b_{ki}$$

Hence it follows that

$$(AB)^T = B^T A^T$$

1-6. Special Matrices

A. Diagonal matrices

Let $D = [d_{ij}]_{n \times n}$ be an $n \times n$ matrix. We call D a **diagonal matrix** if all its nondiagonal entries are zeros. That is,

$$d_{ij} = 0 \qquad \text{whenever } i \neq j \tag{1.46}$$

If $d_{11} = d_1, d_{22} = d_2, \ldots, d_{nn} = d_n$, then the Kronecker delta (1.32) can be used to express D as

DIAGONAL MATRIX

$$D = [d_i \delta_{ij}] = \begin{bmatrix} d_1 & 0 & 0 & \cdots & 0 \\ 0 & d_2 & 0 & \cdots & 0 \\ 0 & 0 & d_3 & \cdots & 0 \\ \vdots & \vdots & \vdots & \ddots & \vdots \\ 0 & 0 & 0 & \cdots & d_n \end{bmatrix} \tag{1.47}$$

There are three properties of diagonal matrices we should know:

(a) The sum of two diagonal matrices is also a diagonal matrix.
(b) The product of two diagonal matrices is also a diagonal matrix.
(c) The product of two diagonal matrices is commutative.

EXAMPLE 1-24: Compute **(a)** $D_1 + D_2$, **(b)** $D_1 D_2$, and **(c)** $D_2 D_1$, where

$$D_1 = \begin{bmatrix} 3 & 0 & 0 \\ 0 & 2 & 0 \\ 0 & 0 & 5 \end{bmatrix}, \qquad D_2 = \begin{bmatrix} 1 & 0 & 0 \\ 0 & 4 & 0 \\ 0 & 0 & 2 \end{bmatrix}$$

Solution

(a)
$$D_1 + D_2 = \begin{bmatrix} 3 & 0 & 0 \\ 0 & 2 & 0 \\ 0 & 0 & 5 \end{bmatrix} + \begin{bmatrix} 1 & 0 & 0 \\ 0 & 4 & 0 \\ 0 & 0 & 2 \end{bmatrix} = \begin{bmatrix} 4 & 0 & 0 \\ 0 & 6 & 0 \\ 0 & 0 & 7 \end{bmatrix}$$

(b)
$$D_1 D_2 = \begin{bmatrix} 3 & 0 & 0 \\ 0 & 2 & 0 \\ 0 & 0 & 5 \end{bmatrix} \begin{bmatrix} 1 & 0 & 0 \\ 0 & 4 & 0 \\ 0 & 0 & 2 \end{bmatrix} = \begin{bmatrix} 3 & 0 & 0 \\ 0 & 8 & 0 \\ 0 & 0 & 10 \end{bmatrix}$$

(c)
$$D_2 D_1 = \begin{bmatrix} 1 & 0 & 0 \\ 0 & 4 & 0 \\ 0 & 0 & 2 \end{bmatrix} \begin{bmatrix} 3 & 0 & 0 \\ 0 & 2 & 0 \\ 0 & 0 & 5 \end{bmatrix} = \begin{bmatrix} 3 & 0 & 0 \\ 0 & 8 & 0 \\ 0 & 0 & 10 \end{bmatrix}$$

EXAMPLE 1-25: Let

$$D = [d_i \delta_{ij}]_{n \times n} = \begin{bmatrix} d_1 & 0 & \cdots & 0 \\ 0 & d_2 & \cdots & 0 \\ \vdots & \vdots & \ddots & \vdots \\ 0 & 0 & \cdots & d_n \end{bmatrix} \qquad \text{where } d_i \neq 0 \text{ for all } i.$$

Show that D is invertible and that

$$D^{-1} = \left[\frac{1}{d_i}\delta_{ij}\right]_{n \times n} = \begin{bmatrix} \frac{1}{d_1} & 0 & \cdots & 0 \\ 0 & \frac{1}{d_2} & \cdots & 0 \\ \vdots & \vdots & \ddots & \vdots \\ 0 & 0 & \cdots & \frac{1}{d_n} \end{bmatrix} \tag{1.48}$$

Solution: Since

$$\begin{bmatrix} d_1 & 0 & \cdots & 0 \\ 0 & d_2 & \cdots & 0 \\ \vdots & \vdots & \ddots & \vdots \\ 0 & 0 & \cdots & d_n \end{bmatrix} \begin{bmatrix} \frac{1}{d_1} & 0 & \cdots & 0 \\ 0 & \frac{1}{d_2} & \cdots & 0 \\ \vdots & \vdots & \ddots & \vdots \\ 0 & 0 & \cdots & \frac{1}{d_n} \end{bmatrix} = \begin{bmatrix} 1 & 0 & \cdots & 0 \\ 0 & 1 & \cdots & 0 \\ \vdots & \vdots & \ddots & \vdots \\ 0 & 0 & \cdots & 1 \end{bmatrix} = I_n$$

$$\begin{bmatrix} \frac{1}{d_1} & 0 & \cdots & 0 \\ 0 & \frac{1}{d_2} & \cdots & 0 \\ \vdots & \vdots & \ddots & \vdots \\ 0 & 0 & \cdots & \frac{1}{d_n} \end{bmatrix} \begin{bmatrix} d_1 & 0 & \cdots & 0 \\ 0 & d_2 & \cdots & 0 \\ \vdots & \vdots & \ddots & \vdots \\ 0 & 0 & \cdots & d_n \end{bmatrix} = \begin{bmatrix} 1 & 0 & \cdots & 0 \\ 0 & 1 & \cdots & 0 \\ \vdots & \vdots & \ddots & \vdots \\ 0 & 0 & \cdots & 1 \end{bmatrix} = I_n$$

thus

$$D^{-1} = \left[\frac{1}{d_i}\delta_{ij}\right]$$

B. Scalar matrices

A diagonal matrix $D = [d_i\delta_{ij}]$ of order n is called a **scalar matrix** if all its diagonal elements d_i are equal:

$$d_1 = d_2 = \cdots = d_n = \lambda$$

that is,

SCALAR MATRIX

$$D = \begin{bmatrix} \lambda & 0 & \cdots & 0 \\ 0 & \lambda & \cdots & 0 \\ \vdots & \vdots & \ddots & \vdots \\ 0 & 0 & \cdots & \lambda \end{bmatrix} \tag{1.49}$$

Note that (1.49) can be rewritten as

$$D = \lambda I_n \tag{1.50}$$

EXAMPLE 1-26: Compute (**a**) $3A$ and (**b**) DA, where

$$A = \begin{bmatrix} 1 & 2 & -1 & 0 \\ 3 & 1 & 2 & 4 \\ 0 & 3 & 5 & 1 \end{bmatrix}, \qquad D = \begin{bmatrix} 3 & 0 & 0 \\ 0 & 3 & 0 \\ 0 & 0 & 3 \end{bmatrix}$$

Solution

(a) By (1.12),

$$3A = 3\begin{bmatrix} 1 & 2 & -1 & 0 \\ 3 & 1 & 2 & 4 \\ 0 & 3 & 5 & 1 \end{bmatrix} = \begin{bmatrix} 3 & 6 & -3 & 0 \\ 9 & 3 & 6 & 12 \\ 0 & 9 & 15 & 3 \end{bmatrix}$$

(b)

$$DA = \begin{bmatrix} 3 & 0 & 0 \\ 0 & 3 & 0 \\ 0 & 0 & 3 \end{bmatrix}\begin{bmatrix} 1 & 2 & -1 & 0 \\ 3 & 1 & 2 & 4 \\ 0 & 3 & 5 & 1 \end{bmatrix} = \begin{bmatrix} 3 & 6 & -3 & 0 \\ 9 & 3 & 6 & 12 \\ 0 & 9 & 15 & 3 \end{bmatrix}$$

Example 1-26 illustrates why D is called a "scalar" matrix:

- The product of a scalar matrix $D = \lambda I_n$ and an $m \times n$ matrix A is the same as the product of the matrix A and the scalar λ.

note: The identity matrix I_n is a scalar matrix, where $\lambda = 1$.

C. Triangular matrices

Let $A = [a_{ij}]$ be an $n \times n$ matrix:

- A is called **upper-triangular** if all the entries *below* the diagonal are zero; that is,

UPPER-TRIANGULAR MATRIX $\qquad a_{ij} = 0 \qquad \text{for } i > j$ **(1.51)**

- A is called **lower-triangular** if all the entries above the diagonal are zero; that is

LOWER-TRIANGULAR MATRIX $\qquad a_{ij} = 0 \qquad \text{for } i < j$ **(1.52)**

- A is said to be a **triangular matrix** if it is either upper-triangular or lower-triangular.

For example, the 3×3 matrices $\begin{bmatrix} 1 & 2 & 3 \\ 0 & 2 & 1 \\ 0 & 0 & 4 \end{bmatrix}$ and $\begin{bmatrix} 1 & 0 & 0 \\ 4 & 0 & 0 \\ 3 & 1 & 2 \end{bmatrix}$ are both triangular: The first is upper-triangular and the second is lower-triangular.

There are two properties of triangular matrices we should know:

(a) The sum of two upper-triangular matrices is upper-triangular, and the sum of two lower-triangular matrices is lower-triangular.

(b) The product of two upper-triangular matrices is upper-triangular, and the product of two lower-triangular matrices is lower-triangular.

EXAMPLE 1-27: Compute **(a)** $A + B$ and **(b)** AB where

$$A = \begin{bmatrix} 1 & 2 & -1 \\ 0 & 1 & 0 \\ 0 & 0 & 3 \end{bmatrix}, \qquad B = \begin{bmatrix} 2 & 1 & 0 \\ 0 & 1 & 3 \\ 0 & 0 & -1 \end{bmatrix}$$

Solution

(a)

$$A + B = \begin{bmatrix} 1 & 2 & -1 \\ 0 & 1 & 0 \\ 0 & 0 & 3 \end{bmatrix} + \begin{bmatrix} 2 & 1 & 0 \\ 0 & 1 & 3 \\ 0 & 0 & -1 \end{bmatrix} = \begin{bmatrix} 3 & 3 & -1 \\ 0 & 2 & 3 \\ 0 & 0 & 2 \end{bmatrix}$$

(b)

$$AB = \begin{bmatrix} 1 & 2 & -1 \\ 0 & 1 & 0 \\ 0 & 0 & 3 \end{bmatrix}\begin{bmatrix} 2 & 1 & 0 \\ 0 & 1 & 3 \\ 0 & 0 & -1 \end{bmatrix} = \begin{bmatrix} 2 & 3 & 7 \\ 0 & 1 & 3 \\ 0 & 0 & -3 \end{bmatrix}$$

D. Symmetric matrices

An $n \times n$ matrix A is called **symmetric** if

SYMMETRIC MATRIX

$$A^T = A \tag{1.53}$$

Thus, if $A = [a_{ij}]_{n \times n}$ is symmetric, then

$$a_{ij} = a_{ji} \tag{1.54}$$

EXAMPLE 1-28: Let $A = \begin{bmatrix} 1 & -1 & 0 \\ 2 & 3 & 1 \end{bmatrix}$. Compute AA^T and A^TA and show that they are symmetric.

Solution

$$B = AA^T = \begin{bmatrix} 1 & -1 & 0 \\ 2 & 3 & 1 \end{bmatrix} \begin{bmatrix} 1 & 2 \\ -1 & 3 \\ 0 & 1 \end{bmatrix} = \begin{bmatrix} 2 & -1 \\ -1 & 14 \end{bmatrix}$$

$$C = A^TA = \begin{bmatrix} 1 & 2 \\ -1 & 3 \\ 0 & 1 \end{bmatrix} \begin{bmatrix} 1 & -1 & 0 \\ 2 & 3 & 1 \end{bmatrix} = \begin{bmatrix} 5 & 5 & 2 \\ 5 & 10 & 3 \\ 2 & 3 & 1 \end{bmatrix}$$

Since $B^T = B$ and $C^T = C$, B and C are symmetric; that is, AA^T and A^TA are symmetric.

SUMMARY

1. A matrix A is any rectangular array of numbers expressed as

$$A = [a_{ij}]_{m \times n} = \begin{bmatrix} a_{11} & a_{12} & \cdots & a_{1n} \\ a_{21} & a_{22} & \cdots & a_{2n} \\ \vdots & \vdots & & \vdots \\ a_{m1} & a_{m2} & \cdots & a_{mn} \end{bmatrix}$$

   Here, a_{ij} is the ijth entry of A, and $m \times n$ is the size of A, where m is the number of rows and n is the number of columns.

2. Equality of matrices: If $A = [a_{ij}]_{m \times n}$ and $B = [b_{ij}]_{m \times n}$,

$$A = B \Rightarrow a_{ij} = b_{ij}$$

3. Sum of matrices: If $A = [a_{ij}]_{m \times n}$, $B = [b_{ij}]_{m \times n}$, and $C = [c_{ij}]_{m \times n}$,

$$C = A + B \Rightarrow c_{ij} = a_{ij} + b_{ij}$$

4. Multiplication of matrices by a scalar: If $A = [a_{ij}]_{m \times n}$ and α is any scalar,

$$\alpha A = [\alpha a_{ij}]_{m \times n}$$

5. Matrix product: If $A = [a_{ij}]_{m \times n}$, $B = [b_{ij}]_{n \times p}$, and $C = [c_{ij}]_{m \times p}$,

$$C = AB \Rightarrow c_{ij} = \sum_{k=1}^{n} a_{ik} b_{kj}$$

   In general, $AB \neq BA$.

6. Properties of basic matrix operations:

$$A + B = B + A \qquad \alpha(A + B) = \alpha A + \alpha B$$

$$A + (B + C) = (A + B) + C \qquad (\alpha + \beta)A = \alpha A + \beta A$$

$$(AB)C = A(BC) \qquad (\alpha\beta)A = \alpha(\beta A)$$

$$A(B + C) = AB + AC \qquad \alpha(AB) = (\alpha A)B = A(\alpha B)$$

7. If A is a square matrix, then the inverse of A, denoted A^{-1}, is defined by

$$AA^{-1} = A^{-1}A = I_n$$

where I_n is the $n \times n$ identity matrix $I_n = \begin{bmatrix} 1 & 0 & \cdots & 0 \\ 0 & 1 & \cdots & 0 \\ \vdots & \vdots & & \vdots \\ 0 & 0 & \cdots & 1 \end{bmatrix}$.

8. Matrix A is nonsingular (or invertible) if A^{-1} exists.
9. If A and B are both nonsingular matrices, then

$$(AB)^{-1} = B^{-1}A^{-1}$$

10. A system of m linear equations of n unknowns

$$\begin{aligned} a_{11}x_1 + a_{12}x_2 + \cdots + a_{1n}x_n &= b_1 \\ a_{21}x_1 + a_{22}x_2 + \cdots + a_{2n}x_n &= b_2 \\ \vdots \qquad \vdots \qquad\qquad \vdots \quad &\quad \vdots \\ a_{m1}x_1 + a_{m2}x_2 + \cdots + a_{mn}x_n &= b_m \end{aligned}$$

can be expressed in matrix form as

$$A\mathbf{x} = \mathbf{b}$$

where

$A = [a_{ij}]_{m \times n} = \begin{bmatrix} a_{11} & a_{12} & \cdots & a_{1n} \\ a_{21} & a_{22} & \cdots & a_{2n} \\ \vdots & \vdots & & \vdots \\ a_{m1} & a_{m2} & \cdots & a_{mn} \end{bmatrix}$ is the coefficient matrix of order $m \times n$

$\mathbf{x} = [x_j]_{n \times 1} = \begin{bmatrix} x_1 \\ x_2 \\ \vdots \\ x_n \end{bmatrix}$ is an $n \times 1$ matrix (or n-vector) consisting of n unknowns

$\mathbf{b} = [b_i]_{m \times 1} = \begin{bmatrix} b_1 \\ b_2 \\ \vdots \\ b_m \end{bmatrix}$ is an $m \times 1$ matrix (or m-vector) representing inhomogeneous terms.

11. If matrix A is nonsingular, then $A\mathbf{x} = \mathbf{b}$ has a unique solution

$$\mathbf{x} = A^{-1}\mathbf{b}$$

12. If $A = [a_{ij}]_{m \times n}$, then the transpose of A, denoted A^T, is defined as

$$A^T = [a_{ji}]_{n \times m}$$

13. Properties of transpose operations:

$$(A + B)^T = A^T + B^T \qquad (A^T)^T = A$$

$$(\alpha A)^T = \alpha A^T \qquad (A^T)^{-1} = (A^{-1})^T$$

$$(AB)^T = B^T A^T$$

14. Matrix A is diagonal if it can be expressed as

$$\begin{bmatrix} a_1 & 0 & \cdots & 0 \\ 0 & a_2 & \cdots & 0 \\ \vdots & \vdots & \ddots & \vdots \\ 0 & 0 & \cdots & a_n \end{bmatrix}$$

15. Sums and products of diagonal matrices are also diagonal matrices.
16. Matrix A is triangular if it can be expressed as

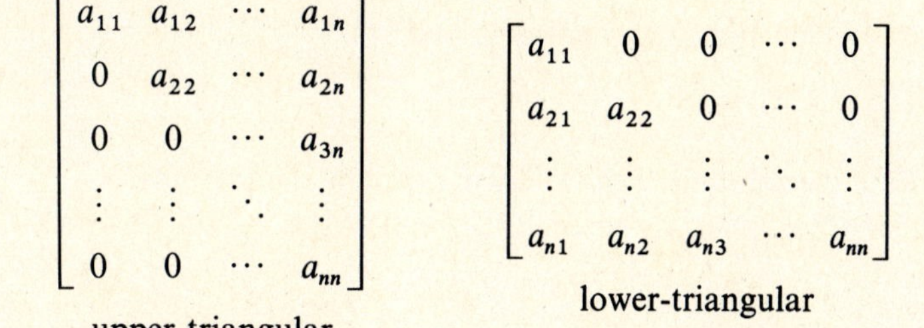

upper-triangular   lower-triangular

17. Matrix A is symmetric if $A^T = A$.

RAISE YOUR GRADES

Can you explain ...?

- ☑ how to add matrices
- ☑ how to multiply two matrices
- ☑ why the product of two matrices is not generally commutative
- ☑ how the inverse of a matrix is defined
- ☑ what the inverse of the product of two matrices is
- ☑ what conditions are necessary for a matrix to be nonsingular (or invertible)
- ☑ what condition is necessary if $A\mathbf{x} = \mathbf{0}$ is to have a nontrivial solution
- ☑ how the transpose of a matrix is defined
- ☑ what the properties of the transpose of a matrix are

SOLVED PROBLEMS

Systems of Linear Equations

PROBLEM 1-1 Find the solution set of (**a**) $2x_1 - x_2 = 1$ and (**b**) $2x_1 + x_2 - x_3 = 2$.

Solution

(**a**) From the equation, we get

$$x_2 = 2x_1 - 1$$

Then, setting $x_1 = t$, we have

$$x_2 = 2t - 1$$

Hence the solution set in terms of the arbitrary parameter t is

$$x_1 = t, \qquad x_2 = 2t - 1$$

(b) From the system, we get

$$x_3 = -2 + 2x_1 + x_2$$

Then setting $x_1 = t$ and $x_2 = s$, we have

$$x_3 = -2 + 2t + s$$

Hence the solution set with arbitrary parameters t and s is

$$x_1 = t, \qquad x_2 = s, \qquad x_3 = -2 + 2t + s$$

PROBLEM 1-2 Find the solution of the following system of equations:

$$3x_1 + 2x_2 = 12$$

$$2x_1 - 5x_2 = -11$$

Solution From the last equation of the system, we get

$$2x_1 = 5x_2 - 11$$

or

$$x_1 = \tfrac{5}{2}x_2 - \tfrac{11}{2}$$

Substituting this value of x_1 into the first equation, we obtain

$$3(\tfrac{5}{2}x_2 - \tfrac{11}{2}) + 2x_2 = 12$$

$$\tfrac{19}{2}x_2 = 12 + \tfrac{33}{2} = \tfrac{57}{2}$$

$$x_2 = \frac{\frac{57}{2}}{\frac{19}{2}} = 3$$

Thus

$$x_1 = \tfrac{5}{2}(3) - \tfrac{11}{2} = \tfrac{15}{2} - \tfrac{11}{2} = \tfrac{4}{2} = 2$$

Hence the solution set is $x_1 = 2$, $x_2 = 3$.

PROBLEM 1-3 Find the solution of the following system of equations:

$$x_1 + 2x_2 - x_3 = 1 \qquad \textbf{(a)}$$

$$3x_1 - 4x_2 + x_3 = -1 \qquad \textbf{(b)}$$

Solution

(1) Adding (a) and (b), we get

$$4x_1 - 2x_2 = 0$$

from which we see that $x_2 = 2x_1$.

(2) Then, from (a), we solve x_3:

$$x_3 = x_1 + 2x_2 - 1$$

(3) Setting $x_1 = t$, we get $x_2 = 2t$ and $x_3 = t + 4t - 1 = 5t - 1$.
Hence the solution set is

$$x_1 = t, \qquad x_2 = 2t, \qquad x_3 = 5t - 1$$

with parameter t.

PROBLEM 1-4 Find the solution of the system

$$x_1 - x_2 = 3 \qquad \textbf{(a)}$$
$$2x_1 + 3x_2 = 2 \qquad \textbf{(b)}$$
$$3x_1 + 2x_2 = 5 \qquad \textbf{(c)}$$

Solution From (a), we have

$$x_1 = 3 + x_2$$

Substituting this value into (b), we have

$$2(3 + x_2) + 3x_2 = 2$$

from which we see that $x_2 = -\frac{4}{5}$; consequently,

$$x_1 = 3 + (-\tfrac{4}{5}) = \tfrac{11}{5}$$

Now, substituting $x_1 = \frac{11}{5}$ and $x_2 = -\frac{4}{5}$ into (c), we have

$$3(\tfrac{11}{5}) + 2(-\tfrac{4}{5}) = \tfrac{25}{5} = 5$$

Thus, the system is consistent and the solution is

$$x_1 = \tfrac{11}{5}, \qquad x_2 = -\tfrac{4}{5}$$

note: (c) can be obtained by adding (a) and (b).

PROBLEM 1-5 Find the solution of the system

$$x_1 + x_2 = 3 \qquad \textbf{(a)}$$
$$2x_1 - 3x_2 = 1 \qquad \textbf{(b)}$$
$$3x_1 + 2x_2 = 4 \qquad \textbf{(c)}$$

Solution From (a), we have

$$x_1 = 3 - x_2$$

Substituting this value into (b), we have

$$2(3 - x_2) - 3x_2 = 1$$

from which we see that $x_2 = 1$; hence

$$x_1 = 3 - 1 = 2$$

Now, substituting $x_1 = 2$ and $x_2 = 1$ into (c), we have

$$3(2) + 2(1) = 8 \neq 4$$

Thus, the system is inconsistent and there is no solution.

Matrix Notation

PROBLEM 1-6 Write the systems of **(a)** $\begin{matrix} 3x_1 + 2x_2 = 12 \\ 2x_1 - 5x_2 = -11 \end{matrix}$ (Problem 1-2) and **(b)** $\begin{matrix} x_1 - x_2 = 3 \\ 2x_1 + 3x_2 = 2 \\ 3x_1 + 2x_2 = 5 \end{matrix}$ (Problem 1-4) in matrix form.

Solution

(a) $A = \begin{bmatrix} 3 & 2 \\ 2 & -5 \end{bmatrix}$, $\mathbf{x} = \begin{bmatrix} x_1 \\ x_2 \end{bmatrix}$, $\mathbf{b} = \begin{bmatrix} 12 \\ -11 \end{bmatrix}$. Thus

$$\begin{bmatrix} 3 & 2 \\ 2 & -5 \end{bmatrix}\begin{bmatrix} x_1 \\ x_2 \end{bmatrix} = \begin{bmatrix} 12 \\ -11 \end{bmatrix}$$

(b) $A = \begin{bmatrix} 1 & -1 \\ 2 & 3 \\ 3 & 2 \end{bmatrix}$, $\mathbf{x} = \begin{bmatrix} x_1 \\ x_2 \end{bmatrix}$, $\mathbf{b} = \begin{bmatrix} 3 \\ 2 \\ 5 \end{bmatrix}$. Thus

$$\begin{bmatrix} 1 & -1 \\ 2 & 3 \\ 3 & 2 \end{bmatrix} \begin{bmatrix} x_1 \\ x_2 \end{bmatrix} = \begin{bmatrix} 3 \\ 2 \\ 5 \end{bmatrix}$$

PROBLEM 1-7 Solve the following matrix equation for x_1, x_2, and x_3:

$$\begin{bmatrix} x_1 + x_2 & 2x_2 + x_3 \\ x_2 - x_3 & x_1 - x_2 \end{bmatrix} = \begin{bmatrix} 4 & 4 \\ -1 & 2 \end{bmatrix}$$

Solution From the definition of matrix equality (1.7), we know that

$$x_1 + x_2 = 4 \qquad \textbf{(a)} \qquad\qquad 2x_2 + x_3 = 4 \qquad \textbf{(c)}$$

$$x_2 - x_3 = -1 \qquad \textbf{(b)} \qquad\qquad x_1 - x_2 = 2 \qquad \textbf{(d)}$$

Then, adding (a) to (d) and subtracting (d) from (a), we get

$$2x_1 = 6 \rightarrow x_1 = 3 \qquad \textbf{(a) + (d)}$$

$$2x_2 = 2 \rightarrow x_2 = 1 \qquad \textbf{(a) − (d)}$$

Substituting $x_2 = 1$ into (b) and (c), we get

$$x_3 = 1 + x_2 = 1 + 1 = 2$$

$$x_3 = 4 - 2x_2 = 4 - 2 = 2 \qquad \text{which is consistent}$$

Thus

$$x_1 = 3, \qquad x_2 = 1, \qquad x_3 = 2$$

Basic Matrix Operations

PROBLEM 1-8 Compute **(a)** $A + B$, **(b)** $A - B$, **(c)** $B + C$, **(d)** $(A + B) + C$, **(e)** $A + (B + C)$, **(f)** $(A + 2B)$, **(g)** $A - D$, where

$$A = \begin{bmatrix} 2 & 3 \\ 4 & 0 \\ -1 & 1 \end{bmatrix}, \quad B = \begin{bmatrix} -1 & 2 \\ 2 & 1 \\ -1 & 1 \end{bmatrix}, \quad C = \begin{bmatrix} 0 & -1 \\ 3 & 2 \\ 2 & -3 \end{bmatrix}, \quad D = \begin{bmatrix} 2 & -1 & 1 \\ 4 & 0 & 2 \end{bmatrix}$$

Solution

(a)
$$A + B = \begin{bmatrix} 2 & 3 \\ 4 & 0 \\ -1 & 1 \end{bmatrix} + \begin{bmatrix} -1 & 2 \\ 2 & 1 \\ -1 & 1 \end{bmatrix} = \begin{bmatrix} 2 + (-1) & 3 + 2 \\ 4 + 2 & 0 + 1 \\ -1 + (-1) & 1 + 1 \end{bmatrix} = \begin{bmatrix} 1 & 5 \\ 6 & 1 \\ -2 & 2 \end{bmatrix}$$

(b)
$$A - B = \begin{bmatrix} 2 & 3 \\ 4 & 0 \\ -1 & 1 \end{bmatrix} - \begin{bmatrix} -1 & 2 \\ 2 & 1 \\ -1 & 1 \end{bmatrix} = \begin{bmatrix} 2 - (-1) & 3 - 2 \\ 4 - 2 & 0 - 1 \\ -1 - (-1) & 1 - 1 \end{bmatrix} = \begin{bmatrix} 3 & 1 \\ 2 & -1 \\ 0 & 0 \end{bmatrix}$$

(c)
$$B + C = \begin{bmatrix} -1 & 2 \\ 2 & 1 \\ -1 & 1 \end{bmatrix} + \begin{bmatrix} 0 & -1 \\ 3 & 2 \\ 2 & -3 \end{bmatrix} = \begin{bmatrix} -1 + 0 & 2 - 1 \\ 2 + 3 & 1 + 2 \\ -1 + 2 & 1 - 3 \end{bmatrix} = \begin{bmatrix} -1 & 1 \\ 5 & 3 \\ 1 & -2 \end{bmatrix}$$

(d)
$$(A+B)+C = \begin{bmatrix} 1 & 5 \\ 6 & 1 \\ -2 & 2 \end{bmatrix} + \begin{bmatrix} 0 & -1 \\ 3 & 2 \\ 2 & -3 \end{bmatrix} = \begin{bmatrix} 1+0 & 5-1 \\ 6+3 & 1+2 \\ -2+2 & 2-3 \end{bmatrix} = \begin{bmatrix} 1 & 4 \\ 9 & 3 \\ 0 & -1 \end{bmatrix}$$

(e)
$$A+(B+C) = \begin{bmatrix} 2 & 3 \\ 4 & 0 \\ -1 & 1 \end{bmatrix} + \begin{bmatrix} -1 & 1 \\ 5 & 3 \\ 1 & -2 \end{bmatrix} = \begin{bmatrix} 2-1 & 3+1 \\ 4+5 & 0+3 \\ -1+1 & 1-2 \end{bmatrix} = \begin{bmatrix} 1 & 4 \\ 9 & 3 \\ 0 & -1 \end{bmatrix}$$

(f) Since $2B = 2\begin{bmatrix} -1 & 2 \\ 2 & 1 \\ -1 & 1 \end{bmatrix} = \begin{bmatrix} -2 & 4 \\ 4 & 2 \\ -2 & 2 \end{bmatrix}$,

$$A+2B = \begin{bmatrix} 2 & 3 \\ 4 & 0 \\ -1 & 1 \end{bmatrix} + \begin{bmatrix} -2 & 4 \\ 4 & 2 \\ -2 & 2 \end{bmatrix} = \begin{bmatrix} 2-2 & 3+4 \\ 4+4 & 0+2 \\ -1-2 & 1+2 \end{bmatrix} = \begin{bmatrix} 0 & 7 \\ 8 & 2 \\ -3 & 3 \end{bmatrix}$$

(g) $A - D$ is undefined, since the sizes of A and D are different.

PROBLEM 1-9 Verify the distributive property (1.26), that $\alpha(A + B) = \alpha A + \alpha B$.

Solution Let $A = [a_{ij}]_{m\times n}$ and $B = [b_{ij}]_{m\times n}$. Then

$$A + B = [a_{ij} + b_{ij}]_{m\times n}$$

and

$$\begin{aligned} \alpha(A+B) &= \alpha[a_{ij} + b_{ij}]_{m\times n} \\ &= [\alpha(a_{ij} + b_{ij})]_{m\times n} \\ &= [\alpha a_{ij} + \alpha b_{ij}]_{m\times n} \\ &= [\alpha a_{ij}]_{m\times n} + [\alpha b_{ij}]_{m\times n} \\ &= \alpha[a_{ij}]_{m\times n} + \alpha[b_{ij}]_{m\times n} \\ &= \alpha A + \alpha B \end{aligned}$$

PROBLEM 1-10 Verify the distributive property (1.27), that $(\alpha + \beta)A = \alpha A + \beta A$.

Solution Let $A = [a_{ij}]_{m\times n}$. Then

$$\begin{aligned} (\alpha+\beta)A &= [(\alpha+\beta)a_{ij}]_{m\times n} \\ &= [\alpha a_{ij} + \beta a_{ij}]_{m\times n} \\ &= [\alpha a_{ij}]_{m\times n} + [\beta a_{ij}]_{m\times n} \\ &= \alpha[a_{ij}]_{m\times n} + \beta[a_{ij}]_{m\times n} \\ &= \alpha A + \beta A \end{aligned}$$

PROBLEM 1-11 Compute **(a)** AB, **(b)** BA, **(c)** AC, **(d)** CA, where

$$A = \begin{bmatrix} 1 & 2 & 4 \\ 3 & 1 & 5 \end{bmatrix}, \qquad B = \begin{bmatrix} 2 & 0 \\ -1 & 4 \end{bmatrix}, \qquad C = \begin{bmatrix} -1 & 2 \\ 1 & 3 \\ 0 & 1 \end{bmatrix}$$

Solution

(a) AB is not defined since A and B are not *conformable*; i.e., A cannot be multiplied by B because A has three columns and B has two rows.

(b) $$BA = \begin{bmatrix} 2 & 0 \\ -1 & 4 \end{bmatrix}\begin{bmatrix} 1 & 2 & 4 \\ 3 & 1 & 5 \end{bmatrix} = \begin{bmatrix} 2(1)+0(3) & 2(2)+0(1) & 2(4)+0(5) \\ -1(1)+4(3) & -1(2)+4(1) & -1(4)+4(5) \end{bmatrix} = \begin{bmatrix} 2 & 4 & 8 \\ 11 & 2 & 16 \end{bmatrix}$$

(c) $$AC = \begin{bmatrix} 1 & 2 & 4 \\ 3 & 1 & 5 \end{bmatrix}\begin{bmatrix} -1 & 2 \\ 1 & 3 \\ 0 & 1 \end{bmatrix} = \begin{bmatrix} 1(-1)+2(1)+4(0) & 1(2)+2(3)+4(1) \\ 3(-1)+1(1)+5(0) & 3(2)+1(3)+5(1) \end{bmatrix} = \begin{bmatrix} 1 & 12 \\ -2 & 14 \end{bmatrix}$$

(d) $$CA = \begin{bmatrix} -1 & 2 \\ 1 & 3 \\ 0 & 1 \end{bmatrix}\begin{bmatrix} 1 & 2 & 4 \\ 3 & 1 & 5 \end{bmatrix} = \begin{bmatrix} -1(1)+2(3) & -1(2)+2(1) & -1(4)+2(5) \\ 1(1)+3(3) & 1(2)+3(1) & 1(4)+3(5) \\ 0(1)+1(3) & 0(2)+1(1) & 0(4)+1(5) \end{bmatrix} = \begin{bmatrix} 5 & 0 & 6 \\ 10 & 5 & 19 \\ 3 & 1 & 5 \end{bmatrix}$$

PROBLEM 1-12 Compute AB and BA, where

$$A = \begin{bmatrix} 1 & -1 \\ -2 & 2 \end{bmatrix} \quad \text{and} \quad B = \begin{bmatrix} 1 & 3 \\ 1 & 3 \end{bmatrix}.$$

Solution

$$AB = \begin{bmatrix} 1 & -1 \\ -2 & 2 \end{bmatrix}\begin{bmatrix} 1 & 3 \\ 1 & 3 \end{bmatrix} = \begin{bmatrix} 1(1)+(-1)(1) & 1(3)+(-1)(3) \\ -2(1)+\ \ 2(1) & -2(3)+\ \ 2(3) \end{bmatrix} = \begin{bmatrix} 0 & 0 \\ 0 & 0 \end{bmatrix}$$

$$BA = \begin{bmatrix} 1 & 3 \\ 1 & 3 \end{bmatrix}\begin{bmatrix} 1 & -1 \\ -2 & 2 \end{bmatrix} = \begin{bmatrix} 1(1)+3(-2) & 1(-1)+3(2) \\ 1(1)+3(-2) & 1(-1)+3(2) \end{bmatrix} = \begin{bmatrix} -5 & 5 \\ -5 & 5 \end{bmatrix} = 5\begin{bmatrix} -1 & 1 \\ -1 & 1 \end{bmatrix}$$

PROBLEM 1-13 Verify that **(a)** $(AB)C = A(BC)$, **(b)** $A(B + C) = AB + AC$, **(c)** $(A + B)C = AC + BC$, where

$$A = \begin{bmatrix} 3 & 1 \\ 2 & 1 \end{bmatrix}, \quad B = \begin{bmatrix} -2 & 1 \\ 0 & 4 \end{bmatrix}, \quad C = \begin{bmatrix} 1 & 2 \\ 2 & -1 \end{bmatrix}$$

Solution

(a) $$AB = \begin{bmatrix} 3 & 1 \\ 2 & 1 \end{bmatrix}\begin{bmatrix} -2 & 1 \\ 0 & 4 \end{bmatrix} = \begin{bmatrix} 3(-2)+1(0) & 3(1)+1(4) \\ 2(-2)+1(0) & 2(1)+1(4) \end{bmatrix} = \begin{bmatrix} -6 & 7 \\ -4 & 6 \end{bmatrix}$$

$$BC = \begin{bmatrix} -2 & 1 \\ 0 & 4 \end{bmatrix}\begin{bmatrix} 1 & 2 \\ 2 & -1 \end{bmatrix} = \begin{bmatrix} -2(1)+1(2) & -2(2)+1(-1) \\ 0(1)+4(2) & 0(2)+4(-1) \end{bmatrix} = \begin{bmatrix} 0 & -5 \\ 8 & -4 \end{bmatrix}$$

$$(AB)C = \begin{bmatrix} -6 & 7 \\ -4 & 6 \end{bmatrix}\begin{bmatrix} 1 & 2 \\ 2 & -1 \end{bmatrix} = \begin{bmatrix} -6(1)+7(2) & -6(2)+7(-1) \\ -4(1)+6(2) & -4(2)+6(-1) \end{bmatrix} = \begin{bmatrix} 8 & -19 \\ 8 & -14 \end{bmatrix}$$

$$A(BC) = \begin{bmatrix} 3 & 1 \\ 2 & 1 \end{bmatrix}\begin{bmatrix} 0 & -5 \\ 8 & -4 \end{bmatrix} = \begin{bmatrix} 3(0)+1(8) & 3(-5)+1(-4) \\ 2(0)+1(8) & 2(-5)+1(-4) \end{bmatrix} = \begin{bmatrix} 8 & -19 \\ 8 & -14 \end{bmatrix}$$

Thus $(AB)C = A(BC)$.

(b) $$AC = \begin{bmatrix} 3 & 1 \\ 2 & 1 \end{bmatrix}\begin{bmatrix} 1 & 2 \\ 2 & -1 \end{bmatrix} = \begin{bmatrix} 3(1)+1(2) & 3(2)+1(-1) \\ 2(1)+1(2) & 2(2)+1(-1) \end{bmatrix} = \begin{bmatrix} 5 & 5 \\ 4 & 3 \end{bmatrix}$$

$$B + C = \begin{bmatrix} -2 & 1 \\ 0 & 4 \end{bmatrix} + \begin{bmatrix} 1 & 2 \\ 2 & -1 \end{bmatrix} = \begin{bmatrix} -1 & 3 \\ 2 & 3 \end{bmatrix}$$

$$A(B + C) = \begin{bmatrix} 3 & 1 \\ 2 & 1 \end{bmatrix}\begin{bmatrix} -1 & 3 \\ 2 & 3 \end{bmatrix} = \begin{bmatrix} 3(-1)+1(2) & 3(3)+1(3) \\ 2(-1)+1(2) & 2(3)+1(3) \end{bmatrix} = \begin{bmatrix} -1 & 12 \\ 0 & 9 \end{bmatrix}$$

From parts **(a)** and **(b)**,

$$AB + AC = \begin{bmatrix} -6 & 7 \\ -4 & 6 \end{bmatrix} + \begin{bmatrix} 5 & 5 \\ 4 & 3 \end{bmatrix} = \begin{bmatrix} -1 & 12 \\ 0 & 9 \end{bmatrix}$$

Thus $A(B + C) = AB + AC$.

(c)

$$A + B = \begin{bmatrix} 3 & 1 \\ 2 & 1 \end{bmatrix} + \begin{bmatrix} -2 & 1 \\ 0 & 4 \end{bmatrix} = \begin{bmatrix} 1 & 2 \\ 2 & 5 \end{bmatrix}$$

$$(A + B)C = \begin{bmatrix} 1 & 2 \\ 2 & 5 \end{bmatrix}\begin{bmatrix} 1 & 2 \\ 2 & -1 \end{bmatrix} = \begin{bmatrix} 5 & 0 \\ 12 & -1 \end{bmatrix}$$

From parts **(b)** and **(a)**,

$$AC + BC = \begin{bmatrix} 5 & 5 \\ 4 & 3 \end{bmatrix} + \begin{bmatrix} 0 & -5 \\ 8 & -4 \end{bmatrix} = \begin{bmatrix} 5 & 0 \\ 12 & -1 \end{bmatrix}$$

Thus $(A + B)C = AC + BC$.

PROBLEM 1-14 Verify the multiplicative associative property (1.23), that $(AB)C = A(BC)$.

Solution Let $A = [a_{ij}]_{m \times n}$, $B = [b_{ij}]_{n \times p}$, $C = [c_{ij}]_{p \times q}$. Then let $D = AB$ and $E = BC$. We must show that $DC = AE$, so we let $D = [d_{ij}]_{m \times p}$, $E = [e_{ij}]_{n \times q}$. Then, by (1.18),

$$d_{il} = \sum_{k=1}^{n} a_{ik}b_{kl} \qquad \text{and} \qquad e_{kj} = \sum_{l=1}^{p} b_{kl}c_{lj}$$

The ijth entry of DC is

$$\sum_{l=1}^{p} d_{il}c_{lj} = \sum_{l=1}^{p}\left(\sum_{k=1}^{n} a_{ik}b_{kl}\right)c_{lj}$$

and the ijth entry of AE is

$$\sum_{k=1}^{n} a_{ik}c_{kj} = \sum_{k=1}^{n} a_{ik}\left(\sum_{l=1}^{p} b_{kl}c_{lj}\right)$$

Since

$$\sum_{l=1}^{p}\left(\sum_{k=1}^{n} a_{ik}b_{kl}\right)c_{lj} = \sum_{l=1}^{p}\sum_{k=1}^{n} a_{ik}b_{kl}c_{lj} = \sum_{k=1}^{n} a_{ik}\left(\sum_{l=1}^{p} b_{kl}c_{lj}\right)$$

it follows that $DC = AE$. Hence, $(AB)C = A(BC)$.

PROBLEM 1-15 Verify the distributive property (1.24), that $A(B + C) = AB + AC$.

Solution Let $A = [a_{ij}]_{m \times n}$, $B = [b_{ij}]_{n \times p}$, $C = [c_{ij}]_{n \times p}$. Then, from definition (1.18), the ijth entries of AB and AC are, respectively,

$$\sum_{k=1}^{n} a_{ik}b_{kj} \qquad \text{and} \qquad \sum_{k=1}^{n} a_{ik}c_{kj}$$

Hence the ijth entry of $AB + AC$ is

$$\sum_{k=1}^{n} a_{ik}b_{kj} + \sum_{k=1}^{n} a_{ik}c_{kj}$$

Next, let $D = B + C$ and $D = [d_{ij}]_{n \times p}$. Then

$$d_{ij} = b_{ij} + c_{ij}$$

Now, the ijth entry of $AD = A(B + C)$ is

$$\sum_{k=1}^{n} a_{ik}d_{kj} = \sum_{k=1}^{n} a_{ik}(b_{kj} + c_{kj})$$

$$= \sum_{k=1}^{n} (a_{ik}b_{kj} + a_{ik}c_{kj})$$

$$= \sum_{k=1}^{n} a_{ik}b_{kj} + \sum_{k=1}^{n} a_{ik}c_{kj}$$

which is exactly the ijth entry of $AB + AC$. Hence, $A(B + C) = AB + AC$.

PROBLEM 1-16 True or false? **(a)** $(AB)^2 = A^2B^2$, **(b)** $(A + B)(A - B) = A^2 - B^2$, where we define

$$A^n = \underbrace{AA\cdots A}_{n \text{ factors}}$$

Solution

(a) Remember that, in manipulating the products of matrices, brackets can be removed [see the multiplicative associative property (1.30)], and powers can be combined as long as the order of products is preserved. Now

$$(AB)^2 = (AB)(AB) = ABAB$$

Thus, **(a)** is not true in general, since $AB \neq BA$.
However, if, $AB = BA$, then

$$(AB)^2 = A(BA)B = A(AB)B = (AA)(BB) = A^2B^2$$

Thus, **(a)** is true if $AB = BA$.

(b) Applying the distributive properties (1.24) and (1.25), we have

$$(A + B)(A - B) = A(A - B) + B(A - B) = A^2 - AB + BA - B^2$$

Again, in general, **(b)** is false, but it will be true if $AB = BA$.

Inverse of a Matrix

PROBLEM 1-17 Verify the identity property (1.34), that $AI_n = I_mA = A$.

Solution Let $A = [a_{ij}]_{m\times n}$. Using the Kronecker delta form (1.33), write the identity matrix as $I_n = [\delta_{ij}]_{n\times n}$. Then by (1.18), the ijth entry of AI_n is

$$\sum_{k=1}^{n} a_{ik}\delta_{kj} = a_{i1}\delta_{1j} + a_{i2}\delta_{2j} + \cdots + a_{in}\delta_{nj} = a_{ij}$$

since $\delta_{kj} = 1$ for $k = j$ and $\delta_{kj} = 0$ for $k \neq j$. Thus, $AI_n = A$.
Similarly, $I_m = [\delta_{ij}]_{m\times m}$, and the ijth entry of I_mA is

$$\sum_{k=1}^{m} \delta_{ik}a_{kj} = \delta_{i1}a_{1j} + \delta_{i2}a_{2j} + \cdots + \delta_{im}a_{mj} = a_{ij}$$

since $\delta_{ik} = 1$ for $k = i$ and $\delta_{ik} = 0$ for $k \neq i$. Thus, $I_mA = A$.
Hence $AI_n = I_mA = A$.

PROBLEM 1-18 Given $A = \begin{bmatrix} 1 & 1 \\ 2 & 3 \end{bmatrix}$, find A^{-1} by definition (1.37): $AA^{-1} = A^{-1}A = I_n$.

Solution Let

$$A^{-1} = \begin{bmatrix} x_{11} & x_{12} \\ x_{21} & x_{22} \end{bmatrix}$$

Then by (1.37),

$$AA^{-1} = \begin{bmatrix} 1 & 1 \\ 2 & 3 \end{bmatrix}\begin{bmatrix} x_{11} & x_{12} \\ x_{21} & x_{22} \end{bmatrix} = \begin{bmatrix} 1 & 0 \\ 0 & 1 \end{bmatrix} = I_2$$

or

$$\begin{bmatrix} x_{11} + x_{21} & x_{12} + x_{22} \\ 2x_{11} + 3x_{21} & 2x_{12} + 3x_{22} \end{bmatrix} = \begin{bmatrix} 1 & 0 \\ 0 & 1 \end{bmatrix}$$

so that

$$\left.\begin{aligned} x_{11} + x_{21} &= 1 \\ 2x_{11} + 3x_{21} &= 0 \end{aligned}\right\} \qquad \textbf{(a)}$$

and

$$\left.\begin{aligned} x_{12} + x_{22} &= 0 \\ 2x_{12} + 3x_{22} &= 1 \end{aligned}\right\} \qquad \textbf{(b)}$$

Solving system (a) for x_{11} and x_{21}, we have

$$x_{11} = 3, \qquad x_{21} = -2$$

Solving system (b) for x_{12} and x_{22}, we have

$$x_{12} = -1, \qquad x_{22} = 1$$

Thus,

$$A^{-1} = \begin{bmatrix} 3 & -1 \\ -2 & 1 \end{bmatrix}$$

(See Example 1-16.)

PROBLEM 1-19 Let $A = \begin{bmatrix} a_{11} & a_{12} \\ a_{21} & a_{22} \end{bmatrix}$. Show that, if $\Delta = a_{11}a_{22} - a_{12}a_{21} \neq 0$, then

$$A^{-1} = \frac{1}{\Delta}\begin{bmatrix} a_{22} & -a_{12} \\ -a_{21} & a_{11} \end{bmatrix}$$

Solution If $\Delta = a_{11}a_{22} - a_{12}a_{21} \neq 0$, then

$$\begin{aligned} AA^{-1} &= \begin{bmatrix} a_{11} & a_{12} \\ a_{21} & a_{22} \end{bmatrix}\left[\frac{1}{\Delta}\begin{bmatrix} a_{22} & -a_{12} \\ -a_{21} & a_{11} \end{bmatrix}\right] \\ &= \frac{1}{\Delta}\begin{bmatrix} a_{11} & a_{12} \\ a_{21} & a_{22} \end{bmatrix}\begin{bmatrix} a_{22} & -a_{12} \\ -a_{21} & a_{11} \end{bmatrix} \\ &= \frac{1}{\Delta}\begin{bmatrix} a_{11}a_{22} - a_{12}a_{21} & -a_{11}a_{12} + a_{12}a_{11} \\ a_{21}a_{22} - a_{22}a_{21} & -a_{21}a_{12} + a_{22}a_{11} \end{bmatrix} \\ &= \frac{1}{\Delta}\begin{bmatrix} \Delta & 0 \\ 0 & \Delta \end{bmatrix} = \begin{bmatrix} 1 & 0 \\ 0 & 1 \end{bmatrix} = I_2 \end{aligned}$$

Similarly,

$$\begin{aligned} A^{-1}A &= \frac{1}{\Delta}\begin{bmatrix} a_{22} & -a_{12} \\ -a_{21} & a_{11} \end{bmatrix}\begin{bmatrix} a_{11} & a_{12} \\ a_{21} & a_{22} \end{bmatrix} \\ &= \frac{1}{\Delta}\begin{bmatrix} a_{22}a_{11} - a_{12}a_{21} & 0 \\ 0 & -a_{21}a_{12} + a_{11}a_{22} \end{bmatrix} \\ &= \frac{1}{\Delta}\begin{bmatrix} \Delta & 0 \\ 0 & \Delta \end{bmatrix} = \begin{bmatrix} 1 & 0 \\ 0 & 1 \end{bmatrix} = I_2 \end{aligned}$$

Therefore,

$$A^{-1} = \frac{1}{\Delta}\begin{bmatrix} a_{22} & -a_{12} \\ -a_{21} & a_{11} \end{bmatrix}$$

PROBLEM 1-20 Using the formula of Problem 1-19, rework Problem 1-18.

Solution $A = \begin{bmatrix} 1 & 1 \\ 2 & 3 \end{bmatrix}$, $\Delta = (1)(3) - (2)(1) = 3 - 2 = 1$. Thus

$$A^{-1} = \frac{1}{\Delta}\begin{bmatrix} 3 & -1 \\ -2 & 1 \end{bmatrix} = \begin{bmatrix} 3 & -1 \\ -2 & 1 \end{bmatrix}$$

PROBLEM 1-21 Using the formula of Problem 1-19, find the inverse of

$$A = \begin{bmatrix} \cos\theta & \sin\theta \\ -\sin\theta & \cos\theta \end{bmatrix}$$

Solution $\Delta = (\cos\theta)(\cos\theta) - (-\sin\theta)(\sin\theta) = \cos^2\theta + \sin^2\theta = 1$. Thus

$$A^{-1} = \frac{1}{\Delta}\begin{bmatrix} \cos\theta & -\sin\theta \\ \sin\theta & \cos\theta \end{bmatrix} = \begin{bmatrix} \cos\theta & -\sin\theta \\ \sin\theta & \cos\theta \end{bmatrix}$$

PROBLEM 1-22 Given that A, B, and C are nonsingular, show that ABC is nonsingular and that

$$(ABC)^{-1} = C^{-1}B^{-1}A^{-1}$$

Solution Let $BC = E$. Then, by Theorem 1-4.2, E is nonsingular and

$$E^{-1} = (BC)^{-1} = C^{-1}B^{-1} \qquad \text{[see property (1.38)]}$$

Next, $ABC = AE$; and, since A and E are both nonsingular, again by Theorem 1-4.2 AE is nonsingular and

$$(AE)^{-1} = E^{-1}A^{-1}$$

That is, ABC is nonsingular, and $(ABC)^{-1} = C^{-1}B^{-1}A^{-1}$.

PROBLEM 1-23 Prove that if $A_1, A_2, \ldots, A_k$ is nonsingular, then $A_1A_2\cdots A_k$ is nonsingular and

$$(A_1A_2\cdots A_k)^{-1} = A_k^{-1}\cdots A_2^{-1}A_1^{-1}$$

Solution We prove this by induction:

Let $A_1A_2\cdots A_r = B_r$ and assume that the above postulate is true for $k = r$. Then, B_r is nonsingular and

$$B_r^{-1} = (A_1A_2\cdots A_r)^{-1} = A_r^{-1}\cdots A_2^{-1}A_1^{-1}$$

Now, consider

$$(A_1A_2\cdots A_rA_{r+1}) = (B_rA_{r+1})$$

Since A_{r+1} and B_r are nonsingular and by Theorem 1-4.2 B_rA_{r+1} is nonsingular and

$$\begin{aligned}(A_1A_2\cdots A_rA_{r+1})^{-1} &= (B_rA_{r+1})^{-1} \\ &= A_{r+1}^{-1}B_r^{-1} \\ &= A_{r+1}^{-1}A_r^{-1}\cdots A_2^{-1}A_1^{-1}\end{aligned}$$

then the postulate is also true for $k = r + 1$.

Now, from property (1.38) the postulate is true for $k = 2$. Thus, for $k = 2 + 1 = 3$, $k = 4$, and so on, the postulate will be true for any k. Thus $(A_1A_2\cdots A_k)^{-1} = A_k^{-1}\cdots A_2^{-1}A_1^{-1}$ is proved.

PROBLEM 1-24 We define

$$A^k = \underbrace{AA\cdots A}_{k \text{ factors}}, \qquad \text{and } A^0 = I_n$$

$$A^{-k} = (A^{-1})^k = \underbrace{A^{-1}A^{-1}\cdots A^{-1}}_{k \text{ factors}}$$

Show that, if A is nonsingular, then A^k is nonsingular and

$$(A^k)^{-1} = (A^{-1})^k = A^{-k} \qquad \text{for } k = 0, 1, 2, \ldots$$

Solution In Problem 1-23, we showed that $(A_1A_2\cdots A_k)^{-1} = A_k^{-1}\cdots A_2^{-1}A_1^{-1}$. Now, let $A_1 = A_2 = \cdots = A_k = A$. Then, $AA\cdots A = A^k$ is nonsingular and

$$(A^k)^{-1} = (AA\cdots A)^{-1} = A^{-1}A^{-1}\cdots A^{-1} = (A^{-1})^k = A^{-k}$$

Transpose of a Matrix

PROBLEM 1-25 Show that $I_n^T = I_n$, where I_n is the identity matrix of order n.

Solution By the definition of the identity matrix [Kronecker delta form (1.33)], we have

$$I_n = [\delta_{ij}]_{n\times n}$$

Now, by the definition of the transpose of a matrix (1.40),

$$I_n^T = [\delta_{ji}]_{n\times n} = [\delta_{ij}]_{n\times n} = I_n$$

$$\text{since } \delta_{ji} = \delta_{ij} = \begin{cases} 1 & j = i \\ 0 & j \neq i \end{cases}.$$

PROBLEM 1-26 Let $A = [2 \;\; -1]$ and $B = \begin{bmatrix} 1 \\ 1 \end{bmatrix}$. **(a)** Show that $AB = I_1$. **(b)** Find A^T and B^T and check that $(AB)^T = B^TA^T$. **(c)** Find A^TB^T.

Solution

(a)

$$AB = [2 \;\; -1]\begin{bmatrix} 1 \\ 1 \end{bmatrix} = 2(1) + (-1)(1) = 1 = I_1$$

(b) Since

$$A^T = \begin{bmatrix} 2 \\ -1 \end{bmatrix} \qquad \text{and} \qquad B^T = [1 \;\; 1],$$

$$B^TA^T = [1 \;\; 1]\begin{bmatrix} 2 \\ -1 \end{bmatrix} = 1(2) + 1(-1) = 1 = I_1 = (AB)^T$$

(c)

$$A^TB^T = \begin{bmatrix} 2 \\ -1 \end{bmatrix}[1 \;\; 1] = \begin{bmatrix} 2 & 2 \\ -1 & -1 \end{bmatrix}$$

PROBLEM 1-27 Verify property (1.41), that $(A + B)^T = A^T + B^T$.

Solution Let $A = [a_{ij}]_{m\times n}$ and $B = [b_{ij}]_{m\times n}$. Then

$$A + B = [a_{ij} + b_{ij}]_{m\times n}$$

and

$$(A + B)^T = [c_{ij}]_{n \times m} \qquad \text{where } c_{ij} = a_{ji} + b_{ji}$$

Now

$$A^T = [a_{ij}^T]_{n \times m} \qquad \text{where } a_{ij}^T = a_{ji}$$

$$B^T = [b_{ij}^T]_{n \times m} \qquad \text{where } b_{ij}^T = b_{ji}$$

and

$$A^T + B^T = [a_{ij}^T + b_{ij}^T]_{n \times m}$$

Since $c_{ij} = a_{ji} + b_{ji} = a_{ij}^T + b_{ij}^T$, we have $(A + B)^T = A^T + B^T$.

PROBLEM 1-28 Show that $(ABC)^T = C^T B^T A^T$.

Solution Let $BC = F$. Then by property (1.43),

$$F^T = (BC)^T = C^T B^T$$

Thus

$$(ABC)^T = (AF)^T = F^T A^T = C^T B^T A^T$$

PROBLEM 1-29 Prove property (1-45), that $(A^T)^{-1} = (A^{-1})^T$ when A is nonsingular.

Solution Let $A^{-1} = B$. Then $AB = BA = I_n$. Taking the transpose and noting $I_n^T = I_n$ (see Problem 1-25), we get

$$(AB)^T = (BA)^T = I_n^T = I_n$$

By property (1.43), we get

$$B^T A^T = A^T B^T = I_n$$

Hence

$$B^T = (A^T)^{-1}$$

that is, $(A^{-1})^T = (A^T)^{-1}$.

Special Matrices

PROBLEM 1-30 Show that the product of two $n \times n$ diagonal matrices is a diagonal matrix and that the product is commutative.

Solution Let

$$D_1 = [a_i \delta_{ij}]_{n \times n} = \begin{bmatrix} a_1 & 0 & \cdots & 0 \\ 0 & a_2 & \cdots & 0 \\ \vdots & \vdots & \ddots & \vdots \\ 0 & 0 & \cdots & a_n \end{bmatrix} \qquad \text{and} \qquad D_2 = [b_i \delta_{ij}]_{n \times n} = \begin{bmatrix} b_1 & 0 & \cdots & 0 \\ 0 & b_2 & \cdots & 0 \\ \vdots & \vdots & \ddots & \vdots \\ 0 & 0 & \cdots & b_n \end{bmatrix}$$

Then

$$D_1 D_2 = \begin{bmatrix} a_1 & 0 & \cdots & 0 \\ 0 & a_2 & \cdots & 0 \\ \vdots & \vdots & \ddots & \vdots \\ 0 & 0 & \cdots & a_n \end{bmatrix} \begin{bmatrix} b_1 & 0 & \cdots & 0 \\ 0 & b_2 & \cdots & 0 \\ \vdots & \vdots & \ddots & \vdots \\ 0 & 0 & \cdots & b_n \end{bmatrix} = \begin{bmatrix} a_1 b_1 & 0 & \cdots & 0 \\ 0 & a_2 b_2 & \cdots & 0 \\ \vdots & \vdots & \ddots & \vdots \\ 0 & 0 & \cdots & a_n b_n \end{bmatrix}$$

which shows that $D_1 D_2$ is diagonal.

Next, since $a_i b_i = b_i a_i$, we have

$$D_1 D_2 = \begin{bmatrix} a_1 b_1 & 0 & \cdots & 0 \\ 0 & a_2 b_2 & \cdots & 0 \\ \vdots & \vdots & \ddots & \vdots \\ 0 & 0 & \cdots & a_n b_n \end{bmatrix} = \begin{bmatrix} b_1 a_1 & 0 & \cdots & 0 \\ 0 & b_2 a_2 & \cdots & 0 \\ \vdots & \vdots & \ddots & \vdots \\ 0 & 0 & \cdots & b_n a_n \end{bmatrix} = D_2 D_1$$

Hence the product is commutative.

PROBLEM 1-31 Show that a diagonal matrix is nonsingular if all its diagonal entries are nonzero, and find its inverse (see Example 1.25).

Solution Let

$$D_1 = [a_i \delta_{ij}]_{n \times n} = \begin{bmatrix} a_1 & 0 & \cdots & 0 \\ 0 & a_2 & \cdots & 0 \\ \vdots & \vdots & \ddots & \vdots \\ 0 & 0 & \cdots & a_n \end{bmatrix} \quad \text{and} \quad D_2 = [b_i \delta_{ij}]_{n \times n} = \begin{bmatrix} b_1 & 0 & \cdots & 0 \\ 0 & b_2 & \cdots & 0 \\ \vdots & \vdots & \ddots & \vdots \\ 0 & 0 & \cdots & b_n \end{bmatrix}$$

Suppose that D_2 is the inverse of D_1; then

$$D_1 D_2 = D_2 D_1 = I_n$$

Now, from Problem 1-30, we have

$$D_1 D_2 = D_2 D_1 = \begin{bmatrix} a_1 b_1 & 0 & \cdots & 0 \\ 0 & a_2 b_2 & \cdots & 0 \\ \vdots & \vdots & \ddots & \vdots \\ 0 & 0 & \cdots & a_n b_n \end{bmatrix} = \begin{bmatrix} 1 & 0 & \cdots & 0 \\ 0 & 1 & \cdots & 0 \\ \vdots & \vdots & \ddots & \vdots \\ 0 & 0 & \cdots & 1 \end{bmatrix}$$

Thus

$$a_i b_i = 1 \quad \text{and} \quad b_i = \frac{1}{a_i}, \quad \text{where } i = 1, 2, \ldots, n$$

and b_i exists only if $a_i \neq 0$.

Hence a diagonal matrix $D = [a_i \delta_{ij}]$ is nonsingular if all a_i's are nonzero, and its inverse is given by

$$D^{-1} = \left[\frac{1}{a_i}\delta_{ij}\right] = \begin{bmatrix} \frac{1}{a_1} & 0 & \cdots & 0 \\ 0 & \frac{1}{a_2} & \cdots & 0 \\ \vdots & \vdots & \ddots & \vdots \\ 0 & 0 & \cdots & \frac{1}{a_n} \end{bmatrix}$$

PROBLEM 1-32 Given $D = [d_i \delta_{ij}]_{n \times n}$, show that $D^k = [d_i^k \delta_{ij}]_{n \times n}$.

Solution Using the result of Problem 1-30 and letting $D_1 = D_2 = D$—that is, $a_i = b_i = d_i$—we obtain

$$D^2 = DD = \begin{bmatrix} d_1^2 & 0 & \cdots & 0 \\ 0 & d_1^2 & \cdots & 0 \\ \vdots & \vdots & \ddots & \vdots \\ 0 & 0 & \cdots & d_n^2 \end{bmatrix}$$

Repeating $D_1 = D^2$, $D_2 = D$, and so on, we obtain

$$D^3 = D^2D = \begin{bmatrix} d_1^3 & 0 & \cdots & 0 \\ 0 & d_1^3 & \cdots & 0 \\ \vdots & \vdots & \ddots & \vdots \\ 0 & 0 & \cdots & d_n^3 \end{bmatrix}$$

and

$$\vdots$$

$$D^k = \begin{bmatrix} d_1^k & 0 & \cdots & 0 \\ 0 & d_2^k & \cdots & 0 \\ \vdots & \vdots & \ddots & \vdots \\ 0 & 0 & \cdots & d_n^k \end{bmatrix} = [d_i^k \delta_{ij}]$$

PROBLEM 1-33 Given $D = \begin{bmatrix} \frac{1}{2} & 0 & 0 \\ 0 & \frac{1}{3} & 0 \\ 0 & 0 & \frac{1}{4} \end{bmatrix}$, find D^{-1} and D^k. Find D^k when $k \to \infty$.

Solution From Problem 1-31, we get

$$D^{-1} = \begin{bmatrix} 2 & 0 & 0 \\ 0 & 3 & 0 \\ 0 & 0 & 4 \end{bmatrix}$$

From Problem 1-32, we have

$$D^k = \begin{bmatrix} (\frac{1}{2})^k & 0 & 0 \\ 0 & (\frac{1}{3})^k & 0 \\ 0 & 0 & (\frac{1}{4})^k \end{bmatrix}$$

If $k \to \infty$, then $(\frac{1}{2})^k \to 0$, $(\frac{1}{3})^k \to 0$, and $(\frac{1}{4})^k \to 0$. Hence

$$\lim_{k \to \infty} D^k = \begin{bmatrix} 0 & 0 & 0 \\ 0 & 0 & 0 \\ 0 & 0 & 0 \end{bmatrix} = O$$

PROBLEM 1-34 Show that $N^3 = O$ but $N^2 \neq O$, where N is the following upper-triangular matrix:

$$N = \begin{bmatrix} 0 & 1 & 2 \\ 0 & 0 & -1 \\ 0 & 0 & 0 \end{bmatrix}$$

Solution

$$N^2 = NN = \begin{bmatrix} 0 & 1 & 2 \\ 0 & 0 & -1 \\ 0 & 0 & 0 \end{bmatrix}\begin{bmatrix} 0 & 1 & 2 \\ 0 & 0 & -1 \\ 0 & 0 & 0 \end{bmatrix} = \begin{bmatrix} 0 & 0 & -1 \\ 0 & 0 & 0 \\ 0 & 0 & 0 \end{bmatrix} \neq O$$

$$N^3 = NN^2 = \begin{bmatrix} 0 & 1 & 2 \\ 0 & 0 & -1 \\ 0 & 0 & 0 \end{bmatrix}\begin{bmatrix} 0 & 0 & -1 \\ 0 & 0 & 0 \\ 0 & 0 & 0 \end{bmatrix} = \begin{bmatrix} 0 & 0 & 0 \\ 0 & 0 & 0 \\ 0 & 0 & 0 \end{bmatrix} = O$$

PROBLEM 1-35 Show that $(N^T)^3 = O$ for the matrix of Problem 1-34.

Solution $N^T = \begin{bmatrix} 0 & 0 & 0 \\ 1 & 0 & 0 \\ 2 & -1 & 0 \end{bmatrix}$, so N^T is the lower-triangular matrix. Now

$$(N^T)^2 = \begin{bmatrix} 0 & 0 & 0 \\ 1 & 0 & 0 \\ 2 & -1 & 0 \end{bmatrix}\begin{bmatrix} 0 & 0 & 0 \\ 1 & 0 & 0 \\ 2 & -1 & 0 \end{bmatrix} = \begin{bmatrix} 0 & 0 & 0 \\ 0 & 0 & 0 \\ -1 & 0 & 0 \end{bmatrix}$$

$$(N^T)^3 = N^T(N^T)^2 = \begin{bmatrix} 0 & 0 & 0 \\ 1 & 0 & 0 \\ 2 & -1 & 0 \end{bmatrix}\begin{bmatrix} 0 & 0 & 0 \\ 0 & 0 & 0 \\ -1 & 0 & 0 \end{bmatrix} = \begin{bmatrix} 0 & 0 & 0 \\ 0 & 0 & 0 \\ 0 & 0 & 0 \end{bmatrix} = O$$

note: A square matrix N is called **nilpotent (of index k)** if $N^k = O$, but $N^{k-1} \neq O$ for some positive integer $k > 1$.

PROBLEM 1-36 Show that AA^T and A^TA are symmetric.

Solution Using formulas (1-43) and (1-44), we have

$$(AA^T)^T = (A^T)^TA^T = AA^T$$

$$(A^TA)^T = A^T(A^T)^T = A^TA$$

Thus, by definition (1.53), AA^T and A^TA are symmetric.

PROBLEM 1-37 Let A and B be symmetric $n \times n$ matrices. Prove that $AB = BA$ if and only if AB is also symmetric.

Solution If A and B are symmetric, then $A^T = A$ and $B^T = B$. Now, if AB is also symmetric, then

$$(AB)^T = AB$$

From (1.43),

$$(AB)^T = B^TA^T = BA$$

Thus $AB = BA$.

Now if $AB = BA$, then

$$(AB)^T = B^TA^T = BA = AB$$

Thus, AB is symmetric.

Hence $AB = BA$ if and only if AB is also symmetric.

PROBLEM 1-38 Show that $A + A^T$ is symmetric for any $n \times n$ matrix A.

Solution

$$\begin{aligned} (A + A^T)^T &= A^T + (A^T)^T && \text{[by (1.41)]} \\ &= A^T + A && \text{[by (1.44)]} \\ &= A + A^T && \text{[by (1.21)]} \end{aligned}$$

Hence, $A + A^T$ is symmetric.

PROBLEM 1-39 A square matrix A is said to be **skew-symmetric** if $A^T = -A$. Show that $A - A^T$ is skew-symmetric.

Solution

$$\begin{aligned}(A - A^T)^T &= A^T - (A^T)^T && \text{[by (1.41)]}\\ &= A^T - A && \text{[by (1.44)]}\\ &= -A + A^T && \text{[by (1.21)]}\\ &= -(A - A^T)\end{aligned}$$

Hence, $A - A^T$ is skew-symmetric.

PROBLEM 1-40 Show that the diagonal elements of a skew-symmetric matrix are all zero.

Solution By definition, if A is skew-symmetric, then

$$A^T = -A \Rightarrow a_{ji} = -a_{ij} \qquad \text{for all } i, j$$

Thus, letting $i = j$,

$$a_{ii} = -a_{ii} \qquad \text{for all } i$$

which is satisfied only if $a_{ii} = 0$ for all i; that is, the diagonal elements of a skew-symmetric matrix are all zero.

PROBLEM 1-41 Show that every square matrix A can be expressed as the sum of a symmetric matrix and a skew-symmetric matrix.

Solution Any square matrix A can be expressed as

$$A = \tfrac{1}{2}(A + A^T) + \tfrac{1}{2}(A - A^T)$$

Then from Problems 1-38 and 1-39, it is seen that $\frac{1}{2}(A + A^T)$ is symmetric and $\frac{1}{2}(A - A^T)$ is skew-symmetric.

PROBLEM 1-42 Given $A = \begin{bmatrix} 1 & 1 & 2 \\ -3 & 0 & 5 \\ 4 & -1 & 4 \end{bmatrix}$, express A as a sum of a symmetric matrix and a skew-symmetric matrix.

Solution For given A, $A^T = \begin{bmatrix} 1 & -3 & 4 \\ 1 & 0 & -1 \\ 2 & 5 & 4 \end{bmatrix}$, then

$$\frac{1}{2}(A + A^T) = \frac{1}{2}\left\{\begin{bmatrix} 1 & 1 & 2 \\ -3 & 0 & 5 \\ 4 & -1 & 4 \end{bmatrix} + \begin{bmatrix} 1 & -3 & 4 \\ 1 & 0 & -1 \\ 2 & 5 & 4 \end{bmatrix}\right\} = \frac{1}{2}\begin{bmatrix} 2 & -2 & 6 \\ -2 & 0 & 4 \\ 6 & 4 & 8 \end{bmatrix} = \begin{bmatrix} 1 & -1 & 3 \\ -1 & 0 & 2 \\ 3 & 2 & 4 \end{bmatrix}$$

$$\frac{1}{2}(A - A^T) = \frac{1}{2}\left\{\begin{bmatrix} 1 & 1 & 2 \\ -3 & 0 & 5 \\ 4 & -1 & 4 \end{bmatrix} - \begin{bmatrix} 1 & -3 & 4 \\ 1 & 0 & -1 \\ 2 & 5 & 4 \end{bmatrix}\right\} = \frac{1}{2}\begin{bmatrix} 0 & 4 & -2 \\ -4 & 0 & 6 \\ 2 & -6 & 0 \end{bmatrix} = \begin{bmatrix} 0 & 2 & -1 \\ -2 & 0 & 3 \\ 1 & -3 & 0 \end{bmatrix}$$

Hence, by the result of Problem 1-41, we get

$$\begin{bmatrix} 1 & 1 & 2 \\ -3 & 0 & 5 \\ 4 & -1 & 4 \end{bmatrix} = \begin{bmatrix} 1 & -1 & 3 \\ -1 & 0 & 2 \\ 3 & 2 & 4 \end{bmatrix} + \begin{bmatrix} 0 & 2 & -1 \\ -2 & 0 & 3 \\ 1 & -3 & 0 \end{bmatrix}$$

Supplementary Exercises

PROBLEM 1-43 Solve the following systems of equations:

(a) $$\begin{aligned} x_1 + 2x_2 + x_3 &= -1 \\ 2x_1 + 4x_2 + x_3 &= 1 \end{aligned}$$

(b) $$\begin{aligned} 3x_1 + 2x_2 + x_3 &= 1 \\ x_2 - x_3 &= 2 \\ 2x_3 &= 4 \end{aligned}$$

(c) $$\begin{aligned} 2x_1 + 3x_2 + x_3 &= 4 \\ x_1 + 2x_2 + x_3 &= 3 \\ 3x_1 - x_2 - 3x_3 &= -1 \end{aligned}$$

Answer **(a)** $(2 - 2t, t, -3)$ **(b)** $(-3, 4, 2)$ **(c)** $(3, -2, 4)$

PROBLEM 1-44 Convert the systems of equations in Problem 1-43 to matrix form.

Answer

(a) $\begin{bmatrix} 1 & 2 & 1 \\ 2 & 4 & 1 \end{bmatrix}\begin{bmatrix} x_1 \\ x_2 \\ x_3 \end{bmatrix} = \begin{bmatrix} -1 \\ 1 \end{bmatrix}$ **(b)** $\begin{bmatrix} 3 & 2 & 1 \\ 0 & 1 & -1 \\ 0 & 0 & 2 \end{bmatrix}\begin{bmatrix} x_1 \\ x_2 \\ x_3 \end{bmatrix} = \begin{bmatrix} 1 \\ 2 \\ 4 \end{bmatrix}$ **(c)** $\begin{bmatrix} 2 & 3 & 1 \\ 1 & 2 & 1 \\ 3 & -1 & -3 \end{bmatrix}\begin{bmatrix} x_1 \\ x_2 \\ x_3 \end{bmatrix} = \begin{bmatrix} 4 \\ 3 \\ -1 \end{bmatrix}$

PROBLEM 1-45 Compute **(a)** $A + BC$ and **(b)** $2A + B$, where

$$A = \begin{bmatrix} 3 & 1 \\ 4 & 2 \end{bmatrix}, \qquad B = \begin{bmatrix} 1 & 2 \\ 3 & 1 \end{bmatrix}, \qquad C = \begin{bmatrix} -2 & 3 \\ 1 & 2 \end{bmatrix}$$

Answer **(a)** $\begin{bmatrix} 3 & 8 \\ -1 & 13 \end{bmatrix}$ **(b)** $\begin{bmatrix} 7 & 4 \\ 11 & 5 \end{bmatrix}$

PROBLEM 1-46 Verify that $3(AB) = (3A)B = A(3B)$, where

$$A = \begin{bmatrix} 1 & 2 \\ 3 & 1 \\ 4 & -1 \end{bmatrix}, \qquad B = \begin{bmatrix} 1 & 3 \\ 2 & -1 \end{bmatrix}$$

PROBLEM 1-47 Compute **(a)** ABC and **(b)** CAB, given

$$A = \begin{bmatrix} 1 & 2 & 1 \\ 1 & -1 & 2 \end{bmatrix}, \qquad B = \begin{bmatrix} 1 \\ 3 \\ -1 \end{bmatrix}, \qquad C = [1 \ \ -1]$$

Answer **(a)** $\begin{bmatrix} 6 & -6 \\ -4 & 4 \end{bmatrix}$ **(b)** 10

PROBLEM 1-48 Find a, b, c, and d, if

$$\begin{bmatrix} a & b \\ c & d \end{bmatrix}\begin{bmatrix} 2 & 1 \\ 3 & 2 \end{bmatrix} = \begin{bmatrix} 5 & 3 \\ 5 & 4 \end{bmatrix}$$

Answer $a = 1 \quad b = 1 \quad c = -2 \quad d = 3$

PROBLEM 1-49 Find all 2×2 matrices A such that $A^2 = I_2$.

Answer A can be any of the following:

$$\pm I_2, \qquad \pm\begin{bmatrix} 1 & b \\ 0 & -1 \end{bmatrix}, \qquad \pm\begin{bmatrix} 1 & 0 \\ c & -1 \end{bmatrix}, \qquad \begin{bmatrix} a & b \\ \dfrac{1 - a^2}{b} & -a \end{bmatrix}$$

PROBLEM 1-50 If $A = \begin{bmatrix} 1 & 1 \\ 1 & 1 \end{bmatrix}$, show that $A^k = \begin{bmatrix} 2^{k-1} & 2^{k-1} \\ 2^{k-1} & 2^{k-1} \end{bmatrix}$. [*Hint:* Use the induction method.]

PROBLEM 1-51 Let $A = \begin{bmatrix} \cos\theta & -\sin\theta \\ \sin\theta & \cos\theta \end{bmatrix}$. Show that

$$A^k = \begin{bmatrix} \cos k\theta & -\sin k\theta \\ \sin k\theta & \cos k\theta \end{bmatrix}$$

[*Hint:* Use the induction method and apply the trigonometric identities $\cos(A + B) = \cos A \cos B - \sin A \sin B$ and $\sin(A + B) = \sin A \cos B + \cos A \sin B$.]

PROBLEM 1-52 If square matrices A and B commute, show that A^m and B^n commute, where m and n are positive integers.

PROBLEM 1-53 Find the inverse of $A = \begin{bmatrix} 1 & 2 & 3 \\ 0 & 1 & 4 \\ 0 & 0 & 1 \end{bmatrix}$.

Answer

$$A^{-1} = \begin{bmatrix} 1 & -2 & 5 \\ 0 & 1 & -4 \\ 0 & 0 & 1 \end{bmatrix}$$

PROBLEM 1-54 Let A and B be any $n \times n$ matrices and let $C = AB$. Prove that if B is singular, then C is also singular. [*Hint:* Use Corollary 1-4.5.]

PROBLEM 1-55 Let A be a nonsingular matrix and k be a nonzero constant. Show that

$$(kA)^{-1} = \frac{1}{k}A^{-1}$$

PROBLEM 1-56 If $A = \begin{bmatrix} 2 & 1 \\ 1 & 4 \\ 3 & -1 \end{bmatrix}$, verify that $(A^T)^T = A$.

PROBLEM 1-57 Let

$$A = \begin{bmatrix} 1 & -2 \\ -2 & 3 \end{bmatrix}, \qquad B = \begin{bmatrix} -2 & 1 \\ 1 & 1 \end{bmatrix}$$

Verify that $(AB)^T = B^T A^T$. Also show that, although A and B are symmetric, AB is not symmetric.

PROBLEM 1-58 Show that if A is symmetric, then A^{-1} is also symmetric. [*Hint:* Use property (1.45).]

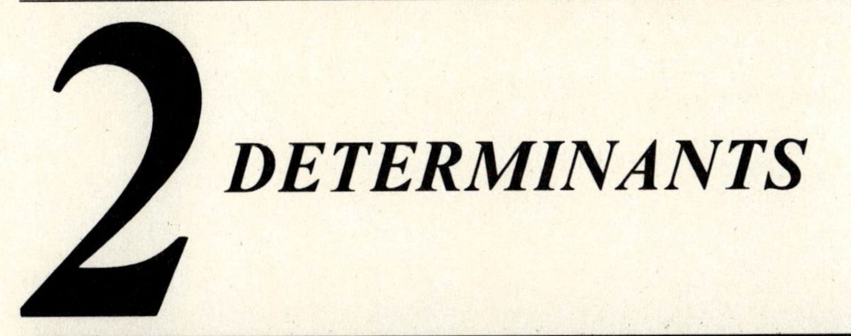

2 DETERMINANTS

THIS CHAPTER IS ABOUT

- ☑ Determinant of a Square Matrix
- ☑ Properties of Determinants
- ☑ Multiplicative Property of Determinants
- ☑ Cofactor Expansion
- ☑ Determinants and the Inverse of a Matrix
- ☑ Systems of Linear Equations and Cramer's Rule

2-1. Determinant of a Square Matrix

Given a square matrix A, we associate with it a certain number called its **determinant**, denoted by det A or $|A|$. The value of this number tells us whether or not the matrix is singular.

A. 1 × 1 Matrices

For a 1×1 matrix $A = [a_{11}]$,

$$\det A = a_{11} \tag{2.1}$$

B. 2 × 2 Matrices

For a 2×2 matrix $A = \begin{bmatrix} a_{11} & a_{12} \\ a_{21} & a_{22} \end{bmatrix}$,

$$\det A = \begin{vmatrix} a_{11} & a_{12} \\ a_{21} & a_{22} \end{vmatrix} = a_{11}a_{22} - a_{12}a_{21} \tag{2.2}$$

Note that this number can be computed by forming the pattern

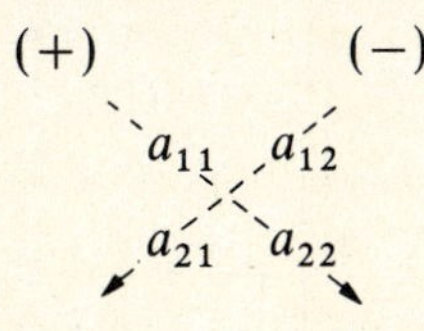

and multiplying the entries on the rightward arrow and then subtracting the product of the entries on the leftward arrow.

EXAMPLE 2-1: Evaluate the determinant of

$$A = \begin{bmatrix} 1 & 2 \\ -1 & 4 \end{bmatrix}$$

Solution: By (2.2),

$$\det A = \begin{vmatrix} 1 & 2 \\ -1 & 4 \end{vmatrix} = (1)(4) - (2)(-1) = 6$$

C. 3 × 3 Matrices

For a 3 × 3 matrix $A = \begin{bmatrix} a_{11} & a_{12} & a_{13} \\ a_{21} & a_{22} & a_{23} \\ a_{31} & a_{32} & a_{33} \end{bmatrix}$,

$$\det A = \begin{vmatrix} a_{11} & a_{12} & a_{13} \\ a_{21} & a_{22} & a_{23} \\ a_{31} & a_{32} & a_{33} \end{vmatrix} \tag{2.3}$$

$$= a_{11}a_{22}a_{33} + a_{12}a_{23}a_{31} + a_{13}a_{21}a_{32} - a_{13}a_{22}a_{31} - a_{11}a_{23}a_{32} - a_{12}a_{21}a_{33}$$

Again, this formula can be obtained by forming the pattern with recopying the first and second columns as shown

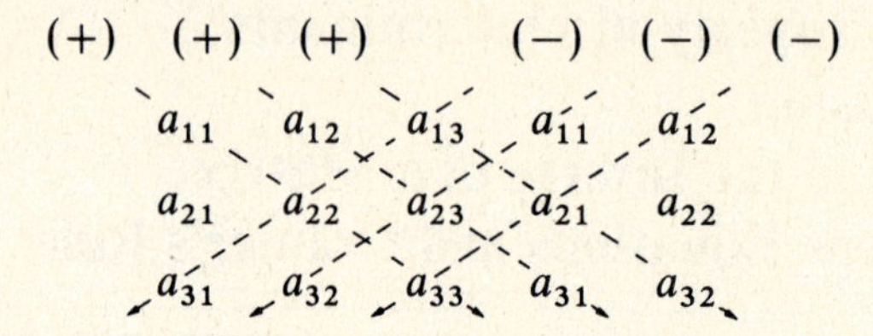

then summing the products on the rightward arrows and subtracting the products on the leftward arrows.

EXAMPLE 2-2: Evaluate the determinant of

$$A = \begin{bmatrix} 1 & 4 & 2 \\ 0 & 1 & 2 \\ 1 & 2 & 3 \end{bmatrix}$$

Solution: Using (2.3) or forming the pattern

$$\begin{array}{ccccc} (+) & (+) & (+) & (-) & (-)\;(-) \\ 1 & 4 & 2 & 1 & 4 \\ 0 & 1 & 2 & 0 & 2 \\ 1 & 2 & 3 & 1 & 2 \end{array}$$

we get

$$\det A = \begin{vmatrix} 1 & 4 & 2 \\ 0 & 1 & 2 \\ 1 & 2 & 3 \end{vmatrix} = 3 + 8 + 0 - 2 - 4 - 0 = 5$$

D. $n \times n$ Matrices

Unfortunately, there is no simple scheme for evaluating the determinant of matrices that are 4 × 4 or larger. The general classical definition of the determinant of an $n \times n$ matrix requires the concept of permutation and combinatorial arguments (and it might give you a headache), and it is of dubious value; so it will not be discussed here. Evaluation of the determinant of an $n \times n$ matrix by the expansion technique is discussed in Section 2-4.

2-2. Properties of Determinants

The simple things about the determinant are not the explicit formulas by which it can be evaluated, but the properties it possesses.

A. Fundamental properties

The determinant can be defined uniquely in terms of the following three properties.

Property 1 The determinant of a matrix A is unchanged if any row or column is replaced by the sum of that row or column and another row or column.

EXAMPLE 2-3: Let

$$A = \begin{bmatrix} a & b \\ c & d \end{bmatrix}, \qquad B = \begin{bmatrix} a+c & b+d \\ c & d \end{bmatrix}$$

Show that $\det A = \det B$ by (2.2).

Solution: By (2.2) we have

$$\det A = \begin{vmatrix} a & b \\ c & d \end{vmatrix} = ad - bc$$

$$\det B = \begin{vmatrix} a+c & b+d \\ c & d \end{vmatrix} = (a+c)d - (b+d)c = ad + cd - bc - dc = ad - bc$$

Hence $\det A = \det B$.

Property 2 The determinant of A is multiplied by the constant k if any row or column is multiplied by k.

EXAMPLE 2-4: Let

$$A = \begin{bmatrix} a & b \\ c & d \end{bmatrix}, \qquad B = \begin{bmatrix} a & b \\ kc & kd \end{bmatrix}$$

Show that $\det B = k(\det A)$ by (2.2).

Solution

$$\det A = \begin{vmatrix} a & b \\ c & d \end{vmatrix} = ad - bc$$

$$\det B = \begin{vmatrix} a & b \\ kc & kd \end{vmatrix} = a(kd) - b(kc)$$

$$= k(ad - bc) = k \det A$$

Property 3 The determinant of the identity matrix I_n is 1, that is,

$$\det I_n = 1 \tag{2.4}$$

EXAMPLE 2-5: Compute $\det I_2$ by (2.2).

Solution

$$\det I_2 = \begin{vmatrix} 1 & 0 \\ 0 & 1 \end{vmatrix} = (1)(1) - (0)(0) = 1 - 0 = 1$$

The above basic properties may be applied to the numerical evaluation of the determinant of matrices, as illustrated in the following example.

EXAMPLE 2-6: Compute det A using the basic properties of determinants for

$$A = \begin{bmatrix} 1 & 2 \\ -1 & 4 \end{bmatrix}$$

Solution: Let

$$|A| = \begin{vmatrix} 1 & 2 \\ -1 & 4 \end{vmatrix}$$

Step 1: Replace the second row by the sum of the second row and the first row (Property 1),

$$|A| = \begin{vmatrix} 1 & 2 \\ 0 & 6 \end{vmatrix}$$

Step 2: Apply Property 2,

$$|A| = (-6)\begin{vmatrix} 1 & 2 \\ 0 & -1 \end{vmatrix}$$

Step 3: Replace the first row by the sum of the first row and twice the second row (Property 1),

$$|A| = (-6)\begin{vmatrix} 1 & 0 \\ 0 & -1 \end{vmatrix}$$

Step 4: Apply Property 2,

$$|A| = (-6)(-1)\begin{vmatrix} 1 & 0 \\ 0 & 1 \end{vmatrix} = 6\begin{vmatrix} 1 & 0 \\ 0 & 1 \end{vmatrix}$$

Step 5: Apply Property 3,

$$|A| = (6)(1) = 6$$

which is the same result obtained in Example 2-1.

B. Derived properties

Other properties of determinants discussed in the following are derived from the three fundamental properties.

Property 4 The determinant of A is unchanged if a multiple of one row or column is added to or subtracted from another row or column.

Property 4 may be seen to follow from Properties 1 and 2 and it can be considered as the extended Property 1.

EXAMPLE 2-7: Let

$$A = \begin{bmatrix} a & b \\ c & d \end{bmatrix}, \qquad B = \begin{bmatrix} a + kc & b + kd \\ c & d \end{bmatrix}$$

Show that det A = det B by (2.2).

Solution: By (2.2) we get

$$\det A = \begin{vmatrix} a & b \\ c & d \end{vmatrix} = ad - bc$$

$$\det B = \begin{vmatrix} a + kc & b + kd \\ c & d \end{vmatrix} = (a + kc)d - (b + kd)c$$

$$= ad - bc = \det A$$

Property 5 The determinant of A changes sign when two rows or columns are interchanged.

Property 5 is a consequence of Properties 1 and 4.

EXAMPLE 2-8: Let

$$A = \begin{bmatrix} a & b \\ c & d \end{bmatrix}, \qquad B = \begin{bmatrix} c & d \\ a & b \end{bmatrix}$$

Show that $\det B = -\det A$ by (2.2).

Solution: By (2.2) we get

$$\det A = \begin{vmatrix} a & b \\ c & d \end{vmatrix} = ad - bc$$

$$\det B = \begin{vmatrix} c & d \\ a & b \end{vmatrix} = cb - da = -(ad - bc) = -\det A$$

Property 6 If A has a zero row or column, then $\det A = 0$.

Property 6 follows from Property 2 for $k = 0$.

Property 7 If any two rows or columns of A are equal or proportional, then $\det A = 0$.

Property 7 follows from Property 5 since, if the equal rows are interchanged, the determinant is supposed to change sign although the matrix stays the same. Thus in this case

$$\det A = -\det A \Rightarrow \det A = 0$$

EXAMPLE 2-9: Compute $\det A$ by (2.2) for

$$A = \begin{bmatrix} a & b \\ ka & kb \end{bmatrix}$$

Solution

$$\det A = \begin{vmatrix} a & b \\ ka & kb \end{vmatrix} = a(kb) - b(ka) = k(ab - ab) = 0$$

Example 2-10: Using Properties 1 through 7 of determinants, evaluate the determinant of

$$A = \begin{bmatrix} 1 & 4 & 2 \\ 0 & 1 & 2 \\ 1 & 2 & 3 \end{bmatrix}$$

Solution: Let

$$|A| = \begin{vmatrix} 1 & 4 & 2 \\ 0 & 1 & 2 \\ 1 & 2 & 3 \end{vmatrix}$$

Step 1: Subtract the first row from the third row,

$$|A| = \begin{vmatrix} 1 & 4 & 2 \\ 0 & 1 & 2 \\ 0 & -2 & 1 \end{vmatrix}$$

Step 2: Add twice the second row to the third row

$$|A| = \begin{vmatrix} 1 & 4 & 2 \\ 0 & 1 & 2 \\ 0 & 0 & 5 \end{vmatrix}$$

Step 3: Subtract 4 times the second row from the first row,

$$|A| = \begin{vmatrix} 1 & 0 & -6 \\ 0 & 1 & 2 \\ 0 & 0 & 5 \end{vmatrix}$$

Step 4: Add $\frac{6}{5}$ times the third row to the first row

$$|A| = \begin{vmatrix} 1 & 0 & 0 \\ 0 & 1 & 2 \\ 0 & 0 & 5 \end{vmatrix}$$

Step 5: Subtract $-\frac{2}{5}$ times the third row from the second row,

$$|A| = \begin{vmatrix} 1 & 0 & 0 \\ 0 & 1 & 0 \\ 0 & 0 & 5 \end{vmatrix}$$

Step 6: Apply fundamental Properties 2 and 3,

$$|A| = (5)\begin{vmatrix} 1 & 0 & 0 \\ 0 & 1 & 0 \\ 0 & 0 & 1 \end{vmatrix} = (5)(1) = 5$$

which is the same value obtained in Example 2-2.

Note that the row modification or row reduction involved in Steps 3 to 6 of this process does not affect the values of the main diagonal entries, the product of which is equal to the value of the determinant. Thus we have the following additional derived property:

Property 8 If A is triangular, then $\det A$ is the product of the diagonal entries; that is

$$\det A = a_{11}a_{22}\ldots a_{nn} \tag{2.5}$$

With Property 8, the determinant of A in Example 2-10 can be found after the completion of the second step.

EXAMPLE 2-11: Evaluate by (2.2) the determinants of

$$A = \begin{bmatrix} a_{11} & a_{12} \\ 0 & a_{22} \end{bmatrix}, \qquad B = \begin{bmatrix} b_{11} & 0 \\ b_{21} & b_{22} \end{bmatrix}$$

Solution: By (2.2), we get

$$\det A = \begin{vmatrix} a_{11} & a_{12} \\ 0 & a_{22} \end{vmatrix} = a_{11}a_{22} - a_{12}(0) = a_{11}a_{22}$$

$$\det B = \begin{vmatrix} b_{11} & 0 \\ b_{21} & b_{22} \end{vmatrix} = b_{11}b_{22} - (0)b_{21} = b_{11}b_{22}$$

In Example 2-10, if the operations with respect to row are replaced by similar operations with respect to columns, the same final value will be arrived at, since the values of the diagonal entries are not affected by interchange of rows and columns. Thus we can add another derived property as follows.

Property 9 The transpose of matrix A has the same determinant as A, that is,

$$\det A = \det A^T \tag{2.6}$$

EXAMPLE 2-12: Show that $\det A = \det A^T$ by (2.2) for

$$A = \begin{bmatrix} a_{11} & a_{12} \\ a_{21} & a_{22} \end{bmatrix}$$

Solution: By (2.2), we get

$$\det A = \begin{vmatrix} a_{11} & a_{12} \\ a_{21} & a_{22} \end{vmatrix} = a_{11}a_{22} - a_{12}a_{21}$$

$$\det A^T = \begin{vmatrix} a_{11} & a_{21} \\ a_{12} & a_{22} \end{vmatrix} = a_{11}a_{22} - a_{21}a_{12} = a_{11}a_{22} - a_{12}a_{21} = \det A$$

2-3. Multiplicative Property of Determinants

THEOREM 2-3.1 If A and B are $n \times n$ matrices, then

$$\det(AB) = (\det A)(\det B) \tag{2.7}$$

EXAMPLE 2-13: Given

$$A = \begin{bmatrix} 2 & 1 \\ 2 & 3 \end{bmatrix}, \qquad B = \begin{bmatrix} 1 & -1 \\ 2 & 3 \end{bmatrix}$$

Show that by (2.2) $\det(AB) = (\det A)(\det B)$.

Solution

$$AB = \begin{bmatrix} 2 & 1 \\ 2 & 3 \end{bmatrix}\begin{bmatrix} 1 & -1 \\ 2 & 3 \end{bmatrix} = \begin{bmatrix} 4 & 1 \\ 8 & 7 \end{bmatrix}$$

Then by (2.2) we get

$$\det A = \begin{vmatrix} 2 & 1 \\ 2 & 3 \end{vmatrix} = 6 - 2 = 4$$

$$\det B = \begin{vmatrix} 1 & -1 \\ 2 & 3 \end{vmatrix} = 3 - (-2) = 5$$

$$\det(AB) = \begin{vmatrix} 4 & 1 \\ 8 & 7 \end{vmatrix} = 28 - 8 = 20 = (4)(5)$$

$$= (\det A)(\det B)$$

THEOREM 2-3.2 If A is nonsingular, then

$$\det A \neq 0 \tag{2.8}$$

EXAMPLE 2-14: Verify Theorem 2-3.2.

Solution: If A is nonsingular, then A^{-1} exists, and

$$AA^{-1} = I_n$$

Thus by Property 3,

$$\det(AA^{-1}) = \det I_n = 1$$

Since by Theorem 2-3.1, (2.7),

$$\det(AA^{-1}) = (\det A)(\det A^{-1})$$

it follows that

$$(\det A)(\det A^{-1}) = 1 \tag{2.9}$$

Therefore

$$\det A \neq 0$$

EXAMPLE 2-15: Show that

$$A = \begin{bmatrix} 1 & 1 & 1 \\ 4 & 2 & 1 \\ 2 & 0 & 2 \end{bmatrix}$$

is nonsingular.

Solution: By (2.3) we get

$$\det A = \begin{vmatrix} 1 & 1 & 1 \\ 4 & 2 & 1 \\ 2 & 0 & 2 \end{vmatrix} = 4 + 2 + 0 - 4 - 0 - 8 = -6 \neq 0$$

Hence, by Theorem 2-3.2, A is nonsingular.

EXAMPLE 2-16: If A is nonsingular, show that

$$\det A^{-1} = \frac{1}{\det A} \tag{2.10}$$

Solution: If A is nonsingular, then $\det A \neq 0$, and from (2.9) it follows that

$$\det A^{-1} = \frac{1}{\det A}$$

2-4. Cofactor Expansion

A. Cofactors

DEFINITION Let $A = [a_{ij}]$ be an $n \times n$ matrix. Let M_{ij} be the $(n-1) \times (n-1)$ matrix obtained from A by deleting the row and column containing a_{ij}, that is, the ith row and the jth column. The det M_{ij} is called the **minor** of a_{ij}. The number A_{ij}, defined by

COFACTOR OF a_{ij}

$$A_{ij} = (-1)^{i+j} \det M_{ij} \tag{2.11}$$

is called the cofactor of a_{ij}. The sign $(-1)^{i+j}$ forms the following pattern:

$$\begin{bmatrix} + & - & + & - & \cdots \\ - & + & - & + & \cdots \\ + & - & + & - & \cdots \\ \vdots & \vdots & \vdots & \vdots & \end{bmatrix}$$

EXAMPLE 2-17: Given

$$A = \begin{bmatrix} 1 & 4 & 2 \\ 0 & 1 & 2 \\ 1 & 2 & 3 \end{bmatrix}$$

find the cofactors A_{11}, A_{23}.

Solution: The minor of a_{11} is

$$|M_{11}| = \begin{vmatrix} 1 & 4 & 2 \\ 0 & 1 & 2 \\ 1 & 2 & 3 \end{vmatrix} = \begin{vmatrix} 1 & 2 \\ 2 & 3 \end{vmatrix} = -1$$

The cofactor of a_{11} is, by (2.11),

$$A_{11} = (-1)^{1+1}|M_{11}| = |M_{11}| = -1$$

Similarly, the minor of a_{23} is

$$|M_{23}| = \begin{vmatrix} 1 & 4 & 2 \\ 0 & 1 & 2 \\ 1 & 2 & 3 \end{vmatrix} = \begin{vmatrix} 1 & 4 \\ 1 & 2 \end{vmatrix} = -2$$

and the cofactor of a_{23} is

$$A_{23} = (-1)^{2+3}|M_{23}| = (-1)|M_{23}| = (-1)(-2) = 2$$

B. Cofactor expansion

THEOREM 2-4.1 If A is an $n \times n$ matrix and A_{ik} denotes the cofactor of a_{ik}, then det A can be expressed as a cofactor expansion using any row or column of A by the following formulas:

$$\det A = a_{i1}A_{i1} + a_{i2}A_{i2} + \cdots + a_{in}A_{in} = \sum_{k=1}^{n} a_{ik}A_{ik} \tag{2.12}$$

for $i = 1, 2, \ldots, n$.

$$\det A = a_{1j}A_{1j} + a_{2j}A_{2j} + \cdots + a_{nj}A_{nj} = \sum_{k=1}^{n} a_{kj}A_{kj} \tag{2.13}$$

for $j = 1, 2, \ldots, n$.

Formula (2.12) is called the **cofactor expansion** (or **Laplace expansion**) **of det A along the ith row**, and formula (2.13) is called the **cofactor expansion of det A along the jth column**. Note that the cofactor expansion technique is most efficient along the row or column that contains the most zeros.

EXAMPLE 2-18: Compute the determinant of

$$A = \begin{bmatrix} a_{11} & a_{12} \\ a_{21} & a_{22} \end{bmatrix}$$

by cofactor expansion.

Solution: Since by (2.1) $\det[a_{11}] = a_{11}$, we have

$$A_{11} = (-1)^{1+1}\det[a_{22}] = a_{22}$$
$$A_{12} = (-1)^{1+2}\det[a_{21}] = -a_{21}$$

so that by formula (2.12) with $i = 1$ (that is, along the first row)

$$\det A = \begin{vmatrix} a_{11} & a_{12} \\ a_{21} & a_{22} \end{vmatrix} = a_{11}A_{11} + a_{12}A_{12} = a_{11}a_{22} - a_{12}a_{21}$$

EXAMPLE 2-19: Compute det A for

$$A = \begin{bmatrix} 2 & 3 & 1 \\ 0 & 3 & 0 \\ -1 & 4 & 1 \end{bmatrix}$$

Solution: Using the cofactor expansion along the second row, we get

$$\det A = \begin{vmatrix} 2 & 3 & 1 \\ 0 & 3 & 0 \\ -1 & 4 & 1 \end{vmatrix} = 3(-1)^{2+2}\begin{vmatrix} 2 & 1 \\ -1 & 1 \end{vmatrix} = 3(3) = 9$$

2-5. Determinants and the Inverse of a Matrix

A. Properties of cofactors

LEMMA 2-5.1 Let A be an $n \times n$ matrix. If A_{jk} denotes the cofactor of a_{jk}, then

$$a_{i1}A_{j1} + a_{i2}A_{j2} + \cdots + a_{in}A_{jn} = \begin{cases} \det A & \text{if } i = j \\ 0 & \text{if } i \neq j \end{cases} \tag{2.14}$$

$$a_{1j}A_{1i} + a_{2j}A_{2i} + \cdots + a_{nj}A_{ni} = \begin{cases} \det A & \text{if } j = i \\ 0 & \text{if } j \neq i \end{cases} \tag{2.15}$$

EXAMPLE 2-20: Verify formula (2.14).

Solution: If $i = j$, formula (2.14) is just the cofactor expansion of det A along the ith row of A [see (2.12)]. In the case $i \neq j$, let A' be the matrix obtained by replacing the jth row of A by the ith row of A.

$$A' = \begin{bmatrix} a_{11} & a_{12} & \cdots & a_{1n} \\ \vdots & \vdots & & \vdots \\ a_{i1} & a_{i2} & \cdots & a_{in} \\ \vdots & \vdots & & \vdots \\ a_{i1} & a_{i2} & \cdots & a_{in} \\ \vdots & \vdots & & \vdots \\ a_{n1} & a_{n2} & \cdots & a_{nn} \end{bmatrix} \begin{matrix} \\ \\ \\ \\ \leftarrow j\text{th row} \\ \\ \\ \end{matrix}$$

Since two rows of A' are the same, then by Property 7 of the determinant

$$\det A' = 0$$

Then it follows from the cofactor expansion of $\det A'$ along the jth row that

$$\begin{aligned}\det A' &= a_{i1}A'_{j1} + a_{i2}A'_{j2} + \cdots + a_{in}A'_{jn}\\ &= a_{i1}A_{j1} + a_{i2}A_{j2} + \cdots + a_{in}A_{jn} = 0\end{aligned}$$

since $A'_{jk} = A_{jk}$. Thus,

$$a_{i1}A_{j1} + a_{i2}A_{j2} + \cdots + a_{in}A_{jn} = \begin{cases}\det A & \text{if } i = j\\ 0 & \text{if } i \neq j\end{cases}$$

B. Adjugate, or classic adjoint, of a matrix

DEFINITION Let $A = [a_{ij}]$ be an $n \times n$ matrix and let A_{ij} denote the cofactor of a_{ij}. The **cofactor matrix** of A, denoted by Cof A, is defined as the $n \times n$ matrix whose ijth entry is A_{ij}, that is

COFACTOR MATRIX

$$\text{Cof } A = [A_{ij}]_{n\times n} \tag{2.16}$$

The transpose of the cofactor matrix of A is called the **adjugate** (or **classic adjoint**) of A, denoted by adj A, that is,

ADJUGATE OF MATRIX A

$$\text{adj } A = [A_{ij}]^T = \begin{bmatrix} A_{11} & A_{21} & \cdots & A_{n1}\\ A_{12} & A_{22} & \cdots & A_{n2}\\ \vdots & \vdots & & \vdots\\ A_{1n} & A_{2n} & \cdots & A_{nn}\end{bmatrix} \tag{2.17}$$

EXAMPLE 2-21: Given

$$A = \begin{bmatrix} 1 & 1\\ 2 & 3\end{bmatrix}$$

find adj A.

Solution: Since $A_{11} = 3$, $A_{12} = -2$, $A_{21} = -1$, $A_{22} = 1$, so

$$\text{adj } A = \begin{bmatrix} 3 & -2\\ -1 & 1\end{bmatrix}^T = \begin{bmatrix} 3 & -1\\ -2 & 1\end{bmatrix}$$

C. Inverse of a matrix

THEOREM 2-5.2 Let A be an $n \times n$ matrix. If A is nonsingular, then

$$A^{-1} = \frac{1}{\det A}\text{adj } A \tag{2.18}$$

EXAMPLE 2-22: Verify (2.18).

Solution: Consider the product $A(\text{adj } A)$, that is,

$$A(\text{adj } A) = \begin{bmatrix} a_{11} & a_{12} & \cdots & a_{1n}\\ a_{21} & a_{22} & \cdots & a_{2n}\\ \vdots & \vdots & & \vdots\\ a_{i1} & a_{i2} & \cdots & a_{in}\\ \vdots & \vdots & & \vdots\\ a_{n1} & a_{n2} & \cdots & a_{nn}\end{bmatrix}\begin{bmatrix} A_{11} & A_{21} & \cdots & A_{j1} & \cdots & A_{n1}\\ A_{12} & A_{22} & \cdots & A_{j2} & \cdots & A_{n2}\\ \vdots & \vdots & & \vdots & & \vdots\\ A_{1n} & A_{2n} & \cdots & A_{jn} & \cdots & A_{nn}\end{bmatrix}$$

We see that the ijth entry of $A(\text{adj } A)$ is

$$a_{i1}A_{j1} + a_{i2}A_{j2} + \cdots + a_{in}A_{jn} \tag{2.19}$$

Now, from formula (2.14) we see that (2.19) is equal to det A if $i = j$ and that (2.19) is equal to zero when $i \neq j$. Thus

$$A(\text{adj } A) = \begin{bmatrix} \det A & 0 & \cdots & 0 \\ 0 & \det A & \cdots & 0 \\ \vdots & \vdots & \ddots & \vdots \\ 0 & 0 & \cdots & \det A \end{bmatrix} = (\det A)I_n \tag{2.20}$$

If A is nonsingular, then by Theorem 2-3.2, det $A \neq 0$ and we may write

$$A\left(\frac{1}{\det A}\text{adj } A\right) = I_n$$

Thus, by definition (1.37), we get

$$A^{-1} = \frac{1}{\det A}\text{adj } A$$

EXAMPLE 2-23: Let

$$\begin{bmatrix} a & b \\ c & d \end{bmatrix}$$

where a, b, c, d are real numbers such that $ad - bc \neq 0$. Show that A^{-1} exists and that

$$A^{-1} = \frac{1}{ad - bc}\begin{bmatrix} d & -b \\ -c & a \end{bmatrix} \tag{2.21}$$

Solution: Since

$$\det A = \begin{vmatrix} a & b \\ c & d \end{vmatrix} = ad - bc \neq 0$$

A^{-1} exists. Now by definition of adj A, we have

$$\text{adj } A = \begin{bmatrix} d & -c \\ -b & a \end{bmatrix}^T = \begin{bmatrix} d & -b \\ -c & a \end{bmatrix}$$

Hence, by Theorem 2-5.2, (2.18),

$$A^{-1} = \frac{1}{\det A}\text{adj } A = \frac{1}{ad - bc}\begin{bmatrix} d & -b \\ -c & a \end{bmatrix}$$

(Cf. Problem 1-19.)

EXAMPLE 2-24: Given

$$A = \begin{bmatrix} 1 & 2 & 3 \\ 2 & 3 & 4 \\ 3 & 4 & 6 \end{bmatrix}$$

find A^{-1}.

Solution: By cofactor expansion along the first column,

$$\det A = \begin{vmatrix} 1 & 2 & 3 \\ 2 & 3 & 4 \\ 3 & 4 & 6 \end{vmatrix} = 1\begin{vmatrix} 3 & 4 \\ 4 & 6 \end{vmatrix} - 2\begin{vmatrix} 2 & 3 \\ 4 & 6 \end{vmatrix} + 3\begin{vmatrix} 2 & 3 \\ 3 & 4 \end{vmatrix}$$

$$= 1(2) - 2(0) + 3(-1) = -1 \neq 0$$

Hence A^{-1} exists. Now, by the definition of cofactors, i.e., (2.11),

$$A_{11} = \begin{vmatrix} 3 & 4 \\ 4 & 6 \end{vmatrix} = 2, \qquad A_{12} = -\begin{vmatrix} 2 & 4 \\ 3 & 6 \end{vmatrix} = 0, \qquad A_{13} = \begin{vmatrix} 2 & 3 \\ 3 & 4 \end{vmatrix} = -1$$

$$A_{21} = -\begin{vmatrix} 2 & 3 \\ 4 & 6 \end{vmatrix} = 0, \qquad A_{22} = \begin{vmatrix} 1 & 3 \\ 3 & 6 \end{vmatrix} = -3, \qquad A_{23} = -\begin{vmatrix} 1 & 2 \\ 3 & 4 \end{vmatrix} = 2$$

$$A_{31} = \begin{vmatrix} 2 & 3 \\ 3 & 4 \end{vmatrix} = -1, \qquad A_{23} = -\begin{vmatrix} 1 & 3 \\ 2 & 4 \end{vmatrix} = 2, \qquad A_{33} = \begin{vmatrix} 1 & 2 \\ 2 & 3 \end{vmatrix} = -1$$

Hence by the definition of adj A,

$$\operatorname{adj} A = \begin{bmatrix} 2 & 0 & -1 \\ 0 & -3 & 2 \\ -1 & 2 & -1 \end{bmatrix}^T = \begin{bmatrix} 2 & 0 & -1 \\ 0 & -3 & 2 \\ -1 & 2 & -1 \end{bmatrix}$$

Thus by (2.18)

$$A^{-1} = \frac{1}{\det A} \operatorname{adj} A = \frac{1}{-1} \begin{bmatrix} 2 & 0 & -1 \\ 0 & -3 & 2 \\ -1 & 2 & -1 \end{bmatrix} = \begin{bmatrix} -2 & 0 & 1 \\ 0 & 3 & -2 \\ 1 & -2 & 1 \end{bmatrix}$$

2-6. Systems of Linear Equations and Cramer's Rule

A. The solution of $A\mathbf{x} = \mathbf{b}$

As shown in Section 1-1, (1.10), a system of n equations in n unknowns

$$\begin{aligned} a_{11}x_1 + a_{12}x_2 + \cdots + a_{1n}x_n &= b_1 \\ a_{21}x_1 + a_{22}x_2 + \cdots + a_{2n}x_n &= b_2 \\ \vdots \qquad \vdots \qquad\quad & \vdots \quad\; \vdots \\ a_{n1}x_1 + a_{n2}x_2 + \cdots + a_{nn}x_n &= b_n \end{aligned} \tag{2.22}$$

can be expressed in matrix form as

$$A\mathbf{x} = \mathbf{b} \tag{2.23}$$

If A is nonsingular, then by Theorem 1-4.3, the system (2.23) has a unique solution given by

$$\mathbf{x} = A^{-1}\mathbf{b} \tag{2.24}$$

Using formula (2.18),

$$\mathbf{x} = A^{-1}\mathbf{b} = \frac{1}{\det A}(\operatorname{adj} A)\mathbf{b} \tag{2.25}$$

B. Cramer's rule

THEOREM 2-6.1 (Cramer's Rule) If $A\mathbf{x} = \mathbf{b}$ is a system of n linear equations in n unknowns such that $\det A \neq 0$, then the system has a unique solution

$$\mathbf{x} = \begin{bmatrix} x_1 \\ x_2 \\ \vdots \\ x_n \end{bmatrix}$$

given by

CRAMER'S RULE

$$x_1 = \frac{\det A_1}{\det A}, \; x_2 = \frac{\det A_2}{\det A}, \ldots, x_n = \frac{\det A_n}{\det A} \tag{2.26}$$

where A_j is the matrix obtained from A by replacing the jth column of A by $\mathbf{b}$.

EXAMPLE 2-25: Verify Cramer's rule (2.26).

Solution: From (2.25), we have

$$\mathbf{x} = \begin{bmatrix} x_1 \\ x_2 \\ \vdots \\ x_n \end{bmatrix} = A^{-1}\mathbf{b} = \frac{1}{\det A}(\text{adj } A)\mathbf{b} \tag{2.27}$$

Using the explicit expression for adj A, (2.17), we have

$$\mathbf{x} = \frac{1}{\det A}(\text{adj } A)\mathbf{b} = \frac{1}{\det A}\begin{bmatrix} A_{11} & A_{21} & \cdots & A_{n1} \\ A_{12} & A_{22} & \cdots & A_{n2} \\ \vdots & \vdots & & \vdots \\ A_{1n} & A_{2n} & \cdots & A_{nn} \end{bmatrix}\begin{bmatrix} b_1 \\ b_2 \\ \vdots \\ b_n \end{bmatrix}$$

Multiplying the matrices gives

$$\begin{bmatrix} x_1 \\ x_2 \\ \vdots \\ x_n \end{bmatrix} = \frac{1}{\det A}\begin{bmatrix} b_1A_{11} + b_2A_{21} + \cdots + b_nA_{n1} \\ b_1A_{12} + b_2A_{22} + \cdots + b_nA_{n2} \\ \vdots \qquad \vdots \qquad \vdots \\ b_1A_{1n} + b_2A_{2n} + \cdots + b_nA_{nn} \end{bmatrix}$$

Therefore the entry in the jth row of $\mathbf{x}$ is

$$x_j = \frac{b_1A_{1j} + b_2A_{2j} + \cdots + b_nA_{nj}}{\det A} \tag{2.28}$$

We now consider the $n \times n$ matrix obtained from A replacing the jth column of A by $\mathbf{b}$:

$$A_j = \begin{bmatrix} a_{11} & a_{12} & \cdots & b_1 & \cdots & a_{1n} \\ a_{21} & a_{22} & \cdots & b_2 & \cdots & a_{2n} \\ \vdots & \vdots & & \vdots & & \vdots \\ a_{n1} & a_{n2} & \cdots & b_n & \cdots & a_{nn} \end{bmatrix}$$

$$\uparrow \\ j\text{th column}$$

If we compute $\det A_j$ using the cofactor expansion along the jth column, we get

$$\det A_j = b_1A_{1j} + b_2A_{2j} + \cdots + b_nA_{nj}$$

Substituting this result in (2.28) yields

$$x_j = \frac{\det A_j}{\det A}, \qquad j = 1, 2, \ldots, n$$

EXAMPLE 2-26: Use Cramer's rule to solve

$$3x_1 + x_2 = 10$$
$$2x_1 - x_2 = 5$$

Solution

$$x_1 = \frac{\begin{vmatrix} 10 & 1 \\ 5 & -1 \end{vmatrix}}{\begin{vmatrix} 3 & 1 \\ 2 & -1 \end{vmatrix}} = \frac{-15}{-5} = 3, \qquad x_2 = \frac{\begin{vmatrix} 3 & 10 \\ 2 & 5 \end{vmatrix}}{\begin{vmatrix} 3 & 1 \\ 2 & -1 \end{vmatrix}} = \frac{-5}{-5} = 1$$

which agree with the result of Example 1-2.

SUMMARY

1. The determinant of a square matrix A is a certain number denoted by $\det A$ or $|A|$.
2. $\det[a_{11}] = a_{11}$

$$\det\begin{bmatrix} a_{11} & a_{12} \\ a_{21} & a_{22} \end{bmatrix} = a_{11}a_{22} - a_{12}a_{21}$$

$$\det\begin{bmatrix} a_{11} & a_{12} & a_{13} \\ a_{21} & a_{22} & a_{23} \\ a_{31} & a_{32} & a_{33} \end{bmatrix} = a_{11}a_{22}a_{33} + a_{12}a_{23}a_{31} + a_{13}a_{21}a_{32} - a_{13}a_{22}a_{31} - a_{11}a_{23}a_{32} - a_{12}a_{21}a_{33}$$

3. The properties of determinants are:
   - The determinant is unchanged if any row or column is replaced by the sum of that row or column and another row or column.
   - The determinant is multiplied by k if any row or column is multiplied by k.
   - The determinant of the identity matrix is 1.
   - The determinant is unchanged if a multiple of one row or column is added to or subtracted from another row or column.
   - The determinant changes sign when two rows or columns are interchanged.
   - The determinant is zero if any row or column consists entirely of zeroes.
   - The determinant is zero if any two rows or columns are equal or proportional.
   - If A is triangular, then $\det A = a_{11}a_{22}\cdots a_{nn}$ (that is, the product of diagonal entries of A).
   - $\det A = \det A^T$
   - $\det(AB) = (\det A)(\det B)$
4. Let $A = [a_{ij}]_{n\times n}$ and M_{ij} be $(n-1)\times(n-1)$ matrix obtained from A by deleting the row and column containing a_{ij}. Then,

   the minor of $a_{ij} = \det M_{ij}$
   the cofactor of $a_{ij} = A_{ij} = (-1)^{i+j}\det M_{ij}$
5. Cofactor expansion of $\det A$ along the ith row is

$$\det A = a_{i1}A_{i1} + a_{i2}A_{i2} + \cdots + a_{in}A_{in}$$

   and along the jth column is

$$\det A = a_{1j}A_{1j} + a_{2j}A_{2j} + \cdots + a_{nj}A_{nj}$$

   where A_{ij} is the cofactor of a_{ij}.
6. A matrix A is nonsingular if $\det A \neq 0$.
7. If A is nonsingular, then its inverse A^{-1} is given by

$$A^{-1} = \frac{1}{\det A}\operatorname{adj} A$$

   where

$$\operatorname{adj} A = [A_{ij}]^T = \begin{bmatrix} A_{11} & A_{21} & \cdots & A_{n1} \\ A_{12} & A_{22} & \cdots & A_{n2} \\ \vdots & \vdots & & \vdots \\ A_{1n} & A_{2n} & \cdots & A_{nn} \end{bmatrix}$$

8. (Cramer's rule) If $A\mathbf{x} = \mathbf{b}$ is a system of n linear equations in n unknowns, and $\det A \neq 0$, then the system has a unique solution

$$\mathbf{x} = \begin{bmatrix} x_1 \\ x_2 \\ \vdots \\ x_n \end{bmatrix}$$

given by

$$x_1 = \frac{\det A_1}{\det A}, x_2 = \frac{\det A_2}{\det A}, \ldots, x_n = \frac{\det A_n}{\det A}$$

where A_j is the matrix obtained from A by replacing the jth column of A by $\mathbf{b}$.

RAISE YOUR GRADES

Can you explain ...?

- ☑ how to evaluate the determinants of 2×2 and 3×3 matrices
- ☑ how to evaluate the determinant of a matrix using the properties of determinants
- ☑ what conditions must obtain for a matrix to be nonsingular
- ☑ how to evaluate a determinant by cofactor expansion
- ☑ how to evaluate the inverse of a matrix by determinant methods
- ☑ Cramer's rule

SOLVED PROBLEMS

Determinant of a Square Matrix A

PROBLEM 2-1 Evaluate the determinants of

$$A = \begin{bmatrix} 3 & 4 \\ 1 & -2 \end{bmatrix}, \qquad B = \begin{bmatrix} 3 & 2 \\ 6 & 4 \end{bmatrix}, \qquad C = \begin{bmatrix} 1 & 0 & 3 \\ 0 & 1 & 2 \\ 1 & 3 & 4 \end{bmatrix}$$

Solution From (2.2),

$$\det A = \begin{vmatrix} 3 & 4 \\ 1 & -2 \end{vmatrix} = 3(-2) - 4(1) = -10$$

$$\det B = \begin{vmatrix} 3 & 2 \\ 6 & 4 \end{vmatrix} = 3(4) - 2(6) = 0$$

From (2.3), or forming the following pattern,

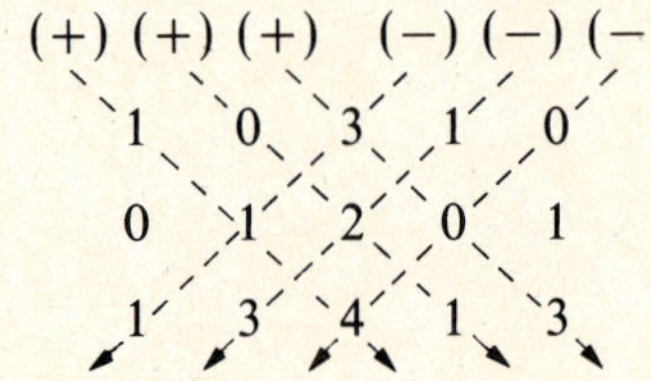

$$\begin{array}{ccccc} (+) & (+) & (+) & (-) & (-)\ (-) \\ 1 & 0 & 3 & 1 & 0 \\ 0 & 1 & 2 & 0 & 1 \\ 1 & 3 & 4 & 1 & 3 \end{array}$$

$$\det C = \begin{vmatrix} 1 & 0 & 3 \\ 0 & 1 & 2 \\ 1 & 3 & 4 \end{vmatrix} = 4 + 0 + 0 - 3 - 6 - 0 = -5$$

PROBLEM 2-2 Find all values of λ for which $\det A = 0$ if

$$A = \begin{bmatrix} \lambda - 3 & -2 \\ 1 & \lambda \end{bmatrix}$$

Solution From (2.2),

$$\det A = \begin{vmatrix} \lambda - 3 & -2 \\ 1 & \lambda \end{vmatrix} = (\lambda - 3)\lambda - (-2)(1) = \lambda^2 - 3\lambda + 2$$

Now

$$\lambda^2 - 3\lambda + 2 = (\lambda - 1)(\lambda - 2)$$

Thus all values of λ for which $\det A = 0$ are

$$\lambda_1 = 1, \qquad \lambda_2 = 2$$

PROBLEM 2-3 Show that

$$\begin{vmatrix} a_1 + a_2 & b_1 + b_2 \\ c & d \end{vmatrix} = \begin{vmatrix} a_1 & b_1 \\ c & d \end{vmatrix} + \begin{vmatrix} a_2 & b_2 \\ c & d \end{vmatrix}$$

Solution By (2.2),

$$\begin{aligned} \begin{vmatrix} a_1 + a_2 & b_1 + b_2 \\ c & d \end{vmatrix} &= (a_1 + a_2)d - (b_1 + b_2)c \\ &= (a_1 d - b_1 c) + (a_2 d - b_2 c) \\ &= \begin{vmatrix} a_1 & b_1 \\ c & d \end{vmatrix} + \begin{vmatrix} a_2 & b_2 \\ c & d \end{vmatrix} \end{aligned}$$

Properties of Determinants

PROBLEM 2-4 Evaluate the determinants of the following matrices by inspection:

$$A = \begin{bmatrix} 1 & 2 & 4 \\ 2 & 6 & -2 \\ 1 & 2 & 4 \end{bmatrix}, \qquad B = \begin{bmatrix} 1 & 2 & 3 \\ 2 & 4 & 6 \\ -1 & 0 & 2 \end{bmatrix}, \qquad C = \begin{bmatrix} 1 & -1 & 2 \\ 2 & 3 & 4 \\ 0 & 0 & 0 \end{bmatrix}$$

Solution Since the first and the third rows of A are equal,

$$\det A = 0$$

Since the first and the second rows of B are proportional,

$$\det B = 0$$

Since the elements in the third row of C are all zeros,

$$\det C = 0$$

PROBLEM 2-5 Compute $\det A$ if

$$A = \begin{bmatrix} 3 & 0 & 1 \\ 1 & 2 & 3 \\ -1 & 4 & 2 \end{bmatrix}$$

Solution

$$\det A = \begin{vmatrix} 3 & 0 & 1 \\ 1 & 2 & 3 \\ -1 & 4 & 2 \end{vmatrix}$$

$$= \begin{vmatrix} 3 & 0 & 1 \\ 0 & 6 & 5 \\ -1 & 4 & 2 \end{vmatrix} \qquad \begin{matrix}\text{(row 2) + (row 3)} \\ \text{[Property 4]}\end{matrix}$$

$$= \begin{vmatrix} 3 & 0 & 1 \\ 0 & 6 & 5 \\ 0 & 4 & \frac{7}{3} \end{vmatrix} \qquad \begin{matrix}\text{(row 3)} + \tfrac{1}{3}\text{(row 1)} \\ \text{[Property 4]}\end{matrix}$$

$$= \begin{vmatrix} 3 & 0 & 1 \\ 0 & 6 & 5 \\ 0 & 0 & -1 \end{vmatrix} \qquad \begin{matrix}\text{(row 3)} - \tfrac{2}{3}\text{(row 2)} \\ \text{[Property 4]}\end{matrix}$$

$$= (3)(6)(-1) \qquad \text{[Apply Property 8]}$$

$$= -18$$

PROBLEM 2-6 Compute $\det A$ if

$$A = \begin{bmatrix} 1 & 2 & 1 & -4 \\ 1 & 1 & 0 & -1 \\ -1 & 2 & -1 & 3 \\ 0 & 4 & 4 & -1 \end{bmatrix}$$

Solution

$$\det A = \begin{vmatrix} 1 & 2 & 1 & -4 \\ 1 & 1 & 0 & -1 \\ -1 & 2 & -1 & 3 \\ 0 & 4 & 4 & -1 \end{vmatrix}$$

$$= \begin{vmatrix} 1 & 2 & 1 & -4 \\ 0 & -1 & -1 & 3 \\ 0 & 4 & 0 & -1 \\ 0 & 4 & 4 & -1 \end{vmatrix} \qquad \begin{matrix}\text{(row 2) - (row 1)} \\ \text{(row 3) + (row 1)} \\ \text{[Property 4]}\end{matrix}$$

$$= \begin{vmatrix} 1 & 2 & 1 & -4 \\ 0 & -1 & -1 & 3 \\ 0 & 0 & -4 & 11 \\ 0 & 0 & 0 & 11 \end{vmatrix} \qquad \begin{matrix}\text{(row 3) + 4(row 2)} \\ \text{(row 4) + 4(row 2)} \\ \text{[Property 4]}\end{matrix}$$

$$= (1)(-1)(-4)(11) \qquad \text{[by Property 8]}$$

$$= 44$$

PROBLEM 2-7 Find $\det A$ if

$$A = \begin{bmatrix} 1 & 1 & 1 \\ a & b & c \\ b+c & c+a & a+b \end{bmatrix}$$

Solution

$$\det A = \begin{vmatrix} 1 & 1 & 1 \\ a & b & c \\ b+c & c+a & a+b \end{vmatrix}$$

$$= \begin{vmatrix} 1 & 1 & 1 \\ a & b & c \\ a+b+c & a+b+c & a+b+c \end{vmatrix} \qquad \text{(row 3) + (row 2)}$$

$$= (a+b+c)\begin{vmatrix} 1 & 1 & 1 \\ a & b & c \\ 1 & 1 & 1 \end{vmatrix} \qquad \text{[Property 2]}$$

$$= (a+b+c)(0) \qquad \text{[by Property 7]}$$

$$= 0$$

PROBLEM 2-8 Let A be an $n \times n$ matrix. Show that

$$\det(kA) = k^n(\det A)$$

Solution Since each row of kA is k multiple of each row of A, by Property 2 we have

$$\det(kA) = \overbrace{(k\ldots k)}^{n \text{ times}}(\det A)$$

$$= k^n(\det A)$$

PROBLEM 2-9 Give an example to show that in general

$$\det(A+B) \neq \det A + \det B.$$

for two $n \times n$ matrices A and B.

Solution Let

$$A = \begin{bmatrix} 1 & 0 \\ 0 & 0 \end{bmatrix}, \qquad B = \begin{bmatrix} 0 & 0 \\ 0 & 1 \end{bmatrix}$$

Then

$$A + B = \begin{bmatrix} 1 & 0 \\ 0 & 0 \end{bmatrix} + \begin{bmatrix} 0 & 0 \\ 0 & 1 \end{bmatrix} = \begin{bmatrix} 1 & 0 \\ 0 & 1 \end{bmatrix}$$

and

$$\det A = \begin{vmatrix} 1 & 0 \\ 0 & 0 \end{vmatrix} = 0, \qquad \det B = \begin{vmatrix} 0 & 0 \\ 0 & 1 \end{vmatrix} = 0, \qquad \det(A+B) = \begin{vmatrix} 1 & 0 \\ 0 & 1 \end{vmatrix} = 1$$

Thus,

$$\det A + \det B = 0 + 0 = 0 \qquad \text{and} \qquad \det(A+B) \neq \det A + \det B$$

Multiplicative Property of Determinants

PROBLEM 2-10 Using 2×2 matrices prove that

$$\det(AB) = (\det A)(\det B)$$

Solution Let

$$A = \begin{bmatrix} a_{11} & a_{12} \\ a_{21} & a_{22} \end{bmatrix}, \qquad B = \begin{bmatrix} b_{11} & b_{12} \\ b_{21} & b_{22} \end{bmatrix}$$

Then

$$AB = \begin{bmatrix} a_{11} & a_{12} \\ a_{21} & a_{22} \end{bmatrix}\begin{bmatrix} b_{11} & b_{12} \\ b_{21} & b_{22} \end{bmatrix} = \begin{bmatrix} a_{11}b_{11} + a_{12}b_{21} & a_{11}b_{12} + a_{12}b_{22} \\ a_{21}b_{11} + a_{22}b_{21} & a_{21}b_{12} + a_{22}b_{22} \end{bmatrix}$$

$$\det A = \begin{vmatrix} a_{11} & a_{12} \\ a_{21} & a_{22} \end{vmatrix} = a_{11}a_{22} - a_{12}a_{21}$$

$$\det B = \begin{vmatrix} b_{11} & b_{12} \\ b_{21} & b_{22} \end{vmatrix} = b_{11}b_{22} - b_{12}b_{21}$$

$$\begin{aligned}
\det(AB) &= \begin{vmatrix} a_{11}b_{11} + a_{12}b_{21} & a_{11}b_{12} + a_{12}b_{22} \\ a_{21}b_{11} + a_{22}b_{21} & a_{21}b_{12} + a_{22}b_{22} \end{vmatrix} \\
&= (a_{11}b_{11} + a_{12}b_{21})(a_{21}b_{12} + a_{22}b_{22}) - (a_{11}b_{12} + a_{12}b_{22})(a_{21}b_{11} + a_{22}b_{21}) \\
&= a_{11}a_{22}(b_{11}b_{22} - b_{12}b_{21}) - a_{12}a_{21}(b_{11}b_{22} - b_{12}b_{21}) + a_{11}a_{21}(b_{11}b_{12} - b_{12}b_{11}) \\
&\quad + a_{12}a_{22}(b_{21}b_{22} - b_{22}b_{21}) \\
&= (a_{11}a_{22} - a_{12}a_{21})(b_{11}b_{22} - b_{12}b_{21}) \\
&= (\det A)(\det B)
\end{aligned}$$

PROBLEM 2-11 Let

$$A = \begin{bmatrix} 1 & 1 & 3 \\ 2 & 1 & 0 \\ 3 & 2 & -1 \end{bmatrix}, \qquad B = \begin{bmatrix} 1 & 2 & -1 \\ -1 & 0 & 1 \\ 1 & 4 & 5 \end{bmatrix}$$

Show that $\det(AB) = (\det A)(\det B)$.

Solution

$$AB = \begin{bmatrix} 1 & 1 & 3 \\ 2 & 1 & 0 \\ 3 & 2 & -1 \end{bmatrix}\begin{bmatrix} 1 & 2 & -1 \\ -1 & 0 & 1 \\ 1 & 4 & 5 \end{bmatrix} = \begin{bmatrix} 3 & 14 & 15 \\ 1 & 4 & -1 \\ 0 & 2 & -6 \end{bmatrix}$$

$$\begin{aligned}
\det A &= \begin{vmatrix} 1 & 1 & 3 \\ 2 & 1 & 0 \\ 3 & 2 & -1 \end{vmatrix} \\
&= \begin{vmatrix} 1 & 1 & 3 \\ 0 & -1 & -6 \\ 0 & -1 & -10 \end{vmatrix} \qquad \begin{matrix} \text{(row 2)} - 2\text{(row 1)} \\ \text{(row 3)} - 3\text{(row 1)} \end{matrix} \\
&= \begin{vmatrix} 1 & 1 & 3 \\ 0 & -1 & -6 \\ 0 & 0 & -4 \end{vmatrix} \qquad \text{(row 3)} - \text{(row 2)} \\
&= (1)(-1)(-4) = 4 \qquad \text{[by Property 8]}
\end{aligned}$$

$$\begin{aligned}
\det B &= \begin{vmatrix} 1 & 2 & -1 \\ -1 & 0 & 1 \\ 1 & 4 & 5 \end{vmatrix} \\
&= \begin{vmatrix} 1 & 2 & -1 \\ 0 & 2 & 0 \\ 0 & 2 & 6 \end{vmatrix} \qquad \begin{matrix} \text{(row 2)} + \text{(row 1)} \\ \text{(row 3)} - \text{(row 1)} \end{matrix}
\end{aligned}$$

$$\det B = \begin{vmatrix} 1 & 2 & -1 \\ 0 & 2 & 0 \\ 0 & 0 & 6 \end{vmatrix} \qquad \text{(row 3)} - \text{(row 2)}$$

$$= (1)(2)(6) = 12 \qquad \text{[by Property 8]}$$

$$\det(AB) = \begin{vmatrix} 3 & 14 & 15 \\ 1 & 4 & -1 \\ 0 & 2 & -6 \end{vmatrix}$$

$$= \begin{vmatrix} 3 & 14 & 15 \\ 0 & -\frac{2}{3} & -6 \\ 0 & 2 & -6 \end{vmatrix} \qquad \text{(row 2)} - \tfrac{1}{3}\text{(row 1)}$$

$$= \begin{vmatrix} 3 & 14 & 15 \\ 0 & -\frac{2}{3} & -6 \\ 0 & 0 & -24 \end{vmatrix} \qquad \text{(row 3)} + 3\text{(row 2)}$$

$$= (3)(-\tfrac{2}{3})(-24) = 48 = (4)(12) = (\det A)(\det B)$$

PROBLEM 2-12 If A and B are two $n \times n$ matrices then $\det(AB) = (\det A)(\det B)$. Using this fact, prove that a matrix A such that $\det A = 0$ does not have an inverse; that is, that A is singular.

Solution If $B = A^{-1}$, then

$$AB = AA^{-1} = I_n \qquad \text{and} \qquad \det(AB) = (\det A)(\det B) = \det I_n = 1$$

Now if $\det A = 0$, then we have $0 = 1$, which is inconsistent. Thus no such B exists and A does not have an inverse; hence A is singular.

PROBLEM 2-13 Show that

$$A = \begin{bmatrix} 1 & 0 & 1 \\ 4 & 2 & 1 \\ 2 & 0 & 2 \end{bmatrix}$$

is singular.

Solution Since the first row and the third row are proportional, $\det A = 0$. Hence A is singular.

PROBLEM 2-14 Let A and B be 3×3 matrices with $\det A = 2$ and $\det B = -3$. Find the value of **(a)** $\det(AB)$, **(b)** $\det(3A)$, **(c)** $\det(A^{-1}B)$.

Solution

(a) By (2.7), we get

$$\det(AB) = (\det A)(\det B) = (2)(-3) = -6$$

(b) From the result of Problem 2-8, we get

$$\det(3A) = 3^3(\det A) = (27)(2) = 54$$

(c) By (2.7) and (2.10) we get

$$\det(A^{-1}B) = (\det A^{-1})(\det B)$$

$$= \frac{1}{\det A}(\det B) = \frac{1}{2}(-3) = -\frac{3}{2}$$

PROBLEM 2-15 Let A and B be $n \times n$ matrices. Prove that AB is nonsingular if and only if A and B are both nonsingular.

Solution Since $\det(AB) = (\det A)(\det B)$, it follows that $\det(AB) \neq 0$ if and only if $\det A \neq 0$ and $\det B \neq 0$. Thus AB is nonsingular if and only if A and B are both nonsingular.

PROBLEM 2-16 Let A and B be $n \times n$ matrices. Prove that if $AB = I_n$, then $BA = I_n$.

Solution If $AB = I_n$, then

$$\det(AB) = (\det A)(\det B) = \det I_n = 1$$

and hence by Problem 2-15 both A and B are nonsingular. It follows then that

$$A^{-1}(AB) = A^{-1}I_n = A^{-1}$$

Since, by (1.23),

$$A^{-1}(AB) = (A^{-1}A)B = I_nB = B$$

we get

$$B = A^{-1}$$

Hence

$$BA = A^{-1}A = I_n$$

PROBLEM 2-17 If A is an $m \times n$ matrix and B is an $n \times m$ matrix, show that

$$\det(AB) = 0 \qquad \text{if } m > n.$$

Solution Let $A = [a_{ij}]_{m\times n}$ and $B = [b_{ij}]_{n\times m}$ and $m > n$. Then

$$AB = \begin{bmatrix} a_{11} & \cdots & a_{1n} \\ a_{21} & \cdots & a_{2n} \\ \vdots & & \vdots \\ a_{m1} & \cdots & a_{mn} \end{bmatrix} \begin{bmatrix} b_{11} & b_{12} & \cdots & b_{1m} \\ \vdots & \vdots & & \vdots \\ b_{n1} & b_{n2} & \cdots & b_{nm} \end{bmatrix}$$

$$= \begin{bmatrix} a_{11} & \cdots & a_{1n} & 0 & \cdots & 0 \\ a_{21} & \cdots & a_{2n} & 0 & \cdots & 0 \\ \vdots & & \vdots & \vdots & & \vdots \\ a_{m1} & \cdots & a_{mn} & \underbrace{0 \quad\;\; 0}_{(m-n)\text{columns}} \end{bmatrix} \begin{bmatrix} b_{11} & b_{12} & \cdots & b_{1m} \\ \vdots & \vdots & & \vdots \\ b_{n1} & b_{n2} & \cdots & b_{nm} \\ 0 & 0 & \cdots & 0 \\ \vdots & \vdots & & \vdots \\ 0 & 0 & \cdots & 0 \end{bmatrix} \left.\vphantom{\begin{matrix}0\\0\\0\end{matrix}}\right\} (m-n)\text{ rows}$$

$$= A'B'$$

Note that AB is an $m \times m$ matrix and A' and B' are both $m \times m$ matrices. Since $\det A' = 0$ and $\det B' = 0$, then by (2.7)

$$\det(AB) = \det(A'B') = (\det A')(\det B') = 0$$

PROBLEM 2-18 Given

$$A = \begin{bmatrix} 1 & 2 \\ -1 & 0 \\ 2 & 1 \end{bmatrix}, \qquad B = \begin{bmatrix} 2 & 3 & 1 \\ -1 & -1 & 2 \end{bmatrix}$$

calculate $\det(AB)$ and $\det(BA)$.

Solution

$$AB = \begin{bmatrix} 1 & 2 \\ -1 & 0 \\ 2 & 1 \end{bmatrix} \begin{bmatrix} 2 & 3 & 1 \\ -1 & -1 & 2 \end{bmatrix} = \begin{bmatrix} 0 & 1 & 5 \\ -2 & -3 & -1 \\ 3 & 5 & 4 \end{bmatrix}$$

$$BA = \begin{bmatrix} 2 & 3 & 1 \\ -1 & -1 & 2 \end{bmatrix} \begin{bmatrix} 1 & 2 \\ -1 & 0 \\ 2 & 1 \end{bmatrix} = \begin{bmatrix} 1 & 5 \\ 4 & 0 \end{bmatrix}$$

$$\det(AB) = \begin{bmatrix} 0 & 1 & 5 \\ -2 & -3 & -1 \\ 3 & 5 & 4 \end{bmatrix} = 0 - 3 - 50 + 45 - 0 + 8 = 0$$

$$\det(BA) = \begin{vmatrix} 1 & 5 \\ 4 & 0 \end{vmatrix} = -20$$

PROBLEM 2-19 If A is an $m \times n$ matrix, B is an $n \times m$ matrix, and $m < n$, then

$$\det(AB) = \sum_i M_i(A) M_i(B) \qquad \textbf{(a)}$$

where $M_i(A)$ and $M_i(B)$ denote the **corresponding *i*th majors** of A and B, respectively. A **major** of an $m \times n$ matrix is any determinant of its maximum order square submatrix of A. A major of A and a major of B are said to be **corresponding majors** of A and B if the columns of A used to form the major of A have the same indices as the rows of B used to form the major of B. Formula (a) is known as the **Binet-Cauchy formula**. Use this formula to find $\det(AB)$ if

$$A = \begin{bmatrix} 2 & 1 & -2 \\ 3 & 3 & 0 \end{bmatrix}, \qquad B = \begin{bmatrix} 2 & 1 \\ -3 & 2 \\ 1 & -2 \end{bmatrix}$$

Solution

$$M_1(A) = \begin{vmatrix} 1 & -2 \\ 3 & 0 \end{vmatrix} = 6, \qquad M_1(B) = \begin{vmatrix} -3 & 2 \\ 1 & -2 \end{vmatrix} = 4$$

$$M_2(A) = \begin{vmatrix} 2 & -2 \\ 3 & 0 \end{vmatrix} = 6, \qquad M_2(B) = \begin{vmatrix} 2 & 1 \\ 1 & -2 \end{vmatrix} = -5$$

$$M_3(A) = \begin{vmatrix} 2 & 1 \\ 3 & 3 \end{vmatrix} = 3, \qquad M_3(B) = \begin{vmatrix} 2 & 1 \\ -3 & 2 \end{vmatrix} = 7$$

Thus by formula (a), we get

$$\begin{aligned} \det(AB) &= M_1(A)M_1(B) + M_2(A)M_2(B) + M_3(A)M_3(B) \\ &= (6)(4) + (6)(-5) + (3)(7) = 15 \end{aligned}$$

Check:

$$AB = \begin{bmatrix} 2 & 1 & -2 \\ 3 & 3 & 0 \end{bmatrix} \begin{bmatrix} 2 & 1 \\ -3 & 2 \\ 1 & -2 \end{bmatrix} = \begin{bmatrix} -1 & 8 \\ -3 & 9 \end{bmatrix}$$

$$\det(AB) = \begin{vmatrix} -1 & 8 \\ -3 & 9 \end{vmatrix} = -9 - (-24) = 15$$

Cofactor Expansion

PROBLEM 2-20 Given

$$A = \begin{bmatrix} 3 & 0 & 1 \\ 1 & 2 & 3 \\ -1 & 4 & 2 \end{bmatrix}$$

compute $\det A$ by cofactor expansion **(a)** along the second row and **(b)** along the second column.

Solution

(a)

$$\det A = \begin{vmatrix} 3 & 0 & 1 \\ 1 & 2 & 3 \\ -1 & 4 & 2 \end{vmatrix} = -1\begin{vmatrix} 0 & 1 \\ 4 & 2 \end{vmatrix} + 2\begin{vmatrix} 3 & 1 \\ -1 & 2 \end{vmatrix} - 3\begin{vmatrix} 3 & 0 \\ -1 & 4 \end{vmatrix}$$

$$= -1(-4) + 2(7) - 3(12) = -18$$

(b)

$$\det A = \begin{vmatrix} 3 & 0 & 1 \\ 1 & 2 & 3 \\ -1 & 4 & 2 \end{vmatrix} = -0\begin{vmatrix} 1 & 3 \\ -1 & 2 \end{vmatrix} + 2\begin{vmatrix} 3 & 1 \\ -1 & 2 \end{vmatrix} - 4\begin{vmatrix} 3 & 1 \\ 1 & 3 \end{vmatrix}$$

$$= 0 + 2(7) - 4(8) = -18$$

PROBLEM 2-21 Compute $\det A$ if

$$A = \begin{vmatrix} 3 & 0 & 1 \\ 1 & 2 & 3 \\ -1 & 4 & 2 \end{vmatrix}$$

Solution We already have a zero in the first row. Subtracting twice the second row from the third row yields

$$\det A = \begin{vmatrix} 3 & 0 & 1 \\ 1 & 2 & 3 \\ -1 & 4 & 2 \end{vmatrix} = \begin{vmatrix} 3 & 0 & 1 \\ 1 & 2 & 3 \\ -3 & 0 & -4 \end{vmatrix}$$

by Property 4. Now, expanding along the second column, we obtain

$$\det A = 2\begin{vmatrix} 3 & 1 \\ -3 & -4 \end{vmatrix} = 2(-9) = -18$$

PROBLEM 2-22 Find $\det A$ if

$$A = \begin{bmatrix} 2 & 0 & 0 & 1 \\ 0 & 1 & 0 & 0 \\ 1 & 6 & 2 & 0 \\ 1 & 1 & -2 & 3 \end{bmatrix}$$

Solution Since there are three zeros in the second row, we expand along that row and get

$$\det A = \begin{vmatrix} 2 & 0 & 0 & 1 \\ 0 & 1 & 0 & 0 \\ 1 & 6 & 2 & 0 \\ 1 & 1 & -2 & 3 \end{vmatrix}$$

$$= (-1)^{2+2}\begin{vmatrix} 2 & 0 & 1 \\ 1 & 2 & 0 \\ 1 & -2 & 3 \end{vmatrix}$$

$$= 2\begin{vmatrix} 2 & 0 \\ -2 & 3 \end{vmatrix} + 1\begin{vmatrix} 1 & 2 \\ 1 & -2 \end{vmatrix}$$ Cofactor expansion along the first row

$$= 2(6) + 1(-4) = 8$$

PROBLEM 2-23 Let A be an $n \times n$ matrix. Using cofactor expansion, verify basic Property 2 of determinants; that is, the determinant of A is multiplied by the constant k if any row or column is multiplied by k.

Solution Let $A = [a_{ij}]_{n\times n}$. Let $B = [b_{ij}]$ be an $n \times n$ matrix that is identical to A except that all entries in the ith row of B are k times the corresponding entries of the ith row of A, that is

$$b_{i1} = ka_{i1},\ b_{i2} = ka_{i2}, \ldots, b_{in} = ka_{in}$$

Since A and B are the same except for the ith row, the cofactors

$$B_{ik} = A_{ik}, \qquad k = 1, 2, \ldots, n$$

then by (2.12)

$$\begin{aligned} \det B &= b_{i1}B_{i1} + b_{i2}B_{i2} + \cdots + b_{in}B_{in} \\ &= (ka_{i1})A_{i1} + (ka_{i2})A_{i2} + \cdots + (ka_{in})A_{in} \\ &= k(a_{i1}A_{i1} + a_{i2}A_{i2} + \cdots + a_{in}A_{in}) \\ &= k(\det A) \end{aligned}$$

The result is also true for the case of the ith column by expanding along the ith column.

PROBLEM 2-24 Using cofactor expansion, prove that if a row or column of an $n \times n$ matrix consists entirely of zeros, then $\det A = 0$.

Solution If the ith row of A consists entirely of zeros, that is,

$$a_{ik} = 0 \qquad \text{for } k = 1, 2, \ldots, n$$

then by (2.12)

$$\det A = a_{i1}A_{i1} + a_{i2}A_{i2} + \cdots + a_{in}A_{in} = 0$$

The result is also true for the ith column by expanding the ith column.

PROBLEM 2-25 Prove that if $A = [a_{ij}]$ is any $n \times n$ upper- (or lower-) triangular matrix, then

$$\det A = a_{11}a_{22}\ldots a_{nn} \qquad \textbf{(a)}$$

Solution We shall prove this by induction. It is trivial for a 1×1 matrix, since $\det[a_{11}] = a_{11}$ by (2.1). For a 2×2 matrix, expanding along the first column we get

$$\det\begin{bmatrix} a_{11} & a_{12} \\ 0 & a_{22} \end{bmatrix} = a_{11}\det[a_{22}] = a_{11}a_{22}$$

Now we assume that (a) is true for any $(n-1) \times (n-1)$ upper-triangular matrix. Consider an $n \times n$ upper-triangular matrix

$$A = \begin{bmatrix} a_{11} & a_{12} & \cdots & a_{1n} \\ 0 & a_{22} & \cdots & a_{2n} \\ \vdots & \vdots & \ddots & \vdots \\ 0 & 0 & \cdots & a_{nn} \end{bmatrix}$$

Expanding along the first column yields

$$\det A = a_{11}\det\begin{bmatrix} a_{22} & a_{23} & \cdots & a_{2n} \\ 0 & a_{33} & \cdots & a_{3n} \\ \vdots & \vdots & \ddots & \vdots \\ 0 & 0 & \cdots & a_{nn} \end{bmatrix} \qquad \textbf{(b)}$$

But the determinant on the right side of (b) is the determinant of an $(n-1) \times (n-1)$ upper-triangular matrix. According to our induction assumption, this determinant is equal to

$$a_{22}a_{33}\dots a_{nn}$$

Therefore

$$\det A = a_{11}a_{22}\dots a_{nn}$$

The result is also true for lower-triangular matrices by expanding along the first row.

PROBLEM 2-26 Let A, B, and C be three $n \times n$ matrices such that for some i,

$$a_{ik} = b_{ik} + c_{ik}, \qquad k = 1, 2, \dots, n$$

and

$$a_{jk} = b_{jk} = c_{jk}, \qquad j \neq i$$

Show that

$$\det A = \det B + \det C$$

Solution Since A, B, and C are the same except for the ith row, the cofactors $A_{ik} = B_{ik} = C_{ik}$, for $k = 1, 2, \dots, n$. Then by cofactor expansion along the ith row, we have

$$\begin{aligned}
\det B + \det C &= (b_{i1}B_{i1} + \cdots + b_{in}B_{in}) + (c_{i1}C_{i1} + \cdots + c_{in}C_{in}) \\
&= (b_{i1} + c_{i1})A_{i1} + \cdots + (b_{in} + c_{in})A_{in} \\
&= a_{i1}A_{i1} + \cdots + a_{in}A_{in} \\
&= \det A
\end{aligned}$$

Determinants and the Inverse of a Matrix

PROBLEM 2-27 Let

$$A = \begin{bmatrix} a & b & c \\ 0 & a & d \\ 0 & 0 & a \end{bmatrix}$$

Suppose that $a \neq 0$. Show that A^{-1} exists and find A^{-1}.

Solution By Property 8,

$$\det A = \begin{vmatrix} a & b & c \\ 0 & a & d \\ 0 & 0 & a \end{vmatrix} = a^3 \neq 0$$

Hence A is nonsingular and A^{-1} exists.

Let

$$A^{-1} = \begin{vmatrix} x_{11} & x_{12} & x_{13} \\ x_{21} & x_{22} & x_{23} \\ x_{31} & x_{32} & x_{33} \end{vmatrix}$$

Then

$$AA^{-1} = \begin{bmatrix} a & b & c \\ 0 & a & d \\ 0 & 0 & a \end{bmatrix} \begin{vmatrix} x_{11} & x_{12} & x_{13} \\ x_{21} & x_{22} & x_{23} \\ x_{31} & x_{32} & x_{33} \end{vmatrix} = \begin{bmatrix} 1 & 0 & 0 \\ 0 & 1 & 0 \\ 0 & 0 & 1 \end{bmatrix} = I_3 \qquad \textbf{(a)}$$

Thus, by multiplying directly and equating each entry for the first column of both sides, we get

$$ax_{11} + bx_{21} + cx_{31} = 1$$
$$ax_{21} + dx_{31} = 0$$
$$ax_{31} = 0$$

Solving the last equation, we get $x_{31} = 0$ since $a \neq 0$. Substituting backward to the second and first equations, we get $x_{21} = 0$, $x_{11} = 1/a$. Hence

$$x_{31} = 0, \qquad x_{21} = 0, \qquad x_{11} = \frac{1}{a}$$

Similarly, equating each entry for the second column of (a), we get

$$ax_{12} + bx_{22} + cx_{32} = 0$$
$$ax_{22} + dx_{32} = 1$$
$$ax_{32} = 0$$

Again solving for x_{32}, x_{22}, x_{12} by back substitution, we get

$$x_{32} = 0, \qquad x_{22} = \frac{1}{a}, \qquad x_{12} = -\frac{b}{a^2}$$

Likewise, equating each entry for the third column of (a), we get

$$ax_{13} + bx_{23} + cx_{33} = 0$$
$$ax_{23} + dx_{33} = 0$$
$$ax_{33} = 1$$

Solving for x_{33}, x_{23}, x_{13} by back substitution, we get

$$x_{33} = \frac{1}{a}, \qquad x_{23} = -\frac{d}{a^2}, \qquad x_{13} = \frac{bd}{a^3} - \frac{c}{a^2}$$

Thus

$$A^{-1} = \begin{bmatrix} \frac{1}{a} & -\frac{b}{a^2} & \frac{bd}{a^3} - \frac{c}{a^2} \\ 0 & \frac{1}{a} & -\frac{d}{a^2} \\ 0 & 0 & \frac{1}{a} \end{bmatrix}$$

PROBLEM 2-28 Find A^{-1} if

$$A = \begin{bmatrix} \cos\theta & -\sin\theta \\ \sin\theta & \cos\theta \end{bmatrix}$$

Solution

$$\det A = \begin{vmatrix} \cos\theta & -\sin\theta \\ \sin\theta & \cos\theta \end{vmatrix} = \cos^2\theta + \sin^2\theta = 1 \neq 0$$

Thus A is nonsingular and A^{-1} exists.

Next, by definition of adj A,

$$\text{adj}\, A = \begin{bmatrix} \cos\theta & -\sin\theta \\ \sin\theta & \cos\theta \end{bmatrix}^T = \begin{bmatrix} \cos\theta & \sin\theta \\ -\sin\theta & \cos\theta \end{bmatrix}$$

Hence, by (2.18),

$$A^{-1} = \frac{1}{\det A}\operatorname{adj} A = \begin{bmatrix} \cos\theta & \sin\theta \\ -\sin\theta & \cos\theta \end{bmatrix}$$

Note that $A^{-1} = A^T$. If $A^{-1} = A^T$, then A is called an **orthogonal matrix** [see (5.40)].

PROBLEM 2-29 Rework Problem 2-27 by formula (2.18).

Solution From Problem 2-27,

$$A = \begin{bmatrix} a & b & c \\ 0 & a & d \\ 0 & 0 & a \end{bmatrix}$$

and

$$\det A = a^3$$

Now, by definition of adj A, we have

$$\operatorname{adj} A = \begin{bmatrix} \begin{vmatrix} a & d \\ 0 & a \end{vmatrix} & -\begin{vmatrix} 0 & d \\ 0 & a \end{vmatrix} & \begin{vmatrix} 0 & a \\ 0 & 0 \end{vmatrix} \\ -\begin{vmatrix} b & c \\ 0 & a \end{vmatrix} & \begin{vmatrix} a & c \\ 0 & a \end{vmatrix} & -\begin{vmatrix} a & b \\ 0 & 0 \end{vmatrix} \\ \begin{vmatrix} b & c \\ a & d \end{vmatrix} & -\begin{vmatrix} a & c \\ 0 & d \end{vmatrix} & \begin{vmatrix} a & b \\ 0 & a \end{vmatrix} \end{bmatrix}^T$$

$$= \begin{bmatrix} a^2 & 0 & 0 \\ -ab & a^2 & 0 \\ bd - ac & -ad & a^2 \end{bmatrix}^T$$

$$= \begin{bmatrix} a^2 & -ab & bd - ac \\ 0 & a^2 & -ad \\ 0 & 0 & a^2 \end{bmatrix}$$

Thus by (2.18),

$$A^{-1} = \frac{1}{\det A}\operatorname{adj} A$$

$$= \frac{1}{a^3}\begin{bmatrix} a^2 & -ab & bd - ac \\ 0 & a^2 & -ad \\ 0 & 0 & a^2 \end{bmatrix}$$

$$= \begin{bmatrix} \frac{1}{a} & -\frac{b}{a^2} & \frac{bd}{a^3} - \frac{c}{a^2} \\ 0 & \frac{1}{a} & -\frac{d}{a^2} \\ 0 & 0 & \frac{1}{a} \end{bmatrix}$$

PROBLEM 2-30 Suppose that A is an $n \times n$ nonsingular matrix. Prove that

$$\det(\operatorname{adj} A) = (\det A)^{n-1}$$

Solution From Example 2-22, (2.20),

$$A \operatorname{adj} A = \begin{bmatrix} \det A & 0 & \cdots & 0 \\ 0 & \det A & \cdots & 0 \\ \vdots & \vdots & \ddots & \vdots \\ 0 & 0 & \cdots & \det A \end{bmatrix}$$

Hence by Property 8, (2.5),

$$\det(A \operatorname{adj} A) = (\det A)^n$$

Then, by Theorem 2-3.1, (2.7),

$$\det(A \operatorname{adj} A) = (\det A)\det(\operatorname{adj} A)$$

Hence

$$(\det A)\det(\operatorname{adj} A) = (\det A)^n$$

Since A is nonsingular, $\det A \neq 0$, hence

$$\det(\operatorname{adj} A) = \frac{1}{\det A}(\det A)^n = (\det A)^{n-1}$$

PROBLEM 2-31 If A is singular, prove that

$$A \operatorname{adj} A = O$$

Solution If A is singular, then $\det A = 0$. Now by (2.20), we have

$$A \operatorname{adj} A = (\det A)I_n = 0I_n = O$$

Systems of Linear Equations and Cramer's Rule

PROBLEM 2-32 Solve the following system of equations:

$$x_1 + x_2 + x_3 = 2$$
$$x_1 + x_2 - x_3 = 1$$
$$x_1 - x_2 - x_3 = -1$$

Solution The system can be expressed as

$$A\mathbf{x} = \mathbf{b}$$

where

$$A = \begin{bmatrix} 1 & 1 & 1 \\ 1 & 1 & -1 \\ 1 & -1 & -1 \end{bmatrix}, \qquad \mathbf{x} = \begin{bmatrix} x_1 \\ x_2 \\ x_3 \end{bmatrix}, \qquad \mathbf{b} = \begin{bmatrix} 2 \\ 1 \\ -1 \end{bmatrix}$$

Then by (2.24),

$$\mathbf{x} = A^{-1}\mathbf{b}$$

Now

$$\det A = \begin{vmatrix} 1 & 1 & 1 \\ 1 & 1 & -1 \\ 1 & -1 & -1 \end{vmatrix} \qquad \begin{matrix}\text{(row 2)} - \text{(row 1)} \\ \text{(row 3)} - \text{(row 1)}\end{matrix}$$

$$= \begin{vmatrix} 1 & 1 & 1 \\ 0 & 0 & -2 \\ 0 & -2 & -2 \end{vmatrix} \qquad \begin{matrix}\text{[cofactor expansion} \\ \text{along the first column]}\end{matrix}$$

$$= \begin{vmatrix} 0 & -2 \\ -2 & -2 \end{vmatrix} = -4$$

Next,

$$\operatorname{adj} A = \begin{bmatrix} \begin{vmatrix} 1 & -1 \\ -1 & -1 \end{vmatrix} & -\begin{vmatrix} 1 & -1 \\ 1 & -1 \end{vmatrix} & \begin{vmatrix} 1 & 1 \\ 1 & -1 \end{vmatrix} \\ -\begin{vmatrix} 1 & 1 \\ -1 & -1 \end{vmatrix} & \begin{vmatrix} 1 & 1 \\ 1 & -1 \end{vmatrix} & -\begin{vmatrix} 1 & 1 \\ 1 & -1 \end{vmatrix} \\ \begin{vmatrix} 1 & 1 \\ 1 & -1 \end{vmatrix} & -\begin{vmatrix} 1 & 1 \\ 1 & -1 \end{vmatrix} & \begin{vmatrix} 1 & 1 \\ 1 & 1 \end{vmatrix} \end{bmatrix}^T$$

$$= \begin{bmatrix} -2 & 0 & -2 \\ 0 & -2 & 2 \\ -2 & 2 & 0 \end{bmatrix}^T = \begin{bmatrix} -2 & 0 & -2 \\ 0 & -2 & 2 \\ -2 & 2 & 0 \end{bmatrix}$$

Thus

$$A^{-1} = \frac{1}{\det A} \operatorname{adj} A = \frac{1}{-4} \begin{bmatrix} -2 & 0 & -2 \\ 0 & -2 & 2 \\ -2 & 2 & 0 \end{bmatrix} = \frac{1}{2} \begin{bmatrix} 1 & 0 & 1 \\ 0 & 1 & -1 \\ 1 & -1 & 0 \end{bmatrix}$$

Then

$$\mathbf{x} = \begin{bmatrix} x_1 \\ x_2 \\ x_3 \end{bmatrix} = A^{-1}\mathbf{b} = \frac{1}{2} \begin{bmatrix} 1 & 0 & 1 \\ 0 & 1 & -1 \\ 1 & -1 & 0 \end{bmatrix} \begin{bmatrix} 2 \\ 1 \\ -1 \end{bmatrix} = \frac{1}{2} \begin{bmatrix} 1 \\ 2 \\ 1 \end{bmatrix} = \begin{bmatrix} \frac{1}{2} \\ 1 \\ \frac{1}{2} \end{bmatrix}$$

and it follows that the solution to the given system is

$$x_1 = \tfrac{1}{2}, \qquad x_2 = 1, \qquad x_3 = \tfrac{1}{2}$$

PROBLEM 2-33 Show that in a homogeneous system of linear equations in which the number of unknowns is greater than the number of equations, a nontrivial solution exists.

Solution Let the homogeneous system be

$$\begin{matrix} a_{11}x_1 + a_{12}x_2 + \cdots + a_{1n}x_n = 0 \\ a_{21}x_1 + a_{22}x_2 + \cdots + a_{2n}x_n = 0 \\ \vdots \qquad \vdots \qquad \qquad \vdots \quad \vdots \\ a_{m1}x_1 + a_{m2}x_2 + \cdots + a_{mn}x_n = 0 \end{matrix} \qquad \textbf{(a)}$$

where $m < n$.

Consider the equation of the form

$$0x_1 + 0x_2 + \cdots + 0x_n = 0 \tag{b}$$

It is seen that any n-tuple of real numbers $(x_1, x_2, \ldots, x_n)$ is a solution of (b).

Next, if we add $n - m$ equations of form (b) to (a), we have the following system:

$$\begin{bmatrix} a_{11} & a_{12} & \cdots & a_{1n} \\ a_{21} & a_{22} & \cdots & a_{2n} \\ \vdots & \vdots & & \vdots \\ a_{m1} & a_{m2} & \cdots & a_{mn} \\ 0 & 0 & \cdots & 0 \\ \vdots & \vdots & & \vdots \\ 0 & 0 & \cdots & 0 \end{bmatrix} \begin{bmatrix} x_1 \\ x_2 \\ \vdots \\ x_n \end{bmatrix} = \begin{bmatrix} 0 \\ 0 \\ \vdots \\ 0 \end{bmatrix} \tag{c}$$

or

$$A'\mathbf{x} = \mathbf{0}$$

where

$$A' = \begin{bmatrix} a_{11} & a_{12} & \cdots & a_{1n} \\ a_{21} & a_{22} & \cdots & a_{2n} \\ \vdots & \vdots & & \vdots \\ a_{m1} & a_{m2} & \cdots & a_{mn} \\ 0 & 0 & \cdots & 0 \\ \vdots & \vdots & & \vdots \\ 0 & 0 & \cdots & 0 \end{bmatrix}$$

Note that A' is an $n \times n$ matrix with $n - m$ rows of zeros. Then by Property 6 of determinants, $\det A' = 0$ and hence A is singular. Thus, by Corollary 1-4.5, system (c)—and consequently, system (a)—will have a nontrivial solution.

PROBLEM 2-34 Rework Problem 2-32 using Cramer's rule.

Solution

$$\det A_1 = \begin{vmatrix} 2 & 1 & 1 \\ 1 & 1 & -1 \\ -1 & -1 & -1 \end{vmatrix} = -2 + 1 - 1 + 1 - 2 + 1 = -2$$

$$\det A_2 = \begin{vmatrix} 1 & 2 & 1 \\ 1 & 1 & -1 \\ 1 & -1 & -1 \end{vmatrix} = -1 - 2 - 1 - 1 - 1 + 2 = -4$$

$$\det A_3 = \begin{vmatrix} 1 & 1 & 2 \\ 1 & 1 & 1 \\ 1 & -1 & -1 \end{vmatrix} = -1 + 1 - 2 - 2 + 1 + 1 = -2$$

Thus

$$x_1 = \frac{\det A_1}{\det A} = \frac{-2}{-4} = \frac{1}{2}, \qquad x_2 = \frac{\det A_2}{\det A} = \frac{-4}{-4} = 1, \qquad x_3 = \frac{\det A_3}{\det A} = \frac{-2}{-4} = \frac{1}{2}$$

PROBLEM 2-35 Use Cramer's rule to derive the law of cosines.

Solution Consider a triangle with sides a, b, and c, and opposite angles α, β, and γ [see Figure 2-1]. Then, using trigonometric definitions, we get

$$\begin{aligned} c(\cos\beta) + b(\cos\gamma) &= a \\ c(\cos\alpha) \qquad\qquad + a(\cos\gamma) &= b \\ b(\cos\alpha) + a(\cos\beta) \qquad\qquad &= c \end{aligned} \tag{a}$$

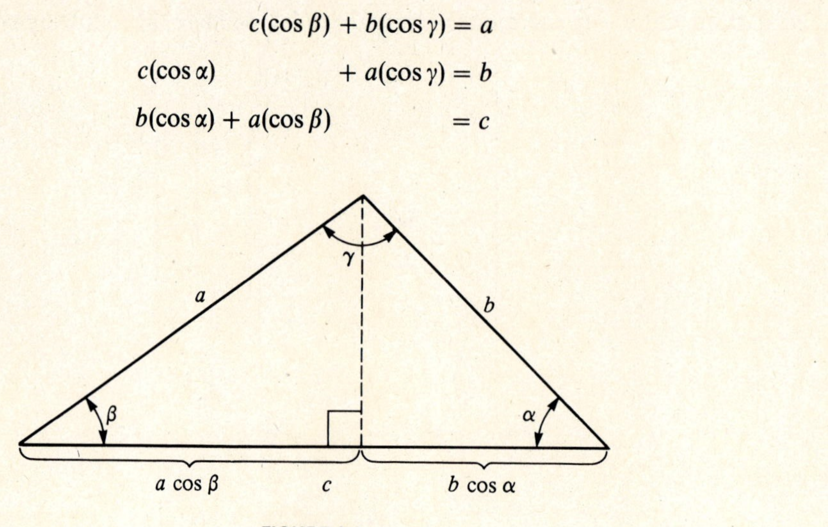

FIGURE 2-1

We wish to solve system (a) for $\cos\alpha$, $\cos\beta$, and $\cos\gamma$. The matrix of coefficients is

$$A = \begin{bmatrix} 0 & c & b \\ c & 0 & a \\ b & a & 0 \end{bmatrix}$$

First

$$\det A = \begin{vmatrix} 0 & c & b \\ c & 0 & a \\ b & a & 0 \end{vmatrix} = 2abc \neq 0$$

if $a \neq 0$, $b \neq 0$, $c \neq 0$, which certainly holds for a triangle. Next

$$\det A_1 = \begin{vmatrix} a & c & b \\ b & 0 & a \\ c & a & 0 \end{vmatrix} = ab^2 + ac^2 - a^3$$

$$\det A_2 = \begin{vmatrix} 0 & a & b \\ c & b & a \\ b & c & 0 \end{vmatrix} = bc^2 + ba^2 - b^3$$

$$\det A_3 = \begin{vmatrix} 0 & c & a \\ c & 0 & b \\ b & a & c \end{vmatrix} = cb^2 + ca^2 - c^3$$

Hence by Cramer's rule

$$\cos\alpha = \frac{\det A_1}{\det A} = \frac{ab^2 + ac^2 - a^3}{2abc} = \frac{b^2 + c^2 - a^2}{2bc}$$

$$\cos\beta = \frac{\det A_2}{\det A} = \frac{bc^2 + ba^2 - b^3}{2abc} = \frac{c^2 + a^2 - b^2}{2ca}$$

$$\cos \gamma = \frac{\det A_3}{\det A} = \frac{cb^2 + ca^2 - c^3}{2abc} = \frac{b^2 + a^2 - c^2}{2ab}$$

which is the desired law of cosines.

*For readers who have studied calculus**

***PROBLEM 2-36** Show that if $f_1(x)$, $f_2(x)$, $g_1(x)$, and $g_2(x)$ are differentiable functions, and if

$$W = \begin{vmatrix} f_1(x) & f_2(x) \\ g_1(x) & g_2(x) \end{vmatrix}$$

then

$$\frac{dW}{dx} = \begin{vmatrix} f_1'(x) & f_2'(x) \\ g_1(x) & g_2(x) \end{vmatrix} + \begin{vmatrix} f_1(x) & f_2(x) \\ g_1'(x) & g_2'(x) \end{vmatrix}$$

Solution

$$W = \begin{vmatrix} f_1(x) & f_2(x) \\ g_1(x) & g_2(x) \end{vmatrix} = f_1(x)g_2(x) - f_2(x)g_1(x)$$

Now according to the derivative rule, $(fg)' = f'g + fg'$, we get

$$\begin{aligned} \frac{dW}{dx} &= f_1'(x)g_2(x) + f_1(x)g_2'(x) - (f_2'(x)g_1(x) + f_2(x)g_1'(x)) \\ &= (f_1'(x)g_2(x) - f_2'(x)g_1(x)) + (f_1(x)g_2'(x) - f_2(x)g_1'(x)) \\ &= \begin{vmatrix} f_1'(x) & f_2'(x) \\ g_1(x) & g_2(x) \end{vmatrix} + \begin{vmatrix} f_1(x) & f_2(x) \\ g_1'(x) & g_2'(x) \end{vmatrix} \end{aligned}$$

***PROBLEM 2-37** Let

$$A = \begin{bmatrix} 1+x & 1-x \\ 1-x & 1+x \end{bmatrix}$$

Find

$$\frac{d}{dx}(\det A)$$

Solution Using the result of Problem 2-36, we have

$$\begin{aligned} \frac{d}{dx}(\det A) &= \frac{d}{dx}\begin{vmatrix} 1+x & 1-x \\ 1-x & 1+x \end{vmatrix} \\ &= \begin{vmatrix} \frac{d}{dx}(1+x) & \frac{d}{dx}(1-x) \\ 1-x & 1+x \end{vmatrix} + \begin{vmatrix} 1+x & 1-x \\ \frac{d}{dx}(1-x) & \frac{d}{dx}(1+x) \end{vmatrix} \\ &= \begin{vmatrix} 1 & -1 \\ 1-x & 1+x \end{vmatrix} + \begin{vmatrix} 1+x & 1-x \\ -1 & 1 \end{vmatrix} \\ &= (1+x) - (-1)(1-x) + (1+x) - (1-x)(-1) \\ &= 4 \end{aligned}$$

Check:

$$\det A = \begin{vmatrix} 1+x & 1-x \\ 1-x & 1+x \end{vmatrix}$$

$$= (1+x)^2 - (1-x)^2$$

$$= (1 + 2x + x^2) - (1 - 2x + x^2) = 4x$$

$$\frac{d}{dx}(\det A) = \frac{d}{dx}(4x) = 4$$

Supplementary Exercises

PROBLEM 2-38 Evaluate the determinants of the matrices

$$A = \begin{bmatrix} 3 & 0 & 0 \\ 2 & 1 & 1 \\ 1 & 2 & 2 \end{bmatrix}, \qquad B = \begin{bmatrix} 1 & 1 & 1 & 3 \\ 0 & 3 & 1 & 1 \\ 0 & 0 & 2 & 2 \\ -1 & -1 & -1 & 2 \end{bmatrix}$$

Answer $|A| = 0 \quad |B| = 30$

PROBLEM 2-39 Find the determinant of A if

$$A = \begin{bmatrix} 0 & a & b \\ -a & 0 & c \\ -b & -c & 0 \end{bmatrix}$$

Answer $|A| = 0$

PROBLEM 2-40 If A is an $n \times n$ skew-symmetric matrix ($A^T = -A$), show that $\det A = 0$ if n is odd.

PROBLEM 2-41 Find the determinant of

$$A = \begin{bmatrix} 1 & a & a^2 \\ 1 & b & b^2 \\ 1 & c & c^2 \end{bmatrix}$$

Answer $|A| = (a - b)(b - c)(c - a)$

PROBLEM 2-42 Suppose A is a 2×1 matrix and B is a 1×2 matrix. Show that $C = AB$ is singular.

PROBLEM 2-43 Does

$$A = \begin{bmatrix} 1 & -2 & 4 \\ 2 & 2 & -2 \\ 3 & 0 & 2 \end{bmatrix}$$

have an inverse?

Answer No

PROBLEM 2-44 Given that A and B are square matrices of the same order and $\det A = 2$, $\det B = 4$. Find $\det(A^{-1}B)$.

Answer 2

PROBLEM 2-45 Given

$$A = \begin{bmatrix} x^2 - 1 & x + 2 \\ x^2 - 2x + 3 & x \end{bmatrix}, \qquad B = \begin{bmatrix} x^2 + x & x + 1 \\ x - 1 & 1 \end{bmatrix}$$

(a) Does the inverse of A exist for any x? **(b)** Does the inverse of B exist for any x?

Answer **(a)** Yes **(b)** No

PROBLEM 2-46 Use cofactor expansion to evaluate the determinants of

$$A = \begin{bmatrix} 1 & -2 & 3 \\ 3 & 2 & 4 \\ 2 & 3 & 2 \end{bmatrix}, \qquad B = \begin{bmatrix} 2 & -1 & 2 & 1 \\ 0 & 1 & -4 & -2 \\ 0 & 0 & 4 & -2 \\ 0 & 0 & 4 & 1 \end{bmatrix}$$

Answer $|A| = 3$ $|B| = 24$

PROBLEM 2-47 Use the classical adjoint formula to compute the inverses of the following matrices:

$$A = \begin{bmatrix} 1 & 0 & 1 \\ 3 & 3 & 4 \\ 2 & 2 & 3 \end{bmatrix}, \qquad B = \begin{bmatrix} \cos\theta & -\sin\theta & 0 \\ \sin\theta & \cos\theta & 0 \\ 0 & 0 & 1 \end{bmatrix}$$

Answer

$$A^{-1} = \begin{bmatrix} 1 & 2 & -3 \\ -1 & 1 & -1 \\ 0 & -2 & 3 \end{bmatrix} \qquad B^{-1} = \begin{bmatrix} \cos\theta & \sin\theta & 0 \\ -\sin\theta & \cos\theta & 0 \\ 0 & 0 & 1 \end{bmatrix}$$

PROBLEM 2-48 Show that if $\det A = 1$, then $\operatorname{adj}(\operatorname{adj} A) = A$. [*Hint:* Use (2.18).]

PROBLEM 2-49 Show that if A is nonsingular, then adj A is also nonsingular and

$$(\operatorname{adj} A)^{-1} = (\det(A^{-1}))A = \operatorname{adj} A^{-1}$$

[*Hint:* Replace A by A^{-1} in (2.18) and find $(\operatorname{adj} A)^{-1}$.]

PROBLEM 2-50 Use Cramer's rule to solve the following systems of linear equations

(a)
$$\begin{aligned} x_1 + x_2 + x_3 &= 1 \\ 3x_1 + 4x_2 + x_3 &= 0 \\ 2x_1 - 3x_2 - x_3 &= 4 \end{aligned}$$

(b)
$$\begin{aligned} 3x_1 - 2x_2 \qquad &= -1 \\ 3x_2 - 2x_3 &= 8 \\ -2x_1 \qquad + 3x_3 &= -5 \end{aligned}$$

Answer **(a)** $(1, -1, 1)$ **(b)** $(1, 2, -1)$

3 VECTOR SPACES

THIS CHAPTER IS ABOUT

- ☑ Vectors in 2-Space and 3-Space
- ☑ Cartesian n-Space
- ☑ Vector Spaces
- ☑ Subspaces
- ☑ Linear Combination and Spanning Sets
- ☑ Linear Independence
- ☑ Basis and Dimension
- ☑ Coordinate Vectors and Change of Basis
- ☑ Column Space and Row Space of a Matrix

3-1. Vectors in 2-Space and 3-Space

A. Representation of points in 2-space and 3-space

We know that a point P on a line can be represented by a number x, once a reference point (origin) and a unit length are selected, and a point P in a plane can be represented by a pair of numbers (x, y). Similarly, a point P in three-dimensional space, or 3-space, can be represented by a triple of numbers (x, y, z). These representations are illustrated in Figure 3-1. Instead of using (x, y, z) we could use (x_1, x_2, x_3). Thus we can define a point in 3-space as an ordered triple of numbers

$$(x_1, x_2, x_3) \tag{3.1}$$

We call x_i the ith **coordinate** of the point. The line could be called **one-dimensional space**, or 1-space, and the plane could be called **two-dimensional space**, or 2-space.

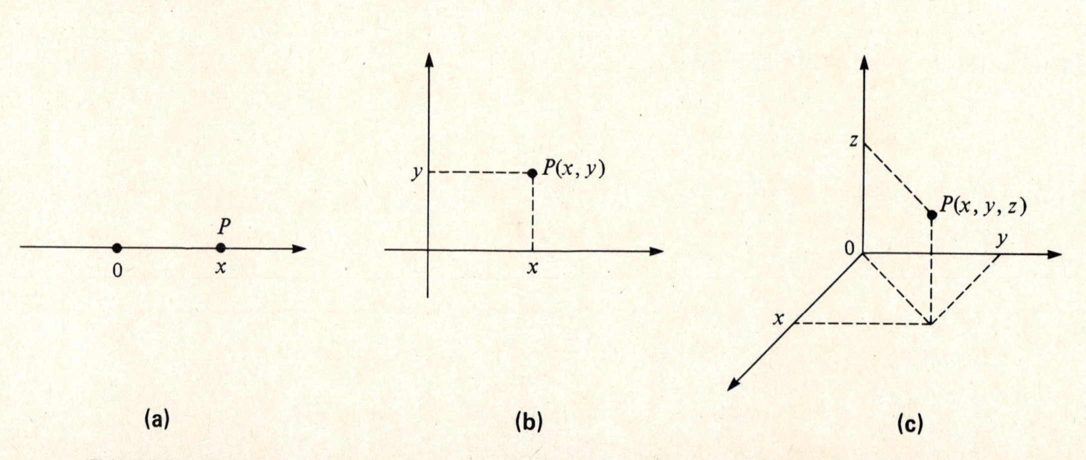

FIGURE 3-1. Representation of a point. (a) Point on a line. (b) Point in a plane. (c) Point in a space.

B. Introduction to geometric vectors

1. Definitions

A **vector** is a quantity characterized by both magnitude and direction; examples are displacement, force, and velocity. We use **boldface letters** such as **u** and **v** to denote vectors. Graphically, a vector **v** is represented by a directed line segment $\overrightarrow{PQ}$, as illustrated in Figure 3-2. The vector **v** has a direction from P to Q. The point P is called the **initial point** and the point Q is called the **terminal point** of **v**. The length $|\overrightarrow{PQ}|$ of the line segment is the magnitude of **v**. We can denote the magnitude of **v** by $\|\mathbf{v}\|$. When the initial point of a vector is fixed, it is called a **fixed**, or **localized**, vector. If the initial point is not fixed, it is called a **free**, or **nonlocalized**, vector. (In this text we assume that all vectors are free vectors.)

FIGURE 3-2

2. Equality of vectors

The vectors **u** and **v** are equal,

$$\mathbf{u} = \mathbf{v} \tag{3.2}$$

when they have the same magnitude and direction (see Figure 3-3).

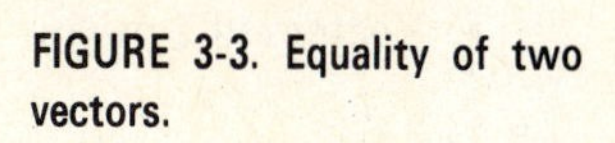
FIGURE 3-3. Equality of two vectors.

3. Addition of vectors

Given two vectors, **u** and **v**, the addition of **u** and **v**

$$\mathbf{u} + \mathbf{v} \tag{3.3}$$

may be performed by using the **triangle rule** or the **parallelogram law**. To perform each of these methods, the terminal point of **u** is placed on the initial point of **v**, and the sum—or **resultant**—of the two vectors is drawn, as shown in Figure 3-4. When the triangle rule is used, the resultant is the vector that can be drawn between the terminal point of **v** and the initial point of **u**, as shown in Figure 3-4b. When the parallelogram law is used, the resultant vector is the diagonal of the parallelogram with sides **u** and **v**, as shown in Figure 3-4c.

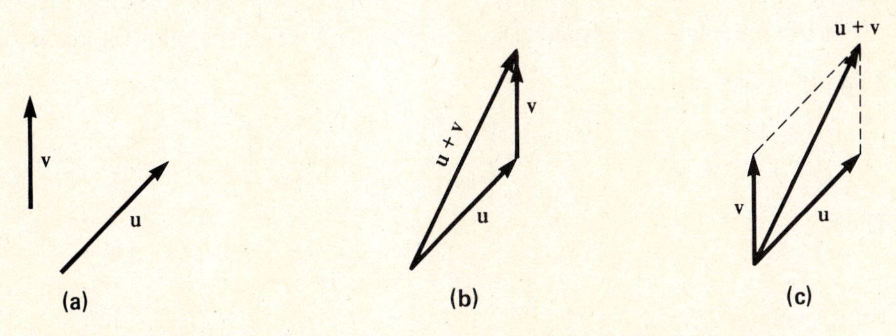

FIGURE 3-4. Addition of vectors. (a) Given vectors. (b) Triangle rule. (c) Parallelogram law.

4. Zero vector: 0

The zero vector, denoted **0**, is a vector that has magnitude 0 and any direction. We define

$$\mathbf{v} + \mathbf{0} = \mathbf{v} \tag{3.4}$$

for any vector **v**.

5. Negative of a vector

If **v** is any nonzero vector, then the negative of **v**, $-\mathbf{v}$, is the vector whose magnitude is equal to that of **v** but whose direction is opposite to that of **v** (see Figure 3-5). This vector has the property

$$\mathbf{v} + (-\mathbf{v}) = \mathbf{0} \tag{3.5}$$

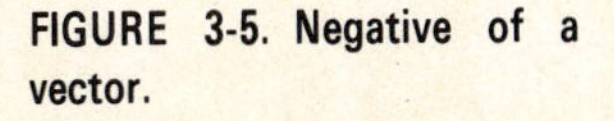
FIGURE 3-5. Negative of a vector.

6. **Scalar multiplication**

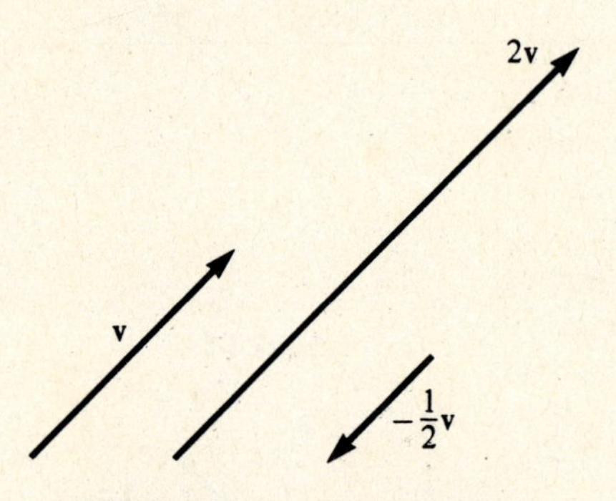

FIGURE 3-6. Scalar multiplication.

Let $\mathbf{v}$ be a nonzero vector and α a scalar. (By a *scalar*, we mean a real number.) Then scalar multiplication of a vector is defined to be a vector $\alpha\mathbf{v}$ whose magnitude is

$$\|\alpha \mathbf{v}\| = |\alpha| \, \|\mathbf{v}\| \tag{3.6}$$

where $|\alpha|$ denotes the absolute value of α. The direction of $\alpha\mathbf{v}$ is the same as that of $\mathbf{v}$ if $\alpha > 0$ and opposite to that of $\mathbf{v}$ if $\alpha < 0$. (See Figure 3-6.)

If we let $\alpha = -1$, we see that

$$(-1)\mathbf{v} = -\mathbf{v} \tag{3.7}$$

Two vectors $\mathbf{u}$ and $\mathbf{v}$ are said to be *parallel* if there is a real number $\alpha \neq 0$ such that $\mathbf{u} = \alpha\mathbf{v}$.

C. Analytical representation of vectors

Problems involving vectors can often be simplified by introducing a rectangular coordinate system.

1. **Vectors in 2-space**

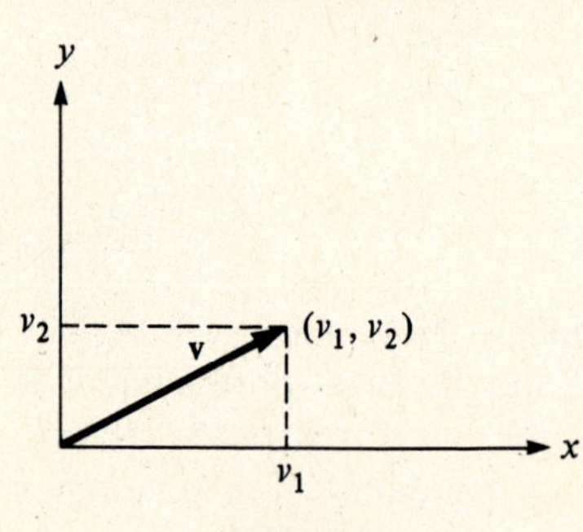

FIGURE 3-7

Let $\mathbf{v}$ be any vector in 2-space. If we place the initial point of $\mathbf{v}$ at the origin of a rectangular coordinate system, we can specify $\mathbf{v}$ by the rectangular coordinate (v_1, v_2) of the terminal point, as shown in Figure 3-7. Thus there is a one-to-one correspondence between a point in 2-space and a vector whose initial point is at the origin. So we can represent $\mathbf{v}$ in 2-space by a 2×1 matrix (see Section 1-2)

$$\mathbf{v} = \begin{bmatrix} v_1 \\ v_2 \end{bmatrix} \tag{3.8}$$

where v_1, v_2 are called the **components** of $\mathbf{v}$ relative to the given coordinate system and are real numbers.

Then from the matrix operation defined in Chapter 1, we get the following vector operations.

(1) *Equality of vectors:* Let

$$\mathbf{u} = \begin{bmatrix} u_1 \\ u_2 \end{bmatrix}, \qquad \mathbf{v} = \begin{bmatrix} v_1 \\ v_2 \end{bmatrix}$$

Then

$$\mathbf{u} = \mathbf{v} \Leftrightarrow u_1 = v_1, u_2 = v_2 \tag{3.9}$$

(2) *Vector addition:* If

$$\mathbf{u} = \begin{bmatrix} u_1 \\ u_2 \end{bmatrix}, \qquad \mathbf{v} = \begin{bmatrix} v_1 \\ v_2 \end{bmatrix}$$

Then

$$\mathbf{u} + \mathbf{v} = \begin{bmatrix} u_1 + v_1 \\ u_2 + v_2 \end{bmatrix} \tag{3.10}$$

(3) *Zero vector* $\mathbf{0}$*:*

$$\mathbf{0} = \begin{bmatrix} 0 \\ 0 \end{bmatrix} \tag{3.11}$$

(4) *Scalar multiplication:* If $\mathbf{v} = \begin{bmatrix} v_1 \\ v_2 \end{bmatrix}$ and α is any scalar, then

$$\alpha \mathbf{v} = \begin{bmatrix} \alpha v_1 \\ \alpha v_2 \end{bmatrix} \tag{3.12}$$

(5) *Negative of* $\mathbf{v}$: If $\mathbf{v} = \begin{bmatrix} v_1 \\ v_2 \end{bmatrix}$, then

$$-\mathbf{v} = \begin{bmatrix} -v_1 \\ -v_2 \end{bmatrix} \tag{3.13}$$

EXAMPLE 3-1: Let $\mathbf{u} = \begin{bmatrix} 3 \\ 1 \end{bmatrix}$ and $\mathbf{v} = \begin{bmatrix} 1 \\ 2 \end{bmatrix}$. Find **(a)** $\mathbf{u} + \mathbf{v}$, **(b)** $2\mathbf{u}$, **(c)** $-\mathbf{v}$, and **(d)** $\mathbf{u} - \mathbf{v} = \mathbf{u} + (-\mathbf{v})$.

Solution

(a)
$$\mathbf{u} + \mathbf{v} = \begin{bmatrix} 3+1 \\ 1+2 \end{bmatrix} = \begin{bmatrix} 4 \\ 3 \end{bmatrix}$$

(b)
$$2\mathbf{u} = \begin{bmatrix} 2(3) \\ 2(1) \end{bmatrix} = \begin{bmatrix} 6 \\ 2 \end{bmatrix}$$

(c)
$$-\mathbf{v} = \begin{bmatrix} -1 \\ -2 \end{bmatrix}$$

(d)
$$\mathbf{u} - \mathbf{v} = \begin{bmatrix} 3-1 \\ 1-2 \end{bmatrix} = \begin{bmatrix} 2 \\ -1 \end{bmatrix}$$

Graphical representations are given in Figure 3-8.

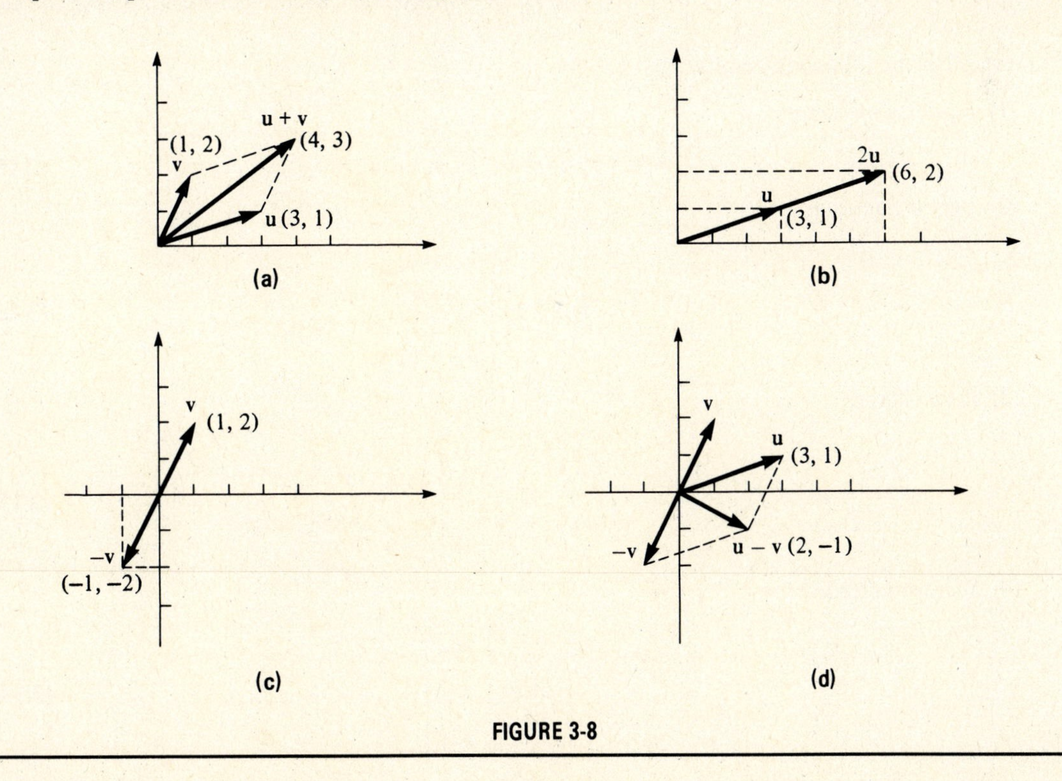

FIGURE 3-8

---

2. **Vectors in 3-space**

Just as for vectors in 2-space, there is a one-to-one correspondence between a point in 3-space (represented by a triple of real numbers) and a vector in 3-space. Thus we can represent **v** in 3-space by a 3×1 matrix

$$\mathbf{v} = \begin{bmatrix} v_1 \\ v_2 \\ v_3 \end{bmatrix} \tag{3.14}$$

where v_1, v_2, v_3 are the components of **v** relative to the given coordinate system (Figure 3-9).

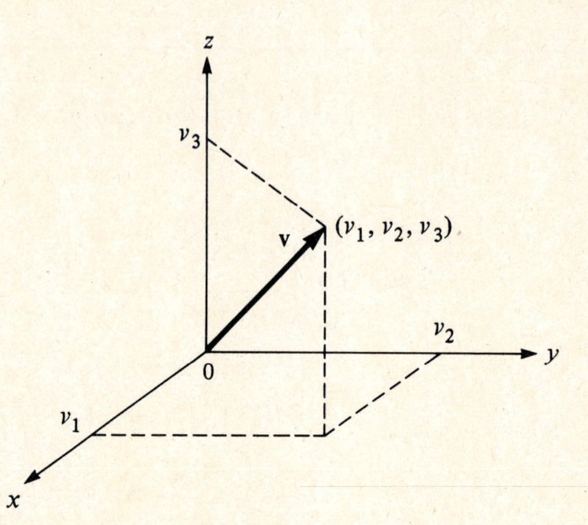

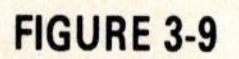

FIGURE 3-9

If $\mathbf{u} = \begin{bmatrix} u_1 \\ u_2 \\ u_3 \end{bmatrix}$ and $\mathbf{v} = \begin{bmatrix} v_1 \\ v_2 \\ v_3 \end{bmatrix}$ are two vectors in 3-space, then, as for vectors in 2-space, we can establish the following results:

(1) *Equality of vectors:*

$$\mathbf{u} = \mathbf{v} \Leftrightarrow u_1 = v_1, u_2 = v_2, u_3 = v_3 \tag{3.15}$$

(2) *Vector addition:*

$$\mathbf{u} + \mathbf{v} = \begin{bmatrix} u_1 + v_1 \\ u_2 + v_2 \\ u_3 + v_3 \end{bmatrix} \tag{3.16}$$

(3) *Zero vector:*

$$\mathbf{0} = \begin{bmatrix} 0 \\ 0 \\ 0 \end{bmatrix} \tag{3.17}$$

(4) *Scalar multiplication:*

$$\alpha\mathbf{v} = \begin{bmatrix} \alpha v_1 \\ \alpha v_2 \\ \alpha v_3 \end{bmatrix} \tag{3.18}$$

(5) *Negative of* **v**:

$$-\mathbf{v} = \begin{bmatrix} -v_1 \\ -v_2 \\ -v_3 \end{bmatrix} \tag{3.19}$$

EXAMPLE 3-2: Let

$$\mathbf{u} = \begin{bmatrix} 1 \\ 2 \\ 3 \end{bmatrix}, \qquad \mathbf{v} = \begin{bmatrix} 2 \\ -1 \\ 3 \end{bmatrix}, \qquad \mathbf{w} = \begin{bmatrix} 3 \\ 2 \\ -1 \end{bmatrix}$$

Find **(a)** $\mathbf{u} + \mathbf{v}$, **(b)** $\mathbf{v} + \mathbf{w}$, **(c)** $(\mathbf{u} + \mathbf{v}) + \mathbf{w}$, and **(d)** $\mathbf{u} + (\mathbf{v} + \mathbf{w})$.

Solution

(a)
$$\mathbf{u} + \mathbf{v} = \begin{bmatrix} 1+2 \\ 2-1 \\ 3+3 \end{bmatrix} = \begin{bmatrix} 3 \\ 1 \\ 6 \end{bmatrix}$$

(b)
$$\mathbf{v} + \mathbf{w} = \begin{bmatrix} 2+3 \\ -1+2 \\ 3-1 \end{bmatrix} = \begin{bmatrix} 5 \\ 1 \\ 2 \end{bmatrix}$$

(c)
$$(\mathbf{u} + \mathbf{v}) + \mathbf{w} = \begin{bmatrix} 3+3 \\ 1+2 \\ 6-1 \end{bmatrix} = \begin{bmatrix} 6 \\ 3 \\ 5 \end{bmatrix}$$

(d)
$$\mathbf{u} + (\mathbf{v} + \mathbf{w}) = \begin{bmatrix} 1+5 \\ 2+1 \\ 3+2 \end{bmatrix} = \begin{bmatrix} 6 \\ 3 \\ 5 \end{bmatrix}$$

3-2. Cartesian *n*-Space

A. Definition of *n*-vector

A point in 1-space is represented by a single number, a point in 2-space is represented by an ordered pair of numbers, and a point in 3-space is represented by an ordered triple of numbers. Thus, a point in n-space can be represented by an ordered n-tuple of real numbers

$$(x_1, x_2, \ldots, x_n) \tag{3.20}$$

where $x_1, x_2, \ldots, x_n$ are called the coordinates of the point. Thus, in the same manner that we can define vectors in 2-space and 3-space, we can define a vector in n-space (***n*-vector** or ***n*-dimensional vector**) by an $n \times 1$ matrix

$$\mathbf{v} = \begin{bmatrix} v_1 \\ v_2 \\ \vdots \\ v_n \end{bmatrix} \tag{3.21}$$

where $v_1, v_2, \ldots, v_n$ are all real numbers and v_i is called the ith component of $\mathbf{v}$.

B. Definition of Cartesian *n*-space

The collection of all ordered n-tuples is called **Cartesian *n*-space** and is denoted by R^n. We can also view R^n as the set of all $n \times 1$ matrices with real entries. That is, $\mathbf{v}$ is an element of R^n, denoted by $\mathbf{v} \in R^n$, if and only if there are n real numbers $v_1, v_2, \ldots, v_n$ such that

$$\mathbf{v} = \begin{bmatrix} v_1 \\ v_2 \\ \vdots \\ v_n \end{bmatrix}$$

C. Definitions

Let

$$\mathbf{u} = \begin{bmatrix} u_1 \\ u_2 \\ \vdots \\ u_n \end{bmatrix}, \qquad \mathbf{v} = \begin{bmatrix} v_1 \\ v_2 \\ \vdots \\ v_n \end{bmatrix}$$

be any two vectors in R^n and α be any scalar. Then

(1) *Equality of vectors:*

$$\mathbf{u} = \mathbf{v} \Leftrightarrow u_i = v_i, \qquad i = 1, 2, \ldots, n \tag{3.22}$$

(2) *Addition of vectors:*

$$\mathbf{u} + \mathbf{v} = \begin{bmatrix} u_1 + v_1 \\ u_2 + v_2 \\ \vdots \\ u_n + v_n \end{bmatrix} \tag{3.23}$$

(3) *Zero vector* $\mathbf{0}$ *in* R^n*:*

$$\mathbf{0} = \begin{bmatrix} 0 \\ 0 \\ \vdots \\ 0 \end{bmatrix} \tag{3.24}$$

(4) *Scalar multiplication:*

$$\alpha \mathbf{v} = \begin{bmatrix} \alpha v_1 \\ \alpha v_2 \\ \vdots \\ \alpha v_n \end{bmatrix} \tag{3.25}$$

(5) *Negative (or additive inverse) of* $\mathbf{v}$*:*

$$-\mathbf{v} = \begin{bmatrix} -v_1 \\ -v_2 \\ \vdots \\ -v_n \end{bmatrix} \tag{3.26}$$

The operations of addition (3.23) and scalar multiplication (3.25) are called the **standard operations** on R^n.

THEOREM 3-2.1 If $\mathbf{u}, \mathbf{v}, \mathbf{w} \in R^n$, and α, β are scalars, then

$$\text{(i) } \mathbf{u} + \mathbf{v} = \mathbf{v} + \mathbf{u}$$
$$\text{(ii) } \mathbf{u} + (\mathbf{v} + \mathbf{w}) = (\mathbf{u} + \mathbf{v}) + \mathbf{w}$$
$$\text{(iii) } \mathbf{v} + \mathbf{0} = \mathbf{v}$$
$$\text{(iv) } \mathbf{v} + (-\mathbf{v}) = \mathbf{0}$$
$$\text{(v) } \alpha(\mathbf{u} + \mathbf{v}) = \alpha\mathbf{u} + \alpha\mathbf{v}$$
$$\text{(vi) } (\alpha + \beta)\mathbf{v} = \alpha\mathbf{v} + \beta\mathbf{v}$$
$$\text{(vii) } (\alpha\beta)\mathbf{v} = \alpha(\beta\mathbf{v})$$
$$\text{(viii) } 1\mathbf{v} = \mathbf{v}$$

EXAMPLE 3-3: Verify properties (i) and (ii) of Theorem 3-2.1.

Solution: Let

$$\mathbf{u} = \begin{bmatrix} u_1 \\ u_2 \\ \vdots \\ u_n \end{bmatrix}, \qquad \mathbf{v} = \begin{bmatrix} v_1 \\ v_2 \\ \vdots \\ v_n \end{bmatrix}, \qquad \mathbf{w} = \begin{bmatrix} w_1 \\ w_2 \\ \vdots \\ w_n \end{bmatrix}$$

Then by the definition of vector addition (3.23)

$$\mathbf{u} + \mathbf{v} = \begin{bmatrix} u_1 + v_1 \\ u_2 + v_2 \\ \vdots \\ u_n + v_n \end{bmatrix} = \begin{bmatrix} v_1 + u_1 \\ v_2 + u_2 \\ \vdots \\ v_n + u_n \end{bmatrix} = \mathbf{v} + \mathbf{u}$$

since $u_i + v_i = v_i + u_i$ for all real numbers. Next,

$$\mathbf{u} + (\mathbf{v} + \mathbf{w}) = \begin{bmatrix} u_1 \\ u_2 \\ \vdots \\ u_n \end{bmatrix} + \begin{bmatrix} v_1 + w_1 \\ v_2 + w_2 \\ \vdots \\ v_n + w_n \end{bmatrix}$$

$$= \begin{bmatrix} u_1 + (v_1 + w_1) \\ u_2 + (v_2 + w_2) \\ \vdots \\ u_n + (v_n + w_n) \end{bmatrix} = \begin{bmatrix} (u_1 + v_1) + w_1 \\ (u_2 + v_2) + w_2 \\ \vdots \\ (u_n + v_n) + w_n \end{bmatrix} = (\mathbf{u} + \mathbf{v}) + \mathbf{w}$$

since $u_i + (v_i + w_i) = (u_i + v_i) + w_i$ for all real numbers.

EXAMPLE 3-4: Verify property (v) of Theorem 3-2.1.

Solution: Let α be any scalar. Then by definition (3.25),

$$\alpha(\mathbf{u} + \mathbf{v}) = \begin{bmatrix} \alpha(u_1 + v_1) \\ \alpha(u_2 + v_2) \\ \vdots \\ \alpha(u_n + v_n) \end{bmatrix}$$

$$\alpha(\mathbf{u}+\mathbf{v}) = \begin{bmatrix} \alpha u_1 + \alpha v_1 \\ \alpha u_2 + \alpha v_2 \\ \vdots \\ \alpha u_n + \alpha v_n \end{bmatrix} = \alpha\mathbf{u} + \alpha\mathbf{v}$$

since $\alpha(u_i + v_i) = \alpha u_i + \alpha v_i$.

---

note: Other properties of Theorem 3-2.1 are verified in Problems 3-5 and 3-6.

3-3. Vector Spaces

A. Definition

Let V be a nonempty set on which the operations of addition and scalar multiplication are defined. By *addition* we mean that, with each pair of elements $\mathbf{u}$ and $\mathbf{v}$ in V, there is a rule determining a unique element $\mathbf{u} + \mathbf{v}$ that is also in V. By *scalar multiplication* we mean that to each element $\mathbf{v}$ in V and to each scalar α we can assign a unique element $\alpha\mathbf{v}$ in V. The set V, together with the operations of addition and scalar multiplication, is said to be a **vector space** if the following axioms are satisfied:

Axioms for Vector Space:

VECTOR SPACE AXIOMS

V1. $\mathbf{u} + \mathbf{v} \in V$

V2. $\mathbf{u} + \mathbf{v} = \mathbf{v} + \mathbf{u}$

V3. $\mathbf{u} + (\mathbf{v} + \mathbf{w}) = (\mathbf{u} + \mathbf{v}) + \mathbf{w}$

V4. There is an element $\mathbf{0}$ in V such that $\mathbf{v} + \mathbf{0} = \mathbf{v}$ for each $\mathbf{v} \in V$

V5. For each $\mathbf{v} \in V$, there is an element $-\mathbf{v}$ in V such that $\mathbf{v} + (-\mathbf{v}) = \mathbf{0}$

V6. $\alpha\mathbf{v} \in V$

V7. $\alpha(\mathbf{u} + \mathbf{v}) = \alpha\mathbf{u} + \alpha\mathbf{v}$

V8. $(\alpha + \beta)\mathbf{v} = \alpha\mathbf{v} + \beta\mathbf{v}$

V9. $(\alpha\beta)\mathbf{v} = \alpha(\beta\mathbf{v})$

V10. $1\mathbf{v} = \mathbf{v}$

The elements of V are called **vectors**. The $\mathbf{0}$ element is called the **zero vector** and $-\mathbf{v}$ is called the **negative** of $\mathbf{v}$.

V is said to be a **real vector space**, or a **vector space over reals**, if the set of scalars is the set of real numbers. If the set of scalars is the set of complex numbers, then V is said to be a **complex vector space**, which will be discussed in Chapter 6. A vector space is also sometimes called a **linear space**.

B. Examples of vector space

1. **The Cartesian n-space: R^n**

EXAMPLE 3-5: Show that R^n is a vector space.

Solution: Since by definition (3.24)

$$\mathbf{0} = \begin{bmatrix} 0 \\ 0 \\ \vdots \\ 0 \end{bmatrix}$$

is an n-vector, $\mathbf{0} \in R^n$, and R^n is not empty. Axioms V1 and V6 follow from the definition of the standard operations on R^n, i.e., (3.23) and (3.25). The remaining axioms follow from Theorem 3-2.1. Thus R^n is a vector space.

---

2. **The space of $m \times n$ matrices: $R_{m\times n}$**

   Let $R_{m\times n}$ denote the set of all $m \times n$ matrices with real entries. The usual addition and scalar multiplication of matrices are defined in Section 1-3, (1.12) and (1.13).

EXAMPLE 3-6: Show that $R_{m\times n}$ is a vector space.

Solution: An $m \times n$ zero matrix with all zero entries belongs to $R_{m\times n}$; thus $R_{m\times n}$ is not empty. If $A, B \in R_{m\times n}$, and α is a scalar, then from the definition of matrix addition (1.12) and scalar multiplication (1.13) it is clear that

$$A + B \in R_{m\times n} \quad \text{and} \quad \alpha A \in R_{m\times n}$$

Thus Axioms V1 and V6 are satisfied. The $m \times n$ zero matrix is the zero vector $\mathbf{0}$ in Axiom V4 and if $\mathbf{v}$ is the $m \times n$ matrix A, then the matrix $-A$ is the vector $-\mathbf{v}$ in Axiom V5. The remaining axioms are satisfied by Theorem 1-3.1 in Section 1-3. Thus we conclude that $R_{m\times n}$ is a vector space.

note: R^n is a special case of $R_{m\times n}$; that is,

$$R^n = R_{n\times 1}$$

---

3. **The space of polynomials of degree less than or equal to n: P_n**

   Let P_n denote the set of all polynomials of degree less than or equal to n, that is, all polynomials expressible in the form

$$a_0 + a_1x + \cdots + a_nx^n$$

   for all x, where $a_0, a_1, \ldots, a_n$ are real numbers. On P_n the following usual addition and scalar multiplication operations are defined:

   Let $\mathbf{p}, \mathbf{q} \in P_n$ and α be any scalar. Then

$$\mathbf{p}(x) = p(x) = a_0 + a_1x + \cdots + a_nx^n \tag{3.27}$$

$$\mathbf{q}(x) = q(x) = b_0 + b_1x + \cdots + b_nx^n \tag{3.28}$$

   and

$$(\mathbf{p} + \mathbf{q})(x) = p(x) + q(x) = (a_0 + b_0) + (a_1 + b_1)x + \cdots + (a_n + b_n)x^n \tag{3.29}$$

$$(\alpha\mathbf{p})(x) = \alpha p(x) = (\alpha a_0) + (\alpha a_1)x + \cdots + (\alpha a_n)x^n \tag{3.30}$$

EXAMPLE 3-7: Show that P_n is a vector space.

Solution: Setting $a_0 = a_1 = \cdots = a_n = 0$ in (3.27), we have the zero polynomial. Thus P_n is not empty. Now, does P_n satisfy all the axioms for a vector space?

V1. By (3.29), $\mathbf{p} + \mathbf{q} \in P_n$.

V2.
$$\begin{aligned}(\mathbf{p} + \mathbf{q})(x) &= p(x) + q(x)\\ &= (a_0 + b_0) + (a_1 + b_1)x + \cdots + (a_n + b_n)x^n\\ &= (b_0 + a_0) + (b_1 + a_1)x + \cdots + (b_n + a_n)x^n\\ &= q(x) + p(x) = (\mathbf{q} + \mathbf{p})(x)\end{aligned}$$

since $a_i + b_i = b_i + a_i$. Thus, $\mathbf{p} + \mathbf{q} = \mathbf{q} + \mathbf{p}$.

V3. Let $\mathbf{r}(x) = r(x) = c_0 + c_1 x + \cdots + c_n x^n$. Then

$$\begin{aligned}[\mathbf{p} + (\mathbf{q} + \mathbf{r})](x) &= p(x) + [q(x) + r(x)] \\ &= a_0 + a_1 x + \cdots + a_n x^n + (b_0 + c_0) + (b_1 + c_1)x + \cdots + (b_n + c_n)x^n \\ &= [a_0 + (b_0 + c_0)] + [a_1 + (b_1 + c_1)]x + \cdots + [a_n + (b_n + c_n)]x^n \\ &= [(a_0 + b_0) + c_0] + [(a_1 + b_1) + c_1]x + \cdots + [(a_n + b_n) + c_n]x^n \\ &= (a_0 + b_0) + (a_1 + b_1)x + \cdots + (a_n + b_n)x^n + c_0 + c_1 x + \cdots + c_n x^n \\ &= [p(x) + q(x)] + r(x) \\ &= [(\mathbf{p} + \mathbf{q}) + \mathbf{r}](x)\end{aligned}$$

since $a_i + (b_i + c_i) = (a_i + b_i) + c_i$. So $\mathbf{p} + (\mathbf{q} + \mathbf{r}) = (\mathbf{p} + \mathbf{q}) + \mathbf{r}$.

V4. Since $\mathbf{z}$, defined by $\mathbf{z}(x) = z(x) = 0$ (that is, the polynomial, all of whose coefficients are zeros) is the zero vector of P_n, we have

$$(\mathbf{p} + \mathbf{z})(x) = p(x) + z(x) = p(x) + 0 = p(x) = \mathbf{p}(x)$$

So $\mathbf{p} + \mathbf{z} = \mathbf{p}$.

V5. Let $-\mathbf{p}$ be defined by $-\mathbf{p}(x) = -p(x)$. Then

$$[\mathbf{p} + (-\mathbf{p})](x) = p(x) + [-p(x)] = p(x) - p(x) = 0$$

Therefore, $-\mathbf{p}$ is the negative of $\mathbf{p}$ for P_n.

V6. By (3.30), $\alpha\mathbf{p} \in P_n$.

V7.

$$\begin{aligned}\alpha(\mathbf{p} + \mathbf{q})(x) &= \alpha[p(x) + q(x)] \\ &= \alpha[(a_0 + b_0) + (a_1 + b_1)x + \cdots + (a_n + b_n)x^n] \\ &= \alpha(a_0 + a_1 x + \cdots + a_n x^n) + \alpha(b_0 + b_1 x + \cdots + b_n x^n) \\ &= \alpha p(x) + \alpha q(x) = (\alpha\mathbf{p} + \alpha\mathbf{q})(x)\end{aligned}$$

Thus $\alpha(\mathbf{p} + \mathbf{q}) = \alpha\mathbf{p} + \alpha\mathbf{q}$.

V8.

$$\begin{aligned}(\alpha + \beta)\mathbf{p}(x) &= (\alpha + \beta)p(x) \\ &= (\alpha + \beta)(a_0 + a_1 x + \cdots + a_n x^n) \\ &= \alpha(a_0 + a_1 x + \cdots + a_n x^n) + \beta(a_0 + a_1 x + \cdots + a_n x^n) \\ &= \alpha p(x) + \beta p(x) \\ &= \alpha\mathbf{p}(x) + \beta\mathbf{p}(x) = (\alpha\mathbf{p} + \beta\mathbf{p})(x)\end{aligned}$$

Thus $(\alpha + \beta)\mathbf{p} = \alpha\mathbf{p} + \beta p$.

V9.

$$\begin{aligned}(\alpha\beta)\mathbf{p}(x) &= (\alpha\beta)p(x) \\ &= (\alpha\beta)(a_0 + a_1 x + \cdots + a_n x^n) \\ &= \alpha(\beta a_0 + \beta a_1 x + \cdots + \beta a_n x^n) \\ &= \alpha[\beta(a_0 + a_1 x + \cdots + a_n x^n)] = \alpha[\beta p(x)] = \alpha(\beta\mathbf{p})(x)\end{aligned}$$

Thus $(\alpha\beta)\mathbf{p} = \alpha(\beta\mathbf{p})$.

V10. Since $(1\mathbf{p})(x) = 1p(x) = p(x) = \mathbf{p}(x)$, then $1\mathbf{p} = \mathbf{p}$.

Thus P_n satisfies all of the axioms for a vector space, and therefore P_n is a vector space.

---

4. **The space of real-valued functions: $F[I]$**

   Let $F[I]$ denote the set of real-valued functions defined on an interval I with the following operations: If $\mathbf{f}(x) = f(x)$ and $\mathbf{g}(x) = g(x)$ are two such functions and α is any scalar, then

$$(\mathbf{f} + \mathbf{g})(x) = f(x) + g(x) \tag{3.31}$$

$$(\alpha \mathbf{f})(x) = \alpha f(x) \tag{3.32}$$

for all x in I.

EXAMPLE 3-8: Show that $F[I]$ is a vector space.

Solution: The function $\mathbf{p}(x) = p(x) = x$ is defined on any interval I. Therefore $F[I]$ is not empty.

V1. Since $(\mathbf{f} + \mathbf{g})(x) = f(x) + g(x)$ for all x in I and $f(x)$ and $g(x)$ are functions defined on I, the function $f(x) + g(x)$ is also defined on I. Therefore $\mathbf{f} + \mathbf{g} \in F[I]$.

V2. Since

$$\begin{aligned} (\mathbf{f} + \mathbf{g})(x) &= f(x) + g(x) \\ &= g(x) + f(x) \qquad [f(x) \text{ and } g(x) \text{ are numbers}] \\ &= (\mathbf{g} + \mathbf{f})(x) \end{aligned}$$

for all x in I, we have $\mathbf{f} + \mathbf{g} = \mathbf{g} + \mathbf{f}$.

V3. Since

$$\begin{aligned} [\mathbf{f} + (\mathbf{g} + \mathbf{h})](x) &= f(x) + (\mathbf{g} + \mathbf{h})(x) \\ &= f(x) + [g(x) + h(x)] \\ &= [f(x) + g(x)] + h(x) \\ &= (\mathbf{f} + \mathbf{g})(x) + h(x) \\ &= [(\mathbf{f} + \mathbf{g}) + \mathbf{h}](x) \end{aligned}$$

for all x in I, $\mathbf{f} + (\mathbf{g} + \mathbf{h}) = (\mathbf{f} + \mathbf{g}) + \mathbf{h}$.

V4. The function $\mathbf{z}$, defined by $\mathbf{z}(x) = z(x) = 0$ for all x in I, is a zero element for $F[I]$ since

$$(\mathbf{f} + \mathbf{z})(x) = f(x) + z(x) = f(x) + 0 = f(x) = \mathbf{f}(x)$$

for all x in I. Thus $\mathbf{f} + \mathbf{z} = \mathbf{f}$.

V5. The function $\mathbf{q}(x)$ defined by $\mathbf{q}(x) = q(x) = -f(x)$ for all x in I is the negative of $\mathbf{f}(x)$ since

$$(\mathbf{f} + \mathbf{q})(x) = f(x) + q(x) = f(x) + [-f(x)] = f(x) - f(x) = 0 = \mathbf{z}(x)$$

for all x in I.

V6. Since $(\alpha \mathbf{f})(x) = \alpha f(x)$, the function $\alpha f(x)$ is also defined in I. Thus $\alpha \mathbf{f} \in F[I]$.

V7. Since

$$\begin{aligned} [\alpha(\mathbf{f} + \mathbf{g})](x) &= \alpha(\mathbf{f} + \mathbf{g})(x) \\ &= \alpha[f(x) + g(x)] \\ &= \alpha f(x) + \alpha g(x) \qquad [f(x) \text{ and } g(x) \text{ are numbers}] \\ &= (\alpha \mathbf{f} + \alpha \mathbf{g})(x) \end{aligned}$$

for all x in I, $\alpha(\mathbf{f} + \mathbf{g}) = \alpha \mathbf{f} + \alpha \mathbf{g}$.

V8. Since

$$\begin{aligned} [(\alpha + \beta)\mathbf{f}](x) &= (\alpha + \beta) f(x) \\ &= \alpha f(x) + \beta f(x) \qquad [f(x) \text{ is a number}] \\ &= (\alpha \mathbf{f} + \beta \mathbf{f})(x) \end{aligned}$$

for all x in I, we have $(\alpha + \beta)\mathbf{f} = \alpha \mathbf{f} + \beta \mathbf{f}$.

V9. Since

$$\begin{aligned}[(\alpha\beta)\mathbf{f}](x) &= (\alpha\beta)f(x) \\ &= \alpha\beta f(x) \\ &= \alpha[\beta f(x)] \qquad [f(x) \text{ is a number}] \\ &= [\alpha(\beta\mathbf{f})](x)\end{aligned}$$

for all x in I, we have $(\alpha\beta)\mathbf{f} = \alpha(\beta\mathbf{f})$.

V10. Since

$$(1\mathbf{f})(x) = 1[f(x)] = \mathbf{f}(x) = \mathbf{f}(x)$$

for all x in I, we have $1\mathbf{f} = \mathbf{f}$.

Thus $F[I]$ satisfies all of the axioms for a vector space, and therefore $F[I]$ is a vector space.

An example of nonvector space is given in Example 3-9.

EXAMPLE 3-9: Let U be the set of all 2-vectors such that the sum of components is equal to 1; that is,

$$U = \left\{ \begin{bmatrix} u_1 \\ u_2 \end{bmatrix} \middle| \, u_1 + u_2 = 1 \right\}$$

Show that U is not a vector space.

Solution: $\begin{bmatrix} 1 \\ 0 \end{bmatrix}$ and $\begin{bmatrix} 0 \\ 1 \end{bmatrix}$ are elements of U, but $\begin{bmatrix} 1 \\ 0 \end{bmatrix} + \begin{bmatrix} 0 \\ 1 \end{bmatrix} = \begin{bmatrix} 1 \\ 1 \end{bmatrix}$ is not an element of U, since $1 + 1 = 2 \neq 1$. Similarly, $2\begin{bmatrix} 1 \\ 0 \end{bmatrix} = \begin{bmatrix} 2 \\ 0 \end{bmatrix}$ is not an element of U either, since $2 + 0 = 2 \neq 1$. Thus Axioms V1 and V6 are not satisfied, and therefore U is not a vector space.

note: Axioms V4 and V5 are not satisfied either. According to Axiom V4, if $\begin{bmatrix} a \\ b \end{bmatrix}$ is a zero element of U, then

$$\begin{bmatrix} 1 \\ 0 \end{bmatrix} + \begin{bmatrix} a \\ b \end{bmatrix} = \begin{bmatrix} 1 + a \\ b \end{bmatrix} = \begin{bmatrix} 1 \\ 0 \end{bmatrix}$$

so that $1 + a = 1$ and $b = 0$; that is, $a = b = 0$. Thus if U has a zero element, then it is $\begin{bmatrix} 0 \\ 0 \end{bmatrix}$. But $\begin{bmatrix} 0 \\ 0 \end{bmatrix}$ is not an element of U, since $0 + 0 \neq 1$. Hence Axiom V4 is not satisfied. And, since U does not have a zero element, the concept of a negative of a vector is meaningless; so Axiom V5 is not satisfied.

C. Additional properties of vector spaces

Let V be a vector space, let $\mathbf{u}, \mathbf{v} \in V$ and let α be a scalar. Then

$$\mathbf{u} + \mathbf{v} = \mathbf{0} \text{ implies } \mathbf{v} = -\mathbf{u} \tag{3.33}$$

$$0\mathbf{v} = \mathbf{0} \tag{3.34}$$

$$\alpha\mathbf{0} = \mathbf{0} \tag{3.35}$$

$$\alpha\mathbf{v} = \mathbf{0} \text{ implies that either } \alpha = 0 \text{ or } \mathbf{v} = \mathbf{0} \tag{3.36}$$

$$(-1)\mathbf{v} = -\mathbf{v} \tag{3.37}$$

EXAMPLE 3-10: Verify property (3.33).

Solution: Suppose that $\mathbf{u} + \mathbf{v} = \mathbf{0}$. Then by Axiom V4,

$$-\mathbf{u} = -\mathbf{u} + \mathbf{0} = -\mathbf{u} + (\mathbf{u} + \mathbf{v})$$

Thus, by Axioms V2, V3, V4, and V5,

$$-\mathbf{u} = (-\mathbf{u} + \mathbf{u}) + \mathbf{v} = \mathbf{0} + \mathbf{v} = \mathbf{v}$$

EXAMPLE 3-11: Verify properties **(a)** (3.34), **(b)** (3.35), and **(c)** (3.36).

Solution

(a) From Axioms V8 and V10,

$$\mathbf{v} = 1\mathbf{v} = (1 + 0)\mathbf{v} = 1\mathbf{v} + 0\mathbf{v} = \mathbf{v} + 0\mathbf{v}$$

Then by Axiom V3,

$$-\mathbf{v} + \mathbf{v} = -\mathbf{v} + (\mathbf{v} + 0\mathbf{v}) = (-\mathbf{v} + \mathbf{v}) + 0\mathbf{v}$$

Next, by Axioms V2, V4, and V5, we have

$$\mathbf{0} = \mathbf{0} + 0\mathbf{v} = 0\mathbf{v}$$

(b) Let $\mathbf{v} = \alpha\mathbf{0}$. Then by Axioms V4 and V7,

$$\mathbf{v} + \mathbf{v} = \alpha\mathbf{0} + \alpha\mathbf{0} = \alpha(\mathbf{0} + \mathbf{0}) = \alpha\mathbf{0} = \mathbf{v}$$

Thus

$$(\mathbf{v} + \mathbf{v}) + (-\mathbf{v}) = \mathbf{v} + (-\mathbf{v})$$

By Axioms V3 and V5,

$$\mathbf{v} + [\mathbf{v} + (-\mathbf{v})] = \mathbf{0}$$
$$\mathbf{v} + \mathbf{0} = \mathbf{0}$$

Then by Axioms V2 and V4 we obtain $\mathbf{v} = \mathbf{0}$; that is $\alpha\mathbf{0} = \mathbf{0}$.

(c) From property (3.34), if $\alpha = 0$, we have

$$\alpha\mathbf{v} = 0\mathbf{v} = \mathbf{0}$$

If $\alpha\mathbf{v} = \mathbf{0}$ and $\alpha \neq 0$, then it follows from property (3.33) and axioms V9 and V10, that

$$\mathbf{0} = \frac{1}{\alpha}\mathbf{0} = \frac{1}{\alpha}(\alpha\mathbf{v}) = \left(\frac{1}{\alpha}\alpha\right)\mathbf{v} = 1\mathbf{v} = \mathbf{v}$$

EXAMPLE 3-12: Verify property (3.37).

Solution: Since $1 + (-1) = 0$, we have

$$[1 + (-1)]\mathbf{v} = 0\mathbf{v} = \mathbf{0} \qquad \text{[by (3.34)]}$$

Thus, by Axiom V8

$$1\mathbf{v} + (-1)\mathbf{v} = \mathbf{0}$$

or, using Axiom V10,

$$\mathbf{v} + (-1)\mathbf{v} = \mathbf{0}$$

Then by property (3.33), we conclude that

$$(-1)\mathbf{v} = -\mathbf{v}$$

3-4. Subspaces

A. Definition

A nonempty subset U of a vector space V is called a **subspace** of V if U is itself a vector space under the same operation of addition and scalar multiplication as V.

B. Closure properties

THEOREM 3-4.1 If U is a nonempty subset of a vector space V, then U is a subspace of V if and only if U satisfies the following conditions:

CLOSURE PROPERTIES

S1. $\mathbf{u} + \mathbf{v} \in U$ whenever $\mathbf{u} \in U$ and $\mathbf{v} \in U$. **(3.38)**

S2. $\alpha\mathbf{u} \in U$ whenever $\mathbf{u} \in U$ and α is any scalar. **(3.39)**

These conditions, often called the **closure properties**, can be restated as follows:

S1. U is closed under addition.

S2. U is closed under scalar multiplication.

Then Theorem 3-4.1 can be restated as follows:

- A nonempty subset U of a vector space V is a subspace of V if and only if it is closed under both addition and scalar multiplication.

EXAMPLE 3-13: Prove Theorem 3-4.1.

Solution: If U is a subspace of a vector space V, then all of the vector space axioms are satisfied. In particular, Conditions S1 and S2 are precisely Axioms V1 and V6, respectively. Conversely, suppose that U is a subset of V and Conditions S1 and S2 are satisfied. Axioms V2, V3, V7, V8, V9, and V10 are satisfied since every element of U is also an element of V. Since Axioms V1 and V6 are precisely Conditions S1 and S2, respectively, we need show only that U satisfies the remaining Axioms, V4 and V5. Setting $\alpha = 0$ and then $\alpha = -1$ in condition (3.39), we obtain the zero element and the negative of each element in U (see properties (3.34) and (3.37)). Thus Axioms V4 and V5 are satisfied. The subset U is a vector space under the same operation of addition and scalar multiplication as V. Therefore U is a subspace.

EXAMPLE 3-14: Consider the subset U of R^2 defined by

$$U = \left\{ \begin{bmatrix} u_1 \\ u_2 \end{bmatrix} \middle| \, u_2 = 2u_1 \right\}$$

Show that U is a subspace of R^2.

Solution: Let $\mathbf{u} = \begin{bmatrix} u_1 \\ u_2 \end{bmatrix} \in U$. Since $u_2 = 2u_1$, we can express any $\mathbf{u} \in U$ as $\mathbf{u} = \begin{bmatrix} a \\ 2a \end{bmatrix}$. Since $\begin{bmatrix} 0 \\ 0 \end{bmatrix} \in U$, the set U is not empty. If $\mathbf{u}, \mathbf{v} \in U$, then

$$\mathbf{u} = \begin{bmatrix} a \\ 2a \end{bmatrix}, \qquad \mathbf{v} = \begin{bmatrix} b \\ 2b \end{bmatrix}$$

and

$$\mathbf{u} + \mathbf{v} = \begin{bmatrix} a + b \\ 2a + 2b \end{bmatrix} = \begin{bmatrix} a + b \\ 2(a + b) \end{bmatrix} \in U$$

Next, if α is any scalar

$$\alpha\mathbf{u} = \begin{bmatrix} \alpha a \\ \alpha 2a \end{bmatrix} = \begin{bmatrix} \alpha a \\ 2(\alpha a) \end{bmatrix} \in U$$

Hence U is closed under addition and scalar multiplication. Therefore by Theorem 3-4.1 U is a subspace of R^2.

note: Geometrically, U consists of all points on a line passing through the origin (expressed as $y = 2x$) in the xy plane (see Figure 3-10).

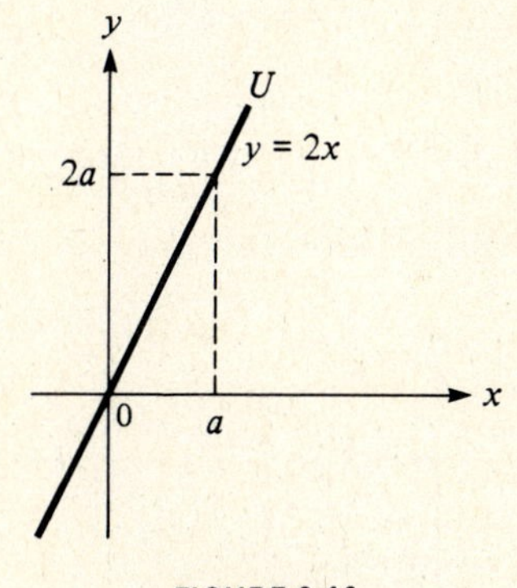

FIGURE 3-10

EXAMPLE 3-15: Consider the subset U of R^2 defined by

$$U = \left\{ \begin{bmatrix} u_1 \\ u_2 \end{bmatrix} \middle| u_1 + u_2 = 1 \right\}$$

Show that U is not a subspace of R^2.

Solution: It is shown in Example 3-9 that U is not a vector space; thus, by the definition of the subspace, U cannot be a subspace of R^2.

---

note: Geometrically, U consists of all points on a line expressed as $x + y = 1$ (or $y = -x + 1$) in the xy plane (see Figure 3-11).

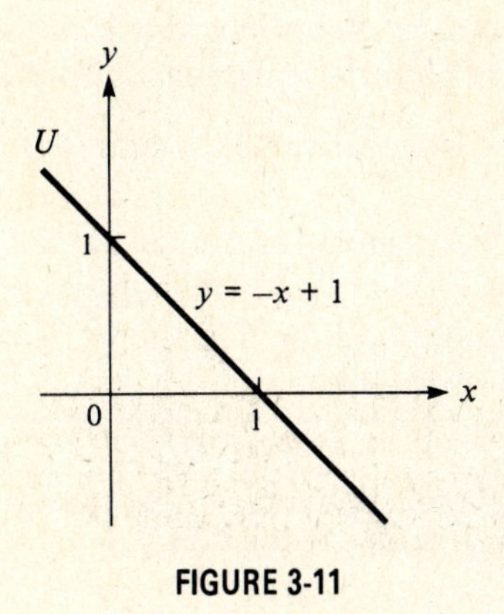

FIGURE 3-11

EXAMPLE 3-16: Consider the subset U of R^2 defined by

$$U = \left\{ \begin{bmatrix} u_1 \\ u_2 \end{bmatrix} \middle| u_1 = 0 \text{ or } u_2 = 0 \right\}$$

Show that U is not a subspace of R^2.

Solution: Let $\mathbf{u} = \begin{bmatrix} u_1 \\ u_2 \end{bmatrix} \in U$. Then $\alpha\mathbf{u} = \begin{bmatrix} \alpha u_1 \\ \alpha u_2 \end{bmatrix}$. If $u_1 = 0$ or $u_2 = 0$, then $\alpha u_1 = 0$ or $\alpha u_2 = 0$ also. Thus $\alpha\mathbf{u} \in U$ and U is closed under scalar multiplication.

But U is not closed under addition since we have

$$\begin{bmatrix} 1 \\ 0 \end{bmatrix} \in U, \qquad \begin{bmatrix} 0 \\ 1 \end{bmatrix} \in U$$

However,

$$\begin{bmatrix}1\\0\end{bmatrix}+\begin{bmatrix}0\\1\end{bmatrix}=\begin{bmatrix}1\\1\end{bmatrix}\notin U$$

Therefore, U is not a subspace of R^2.

note: Geometrically, U consists of all points on the two coordinate axes (see Figure 3-12).

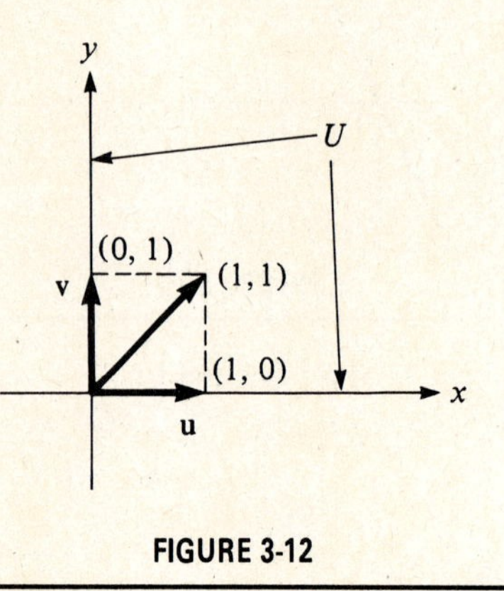

FIGURE 3-12

C. $C[a, b]$ space

Let $C[a,b]$ denote the set of all real-valued functions that are defined and continuous on the closed interval $[a,b]$.

***EXAMPLE 3-17** [*For readers who have studied calculus.*]: Show that $C[a,b]$ is a subspace of the vector space $F[I]$, where $I = [a,b]$, of all real-valued functions in Example 3-8.

Solution: It is obvious that $r(x) = x$ is a real-valued continuous function on $[a,b]$. Thus $C[a,b]$ is a nonempty subset of $F[a,b]$.

Now recall from calculus that, if $f(x)$ and $g(x)$ are real-valued continuous functions on $[a,b]$ and k is a constant, then $f(x) + g(x)$ and $kf(x)$ are also real-valued continuous functions. Thus $C[a,b]$ is closed under addition and scalar multiplication. Therefore $C[a,b]$ is a subspace of $F[a,b]$ of all real-valued functions defined on the closed interval $[a,b]$.

We shall denote the set of all real-valued functions that have a continuous nth derivative on $[a,b]$ by $C^n[a,b]$. In a manner similar to that used in Example 3-17, we can verify that $C^n[a,b]$ is a subspace of $C[a,b]$.

D. Solution space: $S(A)$

DEFINITION Let A be an $m \times n$ matrix and let $S(A)$ denote the set of all solutions to the system

$$A\mathbf{x} = \mathbf{b} \tag{3.40}$$

where $\mathbf{b} \in R^m$ and $\mathbf{x} \in R^n$. Then $\mathbf{x}$ is called the **solution vector** of the system (3.40). Thus,

$$S(A) = \{\mathbf{x} \in R^n | A\mathbf{x} = \mathbf{b}\}$$

Then $S(A)$ is called the **solution space** of $A\mathbf{x} = \mathbf{b}$.

EXAMPLE 3-18: Show that $S(A)$ is not a subspace of R^n if $\mathbf{b} \neq \mathbf{0}$.

Solution: Let $\mathbf{x}_1 \in S(A)$ and $\mathbf{x}_2 \in S(A)$, so that

$$A\mathbf{x}_1 = \mathbf{b} \qquad \text{and} \qquad A\mathbf{x}_2 = \mathbf{b}$$

Then

$$A(\mathbf{x}_1 + \mathbf{x}_2) = A\mathbf{x}_1 + A\mathbf{x}_2 = \mathbf{b} + \mathbf{b} = 2\mathbf{b}$$

Thus $\mathbf{x}_1 + \mathbf{x}_2 \notin S(A)$ and $S(A)$ is not closed under addition. Therefore it follows that $S(A)$ is not a subspace of R^n.

note: $A(\alpha\mathbf{x}_1) = \alpha A\mathbf{x}_1 = \alpha\mathbf{b}$; thus $S(A)$ is not closed under scalar multiplication either.

EXAMPLE 3-19: Show that $S(A)$ is a subspace of R^n if $\mathbf{b} = \mathbf{0}$.

Solution: In this case

$$S(A) = \{\mathbf{x} \in R^n | A\mathbf{x} = \mathbf{0}\}$$

$A\mathbf{0} = \mathbf{0}$; thus $\mathbf{0} \in S(A)$ and $S(A)$ is a nonempty subset of R^n.

Let $\mathbf{x}_1 \in S(A)$ and $\mathbf{x}_2 \in S(A)$, so that

$$A\mathbf{x}_1 = \mathbf{0} \quad \text{and} \quad A\mathbf{x}_2 = \mathbf{0}$$

Then

$$A(\mathbf{x}_1 + \mathbf{x}_2) = A\mathbf{x}_1 + A\mathbf{x}_2 = \mathbf{0} + \mathbf{0} = \mathbf{0}$$

Thus, $\mathbf{x}_1 + \mathbf{x}_2 \in S(A)$.

Next, if α is a scalar and $\mathbf{x} \in S(A)$, then

$$A(\alpha\mathbf{x}) = \alpha A\mathbf{x} = \alpha\mathbf{0} = \mathbf{0} \qquad [\text{by } (3.35)]$$

and hence, $\alpha\mathbf{x} \in S(A)$. Thus $S(A)$ is closed under addition and scalar multiplication and, by Theorem 3-4.1, $S(A)$ is a subspace of R^n.

note: The solution space of the homogeneous system $A\mathbf{x} = \mathbf{0}$ is often called the **null space** of A, denoted by $N(A)$.

3-5. Linear Combination and Spanning Sets

A. Linear combination

DEFINITION Let $\mathbf{v}_1, \mathbf{v}_2, \ldots, \mathbf{v}_n$ be vectors in a vector space V, and $\alpha_1, \alpha_2, \ldots, \alpha_n$ be scalars (real numbers). The vector

LINEAR COMBINATION

$$\mathbf{u} = \alpha_1\mathbf{v}_1 + \alpha_2\mathbf{v}_2 + \cdots + \alpha_n\mathbf{v}_n \tag{3.41}$$

is called a **linear combination** of $\mathbf{v}_1, \mathbf{v}_2, \ldots, \mathbf{v}_n$.

EXAMPLE 3-20: Consider $\mathbf{v}_1 = \begin{bmatrix} 1 \\ 0 \end{bmatrix}$ and $\mathbf{v}_2 = \begin{bmatrix} 1 \\ 1 \end{bmatrix}$ in R^2. Show that $\mathbf{u} = \begin{bmatrix} 4 \\ 3 \end{bmatrix}$ can be expressed as a linear combination of $\mathbf{v}_1$ and $\mathbf{v}_2$.

Solution: In order for $\mathbf{u}$ to be a linear combination of $\mathbf{v}_1$ and $\mathbf{v}_2$, there must be scalars α_1 and α_2 such that

$$\mathbf{u} = \alpha_1\mathbf{v}_1 + \alpha_2\mathbf{v}_2$$

That is,

$$\begin{bmatrix} 4 \\ 3 \end{bmatrix} = \alpha_1\begin{bmatrix} 1 \\ 0 \end{bmatrix} + \alpha_2\begin{bmatrix} 1 \\ 1 \end{bmatrix} = \begin{bmatrix} \alpha_1 + \alpha_2 \\ \alpha_2 \end{bmatrix}$$

Equating the corresponding components, we obtain

$$\alpha_1 + \alpha_2 = 4$$
$$\alpha_2 = 3$$

Solving for α_1, α_2, we obtain $\alpha_1 = 1$ and $\alpha_2 = 3$, and hence we get

$$\mathbf{u} = \mathbf{v}_1 + 3\mathbf{v}_2$$

Check:

$$\mathbf{v}_1 + 3\mathbf{v}_2 = \begin{bmatrix} 1 \\ 0 \end{bmatrix} + 3\begin{bmatrix} 1 \\ 1 \end{bmatrix} = \begin{bmatrix} 1 \\ 0 \end{bmatrix} + \begin{bmatrix} 3 \\ 3 \end{bmatrix} = \begin{bmatrix} 4 \\ 3 \end{bmatrix} = \mathbf{u}$$

EXAMPLE 3-21: Consider $\mathbf{v}_1 = \begin{bmatrix} 2 \\ 1 \\ 1 \end{bmatrix}$ and $\mathbf{v}_2 = \begin{bmatrix} 5 \\ 2 \\ 3 \end{bmatrix}$ in R^3. Show that $\mathbf{u} = \begin{bmatrix} 1 \\ 1 \\ 1 \end{bmatrix}$ cannot be expressed as a linear combination of $\mathbf{v}_1$ and $\mathbf{v}_2$.

Solution: If $\mathbf{u}$ is a linear combination of $\mathbf{v}_1$ and $\mathbf{v}_2$, there must be scalars α_1 and α_2 such that

$$\mathbf{u} = \alpha_1\mathbf{v}_1 + \alpha_2\mathbf{v}_2$$

That is,

$$\begin{bmatrix} 1 \\ 1 \\ 1 \end{bmatrix} = \alpha_1\begin{bmatrix} 2 \\ 1 \\ 1 \end{bmatrix} + \alpha_2\begin{bmatrix} 5 \\ 2 \\ 3 \end{bmatrix} = \begin{bmatrix} 2\alpha_1 + 5\alpha_2 \\ \alpha_1 + 2\alpha_2 \\ \alpha_1 + 3\alpha_2 \end{bmatrix}$$

Equating the corresponding components, we get

$$2\alpha_1 + 5\alpha_2 = 1$$

$$\alpha_1 + 2\alpha_2 = 1$$

$$\alpha_1 + 3\alpha_2 = 1$$

Clearly there is no solution to the above system; therefore there are no scalars α_1 and α_2 such that

$$\mathbf{u} = \alpha_1\mathbf{v}_1 + \alpha_2\mathbf{v}_2$$

Thus $\mathbf{u}$ cannot be expressed as a linear combination of $\mathbf{v}_1$ and $\mathbf{v}_2$.

B. Spanning set

DEFINITION The set of all linear combinations of $\mathbf{v}_1, \mathbf{v}_2, \ldots, \mathbf{v}_n$, that is

$$\alpha_1\mathbf{v}_1 + \alpha_2\mathbf{v}_2 + \cdots + \alpha_n\mathbf{v}_n$$

is called the **span** of $\mathbf{v}_1, \mathbf{v}_2, \ldots, \mathbf{v}_n$ denoted by

$$S(\mathbf{v}_1, \mathbf{v}_2, \ldots, \mathbf{v}_n).$$

The set $\{\mathbf{v}_1, \mathbf{v}_2, \ldots, \mathbf{v}_n\}$ is called a **spanning set** for V if and only if every vector in V can be expressed as a linear combination of $\mathbf{v}_1, \mathbf{v}_2, \ldots, \mathbf{v}_n$.

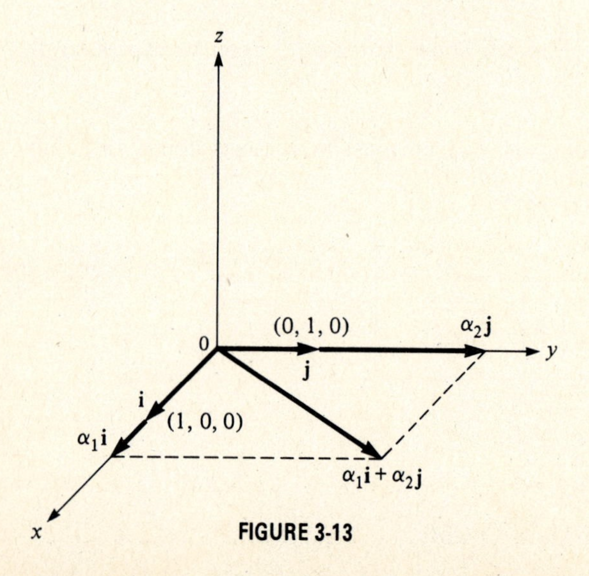

FIGURE 3-13

EXAMPLE 3-22: Let $\mathbf{i} = \begin{bmatrix} 1 \\ 0 \\ 0 \end{bmatrix}$, $\mathbf{j} = \begin{bmatrix} 0 \\ 1 \\ 0 \end{bmatrix}$ be two vectors in R^3. Find the span of $\mathbf{i}$ and $\mathbf{j}$, $S(\mathbf{i}, \mathbf{j})$.

Solution: Let α_1 and α_2 be any scalars. Then in R^3 the span of $\mathbf{i}$ and $\mathbf{j}$ is the set of all vectors of the form

$$\alpha_1\mathbf{i} + \alpha_2\mathbf{j} = \alpha_1\begin{bmatrix} 1 \\ 0 \\ 0 \end{bmatrix} + \alpha_2\begin{bmatrix} 0 \\ 1 \\ 0 \end{bmatrix} = \begin{bmatrix} \alpha_1 \\ \alpha_2 \\ 0 \end{bmatrix}$$

which is the plane determined by $\mathbf{i}$ and $\mathbf{j}$, that is, the xy plane (see Figure 3-13).

EXAMPLE 3-23: Show that $S(\mathbf{i},\mathbf{j})$ of Example 3-22 is a subspace of R^3.

Solution: If $\mathbf{u} \in S(\mathbf{i},\mathbf{j})$ and $\mathbf{v} \in S(\mathbf{i},\mathbf{j})$, then

$$\mathbf{u} = \alpha_1\mathbf{i} + \alpha_2\mathbf{j}$$

and

$$\mathbf{v} = \beta_1\mathbf{i} + \beta_2\mathbf{j}$$

where $\alpha_1, \alpha_2, \beta_1, \beta_2$ are scalars. Therefore,

$$\mathbf{u} + \mathbf{v} = (\alpha_1 + \beta_1)\mathbf{i} + (\alpha_2 + \beta_2)\mathbf{j}$$

and for any scalar λ

$$\lambda\mathbf{u} = (\lambda\alpha_1)\mathbf{i} + (\lambda\alpha_2)\mathbf{j}$$

Thus, $\mathbf{u} + \mathbf{v} \in S(\mathbf{i},\mathbf{j})$, $\lambda\mathbf{u} \in S(\mathbf{i},\mathbf{j})$. Therefore $S(\mathbf{i},\mathbf{j})$ is closed under addition and scalar multiplication and $S(\mathbf{i},\mathbf{j})$ is a subspace of R^3.

EXAMPLE 3-24: Show that the set of 3-vectors $\{\mathbf{i},\mathbf{j},\mathbf{k}\}$ spans R^3, where

$$\mathbf{i} = \begin{bmatrix} 1 \\ 0 \\ 0 \end{bmatrix}, \qquad \mathbf{j} = \begin{bmatrix} 0 \\ 1 \\ 0 \end{bmatrix}, \qquad \mathbf{k} = \begin{bmatrix} 0 \\ 0 \\ 1 \end{bmatrix}$$

Solution: Let $\mathbf{u} = \begin{bmatrix} u_1 \\ u_2 \\ u_3 \end{bmatrix} \in R^3$. Then $\mathbf{u}$ can be expressed as

$$\begin{bmatrix} u_1 \\ u_2 \\ u_3 \end{bmatrix} = u_1\begin{bmatrix} 1 \\ 0 \\ 0 \end{bmatrix} + u_2\begin{bmatrix} 0 \\ 1 \\ 0 \end{bmatrix} + u_3\begin{bmatrix} 0 \\ 0 \\ 1 \end{bmatrix}$$

that is,

$$\mathbf{u} = u_1\mathbf{i} + u_2\mathbf{j} + u_3\mathbf{k}$$

which is a linear combination of **i**, **j**, and **k**. Thus the set $\{\mathbf{i},\mathbf{j},\mathbf{k}\}$ spans R^3.

EXAMPLE 3-25: Show that $S = \{1, x, x^2\}$ spans the vector space P_2 of all polynomials of degree less than or equal to 2.

Solution: Clearly any polynomial $p(x) = a_0 + a_1x + a_2x^2$ in P_2 is a linear combination of the polynomials 1, x, and x^2. Therefore $S = \{1, x, x^2\}$ spans the vector space P_2.

EXAMPLE 3-26: If $\mathbf{v}_1, \mathbf{v}_2, \ldots, \mathbf{v}_m$ are vectors in a vector space V, then show that the span of $\mathbf{v}_1, \mathbf{v}_2, \ldots, \mathbf{v}_m$, $S(\mathbf{v}_1, \mathbf{v}_2, \ldots, \mathbf{v}_m)$, is a subspace of V.

Solution: If **u** and **w** are vectors in $S(\mathbf{v}_1, \mathbf{v}_2, \ldots, \mathbf{v}_m)$, then

$$\mathbf{u} = \alpha_1\mathbf{v}_1 + \alpha_2\mathbf{v}_2 + \cdots + \alpha_m\mathbf{v}_m$$

$$\mathbf{w} = \beta_1\mathbf{v}_1 + \beta_2\mathbf{v}_2 + \cdots + \beta_m\mathbf{v}_m$$

where α_i, β_i $(i = 1, \ldots, m)$ are scalars.

Now

$$\mathbf{u} + \mathbf{w} = (\alpha_1 + \beta_1)\mathbf{v}_1 + (\alpha_2 + \beta_2)\mathbf{v}_2 + \cdots + (\alpha_m + \beta_m)\mathbf{v}_m \in S(\mathbf{v}_1, \mathbf{v}_2, \ldots, \mathbf{v}_m)$$

and for any scalar k,

$$k\mathbf{u} = (k\alpha_1)\mathbf{v}_1 + (k\alpha_2)\mathbf{v}_2 + \cdots + (k\alpha_m)\mathbf{v}_m \in S(\mathbf{v}_1, \mathbf{v}_2, \ldots, \mathbf{v}_m)$$

Thus $S(\mathbf{v}_1, \mathbf{v}_2, \ldots, \mathbf{v}_m)$ is closed under addition and scalar multiplication. Therefore $S(\mathbf{v}_1, \mathbf{v}_2, \ldots, \mathbf{v}_m)$ is a subspace of V.

3-6. Linear Independence

DEFINITION Let V be a vector space and let $\mathbf{v}_1, \mathbf{v}_2, \ldots, \mathbf{v}_n$ be vectors in V. We say that $\mathbf{v}_1, \mathbf{v}_2, \ldots, \mathbf{v}_n$ are *linearly dependent* if there exist scalars $\alpha_1, \alpha_2, \ldots, \alpha_n$ not all equal to zero such that

$$\alpha_1 \mathbf{v}_1 + \alpha_2 \mathbf{v}_2 + \cdots + \alpha_n \mathbf{v}_n = \mathbf{0} \tag{3.42}$$

If no such scalars exist, then we say that $\mathbf{v}_1, \mathbf{v}_2, \ldots, \mathbf{v}_n$ are *linearly independent.*

In other words, vectors $\mathbf{v}_1, \mathbf{v}_2, \ldots, \mathbf{v}_n$ are linearly independent if and only if the following condition is satisfied:

A. Independence criterion

Whenever

$$\alpha_1 \mathbf{v}_1 + \alpha_2 \mathbf{v}_2 + \cdots + \alpha_n \mathbf{v}_n = \mathbf{0}$$

then

$$\alpha_1 = \alpha_2 = \cdots = \alpha_n = 0$$

EXAMPLE 3-27: Show that in P_2, the vectors $\mathbf{p}_1(x) = p_1(x) = 1$, $\mathbf{p}_2(x) = p_2(x) = 1 + x$, $\mathbf{p}_3(x) = p_3(x) = 1 + x + x^2$ are linearly independent.

Solution: If

$$\alpha_1 \mathbf{p}_1 + \alpha_2 \mathbf{p}_2 + \alpha_3 \mathbf{p}_3 = \mathbf{0}$$

where $\alpha_1, \alpha_2, \alpha_3$ are scalars, then

$$\alpha_1 p_1(x) + \alpha_2 p_2(x) + \alpha_3 p_3(x) = 0$$

or

$$\alpha_1(1) + \alpha_2(1 + x) + \alpha_3(1 + x + x^2) = 0$$

or

$$(\alpha_1 + \alpha_2 + \alpha_3) + (\alpha_2 + \alpha_3)x + \alpha_3 x^2 = 0$$

Since a polynomial is zero only if all its coefficients are zero, we must have

$$\alpha_1 + \alpha_2 + \alpha_3 = 0$$
$$\alpha_2 + \alpha_3 = 0$$
$$\alpha_3 = 0$$

This immediately implies $\alpha_3 = \alpha_2 = \alpha_1 = 0$. Thus we conclude that $\{1, 1 + x, 1 + x + x^2\}$ is a linearly independent set of vectors.

***EXAMPLE 3-28** [*For readers who have studied calculus.*]: Show that functions $f(x) = e^x$ and $g(x) = e^{2x}$ in $C[-\infty, \infty]$ are linearly independent.

Solution: Suppose that there are scalars a and b such that

$$ae^x + be^{2x} = 0$$

for all values of x. Differentiating this equation with respect to x, we get

$$ae^x + 2be^{2x} = 0$$

Subtracting the first equation from the second equation, we get

$$be^{2x} = 0$$

and hence $b = 0$. Then from the first equation it follows that

$$ae^x = 0$$

and hence $a = 0$. Therefore, we conclude that e^x and e^{2x} are linearly independent.

B. Independence theorem

THEOREM 3-6.1 Let $\{\mathbf{a}_1, \mathbf{a}_2, \ldots, \mathbf{a}_n\}$ be a set of n vectors in R^n and let A be an $n \times n$ matrix such that $\mathbf{a}_1, \mathbf{a}_2, \ldots, \mathbf{a}_n$ form the column of A. The vector set $\{\mathbf{a}_1, \mathbf{a}_2, \ldots, \mathbf{a}_n\}$ will be linearly independent if and only if A is nonsingular.

EXAMPLE 3-29: Verify Theorem 3-6.1.

Solution: Let $\alpha_1, \alpha_2, \ldots, \alpha_n$ be scalars and consider the equation

$$\alpha_1 \mathbf{a}_1 + \alpha_2 \mathbf{a}_2 + \cdots + \alpha_n \mathbf{a}_n = \mathbf{0} \tag{3.43}$$

Now

$$\begin{aligned}
\alpha_1 \mathbf{a}_1 + \alpha_2 \mathbf{a}_2 + \cdots + \alpha_n \mathbf{a}_n &= \alpha_1 \begin{bmatrix} a_{11} \\ a_{21} \\ \vdots \\ a_{n1} \end{bmatrix} + \alpha_2 \begin{bmatrix} a_{12} \\ a_{22} \\ \vdots \\ a_{n2} \end{bmatrix} + \cdots + \alpha_n \begin{bmatrix} a_{1n} \\ a_{2n} \\ \vdots \\ a_{nn} \end{bmatrix} \\
&= \begin{bmatrix} a_{11}\alpha_1 + a_{12}\alpha_2 + \cdots + a_{1n}\alpha_n \\ a_{21}\alpha_1 + a_{22}\alpha_2 + \cdots + a_{2n}\alpha_n \\ \vdots \\ a_{n1}\alpha_1 + a_{n2}\alpha_2 + \cdots + a_{nn}\alpha_n \end{bmatrix} \\
&= A\boldsymbol{\alpha}
\end{aligned}$$

where

$$\boldsymbol{\alpha} = \begin{bmatrix} \alpha_1 \\ \alpha_2 \\ \vdots \\ \alpha_n \end{bmatrix}$$

Thus (3.43) reduces to the system

$$A\boldsymbol{\alpha} = \mathbf{0} \tag{3.44}$$

Then by Theorem 1-4.4 the system (3.44) will have the trivial solution $\boldsymbol{\alpha} = \mathbf{0}$, that is, $\alpha_1 = \alpha_2 = \cdots = \alpha_n = 0$ if and only if A is nonsingular. Thus, $\mathbf{a}_1, \mathbf{a}_2, \ldots, \mathbf{a}_n$ will be linearly independent if and only if A is nonsingular.

Theorem 3-6.1 offers an easy test for the linear dependency of n vectors in R^n. We simply form a matrix A whose columns are the vectors being tested. Then, by Theorem 2-3.2, if $\det A = 0$, then A is singular and the vectors are linearly dependent. If $\det A \neq 0$, then A is nonsingular and the vectors are linearly independent.

EXAMPLE 3-30: Using Theorem 3-6.1, show that $\{\mathbf{i}, \mathbf{j}, \mathbf{k}\}$ of Example 3-24 is linearly independent.

Solution: Let

$$A = [\mathbf{i} \quad \mathbf{j} \quad \mathbf{k}] = \begin{bmatrix} 1 & 0 & 0 \\ 0 & 1 & 0 \\ 0 & 0 & 1 \end{bmatrix}$$

Now

$$\det A = \begin{vmatrix} 1 & 0 & 0 \\ 0 & 1 & 0 \\ 0 & 0 & 1 \end{vmatrix} = 1 \neq 0$$

Hence **i, j, k** are linearly independent.

EXAMPLE 3-31: Determine whether or not the following vectors are linearly independent in R^3:

$$\mathbf{v}_1 = \begin{bmatrix} 1 \\ 1 \\ 2 \end{bmatrix}, \qquad \mathbf{v}_2 = \begin{bmatrix} 3 \\ 2 \\ 5 \end{bmatrix}, \qquad \mathbf{v}_3 = \begin{bmatrix} -3 \\ 0 \\ -3 \end{bmatrix}$$

Solution: Let

$$A = [\mathbf{v}_1 \quad \mathbf{v}_2 \quad \mathbf{v}_3] = \begin{bmatrix} 1 & 3 & -3 \\ 1 & 2 & 0 \\ 2 & 5 & -3 \end{bmatrix}$$

Now using (2.3) we get

$$\det A = \begin{vmatrix} 1 & 3 & -3 \\ 1 & 2 & 0 \\ 2 & 5 & -3 \end{vmatrix} = -6 + 0 - 15 + 12 - 0 + 9 = 0$$

Hence $\mathbf{v}_1, \mathbf{v}_2, \mathbf{v}_3$ are linearly dependent.

3-7. Basis and Dimension

A. Basis

DEFINITION If V is any vector space and $\mathscr{B} = \{\mathbf{v}_1, \mathbf{v}_2, \ldots, \mathbf{v}_n\}$ is a set of vectors in V, then $\mathscr{B}$ is called a **basis** for V if

B1. $\mathscr{B}$ is linearly independent
B2. $\mathscr{B}$ spans V.

B. Examples

1. **Standard basis for R^n**

EXAMPLE 3-32: Consider $\mathscr{B} = \{\mathbf{e}_1, \mathbf{e}_2, \ldots, \mathbf{e}_n\}$, where

$$\mathbf{e}_1 = \begin{bmatrix} 1 \\ 0 \\ 0 \\ \vdots \\ 0 \end{bmatrix}, \mathbf{e}_2 = \begin{bmatrix} 0 \\ 1 \\ 0 \\ \vdots \\ 0 \end{bmatrix}, \cdots, \mathbf{e}_n = \begin{bmatrix} 0 \\ 0 \\ \vdots \\ 0 \\ 1 \end{bmatrix}$$

Show that $\mathscr{B}$ is a basis for R^n.

Solution: Assume

$$\alpha_1 \mathbf{e}_1 + \alpha_2 \mathbf{e}_2 + \cdots + \alpha_n \mathbf{e}_n = \mathbf{0}$$

Then

$$\alpha_1 \begin{bmatrix} 1 \\ 0 \\ \vdots \\ 0 \end{bmatrix} + \alpha_2 \begin{bmatrix} 0 \\ 1 \\ \vdots \\ 0 \end{bmatrix} + \cdots + \alpha_n \begin{bmatrix} 0 \\ 0 \\ \vdots \\ 1 \end{bmatrix} = \begin{bmatrix} 0 \\ 0 \\ \vdots \\ 0 \end{bmatrix}$$

or

$$\begin{bmatrix} \alpha_1 \\ \alpha_2 \\ \vdots \\ \alpha_n \end{bmatrix} = \begin{bmatrix} 0 \\ 0 \\ \vdots \\ 0 \end{bmatrix}$$

So $\alpha_1 = \alpha_2 = \cdots = \alpha_n = 0$. Thus $\mathscr{B} = \{\mathbf{e}_1, \mathbf{e}_2, \ldots, \mathbf{e}_n\}$ is linearly independent.

Next, for any $\mathbf{v} \in R^n$, we have

$$\mathbf{v} = \begin{bmatrix} v_1 \\ v_2 \\ \vdots \\ v_n \end{bmatrix} = v_1 \begin{bmatrix} 1 \\ 0 \\ \vdots \\ 0 \end{bmatrix} + v_2 \begin{bmatrix} 0 \\ 1 \\ \vdots \\ 0 \end{bmatrix} + \cdots + v_n \begin{bmatrix} 0 \\ 0 \\ \vdots \\ 1 \end{bmatrix}$$

$$= v_1\mathbf{e}_1 + v_2\mathbf{e}_2 + \cdots + v_n\mathbf{e}_n$$

Since every vector $\mathbf{v}$ can be expressed as a suitable linear combination of $\mathbf{e}_1, \mathbf{e}_2, \ldots, \mathbf{e}_n$, we conclude that $\mathscr{B}$ spans R^n. Thus $\mathscr{B}$ satisfies both (B1) and (B2) conditions; hence $\mathscr{B}$ is a basis for R^n. This particular basis is called the *standard basis* for R^n.

---

2. **Standard basis for $R_{2\times 2}$**

EXAMPLE 3-33: Let

$$M_{11} = \begin{bmatrix} 1 & 0 \\ 0 & 0 \end{bmatrix}, \qquad M_{12} = \begin{bmatrix} 0 & 1 \\ 0 & 0 \end{bmatrix}, \qquad M_{21} = \begin{bmatrix} 0 & 0 \\ 1 & 0 \end{bmatrix}, \qquad M_{22} = \begin{bmatrix} 0 & 0 \\ 0 & 1 \end{bmatrix}$$

Show that the set $\mathscr{B} = \{M_{11}, M_{12}, M_{21}, M_{22}\}$ is a basis for the vector space $R_{2\times 2}$ of 2×2 real matrices.

Solution: Assume that

$$\alpha_1 M_{11} + \alpha_2 M_{12} + \alpha_3 M_{21} + \alpha_4 M_{22} = O_{2\times 2}$$

that is,

$$\alpha_1 \begin{bmatrix} 1 & 0 \\ 0 & 0 \end{bmatrix} + \alpha_2 \begin{bmatrix} 0 & 1 \\ 0 & 0 \end{bmatrix} + \alpha_3 \begin{bmatrix} 0 & 0 \\ 1 & 0 \end{bmatrix} + \alpha_4 \begin{bmatrix} 0 & 0 \\ 0 & 1 \end{bmatrix} = \begin{bmatrix} 0 & 0 \\ 0 & 0 \end{bmatrix}$$

Then

$$\begin{bmatrix} \alpha_1 & \alpha_2 \\ \alpha_3 & \alpha_4 \end{bmatrix} = \begin{bmatrix} 0 & 0 \\ 0 & 0 \end{bmatrix}$$

Thus $\alpha_1 = \alpha_2 = \alpha_3 = \alpha_4 = 0$, and we conclude that $\mathscr{B}$ is linearly independent.

Next, any element of $R_{2\times 2}$ can be expressed as

$$\begin{bmatrix} a & b \\ c & d \end{bmatrix} = \begin{bmatrix} a & 0 \\ 0 & 0 \end{bmatrix} + \begin{bmatrix} 0 & b \\ 0 & 0 \end{bmatrix} + \begin{bmatrix} 0 & 0 \\ c & 0 \end{bmatrix} + \begin{bmatrix} 0 & 0 \\ 0 & d \end{bmatrix}$$

$$= a\begin{bmatrix} 1 & 0 \\ 0 & 0 \end{bmatrix} + b\begin{bmatrix} 0 & 1 \\ 0 & 0 \end{bmatrix} + c\begin{bmatrix} 0 & 0 \\ 1 & 0 \end{bmatrix} + d\begin{bmatrix} 0 & 0 \\ 0 & 1 \end{bmatrix}$$

$$= aM_{11} + bM_{12} + cM_{21} + dM_{22}$$

Thus $\mathscr{B}$ spans $R_{2\times 2}$. Hence $\mathscr{B}$ is a basis for $R_{2\times 2}$. This particular basis is called the *standard basis* for $R_{2\times 2}$.

3. **Standard basis for P_n**

EXAMPLE 3-34: Show that the set $\mathscr{B} = \{1, x, x^2, \ldots, x^n\}$ is a basis for the vector space P_n.

Solution: Assume that

$$\alpha_0(1) + \alpha_1 x + \alpha_2 x^2 + \cdots + \alpha_n x^n = 0$$

This equation is an identity and is true for any value of x. This is possible only if $\alpha_0 = \alpha_1 = \cdots = \alpha_n = 0$. Therefore, the set $\mathscr{B}$ is linearly independent.

Next, any element of P_n, any polynomial $p(x)$ of degree less than or equal to n, can be expressed as the linear combination of the elements of $\mathscr{B}$; that is,

$$p(x) = a_0(1) + a_1 x + a_2 x^2 + \cdots + a_n x^n$$

Hence $\mathscr{B}$ spans P_n. Therefore we conclude that $\mathscr{B}$ is a basis for P_n. This particular basis is called the *standard basis* for P_n.

C. Dimension

DEFINITION If V is any vector space and $\mathscr{B} = \{\mathbf{v}_1, \mathbf{v}_2, \ldots, \mathbf{v}_n\}$ is a basis for V, then V is called a **finite-dimensional vector space** and is said to have **dimension** n, denoted by

$$\dim V = n \qquad \textbf{(3.45)}$$

In other words, the number of vectors in a basis is the dimension. A vector space that does not have finite dimension is called an **infinite-dimensional vector space**.

EXAMPLE 3-35: Show that

$$\textbf{(a)}\ \dim R^n = n \qquad \textbf{(3.46)}$$

$$\textbf{(b)}\ \dim R_{2\times 2} = 4 \qquad \textbf{(3.47)}$$

$$\textbf{(c)}\ \dim P_n = n + 1 \qquad \textbf{(3.48)}$$

Solution

(a) The standard basis for R^n, $\{\mathbf{e}_1, \mathbf{e}_2, \ldots, \mathbf{e}_n\}$, contains n vectors (Example 3-32). Thus

$$\dim R^n = n$$

(b) The standard basis for $R_{2\times 2}$ contains 4 elements (matrices; see Example 3-33). Thus

$$\dim R_{2\times 2} = 4$$

(c) The standard basis for P_n, $\{1, x, x^2, \ldots, x^n\}$, contains $n + 1$ elements (Example 3-34). Thus

$$\dim P_n = n + 1$$

D. Additional properties of finite-dimensional vector spaces

THEOREM 3-7.1 If $\mathscr{B} = \{\mathbf{v}_1, \mathbf{v}_2, \ldots, \mathbf{v}_n\}$ is a basis for a finite-dimensional vector space V, then any set of m vectors in V, where $m > n$, is linearly dependent.

EXAMPLE 3-36: Verify Theorem 3-7.1.

Solution: Let $\mathscr{U} = \{\mathbf{u}_1, \mathbf{u}_2, \ldots, \mathbf{u}_m\}$ be any set of m vectors in V, where $m > n$. Since $\mathscr{B} = \{\mathbf{v}_1, \mathbf{v}_2, \ldots, \mathbf{v}_n\}$ is a basis, each $\mathbf{u}_i$ can be expressed as

$$\begin{aligned} \mathbf{u}_1 &= a_{11}\mathbf{v}_1 + a_{21}\mathbf{v}_2 + \cdots + a_{n1}\mathbf{v}_n \\ \mathbf{u}_2 &= a_{12}\mathbf{v}_1 + a_{22}\mathbf{v}_2 + \cdots + a_{n2}\mathbf{v}_n \\ &\vdots \\ \mathbf{u}_m &= a_{1m}\mathbf{v}_1 + a_{2m}\mathbf{v}_2 + \cdots + a_{nm}\mathbf{v}_n \end{aligned} \tag{3.49}$$

To show that $\mathscr{U}$ is linearly dependent, we must find scalars $\alpha_1, \alpha_2, \ldots, \alpha_m$, not all zero, such that

$$\alpha_1\mathbf{u}_1 + \alpha_2\mathbf{u}_2 + \cdots + \alpha_m\mathbf{u}_m = \mathbf{0} \tag{3.50}$$

Substituting the representation for $\mathbf{u}_i$ in (3.49) into (3.50), we obtain

$$\begin{aligned} (\alpha_1 a_{11} + \alpha_2 a_{12} + \cdots + \alpha_m a_{1m})\mathbf{v}_1 + (\alpha_1 a_{21} + \alpha_2 a_{22} + \cdots + \alpha_m a_{2m})\mathbf{v}_2 + \cdots \\ + (\alpha_1 a_{n1} + \alpha_2 a_{n2} + \cdots + \alpha_m a_{nm})\mathbf{v}_n = \mathbf{0} \end{aligned} \tag{3.51}$$

Since $\mathbf{v}_1, \mathbf{v}_2, \ldots, \mathbf{v}_n$ are linearly independent, we must have

$$\begin{aligned} a_{11}\alpha_1 + a_{12}\alpha_2 + \cdots + a_{1m}\alpha_m &= 0 \\ a_{21}\alpha_1 + a_{22}\alpha_2 + \cdots + a_{2m}\alpha_m &= 0 \\ &\vdots \\ a_{n1}\alpha_1 + a_{n2}\alpha_2 + \cdots + a_{nm}\alpha_m &= 0 \end{aligned} \tag{3.52}$$

which is a system of n equations in m unknowns $\alpha_1, \alpha_2, \ldots, \alpha_m$. Since $m > n$, there is a nontrivial solution of this system (see Problem 2-33). Therefore, there are scalars $\alpha_1, \alpha_2, \ldots, \alpha_m$, not all of which are zero, such that condition (3.50) is satisfied. Hence $\mathscr{U}$ is linearly dependent.

Corollary 3-7.2 Any two bases for a finite-dimensional vector space have the same number of vectors.

EXAMPLE 3-37: Verify Corollary 3-7.2.

Solution: Let $\mathscr{B} = \{\mathbf{v}_1, \mathbf{v}_2, \ldots, \mathbf{v}_n\}$ and $\mathscr{U} = \{\mathbf{u}_1, \mathbf{u}_2, \ldots, \mathbf{u}_m\}$ be two bases for a finite-dimensional vector space V. Since $\mathscr{U}$ is a basis and is linearly independent, from Theorem 3-7.1 we conclude that $m < n$. Similarly, since $\mathscr{U}$ is a basis and $\mathscr{B}$ is linearly independent, again we conclude that $n < m$. Therefore $n = m$.

THEOREM 3-7.3 Let V be a vector space of dimension n. Then any set of n linearly independent vectors in V is a basis for V.

EXAMPLE 3-38: Verify Theorem 3-7.3.

Solution: Suppose that $S = \{\mathbf{v}_1, \mathbf{v}_2, \ldots, \mathbf{v}_n\}$ is a set of linearly independent vectors in V. Let $\mathbf{v}$ be any other vector in V. Then it follows from Theorem 3-7.1 that $\{\mathbf{v}, \mathbf{v}_1, \mathbf{v}_2, \ldots, \mathbf{v}_n\}$ are linearly dependent. Thus there exist scalars $\alpha, \alpha_1, \alpha_2, \ldots, \alpha_n$, not all zero, such that

$$\alpha \mathbf{v} + \alpha_1 \mathbf{v}_1 + \alpha_2 \mathbf{v}_2 + \cdots + \alpha_n \mathbf{v}_n = \mathbf{0} \tag{3.53}$$

Now $\alpha \neq 0$. For if $\alpha = 0$, then (3.53) would imply that $\mathbf{v}_1, \mathbf{v}_2, \ldots, \mathbf{v}_n$ are linearly dependent, which contradicts the assumption that they are linearly independent. Thus (3.53) can be solved for $\mathbf{v}$:

$$\mathbf{v} = \left(-\frac{\alpha_1}{\alpha}\right)\mathbf{v}_1 + \left(-\frac{\alpha_2}{\alpha}\right)\mathbf{v}_2 + \cdots + \left(-\frac{\alpha_n}{\alpha}\right)\mathbf{v}_n$$

Since $\mathbf{v}$ is an arbitrary vector in V, it follows that $S = \{\mathbf{v}_1, \mathbf{v}_2, \ldots, \mathbf{v}_n\}$ spans V. Thus S satisfies Conditions (B1) and (B2). Therefore S is a basis for V.

EXAMPLE 3-39: Show that

$$\mathbf{v}_1 = \begin{bmatrix} 1 \\ 1 \\ 1 \end{bmatrix}, \quad \mathbf{v}_2 = \begin{bmatrix} 2 \\ 1 \\ 3 \end{bmatrix}, \quad \mathbf{v}_3 = \begin{bmatrix} 3 \\ -1 \\ 4 \end{bmatrix}, \quad \mathbf{v}_4 = \begin{bmatrix} 1 \\ 0 \\ 1 \end{bmatrix}$$

are linearly dependent.

Solution: Since $\dim R^3 = 3$, from Theorem 3-7.1 it follows that $\{\mathbf{v}_1, \mathbf{v}_2, \mathbf{v}_3, \mathbf{v}_4\}$ are linearly dependent.

EXAMPLE 3-40: Show that $\mathbf{v}_1 = \begin{bmatrix} 1 \\ 0 \end{bmatrix}$ and $\mathbf{v}_2 = \begin{bmatrix} 1 \\ 1 \end{bmatrix}$ in R^2 form a basis for R^2.

Solution: Since $\dim R^2 = 2$, from Theorem 3-7.3 we need show only $\mathbf{v}_1$, $\mathbf{v}_2$ are linearly independent. That follows since

$$\det[\mathbf{v}_1 \mathbf{v}_2] = \begin{vmatrix} 1 & 1 \\ 0 & 1 \end{vmatrix} = 1 \neq 0$$

Thus $\{\mathbf{v}_1, \mathbf{v}_2\}$ forms a basis for R^2.

EXAMPLE 3-41: Show that $\{1, 1 + x, 1 + x + x^2\}$ forms a basis for P_2.

Solution: Since $\dim P_2 = 3$ by (3.48), we need show only that 1, $1 + x$, and $1 + x + x^2$ are linearly independent. This was done in Example 3-27. Thus $\{1, 1 + x, 1 + x + x^2\}$ forms a basis for P_2.

3-8. Coordinate Vectors and Change of Basis

A. Coordinate vectors

DEFINITION Let V be a vector space of dimension n and $\mathscr{B} = \{\mathbf{v}_1, \mathbf{v}_2, \ldots, \mathbf{v}_n\}$ be an **ordered basis** of V. (By ordered basis we mean that the order of $\mathbf{v}_1, \mathbf{v}_2, \ldots, \mathbf{v}_n$ in $\mathscr{B}$ is fixed.) If $\mathbf{v} \in V$, and $\mathbf{v}$ is written as

$$\mathbf{v} = x_1 \mathbf{v}_1 + x_2 \mathbf{v}_2 + \cdots + x_n \mathbf{v}_n \tag{3.54}$$

then we call the ordered n-tuple of numbers $(x_1, x_2, \ldots, x_n)$ the *coordinates* of $\mathbf{v}$ with respect to the basis $\mathscr{B}$ and we call x_i the *ith coordinate.*

Note that the coordinates of any vector in R^n with respect to the standard basis $\{\mathbf{e}_1, \mathbf{e}_2, \ldots, \mathbf{e}_n\}$ are simply the coordinates as defined in (3.20). We say that

$$[\mathbf{v}]_{\mathscr{B}} = \begin{bmatrix} x_1 \\ x_2 \\ \vdots \\ x_n \end{bmatrix}_{\mathscr{B}} \tag{3.55}$$

is the **coordinate vector** of $\mathbf{v}$ with respect to the basis $\mathscr{B} = \{\mathbf{v}_1, \mathbf{v}_2, \ldots, \mathbf{v}_n\}$, where it is understood that the ith component of this vector is the coefficient of $\mathbf{v}_i$ in (3.54). Thus with each element $\mathbf{v}$ of V of dimension n we associate a unique n-vector. Conversely, with a given n-vector

$$\begin{bmatrix} y_1 \\ y_2 \\ \vdots \\ y_n \end{bmatrix}$$

we can associate a unique element

$$y_1\mathbf{v}_1 + y_2\mathbf{v}_2 + \cdots + y_n\mathbf{v}_n$$

of V. Hence every vector space of dimension n can be identified with R^n. Observe also that if

$$\mathbf{v} = x_1\mathbf{v}_1 + x_2\mathbf{v}_2 + \cdots + x_n\mathbf{v}_n \text{ corresponds to } \begin{bmatrix} x_1 \\ x_2 \\ \vdots \\ x_n \end{bmatrix}_{\mathscr{B}}$$

and

$$\mathbf{w} = y_1\mathbf{v}_1 + y_2\mathbf{v}_2 + \cdots + y_n\mathbf{v}_n \text{ corresponds to } \begin{bmatrix} y_1 \\ y_2 \\ \vdots \\ y_n \end{bmatrix}_{\mathscr{B}}$$

then

$$\mathbf{v} + \mathbf{w} = (x_1 + y_1)\mathbf{v}_1 + (x_2 + y_2)\mathbf{v}_2 + \cdots + (x_n + y_n)\mathbf{v}_n \text{ corresponds to } \begin{bmatrix} x_1 + y_1 \\ x_2 + y_2 \\ \vdots \\ x_n + y_n \end{bmatrix}_{\mathscr{B}}$$

and

$$\alpha\mathbf{v} = (\alpha x_1)\mathbf{v}_1 + (\alpha x_2)\mathbf{v}_2 + \cdots + (\alpha x_n)\mathbf{v}_n \text{ corresponds to } \begin{bmatrix} \alpha x_1 \\ \alpha x_2 \\ \vdots \\ \alpha x_n \end{bmatrix}_{\mathscr{B}}$$

for any scalar α. That is,

$$[\mathbf{v} + \mathbf{w}]_{\mathscr{B}} = [\mathbf{v}]_{\mathscr{B}} + [\mathbf{w}]_{\mathscr{B}} \tag{3.56}$$

$$[\alpha\mathbf{v}]_{\mathscr{B}} = \alpha[\mathbf{v}]_{\mathscr{B}} \tag{3.57}$$

Thus the above one-to-one correspondence between V and R^n preserves the vector space operations of vector addition and scalar multiplication. We then say that V and R^n are *isomorphic*.

THEOREM 3-8.1 Let $\mathscr{B} = \{\mathbf{v}_1, \mathbf{v}_2, \ldots, \mathbf{v}_n\}$ be a basis for a vector space V and let $\mathbf{v}$ be any element of V. Then there are unique scalars $\alpha_1, \alpha_2, \ldots, \alpha_n$ such that

$$\mathbf{v} = \alpha_1\mathbf{v}_1 + \alpha_2\mathbf{v}_2 + \cdots + \alpha_n\mathbf{v}_n \tag{3.58}$$

EXAMPLE 3-42: Verify Theorem 3-8.1.

Solution: Suppose that $\mathbf{v} \in V$ can be written as

$$\mathbf{v} = \alpha_1\mathbf{v}_1 + \alpha_2\mathbf{v}_2 + \cdots + \alpha_n\mathbf{v}_n$$

and as

$$\mathbf{v} = \beta_1\mathbf{v}_1 + \beta_2\mathbf{v}_2 + \cdots + \beta_n\mathbf{v}_n$$

Then

$$\begin{aligned}\mathbf{v} - \mathbf{v} = \mathbf{0} &= (\alpha_1\mathbf{v}_1 + \alpha_2\mathbf{v}_2 + \cdots + \alpha_n\mathbf{v}_n) - (\beta_1\mathbf{v}_1 + \beta_2\mathbf{v}_2 + \cdots + \beta_n\mathbf{v}_n)\\ &= (\alpha_1 - \beta_1)\mathbf{v}_1 + (\alpha_2 - \beta_2)\mathbf{v}_2 + \cdots + (\alpha_n - \beta_n)\mathbf{v}_n\end{aligned}$$

Since $\mathscr{B}$ is a basis for V, $\mathscr{B}$ is linearly independent. Then by definition we must have

$$\alpha_1 - \beta_1 = 0, \alpha_2 - \beta_2 = 0, \ldots, \alpha_n - \beta_n = 0$$

so that

$$\alpha_1 = \beta_1, \alpha_2 = \beta_2, \ldots, \alpha_n = \beta_n$$

This completes the proof.

EXAMPLE 3-43: Find the coordinates of $\begin{bmatrix}4\\3\end{bmatrix}$ in R^2 with respect to the basis $\{\mathbf{v}_1, \mathbf{v}_2\}$ where

$$\mathbf{v}_1 = \begin{bmatrix}1\\0\end{bmatrix} \quad \text{and} \quad \mathbf{v}_2 = \begin{bmatrix}1\\1\end{bmatrix}$$

Solution: We have shown in Example 3-40 that $\{\mathbf{v}_1, \mathbf{v}_2\}$ forms a basis for R^2. Also, in Example 3-20, we have expressed

$$\begin{bmatrix}4\\3\end{bmatrix} = \alpha_1\mathbf{v}_1 + \alpha_2\mathbf{v}_2 = 1\begin{bmatrix}1\\0\end{bmatrix} + 3\begin{bmatrix}1\\1\end{bmatrix}$$

Hence the coordinates of $\begin{bmatrix}4\\3\end{bmatrix}$ with respect to the basis $\left\{\begin{bmatrix}1\\0\end{bmatrix}, \begin{bmatrix}1\\1\end{bmatrix}\right\}$ are $(1, 3)$.

EXAMPLE 3-44: Consider a polynomial $p(x) = 1 + 2x + 3x^2$ in P_2.

(a) Find the coordinate vector of $p(x)$ with respect to the standard basis $\{1, x, x^2\}$ for P_2.
(b) Find the coordinate vector of $p(x)$ with respect to the basis $\{1, 1 + x, 1 + x + x^2\}$ for P_2.

Solution

(a) We must find numbers a, b, and c such that

$$p(x) = a(1) + bx + cx^2$$

or

$$1 + 2x + 3x^2 = a + bx + cx^2$$

It is obvious that $a = 1$, $b = 2$, and $c = 3$. Thus the coordinate vector of $1 + 2x + 3x^2$ with respect to the standard basis $\{1, x, x^2\}$ is $\begin{bmatrix}1\\2\\3\end{bmatrix}$.

(b) We have shown in Example 3-41 that $\{1, 1 + x, 1 + x + x^2\}$ forms a basis for P_2. Thus in order to find the coordinates of $p(x)$ with respect to this basis we must find numbers a, b, and c such that

$$p(x) = a(1) + b(1 + x) + c(1 + x + x^2)$$

or

$$1 + 2x + 3x^2 = (a + b + c) + (b + c)x + cx^2$$

Equating the corresponding coefficients, we get

$$a + b + c = 1$$
$$b + c = 2$$
$$c = 3$$

Solving for a, b, and c, we obtain $a = -1, b = -1, c = 3$. Thus the coordinate vector of $1 + 2x + 3x$ with respect to the basis $\{1, 1 + x, 1 + x + x^2\}$ is $\begin{bmatrix} -1 \\ -1 \\ 3 \end{bmatrix}$.

B. Change of basis: The transition matrix

Let $\mathscr{B} = \{\mathbf{v}_1, \mathbf{v}_2, \ldots, \mathbf{v}_n\}$ and $\mathscr{B}' = \{\mathbf{u}_1, \mathbf{u}_2, \ldots, \mathbf{u}_n\}$ be two ordered bases for a vector space V. Let $\mathbf{v} \in V$ and $[\mathbf{v}]_{\mathscr{B}}$ and $[\mathbf{v}]_{\mathscr{B}'}$ be the coordinate vectors of $\mathbf{v}$ with respect to the bases $\mathscr{B}$ and $\mathscr{B}'$, respectively. Let P be the $n \times n$ matrix having $[\mathbf{u}_1]_{\mathscr{B}}, [\mathbf{u}_2]_{\mathscr{B}}, \ldots, [\mathbf{u}_n]_{\mathscr{B}}$ as its columns; that is,

$$P = [[\mathbf{u}_1]_{\mathscr{B}} \vdots [\mathbf{u}_2]_{\mathscr{B}} \vdots \cdots \vdots [\mathbf{u}_n]_{\mathscr{B}}] \tag{3.59}$$

Then

$$[\mathbf{v}]_{\mathscr{B}} = P[\mathbf{v}]_{\mathscr{B}'} \tag{3.60}$$

The matrix P is called the **transition matrix** from $\mathscr{B}'$ to $\mathscr{B}$.

EXAMPLE 3-45: Verify (3.60).

Solution: Let $\mathbf{v} \in V$. Since $\mathscr{B}'$ is a basis for V, $\mathbf{v}$ can be expressed as

$$\mathbf{v} = y_1\mathbf{u}_1 + y_2\mathbf{u}_2 + \cdots + y_n\mathbf{u}_n$$

so that

$$[\mathbf{v}]_{\mathscr{B}'} = \begin{bmatrix} y_1 \\ y_2 \\ \vdots \\ y_n \end{bmatrix} \tag{3.61}$$

Since $\mathscr{B}$ is a basis for V, each vector of $\mathscr{B}'$ can be expressed as

$$\begin{aligned} \mathbf{u}_1 &= p_{11}\mathbf{v}_1 + p_{21}\mathbf{v}_2 + \cdots + p_{n1}\mathbf{v}_n \\ \mathbf{u}_2 &= p_{12}\mathbf{v}_1 + p_{22}\mathbf{v}_2 + \cdots + p_{n2}\mathbf{v}_n \\ &\vdots \\ \mathbf{u}_n &= p_{1n}\mathbf{v}_1 + p_{2n}\mathbf{v}_2 + \cdots + p_{nn}\mathbf{v}_n \end{aligned} \tag{3.62}$$

so that

$$[\mathbf{u}_1]_{\mathscr{B}} = \begin{bmatrix} p_{11} \\ p_{21} \\ \vdots \\ p_{n1} \end{bmatrix}, [\mathbf{u}_2]_{\mathscr{B}} = \begin{bmatrix} p_{12} \\ p_{22} \\ \vdots \\ p_{2n} \end{bmatrix}, \ldots, [\mathbf{u}_n]_{\mathscr{B}} = \begin{bmatrix} p_{1n} \\ p_{2n} \\ \vdots \\ p_{nn} \end{bmatrix} \tag{3.63}$$

and

$$P = [[\mathbf{u}_1]_{\mathscr{B}} \vdots [\mathbf{u}_2]_{\mathscr{B}} \vdots \cdots \vdots [\mathbf{u}_n]_{\mathscr{B}}]$$

$$= \begin{bmatrix} p_{11} & p_{12} & \cdots & p_{1n} \\ p_{21} & p_{22} & \cdots & p_{2n} \\ \vdots & \vdots & \ddots & \vdots \\ p_{n1} & p_{n2} & \cdots & p_{nn} \end{bmatrix} \tag{3.64}$$

Now using (3.62) we get

$$\begin{aligned} \mathbf{v} &= y_1\mathbf{u}_1 + y_2\mathbf{u}_2 + \cdots + y_n\mathbf{u}_n \qquad (3.65) \\ &= y_1(p_{11}\mathbf{v}_1 + p_{21}\mathbf{v}_2 + \cdots + p_{n1}\mathbf{v}_n) \\ &\quad + y_2(p_{12}\mathbf{v}_1 + p_{22}\mathbf{v}_2 + \cdots + p_{n2}\mathbf{v}_n) + \cdots + y_n(p_{1n}\mathbf{v}_1 + p_{2n}\mathbf{v}_2 + \cdots + p_{nn}\mathbf{v}_n) \\ &= (p_{11}y_1 + p_{12}y_2 + \cdots + p_{1n}y_n)\mathbf{v}_1 \\ &\quad + (p_{21}y_1 + p_{22}y_2 + \cdots + p_{2n}y_n)\mathbf{v}_2 + \cdots + (p_{n1}y_1 + p_{n2}y_2 + \cdots + p_{nn}y_n)\mathbf{v}_n \\ &= x_1\mathbf{v}_1 + x_2\mathbf{v}_2 + \cdots + x_n\mathbf{v}_n \end{aligned}$$

where

$$\begin{bmatrix} x_1 \\ x_2 \\ \vdots \\ x_n \end{bmatrix} = [\mathbf{v}]_{\mathscr{B}} \tag{3.66}$$

Thus we have

$$\begin{bmatrix} x_1 \\ x_2 \\ \vdots \\ x_n \end{bmatrix} = \begin{bmatrix} p_{11}y_1 + p_{12}y_2 + \cdots + p_{1n}y_n \\ p_{21}y_1 + p_{22}y_2 + \cdots + p_{2n}y_n \\ \vdots \\ p_{n1}y_1 + p_{n2}y_2 + \cdots + p_{nn}y_n \end{bmatrix} \tag{3.67}$$

$$= \begin{bmatrix} p_{11} & p_{12} & \cdots & p_{1n} \\ p_{21} & p_{22} & \cdots & p_{2n} \\ \vdots & \vdots & \ddots & \vdots \\ p_{n1} & p_{n2} & \cdots & p_{nn} \end{bmatrix} \begin{bmatrix} y_1 \\ y_2 \\ \vdots \\ y_n \end{bmatrix}$$

or

$$[\mathbf{v}]_{\mathscr{B}} = P[\mathbf{v}]_{\mathscr{B}'}$$

which is relation (3.60).

EXAMPLE 3-46: Consider P_2 and let $\mathscr{B} = \{1, x, x^2\}$ and $\mathscr{B}' = \{1, 1 + x, 1 + x + x^2\}$. Find the transition matrix P from $\mathscr{B}'$ to $\mathscr{B}$, and verify the relation (3.60) for the result of Example 3-44.

Solution: We must find the coordinate vectors of each vector in $\mathscr{B}'$ with respect to $\mathscr{B}$. By inspection, we have

$$[1]_{\mathscr{B}} = \begin{bmatrix} 1 \\ 0 \\ 0 \end{bmatrix}, \qquad [1 + x]_{\mathscr{B}} = \begin{bmatrix} 1 \\ 1 \\ 0 \end{bmatrix}, \qquad [1 + x + x^2]_{\mathscr{B}} = \begin{bmatrix} 1 \\ 1 \\ 1 \end{bmatrix}$$

Thus by (3.64) we get

$$P = \begin{bmatrix} 1 & 1 & 1 \\ 0 & 1 & 1 \\ 0 & 0 & 1 \end{bmatrix}$$

Next, for Example 3-44, we have

$$\mathbf{p}(x) = p(x) = 1 + 2x + 3x^2$$

and

$$[\mathbf{p}]_{\mathscr{B}} = \begin{bmatrix} 1 \\ 2 \\ 3 \end{bmatrix}, \qquad [\mathbf{p}]_{\mathscr{B}'} = \begin{bmatrix} -1 \\ -1 \\ 3 \end{bmatrix}$$

Now

$$P[\mathbf{p}]_{\mathscr{B}'} = \begin{bmatrix} 1 & 1 & 1 \\ 0 & 1 & 1 \\ 0 & 0 & 1 \end{bmatrix} \begin{bmatrix} -1 \\ -1 \\ 3 \end{bmatrix} = \begin{bmatrix} 1 \\ 2 \\ 3 \end{bmatrix} = [\mathbf{p}]_{\mathscr{B}}$$

Thus relation (3.60) is verified for this example.

THEOREM 3-8.2 Let $\mathscr{B}$ and $\mathscr{B}'$ be two ordered bases for a finite-dimensional vector space V. If P is the transition matrix from $\mathscr{B}'$ to $\mathscr{B}$, then

(a) P is nonsingular

(b) P^{-1} is the transition matrix from $\mathscr{B}$ to $\mathscr{B}'$, that is, if $\mathbf{v} \in V$, then

$$[\mathbf{v}]_{\mathscr{B}} = P[\mathbf{v}]_{\mathscr{B}'} \tag{3.68}$$

$$[\mathbf{v}]_{\mathscr{B}'} = P^{-1}[\mathbf{v}]_{\mathscr{B}} \tag{3.69}$$

EXAMPLE 3-47: Verify Theorem 3-8.2.

Solution: First we prove that P is nonsingular.

Let $\mathscr{B} = \{\mathbf{v}_1, \mathbf{v}_2, \ldots, \mathbf{v}_n\}$ and $\mathscr{B}' = \{\mathbf{u}_1, \mathbf{u}_2, \ldots, \mathbf{u}_n\}$ be two ordered bases for V. Let $\mathbf{v} \in V$. Then there exist $x_1, x_2, \ldots, x_n$, such that

$$\mathbf{v} = x_1\mathbf{u}_1 + x_2\mathbf{u}_2 + \cdots + x_n\mathbf{u}_n$$

and

$$\mathbf{x} = \begin{bmatrix} x_1 \\ x_2 \\ \vdots \\ x_n \end{bmatrix} = [\mathbf{v}]_{\mathscr{B}'}$$

Now let

$$\mathbf{b} = \begin{bmatrix} b_1 \\ b_2 \\ \vdots \\ b_n \end{bmatrix}$$

be an n-vector and let

$$\mathbf{v} = b_1\mathbf{v}_1 + b_2\mathbf{v}_2 + \cdots + b_n\mathbf{v}_n$$

Then, by (3.60),

$$\mathbf{b} = [\mathbf{v}]_{\mathscr{B}} = P[\mathbf{v}]_{\mathscr{B}'}$$

so that $P\mathbf{x} = \mathbf{b}$ has a solution for every n-vector $\mathbf{b}$. Therefore, by Theorem 1-4.3, P is nonsingular.

Next, let Q be the transition matrix from $\mathscr{B}$ to $\mathscr{B}'$. From (3.60),

$$[\mathbf{v}]_{\mathscr{B}} = P[\mathbf{v}]_{\mathscr{B}'}$$

and

$$[\mathbf{v}]_{\mathscr{B}'} = Q[\mathbf{v}]_{\mathscr{B}}$$

Thus, we have

$$[\mathbf{v}]_{\mathscr{B}} = PQ[\mathbf{v}]_{\mathscr{B}}$$

It follows that $PQ = I_n$, so that $Q = P^{-1}$.

Note that, according to (3.59), Q is given by

$$Q = [[\mathbf{v}_1]_{\mathscr{B}'} \vdots [\mathbf{v}_2]_{\mathscr{B}'} \vdots \cdots \vdots [\mathbf{v}_n]_{\mathscr{B}'}] \tag{3.70}$$

EXAMPLE 3-48: Find the transition matrix Q from $\mathscr{B}'$ to $\mathscr{B}$ of Example 3-46 according to (3.70) and verify that $Q = P^{-1}$.

Solution: We must find the coordinate vectors of each vector in $\mathscr{B}$ with respect to $\mathscr{B}'$. Let

$$\begin{aligned} 1 &= a(1) + b(1 + x) + c(1 + x + x^2) \\ &= (a + b + c) + (b + c)x + cx^2 \end{aligned}$$

Equating the corresponding coefficients, we get

$$a = 1, \qquad b = 0, \qquad c = 0$$

Similarly, let

$$\begin{aligned} x &= a(1) + b(1 + x) + c(1 + x + x^2) \\ &= (a + b + c) + (b + c)x + cx^2 \end{aligned}$$

Equating the corresponding coefficients, we get

$$a = -1, \qquad b = 1, \qquad c = 0$$

Again, let

$$\begin{aligned} x^2 &= a(1) + b(1 + x) + c(1 + x + x^2) \\ &= (a + b + c) + (b + c)x + cx^2 \end{aligned}$$

Equating the corresponding coefficients, we get

$$a = 0, \qquad b = -1, \qquad c = 1$$

Thus, we obtain

$$[1]_{\mathscr{B}'} = \begin{bmatrix} 1 \\ 0 \\ 0 \end{bmatrix}, \qquad [x]_{\mathscr{B}'} = \begin{bmatrix} -1 \\ 1 \\ 0 \end{bmatrix}, \qquad [x^2]_{\mathscr{B}'} = \begin{bmatrix} 0 \\ -1 \\ 1 \end{bmatrix}$$

and by (3.70) we get

$$Q = \begin{bmatrix} 1 & -1 & 0 \\ 0 & 1 & -1 \\ 0 & 0 & 1 \end{bmatrix}$$

Now from Example 3-46 we have

$$P = \begin{bmatrix} 1 & 1 & 1 \\ 0 & 1 & 1 \\ 0 & 0 & 1 \end{bmatrix}$$

and

$$PQ = \begin{bmatrix} 1 & 1 & 1 \\ 0 & 1 & 1 \\ 0 & 0 & 1 \end{bmatrix} \begin{bmatrix} 1 & -1 & 0 \\ 0 & 1 & -1 \\ 0 & 0 & 1 \end{bmatrix} = \begin{bmatrix} 1 & 0 & 0 \\ 0 & 1 & 0 \\ 0 & 0 & 1 \end{bmatrix} = I_3$$

Thus

$$Q = P^{-1}$$

3-9. Column Space and Row Space of a Matrix

A. Column space and row space

DEFINITIONS Consider an $m \times n$ matrix

$$A = \begin{bmatrix} a_{11} & a_{12} & \cdots & a_{1n} \\ a_{21} & a_{22} & \cdots & a_{2n} \\ \vdots & \vdots & \ddots & \vdots \\ a_{m1} & a_{m2} & \cdots & a_{mn} \end{bmatrix}$$

- The vectors

$$\mathbf{c}_1 = \begin{bmatrix} a_{11} \\ a_{21} \\ \vdots \\ a_{m1} \end{bmatrix}, \mathbf{c}_2 = \begin{bmatrix} a_{12} \\ a_{22} \\ \vdots \\ a_{m2} \end{bmatrix}, \ldots, \mathbf{c}_n = \begin{bmatrix} a_{1n} \\ a_{2n} \\ \vdots \\ a_{mn} \end{bmatrix} \tag{3.71}$$

  formed from the columns of A are called the **column vectors** of A, and $\mathbf{c}_i$ can be considered as a vector in R^m.
- The vectors

$$\begin{aligned} \mathbf{r}_1 &= [a_{11}, a_{12}, \ldots, a_{1n}] \\ \mathbf{r}_2 &= [a_{21}, a_{22}, \ldots, a_{2n}] \\ &\vdots \\ \mathbf{r}_m &= [a_{m1}, a_{m2}, \ldots, a_{mn}] \end{aligned} \tag{3.72}$$

  formed from the rows of A are called the **row vectors** of A, and $\mathbf{r}_j$ can be considered as a vector in $R_{1 \times n}$.
- The subspace of R^m spanned by the column vectors of A is called the **column space** of A. The dimension of the column space of A is called the **column rank** of A.
- The subspace of $R_{1 \times n}$ spanned by the row vectors of A is called the **row space** of A. The dimension of the row space of A is called the **row rank** of A.

EXAMPLE 3-49: Find the column space and column rank of the matrix

$$A = \begin{bmatrix} 1 & 0 & 0 \\ 0 & 1 & 0 \end{bmatrix}$$

Solution: The column space of A is the set of all vectors of the form

$$\alpha_1 \begin{bmatrix} 1 \\ 0 \end{bmatrix} + \alpha_2 \begin{bmatrix} 0 \\ 1 \end{bmatrix} + \alpha_3 \begin{bmatrix} 0 \\ 0 \end{bmatrix} = \begin{bmatrix} \alpha_1 \\ \alpha_2 \end{bmatrix}$$

Thus the column space of A is R^2 and the column rank of A is 2 since $\dim R^2 = 2$.

EXAMPLE 3-50: Find the row space and row rank of A of Example 3-49.

Solution: The row space of A is the set of all 1×3 matrices of the form

$$\alpha_1[1, 0, 0] + \alpha_2[0, 1, 0] = [\alpha_1, \alpha_2, 0]$$

Now, $\mathbf{r}_1 = [1,0,0]$ and $\mathbf{r}_2 = [0,1,0]$ are linearly independent, since neither of these two vectors is a scalar multiple of the other (see Problem 3-36). Thus $\{\mathbf{r}_1, \mathbf{r}_2\}$ forms a basis for the row space, and its dimension is 2. Therefore the row rank of A is also 2.

B. Row echelon form

DEFINITION A matrix is said to be in **row echelon form** if:

- the first nonzero entry in each row is 1.
- row i does not consist entirely of zeros, such that the number of leading zero entries in the row $(i+1)$ is greater than the number of leading zero entries in row i.
- there are rows whose entries are all zero, such that these rows are below the rows having nonzero entries.

A matrix can be reduced to row echelon form by the following elementary row operations:

- Multiply a row by a nonzero real number.
- Add a multiple of one row to another row.
- Interchange two rows.

If a matrix B can be obtained from a matrix A by performing a finite sequence of elementary row operations, then B is said to be **row equivalent** to A.

EXAMPLE 3-51: Let

$$A = \begin{bmatrix} 1 & 1 & 1 & 0 \\ 3 & 5 & 1 & 2 \\ 0 & 4 & -4 & 4 \end{bmatrix}$$

Reduce A to row echelon form.

Solution

$$A = \begin{bmatrix} 1 & 1 & 1 & 0 \\ 3 & 5 & 1 & 2 \\ 0 & 4 & -4 & 4 \end{bmatrix}$$

$$\downarrow \qquad \text{(row 2)} - 3\text{(row 1)}$$

$$\begin{bmatrix} 1 & 1 & 1 & 0 \\ 0 & 2 & -2 & 2 \\ 0 & 4 & -4 & 4 \end{bmatrix}$$

$$\downarrow \qquad \tfrac{1}{2}\text{(row 2)}$$

$$\begin{bmatrix} 1 & 1 & 1 & 0 \\ 0 & 1 & -1 & 1 \\ 0 & 4 & -4 & 4 \end{bmatrix}$$

$$\downarrow \qquad \text{(row 3)} - 4\text{(row 2)}$$

$$\begin{bmatrix} 1 & 1 & 1 & 0 \\ 0 & 1 & -1 & 1 \\ 0 & 0 & 0 & 0 \end{bmatrix}$$

THEOREM 3-9.1 Let A be an $m \times n$ matrix that is row equivalent to a matrix B that is in row echelon form. Then:

(1) A and B have the same row space.
(2) The nonzero rows of B form a basis for the row space of A.

EXAMPLE 3-52: Verify Theorem 3-9.1.

Solution

(1) Suppose that the row vectors of a matrix A are $\mathbf{r}_1, \mathbf{r}_2, \ldots, \mathbf{r}_m$, and let A' be obtained from A by performing an elementary row operation. If the row operation is a row interchange, then A' and A have the same row vectors, and hence the same row space. If the row operation is multiplication of a row by a scalar or addition of a multiple of one row to another, then the row vectors $\mathbf{r}'_1, \mathbf{r}'_2, \ldots, \mathbf{r}'_m$ of A' are linear combinations of $\mathbf{r}_1, \mathbf{r}_2, \ldots, \mathbf{r}_m$. Since a vector space is closed under addition and scalar multiplication, all linear combinations of $\mathbf{r}_1, \mathbf{r}_2, \ldots, \mathbf{r}_m$ will lie in the row space of A. Hence, the row space of A' is the same as the row space of A. Since B is obtained from A by repeated application of elementary row operations, we conclude that the row spaces of A and B are identical.

(2) The very nature of row echelon form ensures that the nonzero rows of B are linearly independent. Therefore, the nonzero rows of B form a basis for the row space of B and, hence, of the row space of A.

Theorem 3-9.1 offers an easy means to find a basis for the row space of a matrix and its rank.

EXAMPLE 3-53: Find a basis for the row space of

$$A = \begin{bmatrix} 1 & 1 & 1 & 0 \\ 3 & 5 & 1 & 2 \\ 0 & 4 & -4 & 4 \end{bmatrix}$$

and its row rank.

Solution: From Example 3-51, the reduced echelon form of A is given by

$$B = \begin{bmatrix} 1 & 1 & 1 & 0 \\ 0 & 1 & -1 & 1 \\ 0 & 0 & 0 & 0 \end{bmatrix}$$

Thus $\mathbf{r}_1 = [1, 1, 1, 0]$ and $\mathbf{r}_2 = [0, 1, -1, 1]$ form a basis for the row space of B and also for the row space of A; hence the row rank of A is 2.

C. Equivalence of column space of A and row space of A

From the definition of the transpose of a matrix it is clear that the column space of a matrix is the same as the row space of its transpose. Thus

$$\text{column space of } A = \text{row space of } A^T \qquad \textbf{(3.73)}$$

Hence we can find a basis for the column space of a matrix A by finding a basis for the row space of A^T and transposing back to column form.

EXAMPLE 3-54: Find a basis for the column space of

$$A = \begin{bmatrix} 1 & 1 & 1 & 0 \\ 3 & 5 & 1 & 2 \\ 0 & 4 & -4 & 4 \end{bmatrix}$$

and its column rank.

Solution: Transposing A and reducing to row echelon form, we get

$$A^T = \begin{bmatrix} 1 & 3 & 0 \\ 1 & 5 & 4 \\ 1 & 1 & -4 \\ 0 & 2 & 4 \end{bmatrix}$$

$$\downarrow \qquad \begin{array}{l} (\text{row } 2) - (\text{row } 1) \\ (\text{row } 3) - (\text{row } 1) \end{array}$$

$$\begin{bmatrix} 1 & 3 & 0 \\ 0 & 2 & 4 \\ 0 & -2 & -4 \\ 0 & 2 & 4 \end{bmatrix}$$

$$\downarrow \qquad \begin{array}{l} (\text{row } 3) + (\text{row } 2) \\ (\text{row } 4) - (\text{row } 2) \\ \tfrac{1}{2}(\text{row } 2) \end{array}$$

$$\begin{bmatrix} 1 & 3 & 0 \\ 0 & 1 & 2 \\ 0 & 0 & 0 \\ 0 & 0 & 0 \end{bmatrix}$$

Thus the vectors $\mathbf{r}_1 = [1, 3, 0]$ and $\mathbf{r}_2 = [0, 1, 2]$ form a basis for the row space of A^T. Transposing $\mathbf{r}_1$ and $\mathbf{r}_2$, we get

$$\mathbf{c}_1 = \mathbf{r}_1^T = \begin{bmatrix} 1 \\ 3 \\ 0 \end{bmatrix}, \qquad \mathbf{c}_2 = \mathbf{r}_2^T = \begin{bmatrix} 0 \\ 1 \\ 2 \end{bmatrix}$$

as a basis for the column space of A, and again the column rank of A is 2.

D. Rank and nullity of a matrix

THEOREM 3-9.2 If A is any matrix, then

$$\text{row rank of } A = \text{column rank of } A \tag{3.74}$$

note: For the proof of Theorem 3-9.2, see Problem 3-60.

DEFINITION Because of (3.74), we now define the **rank** of a matrix as the dimension of the row space of A.

$$\text{rank of } A = \text{row rank of } A = \text{column rank of } A \tag{3.75}$$

The **nullity** of a matrix A is the dimension of the null space of A, $N(A)$ (see Example 3-19).

EXAMPLE 3-55: Find the rank and nullity of

$$A = \begin{bmatrix} 1 & 1 & 1 & 0 \\ 3 & 5 & 1 & 2 \\ 0 & 4 & -4 & 4 \end{bmatrix}$$

Solution: From Example 3-53 we know that the row rank of A is 2. Thus by definition (3.75), the rank of A is 2.

Next, the null space $N(A)$ of A is the set of all 4-vectors

$$\mathbf{x} = \begin{bmatrix} x_1 \\ x_2 \\ x_3 \\ x_4 \end{bmatrix}$$

such that $A\mathbf{x} = \mathbf{0}$. Reducing A to row echelon form B, this system is equivalent to $B\mathbf{x} = \mathbf{0}$; that is, from Example 3-51,

$$\begin{aligned} x_1 + x_2 + x_3 \quad\quad &= 0 \\ x_2 - x_3 + x_4 &= 0 \end{aligned}$$

The unknown variables corresponding to the first nonzero entries in each row of the reduced row echelon form are called **lead variables**. The remaining variables corresponding to the columns skipped in the reduced echelon form are referred to as **free variables**.

Thus x_1 and x_2 are lead variables and x_3 and x_4 are free variables. Hence we can write

$$\begin{aligned} x_1 &= -x_2 - x_3 = -2x_3 + x_4 \\ x_2 &= \quad\; x_3 - x_4 \end{aligned}$$

If we set $x_3 = \alpha$ and $x_4 = \beta$, then

$$\mathbf{x} = \begin{bmatrix} x_1 \\ x_2 \\ x_3 \\ x_4 \end{bmatrix} = \begin{bmatrix} -2\alpha + \beta \\ \alpha - \beta \\ \alpha \\ \beta \end{bmatrix} = \alpha \begin{bmatrix} -2 \\ 1 \\ 1 \\ 0 \end{bmatrix} + \beta \begin{bmatrix} 1 \\ -1 \\ 0 \\ 1 \end{bmatrix} \in N(A)$$

Thus the null space of A, $N(A)$, is the span of $\{\mathbf{v}_1, \mathbf{v}_2\}$ where

$$\mathbf{v}_1 = \begin{bmatrix} -2 \\ 1 \\ 1 \\ 0 \end{bmatrix} \quad \text{and} \quad \mathbf{v}_2 = \begin{bmatrix} 1 \\ -1 \\ 0 \\ 1 \end{bmatrix}$$

Since $\mathbf{v}_1$ and $\mathbf{v}_2$ are not the scalar multiple of each other, $\mathbf{v}_1$ and $\mathbf{v}_2$ are linearly independent (see Problem 3-36). Thus $\{\mathbf{v}_1, \mathbf{v}_2\}$ forms a basis of $N(A)$. Hence the dimension of $N(A)$ is equal to 2 and the nullity of A is 2.

EXAMPLE 3-56: Let A be an $m \times n$ matrix. Show that the rank of A plus the nullity of A equals the number of columns of A; that is

$$\text{rank} + \text{nullity} = n \tag{3.76}$$

Solution: Let B be the reduced row echelon form of A. Then the system $A\mathbf{x} = \mathbf{0}$ is equivalent to the system $B\mathbf{x} = \mathbf{0}$. If the rank of A is r, then B will have r nonzero rows, and consequently the system $B\mathbf{x} = \mathbf{0}$ will involve r lead variables and $n - r$ free variables. Thus the dimension of the null space of A, $N(A)$, or the nullity of A, is equal to the number of free variables $n - r$ (see Example 3-55). Thus

$$\text{rank} + \text{nullity} = r + (n - r) = n$$

where n is the number of the columns of A.

E. System of linear equations

THEOREM 3-9.3 A system of linear equations

$$A\mathbf{x} = \mathbf{b} \tag{3.77}$$

is consistent if and only if $\mathbf{b}$ is in the column space of A.

EXAMPLE 3-57: Verify Theorem 3-9.3.

Solution: Consider a system of linear equations

$$A\mathbf{x} = \mathbf{b}$$

or

$$\begin{bmatrix} a_{11} & a_{12} & \cdots & a_{1n} \\ a_{21} & a_{22} & \cdots & a_{2n} \\ \vdots & \vdots & & \vdots \\ a_{m1} & a_{m2} & \cdots & a_{mn} \end{bmatrix} \begin{bmatrix} x_1 \\ x_2 \\ \vdots \\ x_n \end{bmatrix} = \begin{bmatrix} b_1 \\ b_2 \\ \vdots \\ b_m \end{bmatrix}$$

or

$$\begin{bmatrix} a_{11}x_1 + a_{12}x_2 + \cdots + a_{1n}x_n \\ a_{21}x_1 + a_{22}x_2 + \cdots + a_{2n}x_n \\ \vdots \\ a_{m1}x_1 + a_{m2}x_2 + \cdots + a_{mn}x_n \end{bmatrix} = \begin{bmatrix} b_1 \\ b_2 \\ \vdots \\ b_m \end{bmatrix}$$

The above equation can be rewritten in the form

$$x_1 \begin{bmatrix} a_{11} \\ a_{21} \\ \vdots \\ a_{m1} \end{bmatrix} + x_2 \begin{bmatrix} a_{12} \\ a_{22} \\ \vdots \\ a_{m2} \end{bmatrix} + \cdots + x_n \begin{bmatrix} a_{1n} \\ a_{2n} \\ \vdots \\ a_{mn} \end{bmatrix} = \begin{bmatrix} b_1 \\ b_2 \\ \vdots \\ b_m \end{bmatrix}$$

or

$$x_1\mathbf{c}_1 + x_2\mathbf{c}_2 + \cdots + x_n\mathbf{c}_n = \mathbf{b} \tag{3.78}$$

where $\mathbf{c}_i$ is the ith column vector of A.

It follows from (3.78) that the system $A\mathbf{x} = \mathbf{b}$ is consistent if and only if $\mathbf{b}$ can be written as a linear combination of the column vectors of A; that is, if $\mathbf{b}$ is in the column space of A.

THEOREM 3-9.4 A homogeneous system $A\mathbf{x} = \mathbf{0}$ will have only the trivial solution $\mathbf{x} = \mathbf{0}$ if and only if the column vectors of A are linearly independent.

Note that Theorem 3-9.4 is equivalent to Theorem 1-4.4.

EXAMPLE 3-58: Verify Theorem 3-9.4.

Solution: Setting $\mathbf{b} = \mathbf{0}$ in (3.78) of Example 3-57, we get

$$x_1\mathbf{c}_1 + x_2\mathbf{c}_2 + \cdots + x_n\mathbf{c}_n = \mathbf{0} \tag{3.79}$$

Thus if the column vectors $\mathbf{c}_1, \mathbf{c}_2, \ldots, \mathbf{c}_n$ of A are linearly independent, then by the definition of linear independence, (3.79) implies that

$$x_1 = x_2 = \cdots = x_n = 0$$

that is, $\mathbf{x} = \mathbf{0}$ is the solution. On the other hand, if $\mathbf{x} = \mathbf{0}$, that is, $x_1 = x_2 = \cdots = x_n = 0$, then (3.79) implies that $\mathbf{c}_1, \mathbf{c}_2, \ldots, \mathbf{c}_n$ are linearly independent. Thus we have shown that $A\mathbf{x} = \mathbf{0}$ will have only the solution $\mathbf{x} = \mathbf{0}$ if and only if the column vectors of A are linearly independent.

note: The above statement is equivalent to the statement that A is nonsingular if the column vectors of A are linearly independent.

SUMMARY

1. A point in 3-space can be represented by a triple of numbers (x_1, x_2, x_3).
2. A geometric vector in 3-space is a quantity characterized by both magnitude and direction; it can be represented by a directed line segment.
3. Addition of geometric vectors may be performed by using the triangle rule or parallelogram law.
4. There is a one-to-one correspondence between a point in 3-space and a geometric vector in 3-space. Thus a vector in 3-space can be defined by a 3×1 matrix as

$$\mathbf{v} = \begin{bmatrix} v_1 \\ v_2 \\ v_3 \end{bmatrix}$$

5. A point in n-space can be represented by an n-tuple of numbers $(x_1, x_2, \ldots, x_n)$.
6. A vector in n-space, an n-vector, can be defined by an $n \times 1$ matrix as

$$\mathbf{v} = \begin{bmatrix} v_1 \\ v_2 \\ \vdots \\ v_n \end{bmatrix}$$

   where v_i is called the ith component of $\mathbf{v}$.
7. A vector space V is a nonempty set on which the operations of addition and scalar multiplication are defined and the following axioms are satisfied:

   Let $\mathbf{u}$, $\mathbf{v}$, $\mathbf{w} \in V$ and α, β be scalars.

   V1. $\mathbf{u} + \mathbf{v} \in V$

   V2. $\mathbf{u} + \mathbf{v} = \mathbf{v} + \mathbf{u}$

   V3. $\mathbf{u} + (\mathbf{v} + \mathbf{w}) = (\mathbf{u} + \mathbf{v}) + \mathbf{w}$

   V4. $\mathbf{0} \in V$ such that $\mathbf{v} + \mathbf{0} = \mathbf{v}$

   V5. $-\mathbf{v} \in V$ such that $\mathbf{v} + (-\mathbf{v}) = \mathbf{0}$

   V6. $\alpha\mathbf{v} \in V$

   V7. $\alpha(\mathbf{u} + \mathbf{v}) = \alpha\mathbf{u} + \alpha\mathbf{v}$

   V8. $(\alpha + \beta)\mathbf{v} = \alpha\mathbf{v} + \beta\mathbf{v}$

   V9. $(\alpha\beta)\mathbf{v} = \alpha(\beta\mathbf{v})$

   V10. $1\mathbf{v} = \mathbf{v}$

8. A subspace U of V is a nonempty subset of V such that U itself is a vector space.
9. A nonempty subset U of V is a subspace of V if and only if it is closed under both addition and scalar multiplication.
10. The set $\{\mathbf{v}_1, \mathbf{v}_2, \ldots, \mathbf{v}_n\}$, where $\mathbf{v}_i \in V$, is called a spanning set for V if any vector $\mathbf{v} \in V$ can be expressed as a linear combination of $\mathbf{v}_i$ as

$$\mathbf{v} = \alpha_1\mathbf{v}_1 + \alpha_2\mathbf{v}_2 + \cdots + \alpha_n\mathbf{v}_n$$

11. The set $\{\mathbf{v}_1, \mathbf{v}_2, \ldots, \mathbf{v}_n\}$ is linearly independent if

$$\alpha_1\mathbf{v}_1 + \alpha_2\mathbf{v}_2 + \cdots + \alpha_n\mathbf{v}_n = \mathbf{0}$$

   implies

$$\alpha_1 = \alpha_2 = \cdots = \alpha_n = 0$$

12. The set $\{\mathbf{v}_1, \mathbf{v}_2, \ldots, \mathbf{v}_n\}$, where $\mathbf{v}_i \in V$, forms a basis for V if it is linearly independent and spans V.
13. Standard basis for R^n, Cartesian n-space, is $\{\mathbf{e}_1, \mathbf{e}_2, \ldots, \mathbf{e}_n\}$ where

$$\mathbf{e}_1 = \begin{bmatrix} 1 \\ 0 \\ 0 \\ \vdots \\ 0 \end{bmatrix}, \mathbf{e}_2 = \begin{bmatrix} 0 \\ 1 \\ 0 \\ \vdots \\ 0 \end{bmatrix}, \ldots \mathbf{e}_n = \begin{bmatrix} 0 \\ 0 \\ \vdots \\ 0 \\ 1 \end{bmatrix}$$

14. Standard basis for P_n of all polynomials of degree less than or equal to n is $\{1, x, x^2, \ldots, x^n\}$.
15. The dimension of V, dim V, is the number of vectors in a basis for V.
16. If dim $V = n$, then any set of n linearly independent vectors in V is a basis for V.
17. The coordinate vector of $\mathbf{v}$ with respect to the ordered basis $\mathscr{B} = \{\mathbf{v}_1, \mathbf{v}_2, \ldots, \mathbf{v}_n\}$ is defined as

$$[\mathbf{v}]_{\mathscr{B}} = \begin{bmatrix} x_1 \\ x_2 \\ \vdots \\ x_n \end{bmatrix}$$

where

$$\mathbf{v} = x_1\mathbf{v}_1 + x_2\mathbf{v}_2 + \cdots + x_n\mathbf{v}_n$$

18. Let $\mathscr{B} = \{\mathbf{v}_1, \mathbf{v}_2, \ldots, \mathbf{v}_n\}$ and $\mathscr{B}' = \{\mathbf{u}_1, \mathbf{u}_2, \ldots, \mathbf{u}_n\}$ be two ordered bases for V. Then $[\mathbf{v}]_{\mathscr{B}}$ and $[\mathbf{v}]_{\mathscr{B}'}$ are related by

$$[\mathbf{v}]_{\mathscr{B}} = P[\mathbf{v}]_{\mathscr{B}'}$$

where P is the transition matrix defined by

$$P = [[\mathbf{u}_1]_{\mathscr{B}} \vdots [\mathbf{u}_2]_{\mathscr{B}} \vdots \cdots \vdots [\mathbf{u}_n]_{\mathscr{B}}]$$

19. Given an $m \times n$ matrix A, the subspace of R^m spanned by the column vectors is called the column space of A and the subspace $R_{1 \times n}$ spanned by the row vectors of A is called the row space of A.
20. The rank r of A is defined as the dimension of the row space of A, which also equals the dimension of the column space of A.
21. The nullity of an $m \times n$ matrix A is the dimension of the null space of A and equal to $n - r$, where r is the rank of A.
22. If A is an $n \times n$ matrix, then the following statements are equivalent.

(a) A is nonsingular (or invertible).
(b) $A\mathbf{x} = \mathbf{0}$ has only the trivial solution $\mathbf{x} = \mathbf{0}$.
(c) $\det A \neq 0$.
(d) The row vectors of A are linearly independent.
(e) The column vectors of A are linearly independent.
(f) A has rank n.
(g) $A\mathbf{x} = \mathbf{b}$ is consistent for every $n \times 1$ matrix $\mathbf{b}$, that is, $\mathbf{b}$ is in the column space of A.

RAISE YOUR GRADES

Can you explain ...?

☑ what axioms a vector space has to satisfy
☑ the conditions for linear independence of a set of vectors
☑ how the basis for a vector space is defined
☑ how to find the coordinate vector of a vector relative to a given basis
☑ the relationship between two coordinate vectors of a vector relative to two different bases
☑ how to find the rank of a given matrix

SOLVED PROBLEMS

Vectors in 2-Space and 3-Space

PROBLEM 3-1 Let

$$\mathbf{u} = \begin{bmatrix} 2 \\ 3 \end{bmatrix} \quad \text{and} \quad \mathbf{v} = \begin{bmatrix} -1 \\ 1 \end{bmatrix}$$

Find **(a)** $\mathbf{u} + \mathbf{v}$, **(b)** $-\mathbf{v}$, **(c)** $\mathbf{u} - \mathbf{v}$, and **(d)** $\mathbf{u} - 2\mathbf{v}$: illustrate these results graphically.

Solution

(a)
$$\mathbf{u} + \mathbf{v} = \begin{bmatrix} 2 \\ 3 \end{bmatrix} + \begin{bmatrix} -1 \\ 1 \end{bmatrix} = \begin{bmatrix} 2-1 \\ 3+1 \end{bmatrix} = \begin{bmatrix} 1 \\ 4 \end{bmatrix}$$

(b)
$$-\mathbf{v} = (-1)\mathbf{v} = (-1)\begin{bmatrix} -1 \\ 1 \end{bmatrix} = \begin{bmatrix} 1 \\ -1 \end{bmatrix}$$

(c)
$$\mathbf{u} - \mathbf{v} = \mathbf{u} + (-\mathbf{v}) = \begin{bmatrix} 2 \\ 3 \end{bmatrix} + \begin{bmatrix} 1 \\ -1 \end{bmatrix} = \begin{bmatrix} 2+1 \\ 3-1 \end{bmatrix} = \begin{bmatrix} 3 \\ 2 \end{bmatrix}$$

(d)
$$\mathbf{u} - 2\mathbf{v} = \mathbf{u} + (-2\mathbf{v}) = \begin{bmatrix} 2 \\ 3 \end{bmatrix} + (-2)\begin{bmatrix} -1 \\ 1 \end{bmatrix} = \begin{bmatrix} 2 \\ 3 \end{bmatrix} + \begin{bmatrix} 2 \\ -2 \end{bmatrix} = \begin{bmatrix} 2+2 \\ 3-2 \end{bmatrix} = \begin{bmatrix} 4 \\ 1 \end{bmatrix}$$

These results are illustrated in Figure 3-14.

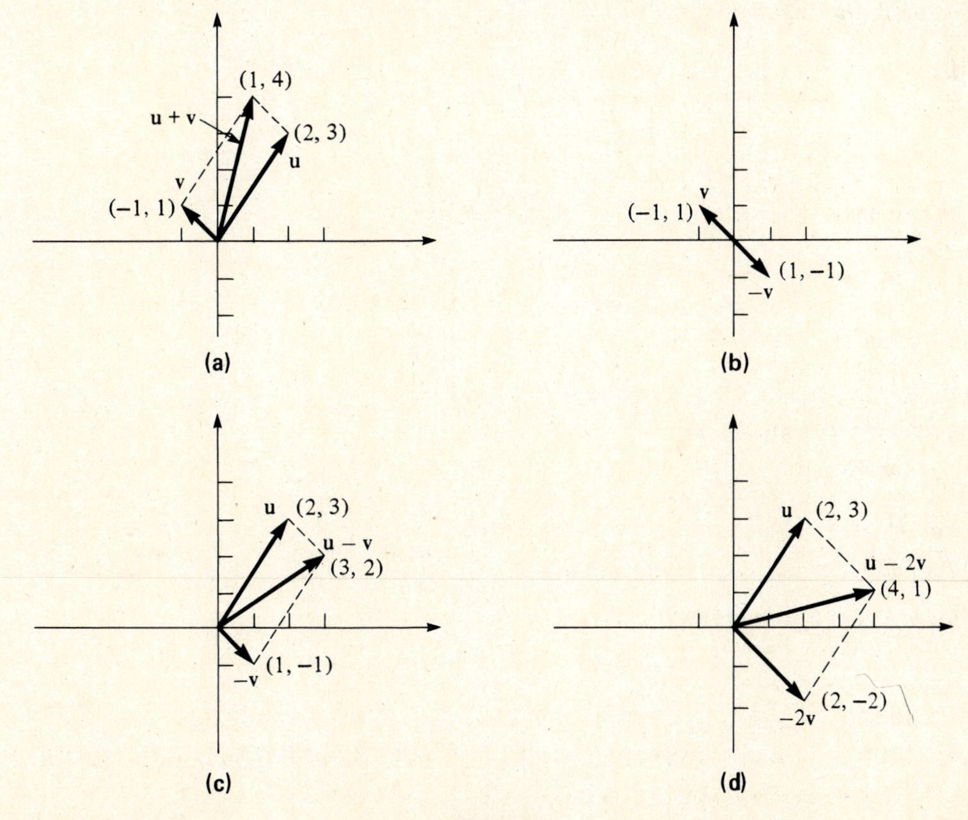

FIGURE 3-14

PROBLEM 3-2 Let $\mathbf{v} = \begin{bmatrix} a \\ b \end{bmatrix}$ be any 2-vector. Then show that if $\mathbf{0} = \begin{bmatrix} 0 \\ 0 \end{bmatrix}$,

(a) $$\mathbf{v} + \mathbf{0} = \mathbf{0} + \mathbf{v} = \mathbf{v}$$

(b) $$\mathbf{v} + (-\mathbf{v}) = \mathbf{0}$$

Solution

(a) $$\mathbf{v} + \mathbf{0} = \begin{bmatrix} a \\ b \end{bmatrix} + \begin{bmatrix} 0 \\ 0 \end{bmatrix} = \begin{bmatrix} a+0 \\ b+0 \end{bmatrix} = \begin{bmatrix} a \\ b \end{bmatrix} = \mathbf{v}$$

$$\mathbf{0} + \mathbf{v} = \begin{bmatrix} 0 \\ 0 \end{bmatrix} + \begin{bmatrix} a \\ b \end{bmatrix} = \begin{bmatrix} 0+a \\ 0+b \end{bmatrix} = \begin{bmatrix} a \\ b \end{bmatrix} = \mathbf{v}$$

Thus,

$$\mathbf{v} + \mathbf{0} = \mathbf{0} + \mathbf{v} = \mathbf{v}.$$

(b) $$-\mathbf{v} = (-1)\mathbf{v} = (-1)\begin{bmatrix} a \\ b \end{bmatrix} = \begin{bmatrix} -a \\ -b \end{bmatrix}$$

Thus

$$\mathbf{v} + (-\mathbf{v}) = \begin{bmatrix} a \\ b \end{bmatrix} + \begin{bmatrix} -a \\ -b \end{bmatrix} = \begin{bmatrix} a-a \\ b-b \end{bmatrix} = \begin{bmatrix} 0 \\ 0 \end{bmatrix} = \mathbf{0}$$

PROBLEM 3-3 Let

$$\mathbf{u} = \begin{bmatrix} 1 \\ 2 \\ k \end{bmatrix}, \qquad \mathbf{v} = \begin{bmatrix} 0 \\ 1 \\ k-1 \end{bmatrix}, \qquad \mathbf{w} = \begin{bmatrix} 3 \\ 4 \\ 3 \end{bmatrix}$$

Find k, α, and β such that $\alpha\mathbf{u} + \beta\mathbf{v} = \mathbf{w}$.

Solution

$$\alpha\mathbf{u} + \beta\mathbf{v} = \alpha\begin{bmatrix} 1 \\ 2 \\ k \end{bmatrix} + \beta\begin{bmatrix} 0 \\ 1 \\ k-1 \end{bmatrix} = \begin{bmatrix} \alpha \\ 2\alpha \\ k\alpha \end{bmatrix} + \begin{bmatrix} 0 \\ \beta \\ \beta(k-1) \end{bmatrix} = \begin{bmatrix} \alpha \\ 2\alpha+\beta \\ (\alpha+\beta)k-\beta \end{bmatrix}$$

Thus we get

$$\begin{bmatrix} \alpha \\ 2\alpha+\beta \\ (\alpha+\beta)k-\beta \end{bmatrix} = \begin{bmatrix} 3 \\ 4 \\ 3 \end{bmatrix}$$

Equating each component, we get

$$\alpha = 3$$
$$2\alpha + \beta = 4$$
$$(\alpha+\beta)k - \beta = 3$$

Solving for α, β, and k, we obtain

$$\alpha = 3, \qquad \beta = -2, \qquad k = 1$$

Check:

$$\alpha\mathbf{u} + \beta\mathbf{v} = 3\begin{bmatrix} 1 \\ 2 \\ 1 \end{bmatrix} + (-2)\begin{bmatrix} 0 \\ 1 \\ 0 \end{bmatrix} = \begin{bmatrix} 3 \\ 6 \\ 3 \end{bmatrix} + \begin{bmatrix} 0 \\ -2 \\ 0 \end{bmatrix} = \begin{bmatrix} 3+0 \\ 6-2 \\ 3+0 \end{bmatrix} = \begin{bmatrix} 3 \\ 4 \\ 3 \end{bmatrix} = \mathbf{w}$$

Cartesian n-Space R^n

PROBLEM 3-4 Let

$$\mathbf{u} = \begin{bmatrix} 2 \\ 0 \\ -1 \\ 3 \end{bmatrix}, \qquad \mathbf{v} = \begin{bmatrix} 3 \\ 1 \\ 2 \\ -1 \end{bmatrix}, \qquad \mathbf{w} = \begin{bmatrix} 2 \\ 3 \\ 4 \\ 0 \end{bmatrix}$$

Find **(a)** $\mathbf{u} + \mathbf{v}$, **(b)** $\mathbf{v} + \mathbf{w}$, **(c)** $(\mathbf{u} + \mathbf{v}) + \mathbf{w}$, **(d)** $\mathbf{u} + (\mathbf{v} + \mathbf{w})$, **(e)** $-\mathbf{u} + 2\mathbf{v}$, **(f)** $3(\mathbf{u} - 2\mathbf{v})$.

Solution

(a)

$$\mathbf{u} + \mathbf{v} = \begin{bmatrix} 2 \\ 0 \\ -1 \\ 3 \end{bmatrix} + \begin{bmatrix} 3 \\ 1 \\ 2 \\ -1 \end{bmatrix} = \begin{bmatrix} 2+3 \\ 0+1 \\ -1+2 \\ 3-1 \end{bmatrix} = \begin{bmatrix} 5 \\ 1 \\ 1 \\ 2 \end{bmatrix}$$

(b)

$$\mathbf{v} + \mathbf{w} = \begin{bmatrix} 3 \\ 1 \\ 2 \\ -1 \end{bmatrix} + \begin{bmatrix} 2 \\ 3 \\ 4 \\ 0 \end{bmatrix} = \begin{bmatrix} 3+2 \\ 1+3 \\ 2+4 \\ -1+0 \end{bmatrix} = \begin{bmatrix} 5 \\ 4 \\ 6 \\ -1 \end{bmatrix}$$

(c) From **(a)**,

$$(\mathbf{u} + \mathbf{v}) + \mathbf{w} = \begin{bmatrix} 5 \\ 1 \\ 1 \\ 2 \end{bmatrix} + \begin{bmatrix} 2 \\ 3 \\ 4 \\ 0 \end{bmatrix} = \begin{bmatrix} 5+2 \\ 1+3 \\ 1+4 \\ 2+0 \end{bmatrix} = \begin{bmatrix} 7 \\ 4 \\ 5 \\ 2 \end{bmatrix}$$

(d) From **(b)**,

$$\mathbf{u} + (\mathbf{v} + \mathbf{w}) = \begin{bmatrix} 2 \\ 0 \\ -1 \\ 3 \end{bmatrix} + \begin{bmatrix} 5 \\ 4 \\ 6 \\ -1 \end{bmatrix} = \begin{bmatrix} 2+5 \\ 0+4 \\ -1+6 \\ 3-1 \end{bmatrix} = \begin{bmatrix} 7 \\ 4 \\ 5 \\ 2 \end{bmatrix}$$

(e)

$$-\mathbf{u} + 2\mathbf{v} = -\begin{bmatrix} 2 \\ 0 \\ -1 \\ 3 \end{bmatrix} + 2\begin{bmatrix} 3 \\ 1 \\ 2 \\ -1 \end{bmatrix} = \begin{bmatrix} -2 \\ 0 \\ 1 \\ -3 \end{bmatrix} + \begin{bmatrix} 6 \\ 2 \\ 4 \\ -2 \end{bmatrix} = \begin{bmatrix} 4 \\ 2 \\ 5 \\ -5 \end{bmatrix}$$

(f) From **(e)**,

$$3(\mathbf{u} - 2\mathbf{v}) = -3(-\mathbf{u} + 2\mathbf{v}) = -3\begin{bmatrix} 4 \\ 2 \\ 5 \\ -5 \end{bmatrix} = \begin{bmatrix} -12 \\ -6 \\ -15 \\ 15 \end{bmatrix}$$

PROBLEM 3-5 Verify properties (iii) and (iv) of Theorem 3-2.1.

Solution Using (3.24) and (3.23),

$$\mathbf{v} + \mathbf{0} = \begin{bmatrix} v_1 \\ v_2 \\ \vdots \\ v_n \end{bmatrix} + \begin{bmatrix} 0 \\ 0 \\ \vdots \\ 0 \end{bmatrix} = \begin{bmatrix} v_1 + 0 \\ v_2 + 0 \\ \vdots \\ v_n + 0 \end{bmatrix} = \begin{bmatrix} v_1 \\ v_2 \\ \vdots \\ v_n \end{bmatrix} = \mathbf{v}$$

Using (3.26) and (3.23),

$$\mathbf{v} + (-\mathbf{v}) = \begin{bmatrix} v_1 \\ v_2 \\ \vdots \\ v_n \end{bmatrix} + \begin{bmatrix} -v_1 \\ -v_2 \\ \vdots \\ -v_n \end{bmatrix} = \begin{bmatrix} v_1 - v_1 \\ v_2 - v_2 \\ \vdots \\ v_n - v_n \end{bmatrix} = \begin{bmatrix} 0 \\ 0 \\ \vdots \\ 0 \end{bmatrix} = \mathbf{0}$$

PROBLEM 3-6 Verify properties (vi) and (vii) of Theorem 3-2.1.

Solution Let α and β be any scalars. Then by definition (3.25)

$$(\alpha + \beta)\mathbf{v} = \begin{bmatrix} (\alpha + \beta)v_1 \\ (\alpha + \beta)v_2 \\ \vdots \\ (\alpha + \beta)v_n \end{bmatrix} = \begin{bmatrix} \alpha v_1 + \beta v_1 \\ \alpha v_2 + \beta v_2 \\ \vdots \\ \alpha v_n + \beta v_n \end{bmatrix} = \alpha\mathbf{v} + \beta\mathbf{v}$$

$$(\alpha\beta)\mathbf{u} = \begin{bmatrix} (\alpha\beta)v_1 \\ (\alpha\beta)v_2 \\ \vdots \\ (\alpha\beta)v_n \end{bmatrix} = \begin{bmatrix} \alpha(\beta v_1) \\ \alpha(\beta v_2) \\ \vdots \\ \alpha(\beta v_n) \end{bmatrix} = \alpha(\beta\mathbf{v})$$

since $(\alpha + \beta)v_i = \alpha v_i + \beta v_i$, $(\alpha\beta)v_i = \alpha(\beta v_i)$ for all real numbers.

PROBLEM 3-7 Show that for every two vectors in R^n, $-(\mathbf{u} + \mathbf{v}) = -\mathbf{u} - \mathbf{v}$ and $-(\mathbf{u} - \mathbf{v}) = -\mathbf{u} + \mathbf{v}$.

Solution Let

$$\mathbf{u} = \begin{bmatrix} u_1 \\ u_2 \\ \vdots \\ u_n \end{bmatrix}, \qquad \mathbf{v} = \begin{bmatrix} v_1 \\ v_2 \\ \vdots \\ v_n \end{bmatrix}$$

Then

$$-(\mathbf{u} + \mathbf{v}) = -\begin{bmatrix} u_1 + v_1 \\ u_2 + v_2 \\ \vdots \\ u_n + v_n \end{bmatrix}$$

$$= \begin{bmatrix} -(u_1 + v_1) \\ -(u_2 + v_2) \\ \vdots \\ -(u_n + v_n) \end{bmatrix}$$

$$= \begin{bmatrix} -u_1 - v_1 \\ -u_2 - v_2 \\ \vdots \\ -u_n - v_n \end{bmatrix} = \begin{bmatrix} -u_1 + (-v_1) \\ -u_2 + (-v_2) \\ \vdots \\ -u_n + (-v_n) \end{bmatrix} = -\mathbf{u} + (-\mathbf{v}) = -\mathbf{u} - \mathbf{v}$$

$$-(\mathbf{u}-\mathbf{v}) = -\begin{bmatrix} u_1 - v_1 \\ u_2 - v_2 \\ \vdots \\ u_n - v_n \end{bmatrix}$$

$$= \begin{bmatrix} -(u_2 - v_1) \\ -(u_2 - v_2) \\ \vdots \\ -(u_n - v_n) \end{bmatrix} = \begin{bmatrix} -u_1 + v_1 \\ -u_2 + v_2 \\ \vdots \\ -u_n + v_n \end{bmatrix} = -\mathbf{u} + \mathbf{v}$$

Vector Space

PROBLEM 3-8 Let V be the set of all 2-vectors such that the sum of components is equal to zero. That is,

$$V = \left\{ \begin{bmatrix} v_1 \\ v_2 \end{bmatrix} \middle| v_1 + v_2 = 0 \right\}$$

Show that V is a vector space.

Solution The set V is nonempty since $\begin{bmatrix} 0 \\ 0 \end{bmatrix}$ is an element of V. The set V satisfies Axioms V2, V3, V7, V8, V9, and V10 because all 2-vectors satisfy them. Thus we need only verify that the remaining axioms (V1, V4, V5, and V6) are satisfied.

Let $\mathbf{u}, \mathbf{v} \in V$; that is,

$$\mathbf{u} = \begin{bmatrix} u_1 \\ u_2 \end{bmatrix}, \qquad \mathbf{v} = \begin{bmatrix} v_1 \\ v_2 \end{bmatrix}$$

so that

$$u_1 + u_2 = 0$$
$$v_1 + v_2 = 0$$

Then

$$\mathbf{u} + \mathbf{v} = \begin{bmatrix} u_1 \\ u_2 \end{bmatrix} + \begin{bmatrix} v_1 \\ v_2 \end{bmatrix} = \begin{bmatrix} u_1 + v_1 \\ u_2 + v_2 \end{bmatrix}$$

and

$$(u_1 + v_1) + (u_2 + v_2) = (u_1 + u_2) + (v_1 + v_2) = 0 + 0 = 0$$

Thus $\mathbf{u} + \mathbf{v} \in V$, and Axiom V1 is satisfied.

Since $\mathbf{0} = \begin{bmatrix} 0 \\ 0 \end{bmatrix} \in V$ and it is the zero element of V, such that $\mathbf{v} + \mathbf{0} = \mathbf{v}$ for each $\mathbf{v} \in V$, Axiom V4 is satisfied.

It is clear that

$$-\mathbf{v} = \begin{bmatrix} -v_1 \\ -v_2 \end{bmatrix}$$

is the negative of $\mathbf{v}$ since

$$\mathbf{v} + (-\mathbf{v}) = \begin{bmatrix} v_1 \\ v_2 \end{bmatrix} + \begin{bmatrix} -v_1 \\ -v_2 \end{bmatrix} = \begin{bmatrix} 0 \\ 0 \end{bmatrix} = \mathbf{0}$$

Furthermore

$$(-v_1) + (-v_2) = (-1)(v_1 + v_2) = (-1)(0) = 0$$

so $-\mathbf{v} \in V$, and Axiom V5 is satisfied.

If α is any scalar, then

$$\alpha v = \alpha \begin{bmatrix} v_1 \\ v_2 \end{bmatrix} = \begin{bmatrix} \alpha v_1 \\ \alpha v_2 \end{bmatrix}$$

and

$$\alpha v_1 + \alpha v_2 = \alpha(v_1 + v_2) = \alpha(0) = 0$$

Thus $\alpha \mathbf{v} \in V$, and Axiom V6 is satisfied.

Therefore V satisfies all the vector space axioms, and we conclude that V is a vector space.

PROBLEM 3-9 Let V be the set of all 2×2 matrices of the form

$$\begin{bmatrix} a & b \\ 0 & c \end{bmatrix}$$

with the usual matrix addition and scalar multiplication [see Section 1-3, (1.12) and (1.13)]. Show that V is a vector space.

Solution The set is not empty, since $\begin{bmatrix} 0 & 0 \\ 0 & 0 \end{bmatrix} \in V$. The set V satisfies Axioms V2, V3, V7, V8, V9, and V10 because all 2×2 matrices satisfy them. Thus again we need only verify that the remaining axioms (V1, V4, V5, and V6) are satisfied.

Let $A, B \in V$, so that

$$A = \begin{bmatrix} a_1 & a_2 \\ 0 & a_4 \end{bmatrix}, \qquad B = \begin{bmatrix} b_1 & b_2 \\ 0 & b_4 \end{bmatrix}$$

Then

$$A + B = \begin{bmatrix} a_1 & a_2 \\ 0 & a_4 \end{bmatrix} + \begin{bmatrix} b_1 & b_2 \\ 0 & b_4 \end{bmatrix} = \begin{bmatrix} a_1 + b_1 & a_2 + b_2 \\ 0 & a_4 + b_4 \end{bmatrix}$$

Thus $A + B \in V$, and Axiom V1 is satisfied.

It is clear that $O = \begin{bmatrix} 0 & 0 \\ 0 & 0 \end{bmatrix} \in V$; and it is the zero element of V since

$$A + O = \begin{bmatrix} a_1 & a_2 \\ 0 & a_4 \end{bmatrix} + \begin{bmatrix} 0 & 0 \\ 0 & 0 \end{bmatrix} = \begin{bmatrix} a_1 & a_2 \\ 0 & a_4 \end{bmatrix} = A$$

Thus Axiom V4 is satisfied.

It is clear that

$$-A = \begin{bmatrix} -a_1 & -a_2 \\ 0 & -a_4 \end{bmatrix} \in V$$

and it is the negative of A since

$$A + (-A) = \begin{bmatrix} a_1 & a_2 \\ 0 & a_4 \end{bmatrix} + \begin{bmatrix} -a_1 & -a_2 \\ 0 & -a_4 \end{bmatrix} = \begin{bmatrix} 0 & 0 \\ 0 & 0 \end{bmatrix} = O$$

Thus Axiom V5 is satisfied.

If α is any scalar, then

$$\alpha A = \alpha \begin{bmatrix} a_1 & a_2 \\ 0 & a_4 \end{bmatrix} = \begin{bmatrix} \alpha a_1 & \alpha a_2 \\ 0 & \alpha a_4 \end{bmatrix} \in V$$

Thus Axiom V6 is satisfied.

Therefore V satisfies all the vector space axioms, and we conclude that V is a vector space.

PROBLEM 3-10 Let U denote the set of all polynomials with real coefficients of exactly degree 2. The operations of addition and scalar multiplication are defined as in (3.29) and (3.30). Show that U is not a vector space.

Solution The sum of two second-degree polynomials need not be a second-degree polynomial. For example, consider two specific polynomials $\mathbf{p}, \mathbf{q} \in U$, where

$$\mathbf{p}(x) = p(x) = x^2$$
$$\mathbf{q}(x) = q(x) = -x^2 + x$$

Then

$$(\mathbf{p} + \mathbf{q})(x) = p(x) + q(x)$$
$$= x^2 + (-x^2 + x) = x$$

which is a first-degree polynomial.

Hence $\mathbf{p} + \mathbf{q} \notin U$, Axiom V1 is not satisfied; therefore U cannot be a vector space. Furthermore, the polynomial that is zero for all x is not a second-degree polynomial, so Axiom V4 is not satisfied either.

PROBLEM 3-11 Let V denote the set of all infinite sequences of real numbers with addition and scalar multiplication defined by

$$\{a_n\} + \{b_n\} = \{a_n + b_n\}$$
$$\alpha\{a_n\} = \{\alpha a_n\}$$

where

$$\{a_n\} = \{a_1, a_2, \ldots\}$$
$$\{b_n\} = \{b_1, b_2, \ldots\}$$

Show that V is a vector space.

Solution The set V is not empty since the zero sequence $\{0, 0, \ldots\}$ is an element of V. If $\{a_n\}$ and $\{b_n\}$ are any two elements of V, then by the definition of addition $\{a_n + b_n\}$ is also an infinite sequence of real numbers. Thus $\{a_n\} + \{b_n\} \in V$, so Axiom V1 holds.

If $\{a_n\}$, $\{b_n\}$, and $\{c_n\}$ are arbitrary elements of V, then for each n

$$a_n + b_n = b_n + a_n$$

and

$$(a_n + b_n) + c_n = a_n + (b_n + c_n)$$

Hence

$$\{a_n\} + \{b_n\} = \{b_n\} + \{a_n\}$$

and

$$(\{a_n\} + \{b_n\}) + \{c_n\} = \{a_n\} + (\{b_n\} + \{c_n\})$$

so Axioms V2 and V3 hold.

The zero vector of V is just the zero sequence $\{0, 0, \ldots\}$, and the negative of $\{a_n\}$ is the sequence $\{-a_n\}$; thus Axioms V4 and V5 hold.

Again, by the definition of scalar multiplication

$$\alpha\{a_n\} = \{\alpha a_n\} \in V$$

so Axiom V6 holds.

The remaining four axioms all hold since

$$\alpha(a_n + b_n) = \alpha a_n + \alpha b_n$$
$$(\alpha + \beta)a_n = \alpha a_n + \beta a_n$$
$$(\alpha\beta)a_n = \alpha(\beta a_n)$$
$$1a_n = a_n$$

for each n.

Thus all vector space axioms hold, therefore, V is a vector space.

PROBLEM 3-12 Let V be a vector space. Show that if $\mathbf{u}, \mathbf{v} \in V$ and

$$\mathbf{u} + \mathbf{v} = \mathbf{u}$$

then

$$\mathbf{v} = \mathbf{0}$$

Solution Suppose that

$$\mathbf{u} + \mathbf{v} = \mathbf{u}$$

By Axiom V5, there is a negative element $-\mathbf{u}$ for $\mathbf{u}$. Adding $-\mathbf{u}$ to both sides, we get

$$-\mathbf{u} + (\mathbf{u} + \mathbf{v}) = -\mathbf{u} + \mathbf{u}$$

or

$$(-\mathbf{u} + \mathbf{u}) + \mathbf{v} = -\mathbf{u} + \mathbf{u} \qquad \text{(by V3)}$$

or

$$(\mathbf{u} + (-\mathbf{u})) + \mathbf{v} = \mathbf{u} + (-\mathbf{u}) \qquad \text{(by V2)}$$

or

$$\mathbf{0} + \mathbf{v} = \mathbf{0} \qquad \text{(by V5)}$$

Thus

$$\mathbf{v} = \mathbf{0} \qquad \text{(by V2 and V4)}$$

PROBLEM 3-13 Let V be a vector space. Show that if $\mathbf{u}, \mathbf{v} \in V$, then there is one and only one $\mathbf{x} \in V$ such that

$$\mathbf{u} + \mathbf{x} = \mathbf{v}$$

Solution First, we show that there is one such $\mathbf{x}$. Let

$$\mathbf{x} = \mathbf{v} + (-\mathbf{u})$$

Then

$$\begin{aligned}
\mathbf{u} + \mathbf{x} &= \mathbf{u} + (\mathbf{v} + (-\mathbf{u})) \\
&= \mathbf{u} + ((-\mathbf{u}) + \mathbf{v}) && \text{(by V2)} \\
&= (\mathbf{u} + (-\mathbf{u})) + \mathbf{v} && \text{(by V3)} \\
&= \mathbf{0} + \mathbf{v} && \text{(by V5)} \\
&= \mathbf{v} && \text{(by V2 and V4)}
\end{aligned}$$

Next, we show that there is only one such $\mathbf{x}$. Suppose $\mathbf{x}_1, \mathbf{x}_2 \in V$ such that

$$\mathbf{u} + \mathbf{x}_1 = \mathbf{v} \qquad \text{and} \qquad \mathbf{u} + \mathbf{x}_2 = \mathbf{v}$$

Then

$$\mathbf{u} + \mathbf{x}_1 = \mathbf{u} + \mathbf{x}_2$$

Adding $(-\mathbf{u})$ to both sides, we get

$$-\mathbf{u} + (\mathbf{u} + \mathbf{x}_1) = -\mathbf{u} + (\mathbf{u} + \mathbf{x}_2)$$

Thus

$$(-\mathbf{u} + \mathbf{u}) + \mathbf{x}_1 = (-\mathbf{u} + \mathbf{u}) + \mathbf{x}_2 \qquad \text{(by V3)}$$

or

$$\mathbf{0} + \mathbf{x}_1 = \mathbf{0} + \mathbf{x}_2 \qquad \text{(by V2 and V5)}$$

or

$$\mathbf{x}_1 = \mathbf{x}_2 \qquad \text{(by V2 and V4)}$$

Note that it is customary to denote the vector $\mathbf{x} = \mathbf{v} + (-\mathbf{u})$ by $\mathbf{v} - \mathbf{u}$, which is the vector obtained from $\mathbf{v}$ by subtracting $\mathbf{u}$.

Subspaces

PROBLEM 3-14 In R^3, let

$$U = \left\{ \begin{bmatrix} u_1 \\ u_2 \\ u_3 \end{bmatrix} \middle| \, u_1 = u_2 \right\}$$

Show that U is a subspace of R^3 and illustrate U geometrically.

Solution It is obvious that

$$\mathbf{0} = \begin{bmatrix} 0 \\ 0 \\ 0 \end{bmatrix} \in U$$

Thus U is not an empty subset of R^3. If

$$\mathbf{u} = \begin{bmatrix} a \\ a \\ c \end{bmatrix} \in U \quad \text{and} \quad \mathbf{v} = \begin{bmatrix} b \\ b \\ d \end{bmatrix} \in U$$

then

$$\mathbf{u} + \mathbf{v} = \begin{bmatrix} a \\ a \\ c \end{bmatrix} + \begin{bmatrix} b \\ b \\ d \end{bmatrix} = \begin{bmatrix} a+b \\ a+b \\ c+d \end{bmatrix} \in U$$

Next, if α is any scalar,

$$\alpha\mathbf{u} = \alpha \begin{bmatrix} a \\ a \\ c \end{bmatrix} = \begin{bmatrix} \alpha a \\ \alpha a \\ \alpha c \end{bmatrix} \in U$$

Thus U is closed under addition and scalar multiplication. Therefore U is a subspace of R^3.

Now, $\mathbf{u}$ can be expressed as

$$\mathbf{u} = \begin{bmatrix} a \\ a \\ c \end{bmatrix} = \begin{bmatrix} a \\ a \\ 0 \end{bmatrix} + \begin{bmatrix} 0 \\ 0 \\ c \end{bmatrix} = a \begin{bmatrix} 1 \\ 1 \\ 0 \end{bmatrix} + c \begin{bmatrix} 0 \\ 0 \\ 1 \end{bmatrix}$$

Thus, geometrically the set U consists of all points on the plane formed by the vectors

$$\begin{bmatrix} 1 \\ 1 \\ 0 \end{bmatrix} \quad \text{and} \quad \begin{bmatrix} 0 \\ 0 \\ 1 \end{bmatrix}$$

as shown in Figure 3-15.

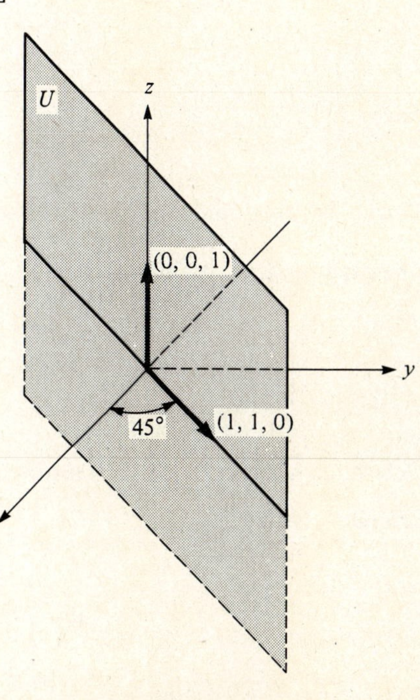

FIGURE 3-15

PROBLEM 3-15 In R^3, let

$$U = \left\{ \begin{bmatrix} u_1 \\ u_2 \\ u_3 \end{bmatrix} \middle| u_1 + u_2 + u_3 = 0 \right\}$$

Show that U is a subspace of R^3.

Solution

$$\mathbf{0} = \begin{bmatrix} 0 \\ 0 \\ 0 \end{bmatrix} \in U$$

since $0 + 0 + 0 = 0$. Thus U is not an empty subset of R^3. If

$$\mathbf{u} = \begin{bmatrix} u_1 \\ u_2 \\ u_3 \end{bmatrix} \in U, \qquad \mathbf{v} = \begin{bmatrix} v_1 \\ v_2 \\ v_3 \end{bmatrix} \in U$$

then

$$u_1 + u_2 + u_3 = 0, \qquad v_1 + v_2 + v_3 = 0$$

and

$$\mathbf{u} + \mathbf{v} = \begin{bmatrix} u_1 + v_1 \\ u_2 + v_2 \\ u_3 + v_3 \end{bmatrix} \in U$$

since

$$(u_1 + v_1) + (u_2 + v_2) + (u_3 + v_3) = (u_1 + u_2 + u_3) + (v_1 + v_2 + v_3) = 0 + 0 = 0$$

Next, if α is any scalar,

$$\alpha\mathbf{u} = \begin{bmatrix} \alpha u_1 \\ \alpha u_2 \\ \alpha u_3 \end{bmatrix} \in U$$

since

$$\alpha u_1 + \alpha u_2 + \alpha u_3 = \alpha(u_1 + u_2 + u_3) = \alpha(0) = 0$$

Thus U is closed under addition and scalar multiplication. Therefore U is a subspace of R^3.

PROBLEM 3-16 In R^3, let

$$U = \left\{ \begin{bmatrix} u_1 \\ u_2 \\ u_3 \end{bmatrix} \middle| u_1^2 + u_2^2 + u_3^2 \le 1 \right\}$$

Determine whether or not U is a subspace of R^3. Illustrate U geometrically.

Solution Now

$$\mathbf{u} = \begin{bmatrix} 1 \\ 0 \\ 0 \end{bmatrix} \in U \quad \text{and} \quad \mathbf{v} = \begin{bmatrix} 0 \\ 1 \\ 0 \end{bmatrix} \in U$$

But

$$\mathbf{u} + \mathbf{v} = \begin{bmatrix} 1 \\ 1 \\ 0 \end{bmatrix} \notin U$$

since $1^2 + 1^2 + 0^2 = 2 > 1$. Thus U is not closed under addition. Therefore U is not a subspace of R^3. U consists of all points on and within a sphere with radius 1 centered at the origin, as shown in Figure 3-16.

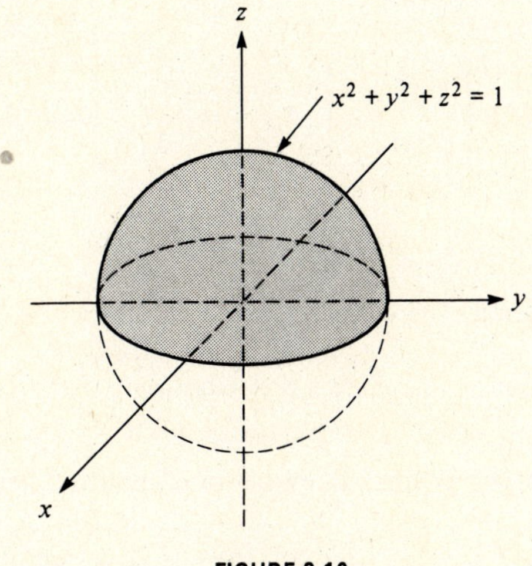

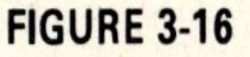

FIGURE 3-16

PROBLEM 3-17 Let V be the set of all 2×2 matrices of the form

$$\begin{bmatrix} a & b \\ 0 & c \end{bmatrix}$$

where a, b, and c are real numbers. Show that V is a subspace of $R_{2\times 2}$.

Solution Since $\begin{bmatrix} 0 & 0 \\ 0 & 0 \end{bmatrix} \in V$, V is a nonempty subset of $R_{2\times 2}$. We have shown that V is a vector space (see Problem 3-9). Thus, by the definition of a subspace, V is a subspace of $R_{2\times 2}$.

PROBLEM 3-18 Let W consist of all 2×2 matrices P for which $P^2 = P$. Show that W is not a subspace of $R_{2\times 2}$.

Solution The identity matrix

$$I_2 = \begin{bmatrix} 1 & 0 \\ 0 & 1 \end{bmatrix} \in W$$

since

$$I_2^2 = \begin{bmatrix} 1 & 0 \\ 0 & 1 \end{bmatrix}\begin{bmatrix} 1 & 0 \\ 0 & 1 \end{bmatrix} = \begin{bmatrix} 1 & 0 \\ 0 & 1 \end{bmatrix} = I_2$$

Thus W is not empty. But

$$\alpha I_2 = \begin{bmatrix} \alpha & 0 \\ 0 & \alpha \end{bmatrix} \notin W$$

since

$$(\alpha I_2)^2 = \begin{bmatrix} \alpha & 0 \\ 0 & \alpha \end{bmatrix}\begin{bmatrix} \alpha & 0 \\ 0 & \alpha \end{bmatrix} = \begin{bmatrix} \alpha^2 & 0 \\ 0 & \alpha^2 \end{bmatrix} = \alpha^2 I_2 \neq \alpha I_2$$

Thus W is not closed under scalar multiplication. W is not a subspace of $R_{2\times 2}$.

PROBLEM 3-19 Let U be a set of all 2×2 matrices whose determinants are 0. Show that U is not a subspace of $R_{2\times 2}$.

Solution

$$A = \begin{bmatrix} 1 & 0 \\ 0 & 0 \end{bmatrix} \in U \qquad \text{since } \det A = 0$$

$$B = \begin{bmatrix} 0 & 0 \\ 0 & 1 \end{bmatrix} \in U \qquad \text{since } \det B = 0$$

But

$$A + B = \begin{bmatrix} 1 & 0 \\ 0 & 0 \end{bmatrix} + \begin{bmatrix} 0 & 0 \\ 0 & 1 \end{bmatrix} = \begin{bmatrix} 1 & 0 \\ 0 & 1 \end{bmatrix} = I_2 \notin U$$

since

$$\det(A + B) = \det I_2 = 1 \neq 0$$

Thus U is not closed under addition. Therefore U is not a subspace of $R_{2\times 2}$.

PROBLEM 3-20 Let W be a set of all square $n \times n$ symmetric matrices. Show that W is a subspace of $R_{n\times n}$.

Solution The $n \times n$ zero matrix $O \in W$ since all entries of O are 0 and hence symmetric. Thus W is not an empty subset of $R_{n\times n}$.

Now suppose

$$A = [a_{ij}]_{n\times n} \in W, \qquad B = [b_{ij}]_{n\times n} \in W$$

so that

$$a_{ji} = a_{ij}, \qquad b_{ji} = b_{ij}$$

Then

$$A + B = [a_{ij} + b_{ij}]_{n\times n} \in W$$

since $a_{ij} + b_{ij} = a_{ji} + b_{ji}$

$$\alpha A = [\alpha a_{ij}]_{n\times n} \in W$$

since $\alpha a_{ji} = \alpha a_{ij}$.

Thus W is closed under both addition and scalar multiplication; therefore W is a subspace of $R_{n\times n}$.

PROBLEM 3-21 Consider the subset of P_2 defined by

$$U = \{a_0 + a_1 x + a_2 x^2 \mid \ a_0 \text{ is an integer}\}$$

Show that U is not a subspace of P_2.

Solution It is obvious that $\mathbf{z}(x) = z(x) = 0 \in U$. Thus U is not empty.

Let

$$\mathbf{p}(x) = p(x) = a_0 + a_1 x + a_2 x^2 \in U$$

$$\mathbf{q}(x) = q(x) = b_0 + b_1 x + b_2 x^2 \in U$$

Now

$$(\mathbf{p} + \mathbf{q})(x) = p(x) + q(x)$$
$$= (a_0 + b_0) + (a_1 + b_1)x + (a_2 + b_2)x^2$$

Since a_0 and b_0 are integers, $a_0 + b_0$ is also an integer and

$$\mathbf{p} + \mathbf{q} \in U$$

Hence U is closed under addition.

Next, if α is any real number, then

$$(\alpha\mathbf{p})(x) = \alpha p(x)$$
$$= \alpha a_0 + \alpha a_1 x + \alpha a_2 x^2$$

If a_0 is an integer, it is not necessarily true that αa_0 is an integer. For example, if $\alpha = \frac{1}{2}$ and $a_0 = 1$, then $\alpha a_0 = \frac{1}{2}$ is not an integer. Therefore U is not closed under scalar multiplication. Hence U is not a subspace of P_2.

***PROBLEM 3-22** [*For readers who have studied calculus.*] In the space P_n, the set of polynomials of degree less than or equal to n with real coefficients, consider the subset

$$Q_n = \left\{\mathbf{p}|\mathbf{p} \in P_n \text{ and } \int_0^1 \mathbf{p}(x)\,dx = 0\right\}$$

that is, the set of polynomials that also satisfy the additional condition of $\int_0^1 p(x)\,dx = 0$. Show that Q_n is a subspace of P_n.

Solution It is obvious that $\mathbf{z}(x) = z(x) = 0 \in Q_n$. Thus Q_n is not empty.

Now, suppose $\mathbf{p} \in Q_n$, $\mathbf{q} \in Q_n$; then

$$\int_0^1 (\mathbf{p} + \mathbf{q})(x)\,dx = \int_0^1 [p(x) + q(x)]\,dx$$
$$= \int_0^1 p(x)\,dx + \int_0^1 q(x)\,dx$$
$$= 0 + 0 = 0$$

So $\mathbf{p} + \mathbf{q} \in Q_n$.

Next, suppose $\mathbf{p} \in Q_n$ and α is a scalar; then

$$\int_0^1 (\alpha\mathbf{p})(x)\,dx = \int_0^1 \alpha p(x)\,dx$$
$$= \alpha \int_0^1 p(x)\,dx$$
$$= \alpha 0 = 0$$

So $\alpha\mathbf{p} \in Q_n$.

Thus we have verified that Q_n is closed under vector addition and scalar multiplication. Therefore, Q_n is a subspace of P_n.

PROBLEM 3-23 Let U be a subset of the real-valued function space $F[I]$ such that U consists of all nonnegative functions, i.e., all functions $\mathbf{f}$ for which $\mathbf{f}(x) > 0$. Show that U is not a subspace of $F[I]$.

Solution Let $\mathbf{f} \in F$ be defined by $\mathbf{f} = f(x) = x^2$. Then $\mathbf{f} \in U$ since $f(x) = x^2 > 0$ for all x.

Now let α be any scalar such that $\alpha < 0$. Then $\alpha\mathbf{f} \notin U$ since $\alpha f(x) = \alpha x^2 < 0$ for all x since $\alpha < 0$. For instance, if $\alpha = -2$, $x = 1$, then

$$\alpha f(x) = \alpha x^2 = -2(1^2) = -2 < 0$$

Thus U is not closed under scalar multiplication and is therefore not a subspace of $F[I]$.

PROBLEM 3-24 Let U and W be subspaces of a vector space V. Define

$$U + W = \{\mathbf{z}|\mathbf{z} = \mathbf{u} + \mathbf{w} \quad \text{where } \mathbf{u} \in U \text{ and } \mathbf{w} \in W\}$$

Show that $U + W$ is a subspace of V.

Solution If $\mathbf{z}_1, \mathbf{z}_2 \in U + W$, then

$$\mathbf{z}_1 = \mathbf{u}_1 + \mathbf{w}_1 \quad \text{and} \quad \mathbf{z}_2 = \mathbf{u}_2 + \mathbf{w}_2$$

where $\mathbf{u}_1, \mathbf{u}_2 \in U$ and $\mathbf{w}_1, \mathbf{w}_2 \in W$.

Since U and W are subspaces, it follows that

$$\mathbf{u}_1 + \mathbf{u}_2 \in U \qquad \text{and} \qquad \mathbf{w}_1 + \mathbf{w}_2 \in W$$

Then

$$\mathbf{z}_1 + \mathbf{z}_2 = (\mathbf{u}_1 + \mathbf{w}_1) + (\mathbf{u}_2 + \mathbf{w}_2) = (\mathbf{u}_1 + \mathbf{u}_2) + (\mathbf{w}_1 + \mathbf{w}_2) \in U + W$$

Thus $U + W$ is closed under addition.

Next, since U and W are subspaces, it follows that if $\mathbf{u} \in U$, then $\alpha\mathbf{u} \in U$; and if $\mathbf{w} \in W$, then $\alpha\mathbf{w} \in W$ for any scalar α. Thus if $\mathbf{z} \in U + W$, then $\mathbf{z} = \mathbf{u} + \mathbf{w}$ and

$$\alpha\mathbf{z} = \alpha(\mathbf{u} + \mathbf{w}) = \alpha\mathbf{u} + \alpha\mathbf{w} \in U + W$$

Thus $U + W$ is closed under scalar multiplication. Therefore $U + W$ is a subspace of V.

Linear Combination and Spanning Sets

PROBLEM 3-25 Consider

$$\mathbf{v}_1 = \begin{bmatrix} 1 \\ 1 \\ 1 \end{bmatrix} \qquad \text{and} \qquad \mathbf{v}_2 = \begin{bmatrix} 1 \\ 2 \\ -1 \end{bmatrix}$$

in R^3. Show that

$$\mathbf{u} = \begin{bmatrix} 3 \\ 4 \\ 1 \end{bmatrix}$$

is a linear combination of $\mathbf{v}_1$ and $\mathbf{v}_2$ and that

$$\mathbf{w} = \begin{bmatrix} 1 \\ 0 \\ 0 \end{bmatrix}$$

is not a linear combination of $\mathbf{v}_1$ and $\mathbf{v}_2$.

Solution In order for $\mathbf{u}$ to be a linear combination of $\mathbf{v}_1$ and $\mathbf{v}_2$ there must be scalars α_1 and α_2, such that $\mathbf{u} = \alpha_1\mathbf{v}_1 + \alpha_2\mathbf{v}_2$; that is,

$$\begin{bmatrix} 3 \\ 4 \\ 1 \end{bmatrix} = \alpha_1 \begin{bmatrix} 1 \\ 1 \\ 1 \end{bmatrix} + \alpha_2 \begin{bmatrix} 1 \\ 2 \\ -1 \end{bmatrix} = \begin{bmatrix} \alpha_1 + \alpha_2 \\ \alpha_1 + 2\alpha_2 \\ \alpha_1 - \alpha_2 \end{bmatrix}$$

Equating the corresponding components, we obtain

$$\begin{aligned} \alpha_1 + \alpha_2 &= 3 \\ \alpha_1 + 2\alpha_2 &= 4 \\ \alpha_1 - \alpha_2 &= 1 \end{aligned}$$

Solving this system yields $\alpha_1 = 2$, $\alpha_2 = 1$, and hence $\mathbf{u} = 2\mathbf{v}_1 + \mathbf{v}_2$. Similarly, if $\mathbf{w}$ is a linear combination of $\mathbf{v}_1$ and $\mathbf{v}_2$, there must be scalars β_1 and β_2 such that $\mathbf{w} = \beta_1\mathbf{v}_1 + \beta_2\mathbf{v}_2$; that is,

$$\begin{bmatrix} 1 \\ 0 \\ 0 \end{bmatrix} = \beta_1 \begin{bmatrix} 1 \\ 1 \\ 1 \end{bmatrix} + \beta_2 \begin{bmatrix} 1 \\ 2 \\ -1 \end{bmatrix} = \begin{bmatrix} \beta_1 + \beta_2 \\ \beta_1 + 2\beta_2 \\ \beta_1 - \beta_2 \end{bmatrix}$$

Again, equating the corresponding components, we get

$$\begin{aligned}\beta_1 + \beta_2 &= 1\\ \beta_1 + 2\beta_2 &= 0\\ \beta_1 - \beta_2 &= 0\end{aligned}$$

This system of equations is inconsistent, so no such scalars exist. (From the first and the third equations we get $\beta_1 = \beta_2 = \frac{1}{2}$. Substituting these values into the second equation gives $\frac{1}{2} + 1 = \frac{3}{2} \neq 0$.) Therefore $\mathbf{w}$ is not a linear combination of $\mathbf{v}_1$ and $\mathbf{v}_2$.

PROBLEM 3-26 Express the matrix

$$D = \begin{bmatrix} 3 & -1 \\ 1 & 1 \end{bmatrix}$$

as a linear combination of

$$A = \begin{bmatrix} 1 & 0 \\ 1 & 1 \end{bmatrix}, \qquad B = \begin{bmatrix} 0 & -1 \\ 0 & 2 \end{bmatrix}, \qquad C = \begin{bmatrix} 0 & 1 \\ 1 & 0 \end{bmatrix}$$

Solution In order for D to be a linear combination of A, B, and C, there must be scalars α_1, α_2, and α_3 such that $D = \alpha_1 A + \alpha_2 B + \alpha_3 C$ or

$$\begin{aligned}\begin{bmatrix} 3 & -1 \\ 1 & 1 \end{bmatrix} &= \alpha_1 \begin{bmatrix} 1 & 0 \\ 1 & 1 \end{bmatrix} + \alpha_2 \begin{bmatrix} 0 & -1 \\ 0 & 2 \end{bmatrix} + \alpha_3 \begin{bmatrix} 0 & 1 \\ 1 & 0 \end{bmatrix}\\ &= \begin{bmatrix} \alpha_1 & -\alpha_2 + \alpha_3 \\ \alpha_1 + \alpha_3 & \alpha_1 + 2\alpha_2 \end{bmatrix}\end{aligned}$$

Equating the corresponding entries of the matrices, we get

$$\alpha_1 = 3, \qquad -\alpha_2 + \alpha_3 = -1, \qquad \alpha_1 + \alpha_3 = 1, \qquad \alpha_1 + 2\alpha_2 = 1$$

from which we obtain $\alpha_1 = 3$, $\alpha_2 = -1$, $\alpha_3 = -2$. Hence $D = 3A - B - 2C$.

Check:

$$\begin{aligned}3A - B - 2C &= 3\begin{bmatrix} 1 & 0 \\ 1 & 1 \end{bmatrix} - \begin{bmatrix} 0 & -1 \\ 0 & 2 \end{bmatrix} - 2\begin{bmatrix} 0 & 1 \\ 1 & 0 \end{bmatrix}\\ &= \begin{bmatrix} 3 & 0 \\ 3 & 3 \end{bmatrix} + \begin{bmatrix} 0 & 1 \\ 0 & -2 \end{bmatrix} + \begin{bmatrix} 0 & -2 \\ -2 & 0 \end{bmatrix} = \begin{bmatrix} 3 & -1 \\ 1 & 1 \end{bmatrix} = D\end{aligned}$$

PROBLEM 3-27 Find a value of k for which the vector

$$\mathbf{u} = \begin{bmatrix} -2 \\ 1 \\ k \end{bmatrix}$$

in R^3 will be a linear combination of the vectors

$$\mathbf{v}_1 = \begin{bmatrix} 0 \\ 3 \\ -2 \end{bmatrix} \quad \text{and} \quad \mathbf{v}_2 = \begin{bmatrix} -1 \\ 2 \\ -5 \end{bmatrix}$$

Solution Let $\mathbf{u} = \alpha_1 \mathbf{v}_1 + \alpha_2 \mathbf{v}_2$. That is,

$$\begin{bmatrix} -2 \\ 1 \\ k \end{bmatrix} = \alpha_1 \begin{bmatrix} 0 \\ 3 \\ -2 \end{bmatrix} + \alpha_2 \begin{bmatrix} -1 \\ 2 \\ -5 \end{bmatrix} = \begin{bmatrix} -\alpha_2 \\ 3\alpha_1 + 2\alpha_2 \\ -2\alpha_1 - 5\alpha_2 \end{bmatrix}$$

Equating the corresponding components, we get

$$\begin{aligned} -\alpha_2 &= -2 \\ 3\alpha_1 + 2\alpha_2 &= 1 \\ -2\alpha_1 - 5\alpha_2 &= k \end{aligned}$$

From the first and the second equations, $\alpha_2 = 2$, $\alpha_1 = -1$. Substituting these values into the last equation, we get $k = -8$.

PROBLEM 3-28 Show that $\sin(1 + x)$ is a linear combination of $\cos x$ and $\sin x$ for all x, but $\sin 2x$ is not a linear combination of $\cos x$ and $\sin x$.

Solution Using the trigonometric identity $\sin(a + b) = \sin a \cos b + \cos a \sin b$, we have $\sin(1 + x) = (\sin 1)\cos x + (\cos 1)\sin x$ for all x. Thus $\sin(1 + x)$ is a linear combination of $\cos x$ and $\sin x$.

Next, suppose $\sin 2x$ is a linear combination of $\cos x$ and $\sin x$, that is, $\sin 2x = \alpha_1 \cos x + \alpha_2 \sin x$ for all x. Setting $x = 0$, we have (since $\cos 0 = 1$, $\sin 0 = 0$) $0 = \alpha_1(1) + \alpha_2(0)$, so that $\alpha_1 = 0$. Then $\sin 2x = \alpha_2 \sin x$.

But from the trigonometric identity $\sin 2x = 2\cos x \sin x$, we have $\alpha_2 = 2\cos x$ for all x. This is impossible since α_2 is a fixed number. Therefore, there are no such α_1, α_2 such that $\sin 2x = \alpha_1 \cos x + \alpha_2 \sin x$. Thus the function $\sin 2x$ is not a linear combination of $\cos x$ and $\sin x$.

PROBLEM 3-29 Show that

$$\mathbf{v}_1 = \begin{bmatrix} 1 \\ 0 \\ 0 \end{bmatrix}, \qquad \mathbf{v}_2 = \begin{bmatrix} 1 \\ 1 \\ 0 \end{bmatrix}, \qquad \mathbf{v}_3 = \begin{bmatrix} 1 \\ 1 \\ 1 \end{bmatrix}$$

span R^3; then express

$$\mathbf{u} = \begin{bmatrix} 5 \\ 3 \\ 1 \end{bmatrix}$$

as a linear combination of $\mathbf{v}_1$, $\mathbf{v}_2$, and $\mathbf{v}_3$.

Solution Let

$$\mathbf{u} = \begin{bmatrix} u_1 \\ u_2 \\ u_3 \end{bmatrix}$$

be any vector in R^3, and assume that

$$\mathbf{u} = \alpha_1\mathbf{v}_1 + \alpha_2\mathbf{v}_2 + \alpha_3\mathbf{v}_3$$

or

$$\begin{bmatrix} u_1 \\ u_2 \\ u_3 \end{bmatrix} = \alpha_1 \begin{bmatrix} 1 \\ 0 \\ 0 \end{bmatrix} + \alpha_2 \begin{bmatrix} 1 \\ 1 \\ 0 \end{bmatrix} + \alpha_3 \begin{bmatrix} 1 \\ 1 \\ 1 \end{bmatrix} = \begin{bmatrix} \alpha_1 + \alpha_2 + \alpha_3 \\ \alpha_2 + \alpha_3 \\ \alpha_3 \end{bmatrix}$$

Equating the corresponding components yields

$$\begin{aligned} \alpha_1 + \alpha_2 + \alpha_3 &= u_1 \\ \alpha_2 + \alpha_3 &= u_2 \\ \alpha_3 &= u_3 \end{aligned}$$

Solving this system, we get

$$\alpha_1 = u_1 - u_2, \qquad \alpha_2 = u_2 - u_3, \qquad \alpha_3 = u_3$$

Thus

$$\mathbf{u} = (u_1 - u_2)\mathbf{v}_1 + (u_2 - u_3)\mathbf{v}_2 + u_3\mathbf{v}_3 \quad \textbf{(a)}$$

so $\{\mathbf{v}_1, \mathbf{v}_2, \mathbf{v}_3\}$ spans R^3.

Now substituting $u_1 - u_2 = 2$, $u_2 - u_3 = 2$, $u_3 = 1$ into (a) for the given $\mathbf{u} = \begin{bmatrix} u_1 \\ u_2 \\ u_3 \end{bmatrix} = \begin{bmatrix} 5 \\ 3 \\ 1 \end{bmatrix}$, we get

$$\begin{bmatrix} 5 \\ 3 \\ 1 \end{bmatrix} = 2\begin{bmatrix} 1 \\ 0 \\ 0 \end{bmatrix} + 2\begin{bmatrix} 1 \\ 1 \\ 0 \end{bmatrix} + 1\begin{bmatrix} 1 \\ 1 \\ 1 \end{bmatrix}$$

PROBLEM 3-30 Show that the set $S = \{1 + x, 1 + x^2, x + x^2\}$ spans the vector space P_2.

Solution The set S spans P_2 if and only if any element $a_0 + a_1x + a_2x^2$ of P_2 can be expressed as a linear combination of the elements of S. Hence S spans P_2 if and only if we can find scalars α_1, α_2, and α_3 such that

$$\alpha_1(1 + x) + \alpha_2(1 + x^2) + \alpha_3(x + x^2) = a_0 + a_1x + a_2x^2$$

for arbitrary a_0, a_1, and a_2. This equation can be rewritten as

$$(\alpha_1 + \alpha_2) + (\alpha_1 + \alpha_3)x + (\alpha_2 + \alpha_3)x^2 = a_0 + a_1x + a_2x^2$$

Equating the coefficients of like powers of x yields

$$\begin{aligned} \alpha_1 + \alpha_2 \qquad &= a_0 \\ \alpha_1 \qquad + \alpha_3 &= a_1 \\ \alpha_2 + \alpha_3 &= a_2 \end{aligned} \quad \textbf{(a)}$$

Using the matrix notation, this system of equations becomes

$$\begin{bmatrix} 1 & 1 & 0 \\ 1 & 0 & 1 \\ 0 & 1 & 1 \end{bmatrix}\begin{bmatrix} \alpha_1 \\ \alpha_2 \\ \alpha_3 \end{bmatrix} = \begin{bmatrix} a_0 \\ a_1 \\ a_2 \end{bmatrix} \quad \textbf{(b)}$$

Since

$$\det\begin{bmatrix} 1 & 1 & 0 \\ 1 & 0 & 1 \\ 0 & 1 & 1 \end{bmatrix} = \begin{vmatrix} 1 & 1 & 0 \\ 1 & 0 & 1 \\ 0 & 1 & 1 \end{vmatrix} = -2 \neq 0$$

by Theorem 2-3.2, the matrix

$$\begin{bmatrix} 1 & 1 & 0 \\ 1 & 0 & 1 \\ 0 & 1 & 1 \end{bmatrix}$$

is nonsingular so we know that Eq. (b), and consequently Eq. (a), has a nonzero solution α_1, α_2, and α_3 for arbitrary a_0, a_1, and a_2 (Theorem 1-4.3). Thus S spans P_2.

Linear Independence

PROBLEM 3-31 Are the vectors

$$\mathbf{v}_1 = \begin{bmatrix} 1 \\ 2 \\ -1 \end{bmatrix}, \qquad \mathbf{v}_2 = \begin{bmatrix} 1 \\ -4 \\ 6 \end{bmatrix}, \qquad \mathbf{v}_3 = \begin{bmatrix} 3 \\ 0 \\ 4 \end{bmatrix}$$

in R^3 linearly independent or linearly dependent?

Solution Consider the equation

$$\alpha_1\mathbf{v}_1 + \alpha_2\mathbf{v}_2 + \alpha_3\mathbf{v}_3 = \mathbf{0}$$

that is,

$$\alpha_1\begin{bmatrix}1\\2\\-1\end{bmatrix} + \alpha_2\begin{bmatrix}1\\-4\\6\end{bmatrix} + \alpha_3\begin{bmatrix}3\\0\\4\end{bmatrix} = \begin{bmatrix}0\\0\\0\end{bmatrix}$$

or

$$\begin{bmatrix}\alpha_1 + \alpha_2 + 3\alpha_3\\2\alpha_1 - 4\alpha_2\\-\alpha_1 + 6\alpha_2 + 4\alpha_3\end{bmatrix} = \begin{bmatrix}0\\0\\0\end{bmatrix}$$

Equating the corresponding components gives

$$\begin{aligned}\alpha_1 + \alpha_2 + 3\alpha_3 &= 0\\2\alpha_1 - 4\alpha_2 \quad\quad &= 0\\-\alpha_1 + 6\alpha_2 + 4\alpha_3 &= 0\end{aligned}$$

Solving this system yields $\alpha_1 = 2$, $\alpha_2 = 1$, $\alpha_3 = -1$, and we get

$$2\mathbf{v}_1 + \mathbf{v}_2 - \mathbf{v}_3 = \mathbf{0}$$

Therefore, the given vectors are linearly dependent.

Alternate Solution Since

$$\det[\mathbf{v}_1, \mathbf{v}_2, \mathbf{v}_3] = \begin{vmatrix}1 & 1 & 3\\2 & -4 & 0\\-1 & 6 & 4\end{vmatrix} = 0$$

by Theorem 3-6.1, $\mathbf{v}_1$, $\mathbf{v}_2$, $\mathbf{v}_3$ are linearly dependent.

PROBLEM 3-32 Determine whether or not the following vectors are linearly independent in R^3.

(a) $\begin{bmatrix}2\\1\\-2\end{bmatrix}, \begin{bmatrix}2\\2\\0\end{bmatrix}, \begin{bmatrix}3\\2\\-2\end{bmatrix}$

(b) $\begin{bmatrix}1\\0\\1\end{bmatrix}, \begin{bmatrix}0\\1\\3\end{bmatrix}, \begin{bmatrix}3\\2\\4\end{bmatrix}$

Solution It will be easier to apply Theorem 3-6.1 to solve this problem.

(a) Since

$$\begin{vmatrix}2 & 2 & 3\\1 & 2 & 2\\-2 & 0 & -2\end{vmatrix} = 0$$

the vectors are linearly dependent.

(b) Since

$$\begin{vmatrix}1 & 0 & 3\\0 & 1 & 2\\1 & 3 & 4\end{vmatrix} = -5 \neq 0$$

the vectors are linearly independent.

PROBLEM 3-33 Determine whether or not the matrices $A, B, C \in R_{2\times 2}$ are independent where

$$A = \begin{bmatrix} 1 & 1 \\ 1 & 1 \end{bmatrix}, \qquad B = \begin{bmatrix} 0 & 1 \\ 1 & 0 \end{bmatrix}, \qquad C = \begin{bmatrix} 1 & 1 \\ 0 & 0 \end{bmatrix}$$

Solution Let $\alpha_1, \alpha_2, \alpha_3$ be any scalars, and suppose

$$\alpha_1 A + \alpha_2 B + \alpha_3 C = O$$

That is,

$$\alpha_1 \begin{bmatrix} 1 & 1 \\ 1 & 1 \end{bmatrix} + \alpha_2 \begin{bmatrix} 0 & 1 \\ 1 & 0 \end{bmatrix} + \alpha_3 \begin{bmatrix} 1 & 1 \\ 0 & 0 \end{bmatrix} = \begin{bmatrix} 0 & 0 \\ 0 & 0 \end{bmatrix}$$

or

$$\begin{bmatrix} \alpha_1 & \alpha_1 \\ \alpha_1 & \alpha_1 \end{bmatrix} + \begin{bmatrix} 0 & \alpha_2 \\ \alpha_2 & 0 \end{bmatrix} + \begin{bmatrix} \alpha_3 & \alpha_3 \\ 0 & 0 \end{bmatrix} = \begin{bmatrix} 0 & 0 \\ 0 & 0 \end{bmatrix}$$

or

$$\begin{bmatrix} \alpha_1 + \alpha_3 & \alpha_1 + \alpha_2 + \alpha_3 \\ \alpha_1 + \alpha_2 & \alpha_1 \end{bmatrix} = \begin{bmatrix} 0 & 0 \\ 0 & 0 \end{bmatrix}$$

Equating the corresponding entries, we get

$$\begin{aligned} \alpha_1 + \ \alpha_3 &= 0 \\ \alpha_1 + \alpha_2 + \alpha_3 &= 0 \\ \alpha_1 + \alpha_2 &= 0 \\ \alpha_1 &= 0 \end{aligned}$$

Solving the above system, we get $\alpha_1 = \alpha_2 = \alpha_3 = 0$. Therefore A, B, C are linearly independent.

PROBLEM 3-34 Show that the polynomials $1, x, x^2$, and x^3 are linearly independent in P_4.

Solution Suppose that there are scalars a_0, a_1, a_2, and a_3 such that

$$a_0(1) + a_1 x + a_2 x^2 + a_3 x^3 = 0$$

Since a polynomial is zero only if all its coefficients are zero, then $a_0 = a_1 = a_2 = a_3 = 0$. Thus, we conclude that $1, x, x^2, x^3$, are linearly independent.

***PROBLEM 3-35** [*For readers who have studied calculus.*] Show that $f(x) = \sin x$, $g(x) = \cos x$, and $h(x) = x$ are linearly independent in $F[I]$.

Solution Suppose that there are scalars a, b, and c such that $af(x) + bg(x) + ch(x) = 0$ or

$$a \sin x + b \cos x + cx = 0 \qquad \textbf{(a)}$$

Differentiating Eq. (a) with respect to x, we get

$$a \cos x - b \sin x + c = 0 \qquad \textbf{(b)}$$

Differentiating Eq. (b) with respect to x, we get

$$-a \sin x - b \cos x = 0 \qquad \textbf{(c)}$$

Differentiating Eq. (c) with respect to x, we get

$$-a \cos x + b \sin x = 0 \qquad \textbf{(d)}$$

Adding Eqs. (b) and (d), we have $c = 0$

Multiplying Eq. (c) by $\sin x$ and Eq. (d) by $\cos x$ and adding,

$$\begin{array}{ll} (\sin x) \times (\text{c}) & -a \sin^2 x - b \cos x \sin x = 0 \\ (\cos x) \times (\text{d}) & \underline{-a \cos^2 x + b \sin x \cos x = 0} \quad (+ \\ & -a(\sin^2 x + \cos^2 x) = 0 \end{array}$$

we obtain $a = 0$, since $\sin^2 x + \cos^2 x = 1$. Then, multiplying Eq. (c) by $\cos x$ and Eq. (d) by $(-\sin x)$ and adding,

$$\begin{array}{lr} (\cos x) \times (c) & -a \sin x \cos x - b \cos^2 x = 0 \\ (-\sin x) \times (d) & a \cos x \sin x - b \sin^2 x = 0 \\ \hline & -b(\cos^2 x + \sin^2 x) = 0 \end{array} \quad (+$$

we obtain $b = 0$.

Since $a \sin x + b \cos x + cx = 0$ implies $a = b = c = 0$, we conclude that $\sin x$, $\cos x$, and x are linearly independent.

Alternate Solution Suppose that there are scalars a, b, and c such that

$$a \sin x + b \cos x + cx = 0 \qquad \textbf{(a)}$$

Setting $x = 0, \pi/2, \pi$, respectively, we have

$$a \sin 0 + b \cos 0 + c(0) = 0 \qquad \textbf{(b)}$$

$$a \sin \frac{\pi}{2} + b \cos \frac{\pi}{2} + c\left(\frac{\pi}{2}\right) = 0 \qquad \textbf{(c)}$$

$$a \sin \pi + b \cos \pi + c(\pi) = 0 \qquad \textbf{(d)}$$

Since $\sin 0 = 0$, $\cos 0 = 1$, $\sin(\pi/2) = 1$, $\cos(\pi/2) = 0$, $\sin \pi = 0$, $\cos \pi = -1$, from Eqs. (b), (c), and (d), we obtain the system

$$b = 0$$

$$a + \left(\frac{\pi}{2}\right)c = 0$$

$$-b + (\pi)c = 0$$

Solving for a, b, and c we get $a = b = c = 0$. Thus $\sin x$, $\cos x$, and x are linearly independent.

PROBLEM 3-36 Show that two vectors $\mathbf{v}_1$ and $\mathbf{v}_2$ are linearly dependent if and only if one of the vectors is a scalar multiple of the other.

Solution Let's assume that $\mathbf{v}_1$ and $\mathbf{v}_2$ are linearly dependent; then there are scalars α_1 and α_2, not both 0, such that

$$\alpha_1 \mathbf{v}_1 + \alpha_2 \mathbf{v}_2 = \mathbf{0}$$

This equation can be rewritten as

$$\mathbf{v}_1 = -\left(\frac{\alpha_2}{\alpha_1}\right)\mathbf{v}_2 \qquad \text{or} \qquad \mathbf{v}_2 = -\left(\frac{\alpha_1}{\alpha_2}\right)\mathbf{v}_1$$

which shows that $\mathbf{v}_1$ is a scalar multiple of $\mathbf{v}_2$, or $\mathbf{v}_2$ is a scalar multiple of $\mathbf{v}_1$. Conversely, if one of the vectors is a scalar multiple of the other, say,

$$\mathbf{v}_1 = k\mathbf{v}_2$$

where k is any nonzero scalar, then this equation can be rewritten as

$$\mathbf{v}_1 - k\mathbf{v}_2 = \mathbf{0}$$

which indicates that in the expression

$$\alpha_1 \mathbf{v}_1 + \alpha_2 \mathbf{v}_2 = \mathbf{0}$$

$\alpha_1 = 1 \neq 0$, $\alpha_2 = -k \neq 0$. Thus $\mathbf{v}_1$ and $\mathbf{v}_2$ are linearly dependent.

PROBLEM 3-37 Let $S = \{\mathbf{v}_1, \mathbf{v}_2, \ldots, \mathbf{v}_n\}$ be a set of vectors in a vector space V. Then show that S is linearly dependent if and only if one of the vectors is a linear combination of the remaining vectors in the set.

Solution Assume that S is a linearly dependent set. Then

$$\alpha_1 \mathbf{v}_1 + \alpha_2 \mathbf{v}_2 + \cdots + \alpha_n \mathbf{v}_n = \mathbf{0}$$

where $\alpha_1, \alpha_2, \ldots, \alpha_n$ are not all zero. Assume $\alpha_i \neq 0$. Then

$$\alpha_i \mathbf{v}_i = (-\alpha_1)\mathbf{v}_1 + \cdots + (-\alpha_{i-1})\mathbf{v}_{i-1} + (-\alpha_{i+1})\mathbf{v}_{i+1} + \cdots + (-\alpha_n)\mathbf{v}_n$$

or

$$\mathbf{v}_i = \left(-\frac{\alpha_1}{\alpha_i}\right)\mathbf{v}_1 + \cdots + \left(-\frac{\alpha_{i-1}}{\alpha_i}\right)\mathbf{v}_{i-1} + \left(-\frac{\alpha_{i+1}}{\alpha_i}\right)\mathbf{v}_{i+1} + \cdots + \left(-\frac{\alpha_n}{\alpha_i}\right)\mathbf{v}_n$$

Hence, $\mathbf{v}_i$ has been expressed as a linear combination of the remaining vectors $\mathbf{v}_1, \ldots, \mathbf{v}_{i-1}, \mathbf{v}_{i+1}, \ldots, \mathbf{v}_n$.

On the other hand, assume that one of the vectors, say $\mathbf{v}_i$, is a linear combination of $\{\mathbf{v}_1, \ldots, \mathbf{v}_{i-1}, \mathbf{v}_{i+1}, \ldots, \mathbf{v}_n\}$. Then

$$\mathbf{v}_i = \beta_1 \mathbf{v}_1 + \cdots + \beta_{i-1}\mathbf{v}_{i-1} + \beta_{i+1}\mathbf{v}_{i+1} + \cdots + \beta_n \mathbf{v}_n$$

or

$$\beta_1 \mathbf{v}_1 + \cdots + \beta_{i-1}\mathbf{v}_{i-1} + (-1)\mathbf{v}_i + \beta_{i+1}\mathbf{v}_{i+1} + \cdots + \beta_n \mathbf{v}_n = \mathbf{0}$$

Since the coefficient of $\mathbf{v}_i$ is $-1 \neq 0$, the vector set $S = \{\mathbf{v}_1, \mathbf{v}_2, \ldots, \mathbf{v}_n\}$ is linearly dependent.

PROBLEM 3-38 Let $S = \{\mathbf{a}_1, \mathbf{a}_2, \ldots, \mathbf{a}_m\}$ be a set of vectors in R^n. Show that if $m > n$, then S is linearly dependent.

Solution As $\dim R^n = n$, by Theorem 3-7.1 S is linearly dependent since $m > n$.

***PROBLEM 3-39** [*For readers who have studied calculus.*] If $\mathbf{f}_1, \mathbf{f}_2, \ldots, \mathbf{f}_n$ are vectors in $C^{(n-1)}[a, b]$, then the function $W[\mathbf{f}_1, \mathbf{f}_2, \ldots, \mathbf{f}_n](x)$ on $[a, b]$ defined by

$$W[\mathbf{f}_1, \mathbf{f}_2, \ldots, \mathbf{f}_n](x) = \begin{vmatrix} f_1(x) & f_2(x) & \cdots & f_n(x) \\ f_1'(x) & f_2'(x) & \cdots & f_n'(x) \\ \vdots & \vdots & & \vdots \\ f_1^{(n-1)}(x) & f_2^{(n-1)}(x) & \cdots & f_n^{(n-1)}(x) \end{vmatrix}$$

is called the Wronskian of $\mathbf{f}_1, \mathbf{f}_2, \ldots, \mathbf{f}_n$.

Prove that if there exists a point x_0 in $[a, b]$ such that $W[\mathbf{f}_1, \mathbf{f}_2, \ldots, \mathbf{f}_n](x_0) \neq 0$, then $\mathbf{f}_1, \mathbf{f}_2, \ldots, \mathbf{f}_n$ are linearly independent.

Solution Let $\alpha_1, \alpha_2, \ldots, \alpha_n$ be scalars and $\mathbf{f}_1, \mathbf{f}_2, \ldots, \mathbf{f}_n$ be vectors in $C^{(n-1)}[a, b]$. Consider the equation

$$\alpha_1 \mathbf{f}_1 + \alpha_2 \mathbf{f}_2 + \cdots + \alpha_n \mathbf{f}_n = \mathbf{0}$$

or

$$\alpha_1 f_1(x) + \alpha_2 f_2(x) + \cdots + \alpha_n f_n(x) = 0 \tag{a}$$

Taking the derivative with respect to x of both sides of (a) yields

$$\alpha_1 f_1'(x) + \alpha_2 f_2'(x) + \cdots + \alpha_n f_n' = 0$$

If we continue taking derivatives of both sides up to the $(n-1)$th derivative, we obtain the system

$$\begin{aligned} \alpha_1 f_1(x) + \alpha_1 f_2(x) + \cdots + \alpha_n f_n(x) &= 0 \\ \alpha_1 f_1'(x) + \alpha_2 f_2'(x) + \cdots + \alpha_n f_n'(x) &= 0 \\ &\vdots \\ \alpha_1 f_1^{(n-1)}(x) + \alpha_2 f_2^{(n-1)}(x) + \cdots + \alpha_n f_n^{(n-1)}(x) &= 0 \end{aligned} \tag{b}$$

which can be rewritten as the matrix equation

$$\begin{bmatrix} f_1(x) & f_2(x) & \cdots & f_n(x) \\ f_1'(x) & f_2'(x) & \cdots & f_n'(x) \\ \vdots & \vdots & & \vdots \\ f_1^{(n-1)}(x) & f_2^{(n-1)}(x) & \cdots & f_n^{(n-1)}(x) \end{bmatrix} \begin{bmatrix} \alpha_1 \\ \alpha_2 \\ \vdots \\ \alpha_n \end{bmatrix} = \begin{bmatrix} 0 \\ 0 \\ \vdots \\ 0 \end{bmatrix} \tag{c}$$

From Theorem 1-4.4 we see that, when the coefficient matrix of (c) is nonsingular for each x in $[a, b]$, then $\alpha_1 = \alpha_2 = \cdots = \alpha_n = 0$ is the only solution to (c). Thus, by the definition of linear independence, $\mathbf{f}_1, \mathbf{f}_2, \ldots, \mathbf{f}_n$ are linearly independent. Next, by Theorem 2-3.2, if the matrix is nonsingular, its determinant is nonzero. Thus we conclude that if the Wronskian of $\mathbf{f}_1, \mathbf{f}_2, \ldots, \mathbf{f}_n$ is nonzero, then $\mathbf{f}_1, \mathbf{f}_2, \ldots, \mathbf{f}_n$ are linearly independent.

*For readers who have studied calculus**

***PROBLEM 3-40** Using the Wronskian, rework Example 3-27.

Solution Since

$$\mathbf{p}_1(x) = p_1(x) = 1, \qquad \mathbf{p}_2(x) = p_2(x) = 1 + x, \qquad \mathbf{p}_3(x) = p_3(x) = 1 + x + x^2,$$

the Wronskian of $\mathbf{p}_1, \mathbf{p}_2, \mathbf{p}_3$ is given by

$$W[\mathbf{p}_1, \mathbf{p}_2, \mathbf{p}_3](x) = \begin{vmatrix} p_1(x) & p_2(x) & p_3(x) \\ p_1'(x) & p_2'(x) & p_3'(x) \\ p_1''(x) & p_2''(x) & p_3''(x) \end{vmatrix}$$

$$= \begin{vmatrix} 1 & 1 + x & 1 + x + x^2 \\ 0 & 1 & 1 + 2x \\ 0 & 0 & 2 \end{vmatrix} = 2$$

Since $W[\mathbf{p}_1, \mathbf{p}_2, \mathbf{p}_3] \neq 0$, we conclude that $1, 1 + x, 1 + x + x^2$ are linearly independent.

***PROBLEM 3-41** Show that $\mathbf{f}_1 = \cos \pi x$, $\mathbf{f}_2 = \sin \pi x$ are linearly independent in $C[-1, 1]$.

Solution Since

$$W[\mathbf{f}_1, \mathbf{f}_2](x) = \begin{vmatrix} \cos \pi x & \sin \pi x \\ -\pi \sin \pi x & \pi \cos \pi x \end{vmatrix}$$

$$= \pi \cos^2 \pi x + \pi \sin^2 \pi x$$

$$= \pi(\cos^2 \pi x + \sin^2 \pi x) = \pi \neq 0$$

$\cos \pi x$, $\sin \pi x$ are linearly independent.

Basis and Dimension

PROBLEM 3-42 In Problem 3-15 we have shown that

$$U = \left\{ \begin{bmatrix} u_1 \\ u_2 \\ u_3 \end{bmatrix} \middle| \; u_1 + u_2 + u_3 = 0 \right\}$$

is a subspace of R^3. Find a basis for U and the dimension of U.

Solution If we let $u_2 = \alpha$ and $u_3 = \beta$, we have $u_1 = -\alpha - \beta$ so that each vector $\mathbf{u}$ in U can be written as

$$\mathbf{u} = \begin{bmatrix} u_1 \\ u_2 \\ u_3 \end{bmatrix} = \begin{bmatrix} -\alpha - \beta \\ \alpha \\ \beta \end{bmatrix} = \alpha \begin{bmatrix} -1 \\ 1 \\ 0 \end{bmatrix} + \beta \begin{bmatrix} -1 \\ 0 \\ 1 \end{bmatrix}$$

Therefore if we let

$$\mathbf{v}_1 = \begin{bmatrix} -1 \\ 1 \\ 0 \end{bmatrix} \quad \text{and} \quad \mathbf{v}_2 = \begin{bmatrix} -1 \\ 0 \\ 1 \end{bmatrix}$$

then

$$\mathbf{u} = \alpha \mathbf{v}_1 + \beta \mathbf{v}_2$$

so that $U = S\{\mathbf{v}_1, \mathbf{v}_2\}$, the span of $\mathbf{v}_1$ and $\mathbf{v}_2$.

Furthermore, $\mathbf{v}_1$ and $\mathbf{v}_2$ are not scalar multiples of each other; thus they are independent (see Problem 3-36). Hence the vectors $\mathbf{v}_1$ and $\mathbf{v}_2$ form a basis for U and $\dim U = 2$.

PROBLEM 3-43 Let $R_{2\times 3}$ be the vector space of all 2×3 matrices. Then show that the matrices

$$M_{11} = \begin{bmatrix} 1 & 0 & 0 \\ 0 & 0 & 0 \end{bmatrix}, \qquad M_{12} = \begin{bmatrix} 0 & 1 & 0 \\ 0 & 0 & 0 \end{bmatrix}, \qquad M_{13} = \begin{bmatrix} 0 & 0 & 1 \\ 0 & 0 & 0 \end{bmatrix}$$

$$M_{21} = \begin{bmatrix} 0 & 0 & 0 \\ 1 & 0 & 0 \end{bmatrix}, \qquad M_{22} = \begin{bmatrix} 0 & 0 & 0 \\ 0 & 1 & 0 \end{bmatrix}, \qquad M_{23} = \begin{bmatrix} 0 & 0 & 0 \\ 0 & 0 & 1 \end{bmatrix}$$

form a basis and $\dim R_{2\times 3} = 6$.

Solution We need to show that $\{M_{ij}\}$ $(i = 1, 2; j = 1, 2, 3)$ is independent and spans $R_{2\times 3}$.

Now suppose that there are scalars α_{ij} such that

$$\alpha_{11}M_{11} + \alpha_{12}M_{12} + \alpha_{13}M_{13} + \alpha_{21}M_{21} + \alpha_{22}M_{22} + \alpha_{23}M_{23} = O$$

or

$$\begin{bmatrix} \alpha_{11} & 0 & 0 \\ 0 & 0 & 0 \end{bmatrix} + \begin{bmatrix} 0 & \alpha_{12} & 0 \\ 0 & 0 & 0 \end{bmatrix} + \begin{bmatrix} 0 & 0 & \alpha_{13} \\ 0 & 0 & 0 \end{bmatrix} + \begin{bmatrix} 0 & 0 & 0 \\ \alpha_{21} & 0 & 0 \end{bmatrix} + \begin{bmatrix} 0 & 0 & 0 \\ 0 & \alpha_{22} & 0 \end{bmatrix} + \begin{bmatrix} 0 & 0 & 0 \\ 0 & 0 & \alpha_{23} \end{bmatrix} = \begin{bmatrix} 0 & 0 & 0 \\ 0 & 0 & 0 \end{bmatrix}$$

or

$$\begin{bmatrix} \alpha_{11} & \alpha_{12} & \alpha_{13} \\ \alpha_{21} & \alpha_{22} & \alpha_{23} \end{bmatrix} = \begin{bmatrix} 0 & 0 & 0 \\ 0 & 0 & 0 \end{bmatrix}$$

Equating the corresponding entries, we obtain

$$\alpha_{11} = \alpha_{12} = \alpha_{13} = \alpha_{21} = \alpha_{22} = \alpha_{23} = 0$$

Thus $\{M_{ij}\}$ is independent.

Next, let $A = [a_{ij}]_{2\times 3}$ be any matrix in $R_{2\times 3}$. Then

$$A = a_{11}M_{11} + a_{12}M_{12} + a_{13}M_{13} + a_{21}M_{21} + a_{22}M_{22} + a_{23}M_{23}$$

Hence $\{M_{ij}\}$ spans $R_{2\times 3}$. Thus $\{M_{ij}\}$ is a basis for $R_{2\times 3}$ and $\dim R_{2\times 3} = 6$.

PROBLEM 3-44 Let V be the vector space of 2×2 symmetric matrices. Show that $\dim V = 3$.

Solution Recall that $A = [a_{ij}]$ is symmetric if $A^T = A$ or equivalently $a_{ji} = a_{ij}$ [see (1.53) and (1.54)]. Thus an arbitrary 2×2 symmetric matrix is of the form

$$A = \begin{bmatrix} a & b \\ b & c \end{bmatrix}$$

Note that there are three variables; thus setting (1) $a = 1, b = 0, c = 0$, (2) $a = 0, b = 1, c = 0$, and (3) $a = 0, b = 0, c = 1$, we have

$$M_1 = \begin{bmatrix} 1 & 0 \\ 0 & 0 \end{bmatrix}, \qquad M_2 = \begin{bmatrix} 0 & 1 \\ 1 & 0 \end{bmatrix}, \qquad M_3 = \begin{bmatrix} 0 & 0 \\ 0 & 1 \end{bmatrix}$$

Now we show that M_1, M_2, and M_3 form a basis for V. First, suppose there are scalars α_1, α_2, and α_3 such that

$$\alpha_1 M_1 + \alpha_2 M_2 + \alpha_3 M_3 = O$$

or

$$\alpha_1 \begin{bmatrix} 1 & 0 \\ 0 & 0 \end{bmatrix} + \alpha_2 \begin{bmatrix} 0 & 1 \\ 1 & 0 \end{bmatrix} + \alpha_3 \begin{bmatrix} 0 & 0 \\ 0 & 1 \end{bmatrix} = \begin{bmatrix} 0 & 0 \\ 0 & 0 \end{bmatrix}$$

or

$$\begin{bmatrix} \alpha_1 & \alpha_2 \\ \alpha_2 & \alpha_3 \end{bmatrix} = \begin{bmatrix} 0 & 0 \\ 0 & 0 \end{bmatrix}$$

Equating the corresponding entries, we get $\alpha_1 = \alpha_2 = \alpha_3 = 0$. Thus M_1, M_2, and M_3 are linearly independent.

Next, for any arbitrary 2×2 symmetric matrix A, we can write

$$A = \begin{bmatrix} a & b \\ b & c \end{bmatrix} = a \begin{bmatrix} 1 & 0 \\ 0 & 0 \end{bmatrix} + b \begin{bmatrix} 0 & 1 \\ 1 & 0 \end{bmatrix} + c \begin{bmatrix} 0 & 0 \\ 0 & 1 \end{bmatrix} = aM_1 + bM_2 + cM_3$$

Thus $\{M_1, M_2, M_3\}$ spans V. Hence $\{M_1, M_2, M_3\}$ is a basis for V and $\dim V = 3$.

PROBLEM 3-45 Show that $\mathscr{B} = \{1 + x, 1 + x^2, x + x^2\}$ is a basis for P_2.

Solution Since $\dim P_2 = 3$, we need show only that $\mathscr{B}$ is linearly independent. Let α_1, α_2, and α_3 be scalars such that

$$\alpha_1(1 + x) + \alpha_2(1 + x^2) + \alpha_3(x + x^2) = 0$$

for all x. Rearranging the above equation,

$$(\alpha_1 + \alpha_2) + (\alpha_1 + \alpha_3)x + (\alpha_2 + \alpha_3)x^2 = 0$$

for all x. Thus we have

$$\begin{aligned} \alpha_1 + \alpha_2 \qquad &= 0 \\ \alpha_1 \qquad + \alpha_3 &= 0 \\ \alpha_2 + \alpha_3 &= 0 \end{aligned}$$

Solving for α_1, α_2, and α_3, we get $\alpha_1 = \alpha_2 = \alpha_3 = 0$. Thus $\mathscr{B}$ is linearly independent and $\mathscr{B}$ is a basis for P_2.

PROBLEM 3-46 Let U be a subspace of an n-dimensional vector space V. Show that U is finite-dimensional and

$$\dim U < \dim V = n$$

In particular, if $\dim U = n$, then $U = V$.

Solution Since $\dim V = n$, any $n + 1$ or more vectors are linearly dependent by Theorem 3-7.1. Furthermore, since a basis of U consists of linearly independent vectors, it cannot contain more than n elements. Thus $\dim U < n$. In particular, if $\{\mathbf{u}_1, \mathbf{u}_2, \ldots, \mathbf{u}_n\}$ is a basis for U, then since it is an independent set with n elements it is also a basis for V. Therefore $U = V$ if $\dim U = n$.

PROBLEM 3-47 For each positive integer n the vector space P_n is a subspace of the vector space V of all polynomial functions. Show that V is not a finite-dimensional vector space.

Solution From (3.48) we have

$$\dim P_n = n + 1$$

Now

$$\lim_{n \to \infty} \dim P_n = \lim_{n \to \infty} (n + 1) = \infty$$

Thus, we conclude that V is not finite-dimensional.

PROBLEM 3-48 Show that the vector space V of Problem 3-11, that is, the set of all infinite sequences of real numbers, is not a finite-dimensional vector space.

Solution It is seen that V is equivalent to R^n with $n \to \infty$. Thus

$$\dim V = \lim_{n\to\infty} \dim R^n = \lim_{n\to\infty} n = \infty$$

Hence V is not a finite-dimensional vector space.

Coordinate Vectors and Change of Basis

PROBLEM 3-49 Show that the vectors

$$\mathbf{v}_1 = \begin{bmatrix} 0 \\ 1 \\ 1 \end{bmatrix}, \qquad \mathbf{v}_2 = \begin{bmatrix} 1 \\ 0 \\ 1 \end{bmatrix}, \qquad \mathbf{v}_3 = \begin{bmatrix} 1 \\ 1 \\ 0 \end{bmatrix}$$

are a basis for R^3. Then find the coordinate vector of $\mathbf{u} = \begin{bmatrix} 2 \\ -2 \\ 4 \end{bmatrix}$ with respect to this basis.

Solution Since $\dim R^3 = 3$, we need show only that $\{\mathbf{v}_1, \mathbf{v}_2, \mathbf{v}_3\}$ is linearly independent. That follows since

$$\begin{vmatrix} 0 & 1 & 1 \\ 1 & 0 & 1 \\ 1 & 1 & 0 \end{vmatrix} = 2 \neq 0$$

Thus $\{\mathbf{v}_1, \mathbf{v}_2, \mathbf{v}_3\}$ is a basis for R^3.

Next, if (x_1, x_2, x_3) are the coordinates of $\mathbf{u}$ with respect to this basis, we must have

$$\mathbf{u} = x_1\mathbf{v}_1 + x_2\mathbf{v}_2 + x_3\mathbf{v}_3$$

or

$$\begin{bmatrix} 2 \\ -2 \\ 4 \end{bmatrix} = x_1 \begin{bmatrix} 0 \\ 1 \\ 1 \end{bmatrix} + x_2 \begin{bmatrix} 1 \\ 0 \\ 1 \end{bmatrix} + x_3 \begin{bmatrix} 1 \\ 1 \\ 0 \end{bmatrix} = \begin{bmatrix} x_2 + x_3 \\ x_1 + x_3 \\ x_1 + x_2 \end{bmatrix}$$

Equating the corresponding components,

$$\begin{aligned} x_2 + x_3 &= 2 \\ x_1 + x_3 &= -2 \\ x_1 + x_2 &= 4 \end{aligned}$$

Solving for x_1, x_2, x_3, we get $x_1 = 0$, $x_2 = 4$, $x_3 = -2$. Thus the coordinate vector of $\mathbf{u}$ with respect to this basis is $\begin{bmatrix} 0 \\ 4 \\ -2 \end{bmatrix}$.

PROBLEM 3-50 Consider a polynomial $p(x) = 1 + 2x + 3x^2$ in P_2 (see Example 3-44). Find the coordinate vector of $p(x)$ with respect to the basis $\mathscr{B} = \{1 + x, 1 + x^2, x + x^2\}$ of Problem 3-45.

Solution In order to find the coordinate vector of $p(x)$ with respect to $\mathscr{B}$, we need to find numbers a, b, and c such that

$$p(x) = a(1 + x) + b(1 + x^2) + c(x + x^2)$$

or

$$\begin{aligned} 1 + 2x + 3x^2 &= a(1 + x) + b(1 + x^2) + c(x + x^2) \\ &= (a + b) + (a + c)x + (b + c)x^2 \end{aligned}$$

Equating the coefficients of like powers of x, we get

$$a + b = 1$$
$$a + c = 2$$
$$b + c = 3$$

Solving for a, b, and c, we get $a = 0, b = 1, c = 2$. Thus the coordinate vector of $p(x)$ with respect to this basis $\mathscr{B}$ is $\begin{bmatrix} 0 \\ 1 \\ 2 \end{bmatrix}$

note: The coordinate vector of $p(x)$ must be a vector in R^3 since $\dim P_2 = 3$.

PROBLEM 3-51 Let $A = \begin{bmatrix} 4 & 2 \\ 3 & -7 \end{bmatrix} \in R_{2\times 2}$ Find the coordinate vector of A with respect to the basis

$$\mathscr{B} = \left\{ \begin{bmatrix} 1 & 0 \\ 0 & 0 \end{bmatrix}, \begin{bmatrix} 1 & 1 \\ 0 & 0 \end{bmatrix}, \begin{bmatrix} 1 & 1 \\ 0 & 1 \end{bmatrix}, \begin{bmatrix} 1 & 1 \\ 1 & 1 \end{bmatrix} \right\}$$

Solution In order to find the coordinate vector of A with respect to $\mathscr{B}$, we need to find numbers x_1, x_2, x_3, x_4 such that

$$A = \begin{bmatrix} 4 & 2 \\ 3 & -7 \end{bmatrix} = x_1 \begin{bmatrix} 1 & 0 \\ 0 & 0 \end{bmatrix} + x_2 \begin{bmatrix} 1 & 1 \\ 0 & 0 \end{bmatrix} + x_3 \begin{bmatrix} 1 & 1 \\ 0 & 1 \end{bmatrix} + x_4 \begin{bmatrix} 1 & 1 \\ 1 & 1 \end{bmatrix}$$
$$= \begin{bmatrix} x_1 & 0 \\ 0 & 0 \end{bmatrix} + \begin{bmatrix} x_2 & x_2 \\ 0 & 0 \end{bmatrix} + \begin{bmatrix} x_3 & x_3 \\ 0 & x_3 \end{bmatrix} + \begin{bmatrix} x_4 & x_4 \\ x_4 & x_4 \end{bmatrix}$$
$$= \begin{bmatrix} x_1 + x_2 + x_3 + x_4 & x_2 + x_3 + x_4 \\ x_4 & x_3 + x_4 \end{bmatrix}$$

Equating the corresponding entries yields

$$x_1 + x_2 + x_3 + x_4 = 4$$
$$x_2 + x_3 + x_4 = 2$$
$$x_3 + x_4 = -7$$
$$x_4 = 3$$

Solving for x_1, x_2, x_3, x_4, we obtain $x_1 = 2, x_2 = 9, x_3 = -10, x_4 = 3$. Thus the coordinate vector of A with respect to $\mathscr{B}$ is $\begin{bmatrix} 2 \\ 9 \\ -10 \\ 3 \end{bmatrix}$

note: The coordinate vector of A must be a vector in R^4 since $\dim R_{2\times 2} = 4$.

PROBLEM 3-52 Consider the ordered basis $\mathscr{B} = \{\mathbf{e}_1, \mathbf{e}_2\}$ and $\mathscr{B}' = \{\mathbf{u}_1, \mathbf{u}_2\}$ for R^2 where

$$\mathbf{e}_1 = \begin{bmatrix} 1 \\ 0 \end{bmatrix}, \quad \mathbf{e}_2 = \begin{bmatrix} 0 \\ 1 \end{bmatrix}, \quad \mathbf{u}_1 = \begin{bmatrix} 1 \\ 1 \end{bmatrix}, \quad \mathbf{u}_2 = \begin{bmatrix} 1 \\ 2 \end{bmatrix}$$

Find the transition matrix P from $\mathscr{B}'$ to $\mathscr{B}$ and transition matrix Q from $\mathscr{B}$ to $\mathscr{B}'$.

Solution Notice that

$$\begin{bmatrix} 1 \\ 1 \end{bmatrix} = 1 \begin{bmatrix} 1 \\ 0 \end{bmatrix} + 1 \begin{bmatrix} 0 \\ 1 \end{bmatrix}$$
$$\begin{bmatrix} 1 \\ 2 \end{bmatrix} = 1 \begin{bmatrix} 1 \\ 0 \end{bmatrix} + 2 \begin{bmatrix} 0 \\ 1 \end{bmatrix}$$

so that

$$[\mathbf{u}_1]_{\mathscr{B}} = \begin{bmatrix} 1 \\ 1 \end{bmatrix}_{\mathscr{B}} \qquad [\mathbf{u}_2]_{\mathscr{B}} = \begin{bmatrix} 1 \\ 2 \end{bmatrix}_{\mathscr{B}}$$

Therefore, by (3.59)

$$P = \begin{bmatrix} 1 & 1 \\ 1 & 2 \end{bmatrix}$$

To compute $[\mathbf{e}_1]_{\mathscr{B}'}$ and $[\mathbf{e}_2]_{\mathscr{B}'}$, we need to find scalars $\alpha_1, \alpha_2, \beta_1, \beta_2$ such that

$$\begin{bmatrix} 1 \\ 0 \end{bmatrix} = \alpha_1 \begin{bmatrix} 1 \\ 1 \end{bmatrix} + \alpha_2 \begin{bmatrix} 1 \\ 2 \end{bmatrix}$$

$$\begin{bmatrix} 0 \\ 1 \end{bmatrix} = \beta_1 \begin{bmatrix} 1 \\ 1 \end{bmatrix} + \beta_2 \begin{bmatrix} 1 \\ 2 \end{bmatrix}$$

or

$$\alpha_1 + \alpha_2 = 1, \qquad \alpha_1 + 2\alpha_2 = 0$$

$$\beta_1 + \beta_2 = 0, \qquad \beta_1 + 2\beta_2 = 1$$

Solving for $\alpha_1, \alpha_2, \beta_1, \beta_2$, we get $\alpha_1 = 2, \alpha_2 = -1, \beta_1 = -1$, and $\beta_2 = 1$. Therefore

$$\mathbf{e}_1 = 2\mathbf{u}_1 - \mathbf{u}_2$$

$$\mathbf{e}_2 = -\mathbf{u}_1 + \mathbf{u}_2$$

and

$$[\mathbf{e}_1]_{\mathscr{B}'} = \begin{bmatrix} 2 \\ -1 \end{bmatrix}, \qquad [\mathbf{e}_2]_{\mathscr{B}'} = \begin{bmatrix} -1 \\ 1 \end{bmatrix}$$

Thus, by (3.70)

$$Q = \begin{bmatrix} 2 & -1 \\ -1 & 1 \end{bmatrix}$$

Notice that

$$PQ = \begin{bmatrix} 1 & 1 \\ 1 & 2 \end{bmatrix} \begin{bmatrix} 2 & -1 \\ -1 & 1 \end{bmatrix} = \begin{bmatrix} 1 & 0 \\ 0 & 1 \end{bmatrix} = I_2$$

PROBLEM 3-53 Consider the ordered bases $\mathscr{B} = \{\mathbf{i}, \mathbf{j}, \mathbf{k}\}$ and $\mathscr{B}' = \{\mathbf{v}_1, \mathbf{v}_2, \mathbf{v}_3\}$ for R^3 where

$$\mathbf{i} = \begin{bmatrix} 1 \\ 0 \\ 0 \end{bmatrix}, \qquad \mathbf{j} = \begin{bmatrix} 0 \\ 1 \\ 0 \end{bmatrix}, \qquad \mathbf{k} = \begin{bmatrix} 0 \\ 0 \\ 1 \end{bmatrix}$$

and

$$\mathbf{v}_1 = \begin{bmatrix} 0 \\ 1 \\ 1 \end{bmatrix}, \qquad \mathbf{v}_2 = \begin{bmatrix} 1 \\ 0 \\ 1 \end{bmatrix}, \qquad \mathbf{v}_3 = \begin{bmatrix} 1 \\ 1 \\ 0 \end{bmatrix}$$

Find the transition matrix Q of $\mathscr{B}$ to $\mathscr{B}'$ and rework Problem 3-49 using Q.

Solution In Problem 3-49 we have shown that $\{\mathbf{v}_1, \mathbf{v}_2, \mathbf{v}_3\}$ forms a basis for R^3. Note also that

$$[\mathbf{v}_1]_{\mathscr{B}} = \begin{bmatrix} 0 \\ 1 \\ 1 \end{bmatrix}, \qquad [\mathbf{v}_2]_{\mathscr{B}} = \begin{bmatrix} 1 \\ 0 \\ 1 \end{bmatrix}, \qquad [\mathbf{v}_3]_{\mathscr{B}} = \begin{bmatrix} 1 \\ 1 \\ 0 \end{bmatrix}$$

Thus by (3.59) the transition matrix P from $\mathscr{B}'$ to $\mathscr{B}$ is

$$P = \begin{bmatrix} 0 & 1 & 1 \\ 1 & 0 & 1 \\ 1 & 1 & 0 \end{bmatrix}$$

Now, inverting P, we obtain the transition matrix Q from $\mathscr{B}$ to $\mathscr{B}'$. Thus

$$Q = P^{-1} = \begin{bmatrix} 0 & 1 & 1 \\ 1 & 0 & 1 \\ 1 & 1 & 0 \end{bmatrix}^{-1} = \frac{1}{2}\begin{bmatrix} -1 & 1 & 1 \\ 1 & -1 & 1 \\ 1 & 1 & -1 \end{bmatrix}$$

Now

$$\mathbf{u} = \begin{bmatrix} 2 \\ -2 \\ 4 \end{bmatrix} \quad \text{and} \quad [\mathbf{u}]_{\mathscr{B}} = \begin{bmatrix} 2 \\ -2 \\ 4 \end{bmatrix}$$

Thus, by (3.69)

$$[\mathbf{u}]_{\mathscr{B}'} = Q[\mathbf{u}]_{\mathscr{B}} = \frac{1}{2}\begin{bmatrix} -1 & 1 & 1 \\ 1 & -1 & 1 \\ 1 & 1 & -1 \end{bmatrix}\begin{bmatrix} 2 \\ -2 \\ 4 \end{bmatrix} = \begin{bmatrix} 0 \\ 4 \\ -2 \end{bmatrix}$$

which is exactly the same result obtained in Problem 3-49.

PROBLEM 3-54 Consider the ordered bases $\mathscr{B} = \{1, x, x^2\}$ and $\mathscr{B}' = \{1 + x, 1 + x^2, x + x^2\}$ for P_2. Find the transition matrix Q from $\mathscr{B}$ to $\mathscr{B}'$ and rework Problem 3-50 using Q.

Solution Let P be the transition matrix from $\mathscr{B}'$ to $\mathscr{B}$. Now, by inspection we have

$$[1 + x]_{\mathscr{B}} = \begin{bmatrix} 1 \\ 1 \\ 0 \end{bmatrix}, \qquad [1 + x^2]_{\mathscr{B}} = \begin{bmatrix} 1 \\ 0 \\ 1 \end{bmatrix}, \qquad [x + x^2]_{\mathscr{B}} = \begin{bmatrix} 0 \\ 1 \\ 1 \end{bmatrix}$$

Thus by (3.59) we get

$$P = \begin{bmatrix} 1 & 1 & 0 \\ 1 & 0 & 1 \\ 0 & 1 & 1 \end{bmatrix}$$

Inverting P, we obtain the transition matrix Q from $\mathscr{B}$ to $\mathscr{B}'$. Thus

$$Q = P^{-1} = \begin{bmatrix} 1 & 1 & 0 \\ 1 & 0 & 1 \\ 0 & 1 & 1 \end{bmatrix}^{-1} = \frac{1}{2}\begin{bmatrix} 1 & 1 & -1 \\ 1 & -1 & 1 \\ -1 & 1 & 1 \end{bmatrix}$$

Now $p(x) = 1 + 2x + 3x^2$ and

$$[p(x)]_{\mathscr{B}} = \begin{bmatrix} 1 \\ 2 \\ 3 \end{bmatrix}$$

Thus by (3.69),

$$[p(x)]_{\mathscr{B}'} = Q[p(x)]_{\mathscr{B}} = \frac{1}{2}\begin{bmatrix} 1 & 1 & -1 \\ 1 & -1 & 1 \\ -1 & 1 & 1 \end{bmatrix}\begin{bmatrix} 1 \\ 2 \\ 3 \end{bmatrix} = \begin{bmatrix} 0 \\ 1 \\ 2 \end{bmatrix}$$

which is exactly the same result obtained in Problem 3-50.

Column Space and Row Space of a Matrix

PROBLEM 3-55 Find the column space and column rank of the matrix

$$A = \begin{bmatrix} 1 & 0 & -1 \\ 2 & 1 & 3 \\ 3 & 4 & -2 \end{bmatrix}$$

Solution The column space of A is the span of $\mathbf{c}_1, \mathbf{c}_2, \mathbf{c}_3$, where

$$\mathbf{c}_1 = \begin{bmatrix} 1 \\ 2 \\ 3 \end{bmatrix}, \qquad \mathbf{c}_2 = \begin{bmatrix} 0 \\ 1 \\ 4 \end{bmatrix}, \qquad \mathbf{c}_3 = \begin{bmatrix} -1 \\ 3 \\ -2 \end{bmatrix}$$

Since

$$\begin{vmatrix} 1 & 0 & -1 \\ 2 & 1 & 3 \\ 3 & 4 & -2 \end{vmatrix} = -19 \neq 0$$

$\mathbf{c}_1, \mathbf{c}_2, \mathbf{c}_3$ are linearly independent, and thus $\{\mathbf{c}_1, \mathbf{c}_2, \mathbf{c}_3\}$ is a basis for the column space of A. Therefore the column space of A is R^3, and the column rank of A is 3 since $\dim R^3 = 3$.

PROBLEM 3-56 Find the row space and row rank of the matrix of Problem 3-55.

Solution The row space of A is the span of $\mathbf{r}_1, \mathbf{r}_2, \mathbf{r}_3$, where

$$\mathbf{r}_1 = [1, 0, -1], \qquad \mathbf{r}_2 = [2, 1, 3], \qquad \mathbf{r}_3 = [3, 4, -2]$$

Thus, the row space of A is the set of vector $\mathbf{v}$ of the form

$$\mathbf{v} = \alpha_1 \mathbf{r}_1 + \alpha_2 \mathbf{r}_2 + \alpha_3 \mathbf{r}_3$$

for any scalars α_1, α_2, and α_3.

Next, consider the equation

$$\alpha_1 \mathbf{r}_1 + \alpha_2 \mathbf{r}_2 + \alpha_3 \mathbf{r}_3 = \mathbf{0}$$

or

$$\alpha_1[1, 0, -1] + \alpha_2[2, 1, 3] + \alpha_3[3, 4, -2] = [0, 0, 0]$$

or

$$[\alpha_1 + 2\alpha_2 + 3\alpha_3, \alpha_2 + 4\alpha_3, -\alpha_1 + 3\alpha_2 - 2\alpha_3] = [0, 0, 0]$$

Equating the corresponding components yields

$$\begin{aligned} \alpha_1 + 2\alpha_2 + 3\alpha_3 &= 0 \\ \alpha_2 + 4\alpha_3 &= 0 \\ -\alpha_1 + 3\alpha_2 - 2\alpha_3 &= 0 \end{aligned} \tag{a}$$

Now

$$\begin{vmatrix} 1 & 2 & 3 \\ 0 & 1 & 4 \\ -1 & 3 & -2 \end{vmatrix} = -19 \neq 0$$

Thus the coefficient matrix of (a) is nonsingular and $\alpha_1 = \alpha_2 = \alpha_3 = 0$ is the only solution of (a). Hence $\mathbf{r}_1, \mathbf{r}_2, \mathbf{r}_3$ are linearly independent and $\{\mathbf{r}_1, \mathbf{r}_2, \mathbf{r}_3\}$ is a basis for the row space of A. Thus the row rank of A is again 3.

PROBLEM 3-57 Find a basis for the space spanned by the vectors

$$\mathbf{w}_1 = \begin{bmatrix} 1 \\ -3 \\ 1 \end{bmatrix}, \quad \mathbf{w}_2 = \begin{bmatrix} -2 \\ 6 \\ 0 \end{bmatrix}, \quad \mathbf{w}_3 = \begin{bmatrix} -4 \\ 12 \\ 1 \end{bmatrix}, \quad \mathbf{w}_4 = \begin{bmatrix} 3 \\ -9 \\ 3 \end{bmatrix}$$

Indicate its dimension.

Solution The space spanned by these vectors is the same as the column space of a matrix given by

$$A = \begin{bmatrix} 1 & -2 & -4 & 3 \\ -3 & 6 & 12 & -9 \\ 1 & 0 & 1 & 3 \end{bmatrix}$$

Transposing and reducing to row echelon form, we have

$$A^T = \begin{bmatrix} 1 & -3 & 1 \\ -2 & 6 & 0 \\ -4 & 12 & 1 \\ 3 & -9 & 3 \end{bmatrix}$$

$$\downarrow \qquad \begin{matrix} (\text{row } 2) + 2(\text{row } 1) \\ (\text{row } 3) + 4(\text{row } 1) \\ (\text{row } 4) - 3(\text{row } 1) \end{matrix}$$

$$\begin{bmatrix} 1 & -3 & 1 \\ 0 & 0 & 2 \\ 0 & 0 & 5 \\ 0 & 0 & 0 \end{bmatrix}$$

$$\downarrow \qquad \begin{matrix} (\text{row } 3) - \tfrac{5}{2}(\text{row } 2) \\ \tfrac{1}{2}(\text{row } 2) \end{matrix}$$

$$\begin{bmatrix} 1 & -3 & 1 \\ 0 & 0 & 1 \\ 0 & 0 & 0 \\ 0 & 0 & 0 \end{bmatrix}$$

Thus the vectors $[1, -3, 1]$ and $[0, 0, 1]$ form a basis for the row space of A^T—or, equivalently,

$$\mathbf{v}_1 = \begin{bmatrix} 1 \\ -3 \\ 1 \end{bmatrix}, \quad \mathbf{v}_2 = \begin{bmatrix} 0 \\ 0 \\ 1 \end{bmatrix}$$

form a basis for the column space of A. Thus $\{\mathbf{v}_1, \mathbf{v}_2\}$ forms a basis for the space spanned by $\{\mathbf{w}_1, \mathbf{w}_2, \mathbf{w}_3, \mathbf{w}_4\}$, and its dimension is 2.

PROBLEM 3-58 Let A and B be row equivalent matrices. Are the column spaces of the two matrices necessarily the same?

Solution In Example 3-46 we have shown that if A and B are row equivalent, then A and B have the same row space; however, their column spaces are not necessarily the same. For example, if

$$A = \begin{bmatrix} 1 & 0 \\ 0 & 0 \end{bmatrix} \quad \text{and} \quad B = \begin{bmatrix} 0 & 0 \\ 1 & 0 \end{bmatrix}$$

then A and B are row equivalent but the column space of A consists of all vectors of the form $\begin{bmatrix} a \\ 0 \end{bmatrix}$

while the column space of B consists of all vectors of the form $\begin{bmatrix} 0 \\ b \end{bmatrix}$. Thus the column spaces of A and B are not the same.

PROBLEM 3-59 Let A and B be $n \times n$ matrices. Show that $AB = O$ if and only if the column space of B is a subspace of the null space of A, $N(A)$, where O is the zero matrix of the nth order.

Solution The column space of B will be a subspace of $N(A)$ if and only if

$$A\mathbf{b}_j = \mathbf{0} \qquad \text{for } j = 1, 2, \ldots, n$$

where $\mathbf{b}_j$ is the jth column of B. However, the jth column of AB is

$$AB\mathbf{e}_j = A\mathbf{b}_j, \qquad \text{for } j = 1, 2, \ldots, n$$

where

$$\mathbf{e}_j = \begin{bmatrix} 0 \\ 0 \\ \vdots \\ 1 \\ \vdots \\ 0 \end{bmatrix} \leftarrow j\text{th component}$$

is the jth vector of the standard basis $\{\mathbf{e}_1, \mathbf{e}_2, \ldots, \mathbf{e}_n\}$ of R^n. Thus the column space of B will be a subspace of $N(A)$ if and only if all the column vectors of AB are $\mathbf{0}$, or, equivalently $AB = O$.

PROBLEM 3-60 Verify Theorem 3-9.2, that is, row rank of A = column rank of A, for any matrix A.

Solution Let

$$A = \begin{bmatrix} a_{11} & a_{12} & \cdots & a_{1n} \\ a_{21} & a_{22} & \cdots & a_{2n} \\ \vdots & \vdots & & \vdots \\ a_{m1} & a_{m2} & \cdots & a_{mn} \end{bmatrix}$$

Denote the row vectors of A by $\mathbf{r}_1, \mathbf{r}_2, \ldots, \mathbf{r}_m$, and assume that the row space of A has dimension k with a basis $\mathscr{B} = \{\mathbf{b}_1, \mathbf{b}_2, \ldots, \mathbf{b}_k\}$ where $\mathbf{b}_i = [b_{i1}, b_{i2}, \ldots, b_{in}]$.

Since $\mathscr{B}$ is a basis for row space, each row vector can be expressed as a linear combination of the vectors in $\mathscr{B}$. That is,

$$\begin{aligned} \mathbf{r}_1 &= c_{11}\mathbf{b}_1 + c_{12}\mathbf{b}_2 + \cdots + c_{1k}\mathbf{b}_k \\ \mathbf{r}_2 &= c_{21}\mathbf{b}_1 + c_{22}\mathbf{b}_2 + \cdots + c_{2k}\mathbf{b}_k \\ &\vdots \\ \mathbf{r}_m &= c_{m1}\mathbf{b}_1 + c_{m2}\mathbf{b}_2 + \cdots + c_{mk}\mathbf{b}_k \end{aligned} \qquad \textbf{(a)}$$

But two vectors are equal if and only if their corresponding components are equal. Thus, equating the jth components on each side of (a), we get

$$\begin{aligned} a_{1j} &= c_{11}b_{1j} + c_{12}b_{2j} + \cdots + c_{1k}b_{kj} \\ a_{2j} &= c_{21}b_{1j} + c_{22}b_{2j} + \cdots + c_{2k}b_{kj} \\ &\vdots \\ a_{mj} &= c_{m1}b_{1j} + c_{m2}b_{2j} + \cdots + c_{mk}b_{kj} \end{aligned} \qquad \textbf{(b)}$$

System (b) can be rewritten in the column vector form

$$\begin{bmatrix} a_{1j} \\ a_{2j} \\ \vdots \\ a_{mj} \end{bmatrix} = b_{1j} \begin{bmatrix} c_{11} \\ c_{21} \\ \vdots \\ c_{m1} \end{bmatrix} + b_{2j} \begin{bmatrix} c_{12} \\ c_{22} \\ \vdots \\ c_{m2} \end{bmatrix} + \cdots + b_{kj} \begin{bmatrix} c_{1k} \\ c_{2k} \\ \vdots \\ c_{mk} \end{bmatrix} \tag{c}$$

but the column vector on the left side of (c) is the jth column vector of A. Since (c) holds for each $j = 1, 2, \ldots, n$, we see that all column vectors of A lie in the subspace of R^m spanned by the k vectors

$$\mathbf{c}_1 = \begin{bmatrix} c_{11} \\ c_{21} \\ \vdots \\ c_{m1} \end{bmatrix}, \mathbf{c}_2 = \begin{bmatrix} c_{12} \\ c_{22} \\ \vdots \\ c_{m2} \end{bmatrix}, \ldots, \mathbf{c}_k = \begin{bmatrix} c_{1k} \\ c_{2k} \\ \vdots \\ c_{mk} \end{bmatrix}$$

Hence the column rank of A is, at most, k and thus

$$\text{column rank of } A \le k = \text{row rank of } A \tag{d}$$

Since the matrix A is arbitrary, this same conclusion applies to A^T, that is,

$$\text{column rank of } A^T \le \text{row rank of } A^T \tag{e}$$

But by (3.73),

$$\text{column rank of } A^T = \text{row rank of } A$$

$$\text{row rank of } A^T = \text{column rank of } A$$

Therefore, (e) becomes

$$k = \text{row rank of } A \le \text{column rank of } A \tag{f}$$

Combining (d) and (f), we conclude that

$$\text{row rank of } A = \text{column rank of } A$$

PROBLEM 3-61 An $m \times n$ matrix A is said to have a **right inverse** if there exists an $n \times m$ matrix C such that $AC = I_m$.

(a) Show that if the column vectors of A span R^m, then A has a right inverse.
(b) Show that if $n < m$, then A cannot have a right inverse.

Solution Let C denote the right inverse of A and let $\mathbf{b} \in R^m$.

(a) If we set $\mathbf{x} = C\mathbf{b}$, then $\mathbf{x} \in R^n$ and $A\mathbf{x} = AC\mathbf{b} = I_m\mathbf{b} = \mathbf{b}$. Thus if A has a right inverse, then $A\mathbf{x} = \mathbf{b}$ will be consistent for each $\mathbf{b} \in R^m$, and consequently the column vectors of A span R^m.
(b) Since $\dim R^m = m$, no set of less than m vectors can span R^m. Thus if $n < m$, and since

$$\text{column rank of } A = \text{row rank of } A \le n < m$$

then the column vectors of A cannot span R^m—and consequently from the result of (a) A cannot have a right inverse.

PROBLEM 3-62 An $m \times n$ matrix A is said to have a **left inverse** if there exists an $n \times m$ matrix F such that $FA = I_n$. Show that a matrix A has a left inverse if and only if A^T has a right inverse.

Solution Assume that A has a left inverse F. Then

$$FA = I_n$$

Transposing, we get

$$(FA)^T = A^T F^T = I_n^T = I_n$$

Thus, by Problem 3-61, it follows that F is a left inverse of A if and only if F^T is a right inverse of A^T. Thus A will have a left inverse if and only if A^T has a right inverse.

PROBLEM 3-63 Find the right inverse of

$$A = \begin{bmatrix} 1 & 2 & 0 \\ -1 & 0 & -1 \end{bmatrix}$$

Solution Since the column vectors of A span R^2, there exists a matrix $C_{3\times 2}$ such that $AC = I_2$. Let

$$C = \begin{bmatrix} c_{11} & c_{12} \\ c_{21} & c_{22} \\ c_{31} & c_{32} \end{bmatrix}$$

Then

$$AC = \begin{bmatrix} 1 & 2 & 0 \\ -1 & 0 & -1 \end{bmatrix} \begin{bmatrix} c_{11} & c_{12} \\ c_{21} & c_{22} \\ c_{31} & c_{32} \end{bmatrix} = \begin{bmatrix} 1 & 0 \\ 0 & 1 \end{bmatrix} = I_2$$

or

$$\begin{bmatrix} c_{11} + 2c_{21} & c_{12} + 2c_{22} \\ -c_{11} - c_{31} & -c_{12} - c_{32} \end{bmatrix} = \begin{bmatrix} 1 & 0 \\ 0 & 1 \end{bmatrix}$$

Equating the corresponding entries, we have

$$c_{11} + 2c_{21} = 1$$

$$c_{12} + 2c_{22} = 0$$

$$-c_{11} - c_{31} = 0$$

$$-c_{12} - c_{32} = 1$$

From the above we get

$$2c_{21} = 1 - c_{11}$$

$$2c_{22} = -c_{12}$$

$$c_{31} = -c_{11}$$

$$c_{32} = -1 - c_{12}$$

Setting $c_{11} = 1$, $c_{12} = 0$, we get $c_{21} = 0$, $c_{22} = 0$, $c_{31} = -1$, $c_{32} = -1$. Thus the right inverse of A is given by

$$C = \begin{bmatrix} 1 & 0 \\ 0 & 0 \\ -1 & -1 \end{bmatrix}$$

Check:

$$AC = \begin{bmatrix} 1 & 2 & 0 \\ -1 & 0 & -1 \end{bmatrix} \begin{bmatrix} 1 & 0 \\ 0 & 0 \\ -1 & -1 \end{bmatrix} = \begin{bmatrix} 1 & 0 \\ 0 & 1 \end{bmatrix}$$

Note that there are infinite numbers of the right inverse of A.

Supplementary Exercises

PROBLEM 3-64 Let V be the set of ordered pairs (x_1, x_2) of real numbers with addition and scalar multiplication defined by

$$(x_1, x_2) + (y_1, y_2) = (x_1 + y_1, x_2 + y_2)$$
$$\alpha(x_1, x_2) = (\alpha x_1, \alpha x_2)$$

Is V a vector space?

Answer Yes

PROBLEM 3-65 Let U be the set of ordered pairs (x_1, x_2) of real numbers with addition and scalar multiplication defined by

$$(x_1, x_2) + (y_1, y_2) = (x_1 + y_1, x_2 + y_2)$$
$$\alpha(x_1, x_2) = (\alpha x_1, 0)$$

U is not a vector space. Which of the ten axioms fail(s) to hold?

Answer Axiom V10: $1\mathbf{v} = \mathbf{v}$

PROBLEM 3-66 Determine whether or not U is a subspace of R^3 if

$$\textbf{(a)}\ U = \left\{ \begin{bmatrix} u_1 \\ u_2 \\ u_3 \end{bmatrix} \middle| u_1 = 2u_2 \right\}, \qquad \textbf{(b)}\ U = \left\{ \begin{bmatrix} u_1 \\ u_2 \\ u_3 \end{bmatrix} \middle| u_1 \le u_2 \le u_3 \right\}$$

Answer **(a)** Yes **(b)** No

PROBLEM 3-67 Let U be all $n \times n$ matrices that commute with a given matrix B; that is,

$$U = \{A \in R_{n \times n} | AB = BA\}$$

Is U a subspace of $R_{n \times n}$?

Answer Yes

PROBLEM 3-68 Find the null space of

$$A = \begin{bmatrix} 1 & 2 & 0 & 1 \\ 1 & 1 & 1 & 0 \end{bmatrix}$$

Answer $N(A)$ consists of all vectors of the form

$$\alpha \begin{bmatrix} -2 \\ 1 \\ 1 \\ 0 \end{bmatrix} + \beta \begin{bmatrix} 1 \\ -1 \\ 0 \\ 1 \end{bmatrix}$$

where α and β are scalars

PROBLEM 3-69 Do the polynomials

$$\mathbf{p}_1 = 1 + x, \qquad \mathbf{p}_2 = 2 + 3x - x^2, \qquad \mathbf{p}_3 = 1 + x^2$$

form a linearly dependent or linearly independent set in P_2?

Answer Linearly dependent

PROBLEM 3-70 Is any set of $n + 2$ elements of P_n linearly dependent?

Answer Yes

PROBLEM 3-71 In $C[-\pi, \pi]$ find the dimension of the subspace spanned by $1, \cos 2x, \cos^2 x$.

Answer 2

PROBLEM 3-72 Consider the two bases of P_3: $\mathscr{B} = \{1, x, x^2, x^3\}$ and $\mathscr{B}' = \{1, 1 + x, (1 + x)^2, (1 + x)^3\}$. Find the transition matrix P from $\mathscr{B}'$ to $\mathscr{B}$ and the transition matrix Q from $\mathscr{B}$ to $\mathscr{B}'$ and show that $Q = P^{-1}$.

Answer

$$P = \begin{bmatrix} 1 & 1 & 1 & 1 \\ 0 & 1 & 2 & 3 \\ 0 & 0 & 1 & 3 \\ 0 & 0 & 0 & 1 \end{bmatrix} \qquad Q = \begin{bmatrix} 1 & -1 & 1 & -1 \\ 0 & 1 & -2 & 3 \\ 0 & 0 & 1 & -3 \\ 0 & 0 & 0 & 1 \end{bmatrix}$$

PROBLEM 3-73 If A and B are $n \times n$ matrices and

$$A\mathbf{x} = B\mathbf{x} \qquad \text{for all } \mathbf{x} \in R^n$$

(a) Show that $N(A - B) = R^n$
(b) Show that rank of $A - B$ is zero and thus $A = B$.

PROBLEM 3-74 If A is an $m \times n$ matrix, what is the largest possible value for the rank of A?

Answer $\min\{m, n\}$, that is, the minimum of m and n

PROBLEM 3-75 If A is a real matrix, show that A and A^TA have the same rank. [*Hint:* Consider $N(A)$ and $N(A^TA)$.]

4 INNER PRODUCT SPACE

THIS CHAPTER IS ABOUT

- ☑ Scalar Products in R^2 and R^3
- ☑ Inner Products in R^n
- ☑ Inner Product Spaces
- ☑ Length, Distance, and Angle in Inner Product Spaces
- ☑ Orthonormal Bases: Gram-Schmidt Process

4-1. Scalar Products in R^2 and R^3

A. Length of a vector in R^2 and R^3

The **Euclidean length** (or **magnitude**) of a vector $\mathbf{v}$ is denoted by $\|\mathbf{v}\|$. Thus from the Pythagorean theorem, if $\mathbf{v} = \begin{bmatrix} v_1 \\ v_2 \end{bmatrix} \in R^2$, then

$$\|\mathbf{v}\| = \sqrt{v_1^2 + v_2^2} \tag{4.1}$$

(see Figure 4-1a). And if $\mathbf{v} = \begin{bmatrix} v_1 \\ v_2 \\ v_3 \end{bmatrix} \in R^3$, then

$$\|\mathbf{v}\| = \sqrt{v_1^2 + v_2^2 + v_3^2} \tag{4.2}$$

(see Figure 4-1b).

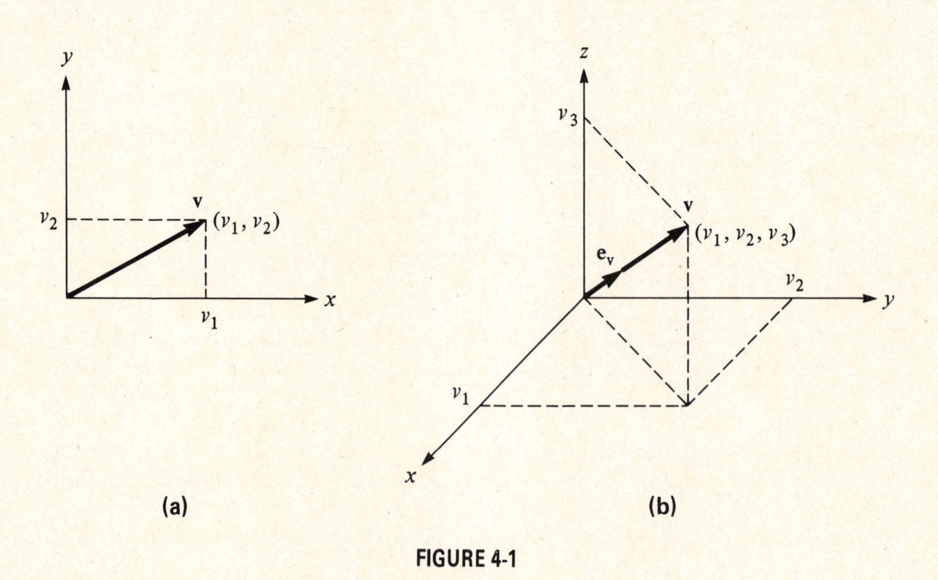

FIGURE 4-1

A vector with length (or magnitude) 1 is called a **unit vector**. Thus a unit vector $\mathbf{e}_\mathbf{v}$ in the direction of $\mathbf{v}$ is given by

UNIT VECTOR

$$\mathbf{e}_\mathbf{v} = \frac{\mathbf{v}}{\|\mathbf{v}\|} \tag{4.3}$$

EXAMPLE 4-1: Find the length of the vector

$$\mathbf{v} = \begin{bmatrix} 1 \\ -2 \\ 2 \end{bmatrix}$$

and a unit vector $\mathbf{e_v}$ in the direction of $\mathbf{v}$.

Solution: From (4.2),

$$\|\mathbf{v}\| = \sqrt{1^2 + (-2)^2 + 2^2} = \sqrt{9} = 3$$

Then by (4.3),

$$\mathbf{e_v} = \frac{\mathbf{v}}{\|\mathbf{v}\|} = \frac{1}{3}\begin{bmatrix} 1 \\ -2 \\ 2 \end{bmatrix} = \begin{bmatrix} \frac{1}{3} \\ -\frac{2}{3} \\ \frac{2}{3} \end{bmatrix}$$

B. Distance between two vectors in R^2 and R^3

Let $\mathbf{u} = \begin{bmatrix} u_1 \\ u_2 \end{bmatrix}$ and $\mathbf{v} = \begin{bmatrix} v_1 \\ v_2 \end{bmatrix}$ be two vectors in R^2 represented by the directed line segments $\overrightarrow{OP}$ and $\overrightarrow{OQ}$, respectively, where $P(u_1, u_2)$ and $Q(v_1, v_2)$ are terminal points of $\mathbf{u}$ and $\mathbf{v}$, respectively. Then the distance between $\mathbf{u}$ and $\mathbf{v}$, denoted by $d(\mathbf{u}, \mathbf{v})$, is the distance between these points; that is, the length of $\overrightarrow{PQ} = \|\mathbf{v} - \mathbf{u}\|$ (see Figure 4-2).

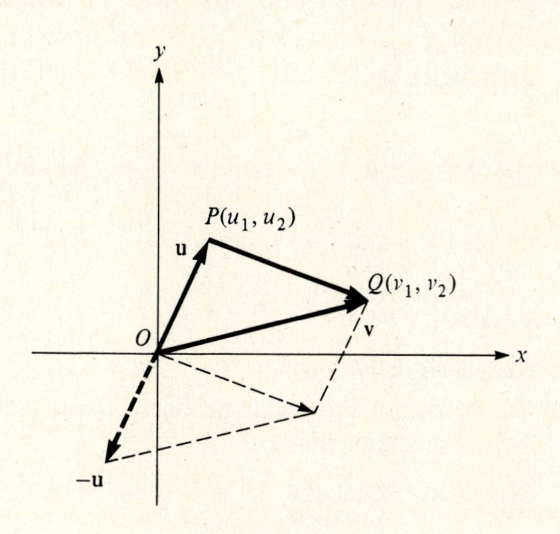

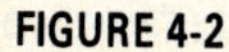
FIGURE 4-2

Since

$$\mathbf{v} - \mathbf{u} = \begin{bmatrix} v_1 - u_1 \\ v_2 - u_2 \end{bmatrix}$$

then

$$d(\mathbf{u}, \mathbf{v}) = \|\mathbf{v} - \mathbf{u}\| = \sqrt{(v_1 - u_1)^2 + (v_2 - u_2)^2} \tag{4.4}$$

Similarly, if $\mathbf{u} = \begin{bmatrix} u_1 \\ u_2 \\ u_3 \end{bmatrix}$ and $\mathbf{v} = \begin{bmatrix} v_1 \\ v_2 \\ v_3 \end{bmatrix}$ are two vectors in R^3, then the distance between them is given by

$$d(\mathbf{u}, \mathbf{v}) = \|\mathbf{v} - \mathbf{u}\| = \sqrt{(v_1 - u_1)^2 + (v_2 - u_2)^2 + (v_3 - u_3)^2} \tag{4.5}$$

EXAMPLE 4-2: Show that if $\mathbf{u}, \mathbf{v} \in R^3$

$$d(\mathbf{u}, \mathbf{v}) = d(\mathbf{v}, \mathbf{u}) \tag{4.6}$$

Solution: From (4.5),

$$\begin{aligned} d(\mathbf{u}, \mathbf{v}) = \|\mathbf{v} - \mathbf{u}\| &= \sqrt{(v_1 - u_1)^2 + (v_2 - u_2)^2 + (v_3 - u_3)^2} \\ &= \sqrt{(u_1 - v_1)^2 + (u_2 - v_2)^2 + (u_3 - v_3)^2} \\ &= \|\mathbf{u} - \mathbf{v}\| = d(\mathbf{v}, \mathbf{u}) \end{aligned}$$

EXAMPLE 4-3: Let

$$\mathbf{u} = \begin{bmatrix} 2 \\ 4 \\ 1 \end{bmatrix} \quad \text{and} \quad \mathbf{v} = \begin{bmatrix} 1 \\ 2 \\ 3 \end{bmatrix}$$

Find the distance $d(\mathbf{u}, \mathbf{v})$ between $\mathbf{u}$ and $\mathbf{v}$.

Solution: From (4.5) we have

$$d(\mathbf{u}, \mathbf{v}) = \sqrt{(1-2)^2 + (2-4)^2 + (3-1)^2} = \sqrt{9} = 3$$

C. Scalar products in R^2 and R^3

1. **Definition**

Let $\mathbf{u} = \begin{bmatrix} u_1 \\ u_2 \end{bmatrix}$ and $\mathbf{v} = \begin{bmatrix} v_1 \\ v_2 \end{bmatrix}$ be two vectors in R^2. The **scalar product** (or **dot product**) of $\mathbf{u}$ and $\mathbf{v}$, denoted by $\mathbf{u} \cdot \mathbf{v}$ (read "$\mathbf{u}$ dot $\mathbf{v}$"), is the scalar

SCALAR PRODUCT IN R^2

$$\mathbf{u} \cdot \mathbf{v} = u_1 v_1 + u_2 v_2 \tag{4.7}$$

Similarly, let $\mathbf{u} = \begin{bmatrix} u_1 \\ u_2 \\ u_3 \end{bmatrix}$ and $\mathbf{v} = \begin{bmatrix} v_1 \\ v_2 \\ v_3 \end{bmatrix}$ be two vectors in R^3. The scalar product of $\mathbf{u}$ and $\mathbf{v}$, $\mathbf{u} \cdot \mathbf{v}$, is the scalar

SCALAR PRODUCT IN R^3

$$\mathbf{u} \cdot \mathbf{v} = u_1 v_1 + u_2 v_2 + u_3 v_3 \tag{4.8}$$

2. **Geometric definition**

Geometrically, the scalar product of two vectors $\mathbf{u}$ and $\mathbf{v}$ in 2-space or 3-space is often defined as

SCALAR PRODUCT

$$\mathbf{u} \cdot \mathbf{v} = \|\mathbf{u}\| \, \|\mathbf{v}\| \cos\theta \tag{4.9}$$

where θ is the angle between $\mathbf{u}$ and $\mathbf{v}$ and $0 \le \theta \le \pi$ (see Figure 4-3).

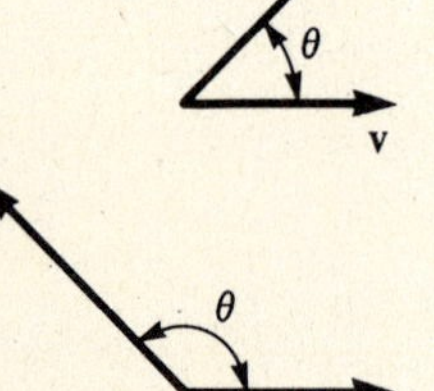
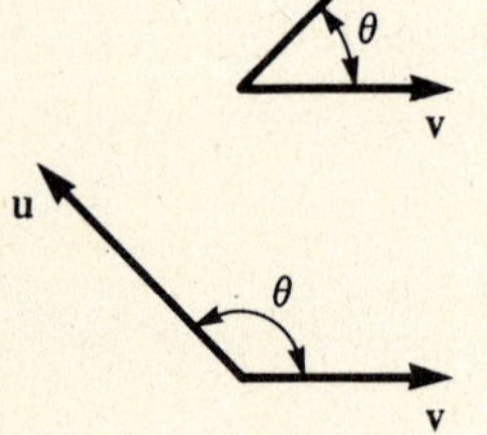
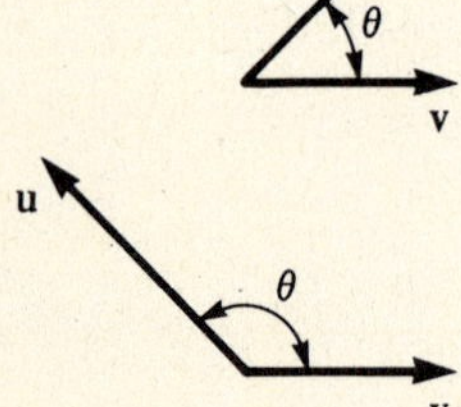
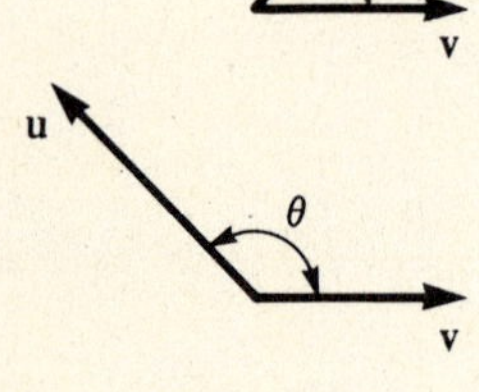

FIGURE 4-3

EXAMPLE 4-4: Let $\mathbf{u} = \begin{bmatrix} 1 \\ 0 \end{bmatrix}$ and $\mathbf{v} = \begin{bmatrix} 1 \\ 1 \end{bmatrix}$. Calculate $\mathbf{u} \cdot \mathbf{v}$ by (4.7) and by (4.9).

Solution: From (4.7),

$$\mathbf{u} \cdot \mathbf{v} = (1)(1) + (0)(1) = 1 + 0 = 1$$

Next, using (4.9) to calculate $\mathbf{u} \cdot \mathbf{v}$, we need to find $\|\mathbf{u}\|$, $\|\mathbf{v}\|$, and angle θ between $\mathbf{u}$ and $\mathbf{v}$.

From (4.1), we get

$$\|\mathbf{u}\| = \sqrt{1^2 + 0^2} = 1 \quad \text{and} \quad \|\mathbf{v}\| = \sqrt{1^2 + 1^2} = \sqrt{2}$$

From Figure 4-4, we see that $\theta = \pi/4$ $(=45°)$. Thus

$$\mathbf{u} \cdot \mathbf{v} = \|\mathbf{u}\| \, \|\mathbf{v}\| \cos\theta = (1)(\sqrt{2}) \cos\left(\frac{\pi}{4}\right) = \sqrt{2} \frac{1}{\sqrt{2}} = 1$$

y
v
(1, 1)
45°
(1, 0)
x
u

FIGURE 4-4

EXAMPLE 4-5: Using the law of cosines, derive (4.8) from (4.9).

Solution: Let

$$\mathbf{u} = \begin{bmatrix} u_1 \\ u_2 \\ u_3 \end{bmatrix} \quad \text{and} \quad \mathbf{v} = \begin{bmatrix} v_1 \\ v_2 \\ v_3 \end{bmatrix}$$

be two nonzero vectors in R^3. If, as shown in Figure 4-5, θ is the angle between $\mathbf{u}$ and $\mathbf{v}$, then the law of

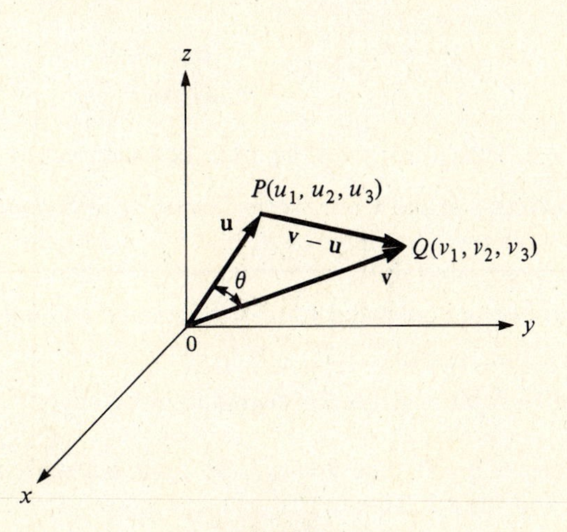

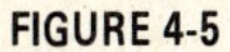
FIGURE 4-5

cosines yields

$$\|\overrightarrow{PQ}\|^2 = \|\mathbf{u}\|^2 + \|\mathbf{v}\|^2 - 2\|\mathbf{u}\|\,\|\mathbf{v}\|\cos\theta \tag{4.10}$$

Since $\overrightarrow{PQ} = \mathbf{v} - \mathbf{u}$, (4.10) can be rewritten as

$$\|\mathbf{u}\|\,\|\mathbf{v}\|\cos\theta = \tfrac{1}{2}(\|\mathbf{u}\|^2 + \|\mathbf{v}\|^2 - \|\mathbf{v} - \mathbf{u}\|^2)$$

Thus, from (4.9), we obtain

$$\mathbf{u}\cdot\mathbf{v} = \tfrac{1}{2}(\|\mathbf{u}\|^2 + \|\mathbf{v}\|^2 - \|\mathbf{v} - \mathbf{u}\|^2)$$

Now from (4.5),

$$\begin{aligned} \|\mathbf{v} - \mathbf{u}\|^2 &= (v_1 - u_1)^2 + (v_2 - u_2)^2 + (v_3 - u_3)^2 \\ &= v_1^2 + v_2^2 + v_3^2 + u_1^2 + u_2^2 + u_3^2 - 2(u_1v_1 + u_2v_2 + u_3v_3) \\ &= \|\mathbf{v}\|^2 + \|\mathbf{u}\|^2 - 2(u_1v_1 + u_2v_2 + u_3v_3) \end{aligned}$$

Thus we obtain

$$\mathbf{u}\cdot\mathbf{v} = u_1v_1 + u_2v_2 + u_3v_3$$

which is precisely (4.8).

EXAMPLE 4-6: Let $\mathbf{v}$ be a vector in R^2 or R^3. Show that

$$\|\mathbf{v}\| = (\mathbf{v}\cdot\mathbf{v})^{1/2} \tag{4.11}$$

Solution: We shall prove (4.11) for vectors in R^3.

Let $\mathbf{u} = \mathbf{v}$ in definition (4.8). Then

$$\mathbf{v}\cdot\mathbf{v} = v_1^2 + v_2^2 + v_3^2 = \|\mathbf{v}\|^2$$

in view of (4.2). Thus

$$\|\mathbf{v}\| = (\mathbf{v}\cdot\mathbf{v})^{1/2}$$

Alternate Solution: Let $\mathbf{u} = \mathbf{v}$ in definition (4.9). Then $\theta = 0$ and

$$\mathbf{v} \cdot \mathbf{v} = \|\mathbf{v}\|^2 \cos 0 = \|\mathbf{v}\|^2$$

since $\cos 0 = 1$. Thus

$$\|\mathbf{v}\| = (\mathbf{v} \cdot \mathbf{v})^{1/2}$$

D. Angle between two vectors in R^2 and R^3

Let $\mathbf{u}$ and $\mathbf{v}$ be two vectors in R^2 or R^3. Then, from definition (4.9), we can write

$$\cos \theta = \frac{\mathbf{u} \cdot \mathbf{v}}{\|\mathbf{u}\| \, \|\mathbf{v}\|} \tag{4.12}$$

or

ANGLE BETWEEN TWO VECTORS

$$\theta = \cos^{-1}\left(\frac{\mathbf{u} \cdot \mathbf{v}}{\|\mathbf{u}\| \, \|\mathbf{v}\|}\right) \tag{4.13}$$

provided $\mathbf{u} \neq \mathbf{0}$ and $\mathbf{v} \neq \mathbf{0}$. Thus (4.13) gives the angle between vectors $\mathbf{u}$ and $\mathbf{v}$ in R^2 or R^3.

DEFINITION: Two vectors, $\mathbf{u}$ and $\mathbf{v}$, are *perpendicular*, or *orthogonal*, to each other (denoted by $\mathbf{u} \perp \mathbf{v}$) if the angle θ between them is a right angle (i.e., $\theta = \pi/2$ radian $= 90°$).

EXAMPLE 4-7: Show that two vectors $\mathbf{u}$ and $\mathbf{v}$ are orthogonal if and only if $\mathbf{u} \cdot \mathbf{v} = 0$.

Solution: If $\mathbf{u} \cdot \mathbf{v} = \|\mathbf{u}\| \, \|\mathbf{v}\| \cos \theta = 0$, we can conclude that $\|\mathbf{u}\| = 0$ or $\|\mathbf{v}\| = 0$ or $\cos \theta = 0$. Since the direction of the zero vector $\mathbf{0}$ is arbitrary, it is considered to be orthogonal to any vector. So if $\mathbf{u} = \mathbf{0}$ or $\mathbf{v} = \mathbf{0}$, then $\mathbf{u} \perp \mathbf{v}$.

When $\mathbf{u} \neq \mathbf{0}$, $\mathbf{v} \neq \mathbf{0}$, and $\cos \theta = 0$, then $\theta = \pi/2$. So again $\mathbf{u} \perp \mathbf{v}$. Conversely, if $\mathbf{u} \perp \mathbf{v}$, $\mathbf{u} \neq \mathbf{0}$, and $\mathbf{v} \neq \mathbf{0}$, the angle between $\mathbf{u}$ and $\mathbf{v}$ is $\pi/2$. Thus

$$\mathbf{u} \cdot \mathbf{v} = \|\mathbf{u}\| \, \|\mathbf{v}\| \cos \frac{\pi}{2} = 0$$

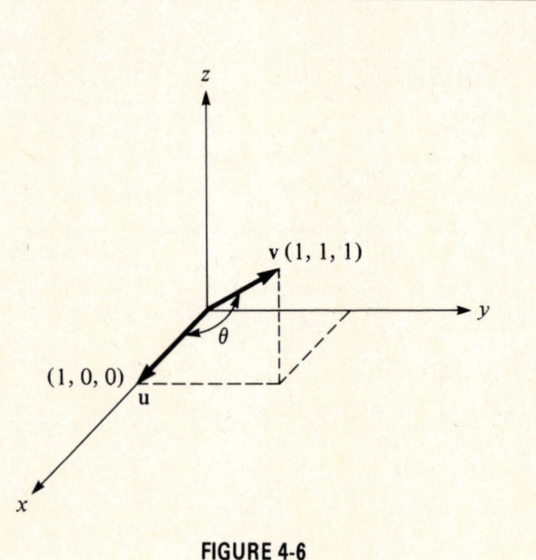

FIGURE 4-6

EXAMPLE 4-8: Find the angle between the vectors

$$\mathbf{u} = \begin{bmatrix} 1 \\ 0 \\ 0 \end{bmatrix} \quad \text{and} \quad \mathbf{v} = \begin{bmatrix} 1 \\ 1 \\ 1 \end{bmatrix}$$

Solution: By (4.7),

$$\mathbf{u} \cdot \mathbf{v} = 1 + 0 + 0 = 1$$

By (4.2),

$$\|\mathbf{u}\| = \sqrt{1^2 + 0^2 + 0^2} = 1$$

$$\|\mathbf{v}\| = \sqrt{1^2 + 1^2 + 1^2} = \sqrt{3}$$

Thus by (4.13) we get

$$\theta = \cos^{-1}\left(\frac{\mathbf{u} \cdot \mathbf{v}}{\|\mathbf{u}\| \, \|\mathbf{v}\|}\right) = \cos^{-1}\left(\frac{1}{\sqrt{3}}\right) \simeq 54°44'$$

This is illustrated in Figure 4-6.

E. Properties of scalar products in R^2 and R^3

The scalar product in R^2 and R^3 has the following properties:

If $\mathbf{u}$, $\mathbf{v}$ and $\mathbf{w}$ are vectors in R^2 or R^3 and α is a scalar, then

$$\mathbf{u} \cdot \mathbf{v} = \mathbf{v} \cdot \mathbf{u} \tag{4.14}$$

$$\mathbf{u} \cdot (\mathbf{v} + \mathbf{w}) = \mathbf{u} \cdot \mathbf{v} + \mathbf{u} \cdot \mathbf{w} \tag{4.15}$$

$$\alpha(\mathbf{u} \cdot \mathbf{v}) = (\alpha\mathbf{u}) \cdot \mathbf{v} = \mathbf{u} \cdot (\alpha\mathbf{v}) \tag{4.16}$$

$$\mathbf{v} \cdot \mathbf{v} \geq 0 \quad \text{and} \quad \mathbf{v} \cdot \mathbf{v} = 0 \quad \text{if and only if } \mathbf{v} = \mathbf{0} \tag{4.17}$$

EXAMPLE 4-9: Prove property (4.16) for vectors in R^3.

Solution: Let

$$\mathbf{u} = \begin{bmatrix} u_1 \\ u_2 \\ u_3 \end{bmatrix}, \qquad \mathbf{v} = \begin{bmatrix} v_1 \\ v_2 \\ v_3 \end{bmatrix}$$

Then

$$\alpha\mathbf{u} = \begin{bmatrix} \alpha u_1 \\ \alpha u_2 \\ \alpha u_3 \end{bmatrix}, \qquad \alpha\mathbf{v} = \begin{bmatrix} \alpha v_1 \\ \alpha v_2 \\ \alpha v_3 \end{bmatrix}$$

Now

$$\begin{aligned} \alpha(\mathbf{u}\cdot\mathbf{v}) &= \alpha(u_1v_1 + u_2v_2 + u_3v_3) \\ &= (\alpha u_1)v_1 + (\alpha u_2)v_2 + (\alpha u_3)v_3 \\ &= (\alpha\mathbf{u})\cdot\mathbf{v} \end{aligned}$$

Similarly,

$$\begin{aligned} \alpha(\mathbf{u}\cdot\mathbf{v}) &= \alpha(u_1v_1 + u_2v_2 + u_3v_3) \\ &= u_1(\alpha v_1) + u_2(\alpha v_2) + u_3(\alpha v_3) \\ &= \mathbf{u}\cdot(\alpha\mathbf{v}) \end{aligned}$$

F. Projection

(1) The **(orthogonal) projection** of a vector **v** onto **a**, denoted by $\text{proj}_{\mathbf{a}}\,\mathbf{v}$, is a vector defined by

PROJECTION $\quad \text{proj}_{\mathbf{a}}\,\mathbf{v} = (\|\mathbf{v}\|\cos\theta)\mathbf{e}_{\mathbf{a}}$ **(4.18)**

where θ is the angle between **v** and **a**, and $\mathbf{e}_{\mathbf{a}}$ is a unit vector in the direction of **a** (see Figure 4-7).

(2) The **component** of vector **v** along nonzero vector **a**, denoted by $\text{comp}_{\mathbf{a}}\,\mathbf{v}$ is a scalar defined by

COMPONENT $\quad \text{comp}_{\mathbf{a}}\,\mathbf{v} = \|\mathbf{v}\|\cos\theta$ **(4.19)**

(3) The vector $\mathbf{v}'$ obtained by subtracting $\text{proj}_{\mathbf{a}}\,\mathbf{v}$ from **v**,

ORTHOGONAL COMPLEMENT $\quad \mathbf{v}' = \mathbf{v} - \text{proj}_{\mathbf{a}}\,\mathbf{v}$ **(4.20)**

is called the **orthogonal complement** of **v** relative to **a**.

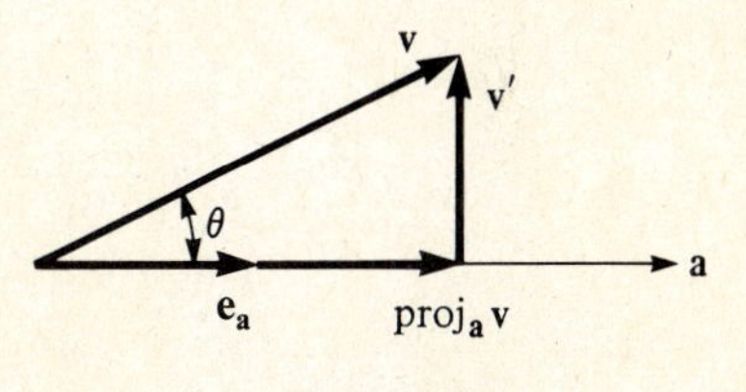

FIGURE 4-7

EXAMPLE 4-10: Show that

$$\text{proj}_{\mathbf{a}}\,\mathbf{v} = \frac{(\mathbf{v}\cdot\mathbf{a})}{\|\mathbf{a}\|^2}\mathbf{a} \tag{4.21}$$

Solution: Let $\mathbf{e}_{\mathbf{a}}$ be a unit vector in the direction of **a**. Then by (4.3)

$$\mathbf{e}_{\mathbf{a}} = \frac{\mathbf{a}}{\|\mathbf{a}\|}$$

Using (4.9) and (4.16), we can write

$$\|\mathbf{v}\|\cos\theta = \mathbf{v}\cdot\mathbf{e}_{\mathbf{a}} = \mathbf{v}\cdot\left(\frac{\mathbf{a}}{\|\mathbf{a}\|}\right) = \left(\frac{\mathbf{v}\cdot\mathbf{a}}{\|\mathbf{a}\|}\right)$$

Thus, substituting these expressions into (4.18), we get

$$\text{proj}_{\mathbf{a}}\mathbf{v} = (\|\mathbf{v}\| \cos\theta)\,\mathbf{e}_{\mathbf{a}} = \frac{(\mathbf{v}\cdot\mathbf{a})}{\|\mathbf{a}\|^2}\mathbf{a}$$

EXAMPLE 4-11: Find the projection of $\mathbf{v} = \begin{bmatrix} 3 \\ 4 \end{bmatrix}$ onto **(a)** $\mathbf{a} = \begin{bmatrix} 1 \\ 0 \end{bmatrix}$ and **(b)** $\mathbf{b} = \begin{bmatrix} 2 \\ 1 \end{bmatrix}$ and illustrate the results geometrically.

Solution

(a)

$$\|\mathbf{a}\| = \sqrt{1^2 + 0^2} = 1$$

$$\mathbf{v}\cdot\mathbf{a} = (3)(1) + (4)(0) = 3$$

Thus, by (4.21),

$$\text{proj}_{\mathbf{a}}\mathbf{v} = \frac{(\mathbf{v}\cdot\mathbf{a})}{\|\mathbf{a}\|^2}\mathbf{a} = \frac{3}{1}\begin{bmatrix} 1 \\ 0 \end{bmatrix} = \begin{bmatrix} 3 \\ 0 \end{bmatrix}$$

(b)

$$\|\mathbf{b}\| = \sqrt{2^2 + 1^2} = \sqrt{5}$$

$$\mathbf{v}\cdot\mathbf{b} = (3)(2) + (4)(1) = 10$$

Thus,

$$\text{proj}_{\mathbf{b}}\mathbf{v} = \frac{(\mathbf{v}\cdot\mathbf{b})}{\|\mathbf{b}\|^2}\mathbf{b} = \frac{10}{5}\begin{bmatrix} 2 \\ 1 \end{bmatrix} = \begin{bmatrix} 4 \\ 2 \end{bmatrix}$$

Geometric illustrations of the results are given in Figure 4-8.

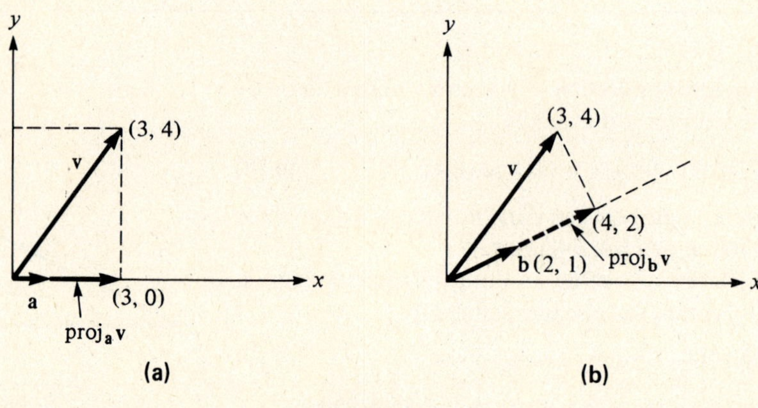

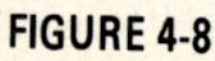
FIGURE 4-8

4-2. Inner Products in R^n

A. Euclidean inner products

1. **Definition**

Let

$$\mathbf{u} = \begin{bmatrix} u_1 \\ u_2 \\ \vdots \\ u_n \end{bmatrix} \quad \text{and} \quad \mathbf{v} = \begin{bmatrix} v_1 \\ v_2 \\ \vdots \\ v_n \end{bmatrix}$$

be vectors in R^n. Then the **Euclidean inner product** of **u** and **v**, denoted by $\langle \mathbf{u}, \mathbf{v} \rangle$, is defined by

EUCLIDEAN INNER PRODUCT IN R^n

$$\langle \mathbf{u}, \mathbf{v} \rangle = u_1 v_1 + u_2 v_2 + \cdots + u_n v_n \tag{4.22}$$

Observe that when $n = 2$ or 3, the Euclidean inner product is the ordinary scalar or dot product (see Section 4-1).

Two vectors **u** and **v** in R^n may be regarded as $n \times 1$ matrices; then the Euclidean inner product of **u** and **v**, $\langle \mathbf{u}, \mathbf{v} \rangle$, (4.22), can be expressed as

$$\langle \mathbf{u}, \mathbf{v} \rangle = \mathbf{u}^T\mathbf{v} = [u_1 u_2 \cdots u_n]\begin{bmatrix} v_1 \\ v_2 \\ \vdots \\ v_n \end{bmatrix} = u_1v_1 + u_2v_2 + \cdots + u_nv_n \tag{4.23}$$

or

$$\langle \mathbf{u}, \mathbf{v} \rangle = \mathbf{v}^T\mathbf{u} = [v_1 v_2 \cdots v_n]\begin{bmatrix} u_1 \\ u_2 \\ \vdots \\ u_n \end{bmatrix} = v_1u_1 + v_2u_2 + \cdots + v_nu_n \tag{4.24}$$

where $\mathbf{u}^T$ and $\mathbf{v}^T$ denote the transposes of **u** and **v**, respectively.

2. **Properties of Euclidean inner products**

The Euclidean inner product in R^n has the following properties:

If **u**, **v**, and **w** are vectors in R^n, and α is a scalar, then

$$\langle \mathbf{u}, \mathbf{v} \rangle = \langle \mathbf{v}, \mathbf{u} \rangle \tag{4.25}$$

$$\langle \mathbf{u}, \mathbf{v} + \mathbf{w} \rangle = \langle \mathbf{u}, \mathbf{v} \rangle + \langle \mathbf{u}, \mathbf{w} \rangle \tag{4.26}$$

$$\alpha\langle \mathbf{u}, \mathbf{v} \rangle = \langle \alpha\mathbf{u}, \mathbf{v} \rangle = \langle \mathbf{u}, \alpha\mathbf{v} \rangle \tag{4.27}$$

$$\langle \mathbf{v}, \mathbf{v} \rangle \geq 0 \quad \text{and} \quad \langle \mathbf{v}, \mathbf{v} \rangle = 0 \quad \text{if and only if } \mathbf{v} = \mathbf{0} \tag{4.28}$$

EXAMPLE 4-12: Prove property (4.26)

Solution: Let

$$\mathbf{u} = \begin{bmatrix} u_1 \\ u_2 \\ \vdots \\ u_n \end{bmatrix}, \qquad \mathbf{v} = \begin{bmatrix} v_1 \\ v_2 \\ \vdots \\ v_n \end{bmatrix}, \qquad \mathbf{w} = \begin{bmatrix} w_1 \\ w_2 \\ \vdots \\ w_n \end{bmatrix}$$

Then

$$\mathbf{v} + \mathbf{w} = \begin{bmatrix} v_1 + w_1 \\ v_2 + w_2 \\ \vdots \\ v_n + w_n \end{bmatrix}$$

and

$$\begin{aligned}\langle \mathbf{u}, \mathbf{v} + \mathbf{w} \rangle &= u_1(v_1 + w_1) + u_2(v_2 + w_2) + \cdots + u_n(v_n + w_n) \\ &= (u_1v_1 + u_2v_2 + \cdots + u_nv_n) + (u_1w_1 + u_2w_2 + \cdots + u_nw_n) \\ &= \langle \mathbf{u}, \mathbf{v} \rangle + \langle \mathbf{u}, \mathbf{w} \rangle\end{aligned}$$

B. Euclidean length of a vector in R^n

1. **Definitions**

By analogy with the familiar formulas in R^2 and R^3, we define the Euclidean length of a vector **v** in R^n, $\|\mathbf{v}\|$, by

EUCLIDEAN LENGTH

$$\|\mathbf{v}\| = \langle \mathbf{v}, \mathbf{v} \rangle^{1/2} = \sqrt{v_1^2 + v_2^2 + \cdots + v_n^2} \tag{4.29}$$

The Euclidean length of **v** is often called the **Euclidean norm** of **v**.

We shall say that a vector **e** is a unit vector if

$$\|\mathbf{e}\| = 1$$

Similarly, the Euclidean distance between two vectors **u** and **v** in R^n, $d(\mathbf{u}, \mathbf{v})$, is defined by

EUCLIDEAN DISTANCE

$$d(\mathbf{u}, \mathbf{v}) = \|\mathbf{v} - \mathbf{u}\| = \sqrt{(v_1 - u_1)^2 + (v_2 - u_2)^2 + \cdots + (v_n - u_n)^2} \tag{4.30}$$

EXAMPLE 4-13: Compute **(a)** $\langle \mathbf{u}, \mathbf{v} \rangle$, **(b)** $\|\mathbf{u}\|$ and $\|\mathbf{v}\|$, and **(c)** $d(\mathbf{u}, \mathbf{v})$, where

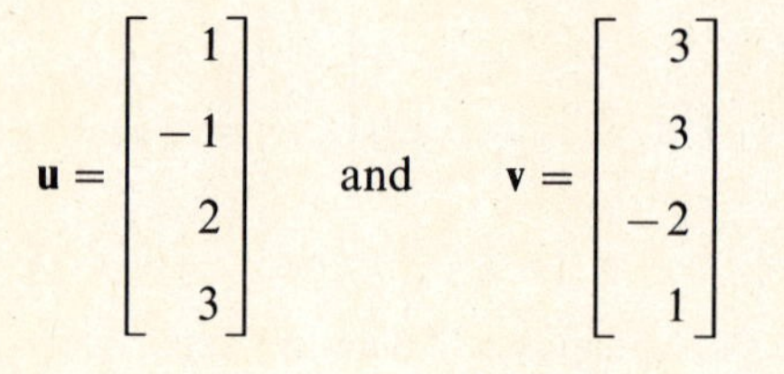

$$\mathbf{u} = \begin{bmatrix} 1 \\ -1 \\ 2 \\ 3 \end{bmatrix} \quad \text{and} \quad \mathbf{v} = \begin{bmatrix} 3 \\ 3 \\ -2 \\ 1 \end{bmatrix}$$

Solution

(a) By (4.22),

$$\langle \mathbf{u}, \mathbf{v} \rangle = (1)(3) + (-1)(3) + (2)(-2) + (3)(1) = -1$$

(b) By (4.29),

$$\|\mathbf{u}\| = \sqrt{1^2 + (-1)^2 + 2^2 + 3^2} = \sqrt{15}$$
$$\|\mathbf{v}\| = \sqrt{3^2 + 3^2 + (-2)^2 + 1^2} = \sqrt{23}$$

(c) By (4.30),

$$d(\mathbf{u}, \mathbf{v}) = \|\mathbf{v} - \mathbf{u}\| = \sqrt{(3-1)^2 + (3-(-1))^2 + (-2-2)^2 + (1-3)^2} = \sqrt{40} = 2\sqrt{10}$$

---

2. **Properties of the Euclidean norm**
   Let $\mathbf{u}, \mathbf{v} \in R^n$ and α be any scalar. Then

$$\|\mathbf{v}\| \geq 0 \quad \text{and} \quad \|\mathbf{v}\| = 0 \quad \text{if and only if } \mathbf{v} = \mathbf{0} \tag{4.31}$$

$$\|\alpha\mathbf{v}\| = |\alpha| \, \|\mathbf{v}\| \tag{4.32}$$

$$\|\mathbf{u} + \mathbf{v}\| \leq \|\mathbf{u}\| + \|\mathbf{v}\| \tag{4.33}$$

EXAMPLE 4-14: Prove (4.32).

Solution: By (4.29) we get

$$\begin{aligned} \|\alpha\mathbf{v}\| &= \sqrt{(\alpha v_1)^2 + (\alpha v_2)^2 + \cdots + (\alpha v_n)^2} \\ &= \sqrt{\alpha^2(v_1^2 + v_2^2 + \cdots + v_n^2)} \\ &= |\alpha|\sqrt{v_1^2 + v_2^2 + \cdots + v_n^2} \\ &= |\alpha| \, \|\mathbf{v}\| \end{aligned}$$

since $\sqrt{\alpha^2} = |\alpha| =$ absolute value of α.

Inequality (4.33) is known as the **triangle inequality**. For the geometric interpretations see Problem 4-12, and for the proof see Problem 4-17.

3. **Cauchy-Schwarz inequality**
   Let $\mathbf{u}, \mathbf{v} \in R^n$. Then

CAUCHY-SCHWARZ INEQUALITY

$$|\langle \mathbf{u}, \mathbf{v} \rangle| \le \|\mathbf{u}\| \, \|\mathbf{v}\| \tag{4.34}$$

This inequality is known as the **Cauchy-Schwarz inequality**.

EXAMPLE 4-15: Verify the Cauchy-Schwarz inequality (4.34).

Solution: If $\mathbf{u} = \mathbf{0}$ or $\mathbf{v} = \mathbf{0}$, then $\langle \mathbf{u}, \mathbf{v} \rangle = 0$ and

$$\langle \mathbf{u}, \mathbf{v} \rangle = \|\mathbf{u}\| \, \|\mathbf{v}\| = 0$$

If $\mathbf{u} \ne \mathbf{0}$, $\mathbf{v} \ne \mathbf{0}$, then for any α, by (4.29), (4.31), and the properties of Euclidean inner products (4.25)–(4.28), we have

$$\begin{aligned} 0 \le \|\mathbf{u} - \alpha\mathbf{v}\|^2 &= \langle \mathbf{u} - \alpha\mathbf{v}, \mathbf{u} - \alpha\mathbf{v} \rangle \\ &= \langle \mathbf{u}, \mathbf{u} \rangle - \langle \mathbf{u}, \alpha\mathbf{v} \rangle - \langle \alpha\mathbf{v}, \mathbf{u} \rangle + \langle \alpha\mathbf{v}, \alpha\mathbf{v} \rangle \\ &= \langle \mathbf{u}, \mathbf{u} \rangle - \alpha\langle \mathbf{u}, \mathbf{v} \rangle - \alpha\langle \mathbf{u}, \mathbf{v} \rangle + \alpha^2\langle \mathbf{v}, \mathbf{v} \rangle \\ &= \|\mathbf{u}\|^2 - 2\alpha\langle \mathbf{u}, \mathbf{v} \rangle + \alpha^2\|\mathbf{v}\|^2 \end{aligned} \tag{4.35}$$

Next, if we let $\alpha = \langle \mathbf{u}, \mathbf{v} \rangle / \|\mathbf{v}\|^2$ in the above inequality, we get

$$0 \le \|\mathbf{u}\|^2 - \frac{2\langle \mathbf{u}, \mathbf{v} \rangle^2}{\|\mathbf{v}\|^2} + \frac{\langle \mathbf{u}, \mathbf{v} \rangle^2}{\|\mathbf{v}\|^4}\|\mathbf{v}\|^2$$

or

$$0 \le \|\mathbf{u}\|^2 - \frac{\langle \mathbf{u}, \mathbf{v} \rangle^2}{\|\mathbf{v}\|^2}$$

Multiplying both sides by $\|\mathbf{v}\|^2$, we get

$$0 \le \|\mathbf{u}\|^2\|\mathbf{v}\|^2 - \langle \mathbf{u}, \mathbf{v} \rangle^2$$

or

$$\langle \mathbf{u}, \mathbf{v} \rangle^2 \le \|\mathbf{u}\|^2\|\mathbf{v}\|^2$$

Taking the square root of both sides, we get

$$|\langle \mathbf{u}, \mathbf{v} \rangle| \le \|\mathbf{u}\| \, \|\mathbf{v}\|$$

C. Orthogonality

1. **Angle between u and v in R^n**

   From (4.34) we have

$$-1 \le \frac{\langle \mathbf{u}, \mathbf{v} \rangle}{\|\mathbf{u}\| \, \|\mathbf{v}\|} \le 1 \tag{4.36}$$

   Then there exists a unique angle θ such that $0 \le \theta \le \pi$ and such that

$$\cos\theta = \frac{\langle \mathbf{u}, \mathbf{v} \rangle}{\|\mathbf{u}\| \, \|\mathbf{v}\|} \tag{4.37}$$

   We define this angle to be the angle between **u** and **v** [see (4.12)].

2. **Orthogonality**

   If $\mathbf{u}, \mathbf{v} \in R^n$, then **u** and **v** are said to be *orthogonal*, denoted by $\mathbf{u} \perp \mathbf{v}$, if

ORTHOGONALITY CONDITION

$$\langle \mathbf{u}, \mathbf{v} \rangle = 0. \tag{4.38}$$

It is seen from (4.37) that if $\langle \mathbf{u}, \mathbf{v} \rangle = 0$, then $\cos\theta = 0$; that is, $\theta = \pi/2$ radian $= 90°$.

EXAMPLE 4-16: Show that

$$\mathbf{u} = \begin{bmatrix} 3 \\ 1 \\ 1 \\ 5 \\ 1 \end{bmatrix} \quad \text{and} \quad \mathbf{v} = \begin{bmatrix} -1 \\ 2 \\ 3 \\ 0 \\ -2 \end{bmatrix}$$

are orthogonal.

Solution: Since

$$\langle \mathbf{u}, \mathbf{v} \rangle = (3)(-1) + (1)(2) + (1)(3) + (5)(0) + (1)(-2) = 0$$

thus, by definition, $\mathbf{u} \perp \mathbf{v}$.

4-3. Inner Product Spaces

A. Inner products

In a general vector space, an inner (or scalar) product is defined axiomatically using the properties of Euclidean inner products described in Section 4-2.

DEFINITION An **inner product** on a vector space V is a function that assigns a real number $\langle \mathbf{u}, \mathbf{v} \rangle$ to each pair $\mathbf{u}$ and $\mathbf{v}$ of vector space V such that the following axioms are satisfied for all $\mathbf{u}, \mathbf{v}, \mathbf{w}$ in V and real scalars (numbers) α:

INNER PRODUCT AXIOMS

$$\text{(I1)} \quad \langle \mathbf{u}, \mathbf{v} \rangle = \langle \mathbf{v}, \mathbf{u} \rangle \tag{4.39}$$

$$\text{(I2)} \quad \langle \mathbf{u}, \mathbf{v} + \mathbf{w} \rangle = \langle \mathbf{u}, \mathbf{v} \rangle + \langle \mathbf{u}, \mathbf{w} \rangle \tag{4.40}$$

$$\text{(I3)} \quad \langle \alpha\mathbf{u}, \mathbf{v} \rangle = \alpha\langle \mathbf{u}, \mathbf{v} \rangle \tag{4.41}$$

$$\text{(I4)} \quad \langle \mathbf{v}, \mathbf{v} \rangle \geq 0 \quad \text{and} \quad \langle \mathbf{v}, \mathbf{v} \rangle = 0 \quad \text{if and only if } \mathbf{v} = \mathbf{0} \tag{4.42}$$

A vector space with an inner product is called the **inner product space.**

EXAMPLE 4-17: Let $\mathbf{u} = \begin{bmatrix} u_1 \\ u_2 \end{bmatrix}$ and $\mathbf{v} = \begin{bmatrix} v_1 \\ v_2 \end{bmatrix}$ be any two vectors in R^2, and consider the function

$$\langle \mathbf{u}, \mathbf{v} \rangle = 2u_1v_1 + 3u_2v_2 \tag{4.43}$$

Show that this function is an inner product on R^2.

Solution

(1)

$$\begin{aligned} \langle \mathbf{u}, \mathbf{v} \rangle &= 2u_1v_1 + 3u_2v_2 \\ &= 2v_1u_1 + 3v_2u_2 \\ &= \langle \mathbf{v}, \mathbf{u} \rangle \end{aligned}$$

So Axiom I1 is satisfied.

(2) Let $\mathbf{w} = \begin{bmatrix} w_1 \\ w_2 \end{bmatrix} \in R^2$; then

$$\begin{aligned} \langle \mathbf{u}, \mathbf{v} + \mathbf{w} \rangle &= 2u_1(v_1 + w_1) + 3u_2(v_2 + w_2) \\ &= (2u_1v_1 + 3u_2v_2) + (2u_1w_1 + 3u_2w_2) \\ &= \langle \mathbf{u}, \mathbf{v} \rangle + \langle \mathbf{u}, \mathbf{w} \rangle \end{aligned}$$

So Axiom I2 is satisfied.

(3) Let α be any scalar; then

$$\begin{aligned}\langle \alpha\mathbf{u}, \mathbf{v}\rangle &= 2(\alpha u_1)v_1 + 3(\alpha u_2)v_2 \\ &= \alpha(2u_1v_1 + 3u_2v_2) \\ &= \alpha\langle \mathbf{u}, \mathbf{v}\rangle\end{aligned}$$

So Axiom I3 is satisfied.

(4) Since

$$\langle \mathbf{v}, \mathbf{v}\rangle = 2v_1^2 + 3v_2^2$$

we see that $\langle \mathbf{v}, \mathbf{v}\rangle > 0$ if either $v_1 \neq 0$ or $v_2 \neq 0$ and $\langle \mathbf{v}, \mathbf{v}\rangle = 0$ if and only if $v_1 = v_2 = 0$. Thus $\langle \mathbf{v}, \mathbf{v}\rangle \geq 0$ and $\langle \mathbf{v}, \mathbf{v}\rangle = 0$ if and only if $\mathbf{v} = \mathbf{0}$. So Axiom I4 is satisfied. Hence, the function defined in (4.43) is an inner product on R^2.

As Example 4.17 indicates, it is possible to define an inner product other than the Euclidean inner product (4.7) or (4.22) on R^n.

EXAMPLE 4-18: Let

$$A = \begin{bmatrix} a_1 & a_2 \\ a_3 & a_4 \end{bmatrix} \quad \text{and} \quad B = \begin{bmatrix} b_1 & b_2 \\ b_3 & b_4 \end{bmatrix}$$

be any two elements (or vectors) of the vector space $R_{2\times 2}$ of all 2×2 matrices with real entries. Show that

$$\langle A, B\rangle = a_1b_1 + a_2b_2 + a_3b_3 + a_4b_4 \tag{4.44}$$

defines an inner product on $R_{2\times 2}$.

Solution

(1)

$$\begin{aligned}\langle A, B\rangle &= a_1b_1 + a_2b_2 + a_3b_3 + a_4b_4 \\ &= b_1a_1 + b_2a_2 + b_3a_3 + b_4a_4 \\ &= \langle B, A\rangle\end{aligned}$$

Thus Axiom I1 is satisfied.

(2) If

$$C = \begin{bmatrix} c_1 & c_2 \\ c_3 & c_4 \end{bmatrix} \in R_{2\times 2}$$

then

$$B + C = \begin{bmatrix} b_1 + c_1 & b_2 + c_2 \\ b_3 + c_3 & b_4 + c_4 \end{bmatrix}$$

and

$$\begin{aligned}\langle A, B + C\rangle &= a_1(b_1 + c_1) + a_2(b_2 + c_2) + a_3(b_3 + c_3) + a_4(b_4 + c_4) \\ &= (a_1b_1 + a_2b_2 + a_3b_3 + a_4b_4) + (a_1c_1 + a_2c_2 + a_3c_3 + a_4c_4) \\ &= \langle A, B\rangle + \langle A, C\rangle\end{aligned}$$

So, Axiom I2 is satisfied.

(3) If α is any scalar, then

$$\alpha A = \begin{bmatrix} \alpha a_1 & \alpha a_2 \\ \alpha a_3 & \alpha a_4 \end{bmatrix}$$

and

$$\begin{aligned}\langle \alpha A, B\rangle &= (\alpha a_1)b_1 + (\alpha a_2)b_2 + (\alpha a_3)b_3 + (\alpha a_4)b_4\\ &= \alpha(a_1 b_1 + a_2 b_2 + a_3 b_3 + a_4 b_4)\\ &= \alpha\langle A, B\rangle\end{aligned}$$

So Axiom I3 is satisfied.

(4) Since

$$\langle A, A\rangle = a_1^2 + a_2^2 + a_3^2 + a_4^2$$

we see that $\langle A, A\rangle \geq 0$ and $\langle A, A\rangle = 0$ if all a_i are zero, that is, if $A = O$. Thus Axiom I4 is satisfied.

Thus the function defined in (4.44) is an inner product of $R_{2\times 2}$.

EXAMPLE 4-19: Consider the vector space P_2 of all polynomials of degree less than or equal to 2 and the function

$$\langle \mathbf{p}, \mathbf{q}\rangle = a_0 b_0 + a_1 b_1 + a_2 b_2 \tag{4.45}$$

where

$$\mathbf{p} = p(x) = a_0 + a_1 x + a_2 x^2$$

$$\mathbf{q} = q(x) = b_0 + b_1 x + b_2 x^2$$

Show that $\langle \mathbf{p}, \mathbf{q}\rangle$ is an inner product on P_2.

Solution

(1)

$$\begin{aligned}\langle \mathbf{p}, \mathbf{q}\rangle &= a_0 b_0 + a_1 b_1 + a_2 b_2\\ &= b_0 a_0 + b_1 a_1 + b_2 a_2\\ &= \langle \mathbf{q}, \mathbf{p}\rangle\end{aligned}$$

So Axiom I1 is satisfied.

(2) If $\mathbf{r} = r(x) = c_0 + c_1 x + c_2 x^2$, then

$$\mathbf{q} + \mathbf{r} = q(x) + r(x) = (b_0 + c_0) + (b_1 + c_1)x + (b_2 + c_2)x^2$$

and

$$\begin{aligned}\langle \mathbf{p}, \mathbf{q} + \mathbf{r}\rangle &= a_0(b_0 + c_0) + a_1(b_1 + c_1) + a_2(b_2 + c_2)\\ &= (a_0 b_0 + a_1 b_1 + a_2 b_2) + (a_0 c_0 + a_1 c_1 + a_2 c_2)\\ &= \langle \mathbf{p}, \mathbf{q}\rangle + \langle \mathbf{p}, \mathbf{r}\rangle\end{aligned}$$

So Axiom I2 is satisfied.

(3) If α is any real number, then

$$\alpha\mathbf{p} = \alpha p(x) = \alpha a_0 + \alpha a_1 x + \alpha a_2 x^2$$

and

$$\begin{aligned}\langle \alpha\mathbf{p}, \mathbf{q}\rangle &= (\alpha a_0)b_0 + (\alpha a_1)b_1 + (\alpha a_2)b_2\\ &= \alpha(a_0 b_0 + a_1 b_1 + a_2 b_2)\\ &= \alpha\langle \mathbf{p}, \mathbf{q}\rangle\end{aligned}$$

So Axiom I3 is satisfied.

(4) Since

$$\langle \mathbf{p}, \mathbf{p}\rangle = a_0^2 + a_1^2 + a_2^2$$

we see that $\langle \mathbf{p}, \mathbf{p}\rangle \geq 0$ and $\langle \mathbf{p}, \mathbf{p}\rangle = 0$ if and only if $a_0 = a_1 = a_2 = 0$. Therefore Axiom I4 is satisfied.

Thus, the function defined in (4.45) is an inner product of P_2.

***EXAMPLE 4-20** [*For readers who have studied calculus.*]: Let $\mathbf{p} = p(x)$ and $\mathbf{q} = q(x)$ be two polynomials in P_n, and consider the function

$$\langle \mathbf{p}, \mathbf{q} \rangle = \int_a^b p(x)q(x)\,dx \tag{4.46}$$

where a and b are any fixed real numbers and $a < b$. Show that $\langle \mathbf{p}, \mathbf{q} \rangle$ is an inner product on P_n.

Solution

(1)
$$\langle \mathbf{p}, \mathbf{q} \rangle = \int_a^b p(x)q(x)\,dx = \int_a^b q(x)p(x)\,dx = \langle \mathbf{q}, \mathbf{p} \rangle$$

So Axiom I1 is satisfied.

(2) If $\mathbf{r} = r(x) \in P_n$, then

$$\begin{aligned}\langle \mathbf{p}, \mathbf{q} + \mathbf{r} \rangle &= \int_a^b p(x)[q(x) + r(x)]\,dx \\ &= \int_a^b p(x)q(x)\,dx + \int_a^b p(x)r(x)\,dx \\ &= \langle \mathbf{p}, \mathbf{q} \rangle + \langle \mathbf{p}, \mathbf{r} \rangle\end{aligned}$$

So Axiom I2 is satisfied.

(3) If α is any scalar, then

$$\begin{aligned}\langle \alpha\mathbf{p}, \mathbf{q} \rangle &= \int_a^b \alpha p(x)q(x)\,dx \\ &= \alpha \int_a^b p(x)q(x)\,dx \\ &= \alpha\langle \mathbf{p}, \mathbf{q} \rangle\end{aligned}$$

So Axiom I3 is satisfied.

(4) If $\mathbf{p} = p(x)$ is any polynomial in P_n, then $[p(x)]^2 \geq 0$ for all x; therefore

$$\langle \mathbf{p}, \mathbf{p} \rangle = \int_a^b [p(x)]^2\,dx \geq 0$$

Further, since $[p(x)]^2 \geq 0$ and $p(x)$ is a continuous function,

$$\int_a^b [p(x)]^2\,dx = 0 \qquad \text{if and only if } p(x) = 0 \qquad \text{for all } x$$

satisfying $a \leq x \leq b$. Thus $\langle \mathbf{p}, \mathbf{p} \rangle = \int_a^b [p(x)]^2\,dx = 0$ if and only if $\mathbf{p} = \mathbf{0}$. Therefore Axiom I4 is satisfied.

Hence the function defined in (4.46) is an inner product on P_n.

note: Example 4-20 indicates that it is possible to define different inner products on P_n. (See also Example 4-19.)

***EXAMPLE 4-21** [*For readers who have studied calculus.*]: Consider the vector space $C[a, b]$ of all continuous functions defined on the interval $[a, b]$.

Let $\mathbf{f} = f(x)$, $\mathbf{g} = g(x)$ be two functions in $C[a, b]$ and consider the function

$$\langle \mathbf{f}, \mathbf{g} \rangle = \int_a^b f(x)g(x)\,dx \tag{4.47}$$

Show that $\langle \mathbf{f}, \mathbf{g} \rangle$ is an inner product on $C[a, b]$.

Solution: Verification follows exactly as in Example 4-20 with **f** and **g** replacing **p** and **q**, respectively.

B. Additional properties of inner products

Let V be an inner product space. Let $\mathbf{u}, \mathbf{v}, \mathbf{w} \in V$ and α be any scalar. Then

$$\langle \mathbf{0}, \mathbf{v} \rangle = 0 \tag{4.48}$$

$$\langle \mathbf{u} + \mathbf{v}, \mathbf{w} \rangle = \langle \mathbf{u}, \mathbf{w} \rangle + \langle \mathbf{v}, \mathbf{w} \rangle \tag{4.49}$$

$$\langle \mathbf{u}, \alpha\mathbf{v} \rangle = \alpha\langle \mathbf{u}, \mathbf{v} \rangle \tag{4.50}$$

EXAMPLE 4-22: Verify (4.49).

Solution

$$\begin{aligned} \langle \mathbf{u} + \mathbf{v}, \mathbf{w} \rangle &= \langle \mathbf{w}, \mathbf{u} + \mathbf{v} \rangle && \text{[by axiom (I1)]} \\ &= \langle \mathbf{w}, \mathbf{u} \rangle + \langle \mathbf{w}, \mathbf{v} \rangle && \text{[by axiom (I2)]} \\ &= \langle \mathbf{u}, \mathbf{w} \rangle + \langle \mathbf{v}, \mathbf{w} \rangle && \text{[by axiom (I1)]} \end{aligned}$$

For the proofs of properties (4.48) and (4.50), see Problems 4-23 and 4-24.

4-4. Length, Distance, and Angle in Inner Product Spaces

A. Definitions

1. Norm of a vector

If V is an inner product space, then the **norm** (or **length**) of a vector $\mathbf{v} \in V$, denoted by $\|\mathbf{v}\|$, is defined by

NORM OF A VECTOR

$$\|\mathbf{v}\| = \langle \mathbf{v}, \mathbf{v} \rangle^{1/2} \tag{4.51}$$

2. Distance between two vectors

The distance between two vectors $\mathbf{u}$ and $\mathbf{v}$ in V, denoted by $d(\mathbf{u}, \mathbf{v})$, is defined by

DISTANCE BETWEEN TWO VECTORS

$$d(\mathbf{u}, \mathbf{v}) = \|\mathbf{v} - \mathbf{u}\| \tag{4.52}$$

B. Basic properties of norm (or length)

If V is an inner product space, then for every $\mathbf{u}, \mathbf{v} \in V$ and any scalar α,

BASIC PROPERTIES OF NORM

L1. $\|\mathbf{v}\| \geq 0$ (4.53)

L2. $\|\mathbf{v}\| = 0$ if and only if $\mathbf{v} = \mathbf{0}$ (4.54)

L3. $\|\alpha\mathbf{v}\| = |\alpha|\,\|\mathbf{v}\|$ (4.55)

L4. $\|\mathbf{u} + \mathbf{v}\| \leq \|\mathbf{u}\| + \|\mathbf{v}\|$ (triangle inequality) (4.56)

C. Basic properties of distance

If V is an inner product space, then for every $\mathbf{u}, \mathbf{v}, \mathbf{w} \in V$,

BASIC PROPERTIES OF DISTANCE

D1. $d(\mathbf{u}, \mathbf{v}) \geq 0$ (4.57)

D2. $d(\mathbf{u}, \mathbf{v}) = 0$ if and only if $\mathbf{u} = \mathbf{v}$ (4.58)

D3. $d(\mathbf{u}, \mathbf{v}) = d(\mathbf{v}, \mathbf{u})$ (4.59)

D4. $d(\mathbf{u}, \mathbf{v}) \leq d(\mathbf{u}, \mathbf{w}) + d(\mathbf{w}, \mathbf{v})$ (triangle inequality) (4.60)

Note that the above properties are considered to be the most fundamental properties of Euclidean length and distance in R^2 and R^3.

Properties (4.53), (4.54), (4.57), and (4.58) follow from definitions (4.51) and (4.52) and Axiom I4. For the proof of (4.56), see Problem 4-17. Property (4.60) follows from (4.56). Properties (4.55) and (4.59) are proved in the following examples.

EXAMPLE 4-23: Prove property (4.55).

Solution: By definition (4.51),

$$\|\alpha\mathbf{v}\| = \langle\alpha\mathbf{v}, \alpha\mathbf{v}\rangle^{1/2}$$
$$= [\alpha^2\langle\mathbf{v}, \mathbf{v}\rangle]^{1/2} \qquad \text{[by (4.41) and (4.50)]}$$
$$= |\alpha|\,\|\mathbf{v}\| \qquad \text{[by definition (4.51)]}$$

since $\sqrt{\alpha^2} = |\alpha|$.

EXAMPLE 4-24: Prove property (4.59).

Solution: By definition (4.52),

$$d(\mathbf{u}, \mathbf{v}) = \|\mathbf{v} - \mathbf{u}\|$$
$$= \|(-1)(\mathbf{u} - \mathbf{v})\|$$
$$= |(-1)|\,\|(\mathbf{u} - \mathbf{v})\| \qquad \text{[by (4.55)]}$$
$$= \|\mathbf{u} - \mathbf{v}\| \qquad \text{[since } |(-1)| = 1\text{]}$$
$$= d(\mathbf{v}, \mathbf{u}) \qquad \text{[by definition (4.52)]}$$

EXAMPLE 4-25: Given

$$A = \begin{bmatrix} 1 & 2 \\ 3 & 4 \end{bmatrix}, \qquad B = \begin{bmatrix} -2 & -1 \\ 1 & -5 \end{bmatrix}$$

Find **(a)** the norm of A and B and **(b)** the distance between A and B with respect to the inner product defined in (4.44).

Solution

(a) By definitions (4.51) and (4.44) we have

$$\|A\| = \langle A, A\rangle^{1/2} = \sqrt{1^2 + 2^2 + 3^2 + 4^2} = \sqrt{30}$$
$$\|B\| = \langle B, B\rangle^{1/2} = \sqrt{(-2)^2 + (-1)^2 + 1^2 + (-5)^2} = \sqrt{31}$$

(b)
$$B - A = \begin{bmatrix} -2 & -1 \\ 1 & -5 \end{bmatrix} - \begin{bmatrix} 1 & 2 \\ 3 & 4 \end{bmatrix} = \begin{bmatrix} -3 & -3 \\ -2 & -9 \end{bmatrix}$$

so by definition (4.52),

$$d(A, B) = \|B - A\| = \langle B - A, B - A\rangle^{1/2}$$
$$= \sqrt{(-3)^2 + (-3)^2 + (-2)^2 + (-9)^2} = \sqrt{103}$$

EXAMPLE 4-26: Let $\mathbf{p} = x$, $\mathbf{q} = x^2$, in P_2. Calculate **(a)** $\|\mathbf{p}\|$, $\|\mathbf{q}\|$, and **(b)** $d(\mathbf{p}, \mathbf{q})$ with respect to the inner product defined in (4.45).

Solution

(a) By (4.51) and (4.45) we have

$$\|\mathbf{p}\| = \langle\mathbf{p}, \mathbf{p}\rangle^{1/2} = \sqrt{1^2} = 1$$
$$\|\mathbf{q}\| = \langle\mathbf{q}, \mathbf{q}\rangle^{1/2} = \sqrt{1^2} = 1$$

(b) $\mathbf{q} - \mathbf{p} = x^2 - x$, so by definition (4.52),

$$d(\mathbf{p}, \mathbf{q}) = \|\mathbf{q} - \mathbf{p}\| = \langle\mathbf{q} - \mathbf{p}, \mathbf{q} - \mathbf{p}\rangle^{1/2} = \sqrt{1^2 + (-1)^2} = \sqrt{2}$$

***EXAMPLE 4-27** [*For readers who have studied calculus.*]: Rework Example 4-26 with respect to the inner product defined in (4.46) with $a = -1$ and $b = 1$; that is

$$\langle \mathbf{p}, \mathbf{q} \rangle = \int_{-1}^{1} p(x)q(x)\,dx$$

Solution

(a) $\mathbf{p} = x$, $\mathbf{q} = x^2$, then

$$\|\mathbf{p}\| = \langle \mathbf{p}, \mathbf{p} \rangle^{1/2} = \left[\int_{-1}^{1} (x)^2\,dx\right]^{1/2} = \left[\frac{x^3}{3}\bigg|_{-1}^{1}\right]^{1/2} = \sqrt{\frac{2}{3}}$$

$$\|\mathbf{q}\| = \langle \mathbf{q}, \mathbf{q} \rangle^{1/2} = \left[\int_{-1}^{1} (x^2)^2\,dx\right]^{1/2} = \left[\frac{x^5}{5}\bigg|_{-1}^{1}\right]^{1/2} = \sqrt{\frac{2}{5}}$$

(b) $\mathbf{q} - \mathbf{p} = x^2 - x$, then

$$\begin{aligned} d(\mathbf{p}, \mathbf{q}) &= \|\mathbf{q} - \mathbf{p}\| \\ &= \langle \mathbf{q} - \mathbf{p}, \mathbf{q} - \mathbf{p} \rangle^{1/2} \\ &= \left[\int_{-1}^{1} (x^2 - x)^2\,dx\right]^{1/2} \\ &= \left[\int_{-1}^{1} (x^4 - 2x^3 + x^2)\,dx\right]^{1/2} \\ &= \left[\left(\frac{x^5}{5} - \frac{x^4}{2} + \frac{x^3}{3}\right)\bigg|_{-1}^{1}\right]^{1/2} = \frac{4}{\sqrt{15}} \end{aligned}$$

D. Cauchy-Schwarz inequality

If **u** and **v** are any elements (or vectors) of an inner product space V, then

CAUCHY-SCHWARZ INEQUALITY

$$|\langle \mathbf{u}, \mathbf{v} \rangle| \le \|\mathbf{u}\|\,\|\mathbf{v}\| \tag{4.61}$$

Inequality (4.61) is known as the **Cauchy-Schwarz inequality**. For the proof of (4.61), see Example 4-14.

E. Orthogonality

1. **Angle between u and v**

From inequality (4.61) we have

$$\frac{|\langle \mathbf{u}, \mathbf{v} \rangle|}{\|\mathbf{u}\|\,\|\mathbf{v}\|} \le 1 \tag{4.62}$$

or equivalently,

$$-1 \le \frac{\langle \mathbf{u}, \mathbf{v} \rangle}{\|\mathbf{u}\|\,\|\mathbf{v}\|} \le 1 \tag{4.63}$$

provided $\|\mathbf{u}\| \ne 0$ and $\|\mathbf{v}\| \ne 0$. Consequently, there is a unique angle θ such that

$$\cos\theta = \frac{\langle \mathbf{u}, \mathbf{v} \rangle}{\|\mathbf{u}\|\,\|\mathbf{v}\|} \quad \text{and} \quad 0 \le \theta \le \pi \tag{4.64}$$

The angle θ is the angle between **u** and **v**.

2. **Orthogonality**

Two vectors **u** and **v** of an inner product space V are said to be *orthogonal* if $\langle \mathbf{u}, \mathbf{v} \rangle = 0$.

THEOREM 4-4.1 Let **u** and **v** be any vectors of an inner product space V. Then **u** and **v** are orthogonal if and only if

$$\|\mathbf{u} + \mathbf{v}\|^2 = \|\mathbf{u}\|^2 + \|\mathbf{v}\|^2 \tag{4.65}$$

EXAMPLE 4-28: Verify Theorem 4-4.1.

Solution: By definition (4.51) and by (4.40) and (4.39), we get

$$\begin{aligned}\|\mathbf{u}+\mathbf{v}\|^2 &= \langle \mathbf{u}+\mathbf{v}, \mathbf{u}+\mathbf{v}\rangle \\ &= \langle \mathbf{u},\mathbf{u}\rangle + 2\langle \mathbf{u},\mathbf{v}\rangle + \langle \mathbf{v},\mathbf{v}\rangle \\ &= \|\mathbf{u}\|^2 + 2\langle \mathbf{u},\mathbf{v}\rangle + \|\mathbf{v}\|^2\end{aligned}$$

Thus the identity in (4.65) holds if and only if $\langle \mathbf{u},\mathbf{v}\rangle = 0$, that is, if and only if **u** and **v** are orthogonal.

Equation (4.65) is the generalized Pythagorean theorem. Note that in R^2 and R^3 with the Euclidean inner product, (4.65) reduces to the ordinary Pythagorean theorem (see Figure 4-9).

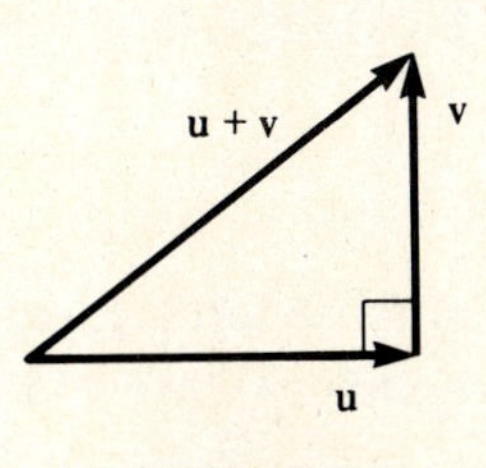

FIGURE 4-9

EXAMPLE 4-29: Find the angle θ between the matrices

$$A = \begin{bmatrix} 1 & 2 \\ 3 & 4 \end{bmatrix} \quad \text{and} \quad B = \begin{bmatrix} -2 & -1 \\ 1 & -5 \end{bmatrix}$$

of $R_{2\times 2}$ with respect to the inner product defined in (4.44).

Solution: From Example 4-25 we have

$$\|A\| = \sqrt{30}$$
$$\|B\| = \sqrt{31}$$

Now, by (4.44),

$$\langle A, B\rangle = (1)(-2) + (2)(-1) + (3)(1) + (4)(-5) = -21$$

Thus, by (4.64), we get

$$\cos\theta = \frac{\langle A, B\rangle}{\|A\|\,\|B\|} = \frac{-21}{\sqrt{30}\sqrt{31}} = -0.6886$$

or

$$\theta = \cos^{-1}(-0.6886) = 133.52°$$

EXAMPLE 4-30: Show that

$$A = \begin{bmatrix} 1 & -1 \\ 3 & 4 \end{bmatrix} \quad \text{and} \quad B = \begin{bmatrix} -5 & 5 \\ 2 & 1 \end{bmatrix}$$

are orthogonal with respect to the inner product of (4.44).

Solution: By (4.44) we get

$$\langle A, B\rangle = (1)(-5) + (-1)(5) + (3)(2) + (4)(1) = 0$$

Thus, by definition, A and B are orthogonal.

***EXAMPLE 4-31** [*For readers who have studied calculus.*]**:** Consider the elements $\sin x$ and $\cos x$ of the vector space $C[0, \pi]$. Show that $\sin x$ and $\cos x$ are orthogonal with respect to the inner product defined by (4.47).

Solution: By (4.47),

$$\langle \sin x, \cos x \rangle = \int_0^{\pi} \sin x \cos x \, dx = \tfrac{1}{2} \sin^2 x \Big|_0^{\pi} = 0$$

Thus by definition, $\sin x$ and $\cos x$ are orthogonal.

EXAMPLE 4-32: Show that the vectors

$$\mathbf{u} = \begin{bmatrix} 1 \\ 1 \end{bmatrix} \quad \text{and} \quad \mathbf{v} = \begin{bmatrix} -1 \\ 1 \end{bmatrix}$$

in R^2 are orthogonal with respect to the Euclidean inner product (4.22) but are not orthogonal with respect to the inner product defined by (4.43).

Solution: From (4.22) we have

$$\langle \mathbf{u}, \mathbf{v} \rangle = u_1 v_1 + u_2 v_2 = (1)(-1) + (1)(1) = 0$$

Thus, by definition, $\mathbf{u}$ and $\mathbf{v}$ are orthogonal with respect to the Euclidean inner product. On the other hand, from (4.43) we get

$$\langle \mathbf{u}, \mathbf{v} \rangle = 2u_1 v_1 + 3u_2 v_2 = 2(1)(-1) + 3(1)(1) = 1 \neq 0$$

Thus, $\mathbf{u}$ and $\mathbf{v}$ are not orthogonal with respect to this inner product.

note: This example shows that care must be taken when stating that two given vectors of a vector space are orthogonal (see also Problem 4-28).

EXAMPLE 4-33: Let $\mathbf{a}$ be a nonzero vector of an inner product space V. Then show that for any vector $\mathbf{v}$ in V there exists a unique number α such that $\mathbf{v} - \alpha\mathbf{a}$ is orthogonal to $\mathbf{a}$.

Solution: For $\mathbf{v} - \alpha\mathbf{a}$ to be orthogonal to $\mathbf{a}$, then by definition we must have

$$\langle \mathbf{v} - \alpha\mathbf{a}, \mathbf{a} \rangle = 0$$

or

$$\begin{aligned} \langle \mathbf{v} - \alpha\mathbf{a}, \mathbf{a} \rangle &= \langle \mathbf{v}, \mathbf{a} \rangle + \langle -\alpha\mathbf{a}, \mathbf{a} \rangle && \text{[by (4.49)]} \\ &= \langle \mathbf{v}, \mathbf{a} \rangle - \alpha\langle \mathbf{a}, \mathbf{a} \rangle && \text{[by (4.41)]} \\ &= 0 \end{aligned}$$

Thus,

$$\langle \mathbf{v}, \mathbf{a} \rangle = \alpha\langle \mathbf{a}, \mathbf{a} \rangle = \alpha\|\mathbf{a}\|^2$$

and

$$\alpha = \frac{\langle \mathbf{v}, \mathbf{a} \rangle}{\|\mathbf{a}\|^2} \tag{4.66}$$

Conversely, if $\alpha = \langle \mathbf{v}, \mathbf{a} \rangle / \|\mathbf{a}\|^2$, then

$$\begin{aligned} \langle \mathbf{v} - \alpha\mathbf{a}, \mathbf{a} \rangle &= \left\langle \mathbf{v} - \frac{\langle \mathbf{v}, \mathbf{a} \rangle}{\|\mathbf{a}\|^2}\mathbf{a}, \mathbf{a} \right\rangle \\ &= \langle \mathbf{v}, \mathbf{a} \rangle + \left\langle -\frac{\langle \mathbf{v}, \mathbf{a} \rangle}{\|\mathbf{a}\|^2}\mathbf{a}, \mathbf{a} \right\rangle \\ &= \langle \mathbf{v}, \mathbf{a} \rangle - \frac{\langle \mathbf{v}, \mathbf{a} \rangle}{\|\mathbf{a}\|^2}\langle \mathbf{a}, \mathbf{a} \rangle \end{aligned}$$

$$\langle \mathbf{v} - \alpha\mathbf{a}, \mathbf{a} \rangle = \langle \mathbf{v}, \mathbf{a} \rangle - \frac{\langle \mathbf{v}, \mathbf{a} \rangle}{\|\mathbf{a}\|^2} \|\mathbf{a}\|^2$$

$$= \langle \mathbf{v}, \mathbf{a} \rangle - \langle \mathbf{v}, \mathbf{a} \rangle = 0$$

Thus, $\mathbf{v} - \alpha\mathbf{a}$ is orthogonal to $\mathbf{a}$.

F. Projection

As with the case of R^2 and R^3, we define the **orthogonal projection** of $\mathbf{v}$ onto $\mathbf{a}$, denoted by $\text{proj}_{\mathbf{a}} \mathbf{v}$, to be the vector $\alpha\mathbf{a}$ of Example 4-33. Thus

ORTHOGONAL PROJECTION

$$\text{proj}_{\mathbf{a}} \mathbf{v} = \alpha\mathbf{a} = \frac{\langle \mathbf{v}, \mathbf{a} \rangle}{\|\mathbf{a}\|^2} \mathbf{a} \tag{4.67}$$

We call

$$\alpha = \frac{\langle \mathbf{v}, \mathbf{a} \rangle}{\|\mathbf{a}\|^2} \tag{4.68}$$

the *component* of $\mathbf{v}$ along $\mathbf{a}$, or the *Fourier coefficient* of $\mathbf{v}$ with respect to $\mathbf{a}$. Note that if $\mathbf{a}$ is a unit vector, then

$$\alpha = \langle \mathbf{v}, \mathbf{a} \rangle \tag{4.69}$$

The vector

ORTHOGONAL COMPONENT

$$\mathbf{v}' = \mathbf{v} - \alpha\mathbf{a} = \mathbf{v} - \text{proj}_{\mathbf{a}} \mathbf{v} = \mathbf{v} - \frac{\langle \mathbf{v}, \mathbf{a} \rangle}{\|\mathbf{a}\|^2} \mathbf{a} \tag{4.70}$$

is called the *orthogonal component* of $\mathbf{v}$ relative to $\mathbf{a}$ (see Figure 4-10).

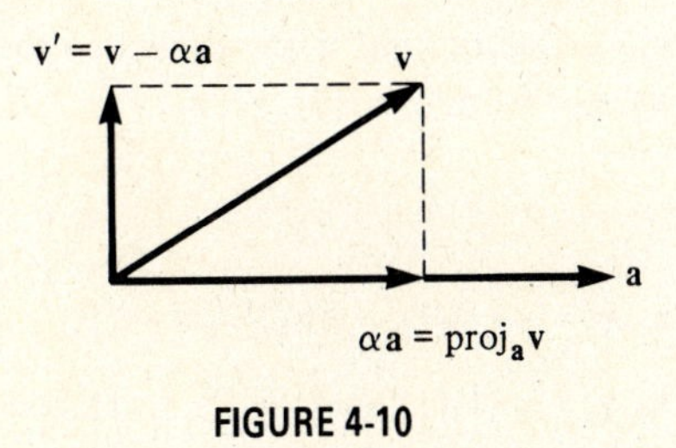

FIGURE 4-10

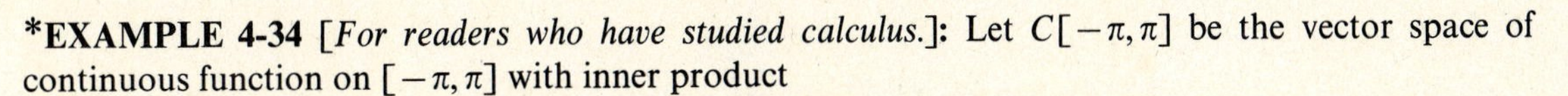

***EXAMPLE 4-34** [*For readers who have studied calculus.*]: Let $C[-\pi, \pi]$ be the vector space of continuous function on $[-\pi, \pi]$ with inner product

$$\langle \mathbf{f}, \mathbf{g} \rangle = \int_{-\pi}^{\pi} f(x)g(x)\,dx$$

Let $\mathbf{f} = f(x) = \sin kx$, where k is some integer and $k > 0$. If $\mathbf{g} = g(x)$ is any continuous function on $[-\pi, \pi]$, then find the Fourier coefficient of $g(x)$ with respect to $f(x)$.

Solution

$$\|\mathbf{f}\| = \langle \mathbf{f}, \mathbf{f} \rangle^{1/2} = \left[\int_{-\pi}^{\pi} \sin^2 kx\,dx \right]^{1/2}$$

$$= \left[\int_{-\pi}^{\pi} \tfrac{1}{2}[1 - \cos 2kx]\,dx \right]^{1/2}$$

$$= \sqrt{\pi}$$

Thus, the Fourier coefficient of $g(x)$ with respect to $f(x)$ is, by (4.68),

$$\frac{\langle \mathbf{g}, \mathbf{f} \rangle}{\|\mathbf{f}\|^2} = \frac{1}{\pi} \int_{-\pi}^{\pi} g(x) \sin kx\,dx$$

4-5. Orthonormal Bases: Gram-Schmidt Process

A. Orthogonal sets

Let $\mathbf{v}_1, \mathbf{v}_2, \ldots, \mathbf{v}_n$ be vectors in an inner product space V. If $\langle \mathbf{v}_i, \mathbf{v}_j \rangle = 0$ whenever $i \neq j$, then $\{\mathbf{v}_1, \mathbf{v}_2, \ldots, \mathbf{v}_n\}$ is said to be an **orthogonal set** of vectors.

If we set

$$\mathbf{u} = \frac{1}{\|\mathbf{v}\|}\mathbf{v}$$

then $\|\mathbf{u}\| = 1$, so that $\mathbf{u}$ is a unit vector [see (4.3)]. When a vector is converted to a unit vector in this manner, we say that it has been *normalized*. An **orthonormal set** of vectors is an orthogonal set of unit vectors. Thus, the set $\{\mathbf{u}_1, \mathbf{u}_2, \ldots, \mathbf{u}_n\}$ will be orthonormal if and only if

$$\langle \mathbf{u}_i, \mathbf{u}_j \rangle = \delta_{ij} \tag{4.71}$$

where

$$\delta_{ij} = \begin{cases} 1 & \text{for } i = j \\ 0 & \text{for } i \neq j \end{cases}$$

EXAMPLE 4-35: Let

$$\mathbf{v}_1 = \begin{bmatrix} 1 \\ 0 \\ -1 \end{bmatrix} \qquad \mathbf{v}_2 = \begin{bmatrix} 1 \\ 1 \\ 1 \end{bmatrix} \qquad \mathbf{v}_3 = \begin{bmatrix} 1 \\ -2 \\ 1 \end{bmatrix}$$

(a) Show that the set $S = \{\mathbf{v}_1, \mathbf{v}_2, \mathbf{v}_3\}$ is orthogonal with respect to the Euclidean inner product.
(b) Find an orthonormal set $\{\mathbf{u}_1, \mathbf{u}_2, \mathbf{u}_3\}$ from $S = \{\mathbf{v}_1, \mathbf{v}_2, \mathbf{v}_3\}$.

Solution

(a) Since

$$\langle \mathbf{v}_1, \mathbf{v}_2 \rangle = (1)(1) + (0)(1) + (-1)(1) = 0$$
$$\langle \mathbf{v}_2, \mathbf{v}_3 \rangle = (1)(1) + (1)(-2) + (1)(1) = 0$$
$$\langle \mathbf{v}_3, \mathbf{v}_1 \rangle = (1)(1) + (-2)(0) + (1)(-1) = 0$$

the set S is orthogonal.

(b) Now

$$\|\mathbf{v}_1\| = \langle \mathbf{v}_1, \mathbf{v}_1 \rangle^{1/2} = \sqrt{1^2 + 0^2 + (-1)^2} = \sqrt{2}$$
$$\|\mathbf{v}_2\| = \langle \mathbf{v}_2, \mathbf{v}_2 \rangle^{1/2} = \sqrt{1^2 + 1^2 + 1^2} = \sqrt{3}$$
$$\|\mathbf{v}_3\| = \langle \mathbf{v}_3, \mathbf{v}_3 \rangle^{1/2} = \sqrt{1^2 + (-2)^2 + 1^2} = \sqrt{6}$$

Let

$$\mathbf{u}_1 = \frac{1}{\|\mathbf{v}_1\|}\mathbf{v}_1 = \frac{1}{\sqrt{2}}\begin{bmatrix} 1 \\ 0 \\ -1 \end{bmatrix} = \begin{bmatrix} \frac{1}{\sqrt{2}} \\ 0 \\ -\frac{1}{\sqrt{2}} \end{bmatrix}$$

$$\mathbf{u}_2 = \frac{1}{\|\mathbf{v}_2\|}\mathbf{v}_2 = \frac{1}{\sqrt{3}}\begin{bmatrix} 1 \\ 1 \\ 1 \end{bmatrix} = \begin{bmatrix} \frac{1}{\sqrt{3}} \\ \frac{1}{\sqrt{3}} \\ \frac{1}{\sqrt{3}} \end{bmatrix}$$

$$\mathbf{u}_3 = \frac{1}{\|\mathbf{v}_3\|}\mathbf{v}_3 = \frac{1}{\sqrt{6}}\begin{bmatrix} 1 \\ -2 \\ 1 \end{bmatrix} = \begin{bmatrix} \frac{1}{\sqrt{6}} \\ -\frac{2}{\sqrt{6}} \\ \frac{1}{\sqrt{6}} \end{bmatrix}$$

Then

$$\|\mathbf{u}_1\| = \|\mathbf{u}_2\| = \|\mathbf{u}_3\| = 1 \quad \text{and} \quad \langle \mathbf{u}_1, \mathbf{u}_2 \rangle = \langle \mathbf{u}_2, \mathbf{u}_3 \rangle = \langle \mathbf{u}_3, \mathbf{u}_1 \rangle = 0$$

Thus the set $\{\mathbf{u}_1, \mathbf{u}_2, \mathbf{u}_3\}$ is an orthonormal set.

EXAMPLE 4-36: Let $S = \{\mathbf{v}_1, \mathbf{v}_2, \ldots, \mathbf{v}_n\}$ be an orthogonal set of nonzero vectors of an inner product space. Show that S is linearly independent.

Solution: Let $\alpha_1, \alpha_2, \ldots, \alpha_n$ be any scalars such that

$$\alpha_1 \mathbf{v}_1 + \alpha_2 \mathbf{v}_2 + \cdots + \alpha_n \mathbf{v}_n = \mathbf{0}$$

Then for any i, $i = 1, 2, \ldots, n$, we have

$$\langle \alpha_1 \mathbf{v}_1 + \alpha_2 \mathbf{v}_2 + \cdots + \alpha_n \mathbf{v}_n, \mathbf{v}_i \rangle = \langle \mathbf{0}, \mathbf{v}_i \rangle = 0$$

or

$$\alpha_1 \langle \mathbf{v}_1, \mathbf{v}_i \rangle + \alpha_2 \langle \mathbf{v}_2, \mathbf{v}_i \rangle + \cdots + \alpha_n \langle \mathbf{v}_n, \mathbf{v}_i \rangle = 0 \tag{4.72}$$

Since S is orthogonal, we have $\langle \mathbf{v}_i, \mathbf{v}_j \rangle = 0$ if $i \neq j$, and (4.72) becomes

$$\alpha_i \langle \mathbf{v}_i, \mathbf{v}_i \rangle = 0$$

Now $\langle \mathbf{v}_i, \mathbf{v}_i \rangle \neq 0$ since $\mathbf{v}_i \neq \mathbf{0}$. Therefore $\alpha_i = 0$ for $i = 1, 2, \ldots, n$; that is, $\alpha_1 = \alpha_2 = \cdots = \alpha_n = 0$. Thus the set S is linearly independent.

B. Orthonormal basis

If $\mathscr{B} = \{\mathbf{u}_1, \mathbf{u}_2, \ldots, \mathbf{u}_n\}$ is an orthonormal set and is also a basis for an inner product space V, then we call $\mathscr{B}$ an **orthonormal basis** for V.

THEOREM 4-5.1 If $\mathscr{B} = \{\mathbf{u}_1, \mathbf{u}_2, \ldots, \mathbf{u}_n\}$ is an orthonormal basis for an inner product space V, and $\mathbf{v}$ is any vector in V, then

$$\mathbf{v} = \langle \mathbf{v}, \mathbf{u}_1 \rangle \mathbf{u}_1 + \langle \mathbf{v}, \mathbf{u}_2 \rangle \mathbf{u}_2 + \cdots + \langle \mathbf{v}, \mathbf{u}_n \rangle \mathbf{u}_n \tag{4.73}$$

EXAMPLE 4-37: Verify Theorem 4-5.1.

Solution: Since $\mathscr{B} = \{\mathbf{u}_1, \mathbf{u}_2, \ldots, \mathbf{u}_n\}$ is a basis for V, a vector $\mathbf{v} \in V$ can be expressed as

$$\mathbf{v} = \alpha_1 \mathbf{u}_1 + \alpha_2 \mathbf{u}_2 + \cdots + \alpha_n \mathbf{u}_n \tag{4.74}$$

Then for any i, $i = 1, 2, \ldots, n$,

$$\begin{aligned} \langle \mathbf{v}, \mathbf{u}_i \rangle &= \langle \alpha_1 \mathbf{u}_1 + \alpha_2 \mathbf{u}_2 + \cdots + \alpha_n \mathbf{u}_n, \mathbf{u}_i \rangle \\ &= \alpha_1 \langle \mathbf{u}_1, \mathbf{u}_i \rangle + \alpha_2 \langle \mathbf{u}_2, \mathbf{u}_i \rangle + \cdots + \alpha_n \langle \mathbf{u}_n, \mathbf{u}_i \rangle \\ &= \alpha_i \end{aligned}$$

since $\mathscr{B}$ is an orthonormal set and $\langle \mathbf{u}_j, \mathbf{u}_i \rangle = 0$ for $i \neq j$ and $\langle \mathbf{u}_i, \mathbf{u}_i \rangle = \|\mathbf{u}_i\| = 1$. Hence (4.74) can be rewritten as

$$\mathbf{v} = \langle \mathbf{v}, \mathbf{u}_1 \rangle \mathbf{u}_1 + \langle \mathbf{v}, \mathbf{u}_2 \rangle \mathbf{u}_2 + \cdots + \langle \mathbf{v}, \mathbf{u}_n \rangle \mathbf{u}_n$$

EXAMPLE 4-38: Write the vector $\mathbf{v} = \begin{bmatrix} 2 \\ 5 \\ 3 \end{bmatrix}$ as a linear combination of

$$\mathbf{u}_1 = \begin{bmatrix} \frac{1}{\sqrt{2}} \\ 0 \\ -\frac{1}{\sqrt{2}} \end{bmatrix}, \qquad \mathbf{u}_2 = \begin{bmatrix} \frac{1}{\sqrt{3}} \\ \frac{1}{\sqrt{3}} \\ \frac{1}{\sqrt{3}} \end{bmatrix}, \qquad \mathbf{u}_3 = \begin{bmatrix} \frac{1}{\sqrt{6}} \\ -\frac{2}{\sqrt{6}} \\ \frac{1}{\sqrt{6}} \end{bmatrix}$$

Solution: In Example 4-35 we have shown that $S = \{\mathbf{u}_1, \mathbf{u}_2, \mathbf{u}_3\}$ is an orthonormal set. Since $\dim R^3 = 3$, S forms an orthonormal basis for R^3. Now,

$$\langle \mathbf{v}, \mathbf{u}_1 \rangle = (2)\left(\frac{1}{\sqrt{2}}\right) + (5)(0) + (3)\left(-\frac{1}{\sqrt{2}}\right) = -\frac{1}{\sqrt{2}}$$

$$\langle \mathbf{v}, \mathbf{u}_2 \rangle = (2)\left(\frac{1}{\sqrt{3}}\right) + (5)\left(\frac{1}{\sqrt{3}}\right) + 3\left(\frac{1}{\sqrt{3}}\right) = \frac{10}{\sqrt{3}}$$

$$\langle \mathbf{v}, \mathbf{u}_3 \rangle = (2)\left(\frac{1}{\sqrt{6}}\right) + 5\left(-\frac{2}{\sqrt{6}}\right) + 3\left(\frac{1}{\sqrt{6}}\right) = -\frac{5}{\sqrt{6}}$$

Hence by (4.73) we have

$$\begin{bmatrix} 2 \\ 5 \\ 3 \end{bmatrix} = \langle \mathbf{v}, \mathbf{u}_1 \rangle \mathbf{u}_1 + \langle \mathbf{v}, \mathbf{u}_2 \rangle \mathbf{u}_2 + \langle \mathbf{v}, \mathbf{u}_3 \rangle \mathbf{u}_3$$

$$= -\frac{1}{\sqrt{2}}\mathbf{u}_1 + \frac{10}{\sqrt{3}}\mathbf{u}_2 - \frac{5}{\sqrt{6}}\mathbf{u}_3$$

EXAMPLE 4-39: Let $\mathscr{B} = \{\mathbf{u}_1, \mathbf{u}_2, \ldots, \mathbf{u}_n\}$ be an orthonormal basis for an inner product space V. If $\mathbf{u}$ and $\mathbf{v}$ are any two vectors in V such that

$$\mathbf{u} = \alpha_1 \mathbf{u}_1 + \alpha_2 \mathbf{u}_2 + \cdots + \alpha_n \mathbf{u}_n \tag{4.75}$$

$$\mathbf{v} = \beta_1 \mathbf{u}_1 + \beta_2 \mathbf{u}_2 + \cdots + \beta_n \mathbf{u}_n \tag{4.76}$$

then show that

$$\langle \mathbf{u}, \mathbf{v} \rangle = \alpha_1 \beta_1 + \alpha_2 \beta_2 + \cdots + \alpha_n \beta_n = \sum_{i=1}^{n} \alpha_i \beta_i \tag{4.77}$$

Solution: Since $\mathscr{B}$ is an orthonormal basis,

$$\langle \mathbf{u}_i, \mathbf{u}_j \rangle = \delta_{ij} = \begin{cases} 1, & i = j \\ 0, & i \neq j \end{cases}$$

Then, using (4.75) and (4.76), we have

$$\begin{aligned} \langle \mathbf{u}, \mathbf{v} \rangle &= \left\langle \sum_{i=1}^{n} \alpha_i \mathbf{u}_i, \sum_{j=1}^{n} \beta_j \mathbf{u}_j \right\rangle \\ &= \sum_{i=1}^{n} \sum_{j=1}^{n} \alpha_i \beta_j \langle \mathbf{u}_i, \mathbf{u}_j \rangle \\ &= \sum_{i=1}^{n} \sum_{j=1}^{n} \alpha_i \beta_j \delta_{ij} \\ &= \sum_{i=1}^{n} \alpha_i \beta_i \\ &= \alpha_1 \beta_1 + \alpha_2 \beta_2 + \cdots + \alpha_n \beta_n \end{aligned}$$

C. Gram-Schmidt orthogonalization process

The **Gram-Schmidt process** is a procedure for obtaining an orthonormal set of vectors from an arbitrary collection of linearly independent vectors:

Let $\{\mathbf{v}_1, \mathbf{v}_2, \ldots, \mathbf{v}_n\}$ be any basis for V.

Step 1. Let $\mathbf{u}_1 = \dfrac{1}{\|\mathbf{v}_1\|}\mathbf{v}_1$

Step 2. Let $\mathbf{w}_2 = \mathbf{v}_2 - \langle \mathbf{v}_2, \mathbf{u}_1 \rangle \mathbf{u}_1$

Step 3. Let $\mathbf{u}_2 = \dfrac{1}{\|\mathbf{w}_2\|}\mathbf{w}_2$

Step 4. Let $\mathbf{w}_3 = \mathbf{v}_3 - \langle \mathbf{v}_3, \mathbf{u}_1 \rangle \mathbf{u}_1 - \langle \mathbf{v}_3, \mathbf{u}_2 \rangle \mathbf{u}_2$

Step 5. Let $\mathbf{u}_3 = \dfrac{1}{\|\mathbf{w}_3\|}\mathbf{w}_3$

Step 6. Continue in this way for $k = 4, 5, \ldots, n$ by letting

$$\mathbf{w}_k = \mathbf{v}_k - \langle \mathbf{v}_k, \mathbf{u}_1 \rangle \mathbf{u}_1 - \langle \mathbf{v}_k, \mathbf{u}_2 \rangle \mathbf{u}_2 - \cdots - \langle \mathbf{v}_k, \mathbf{u}_{k-1} \rangle \mathbf{u}_{k-1}$$

and then let

$$\mathbf{u}_k = \frac{1}{\|\mathbf{w}_k\|}\mathbf{w}_k$$

The set $\{\mathbf{u}_1, \mathbf{u}_2, \ldots, \mathbf{u}_n\}$ so obtained is an orthonormal basis for V (see Problem 4-37). From Example 4-33 we see that in step 2, $\langle \mathbf{v}_2, \mathbf{u}_1 \rangle \mathbf{u}_1$ is the projection of $\mathbf{v}_2$ onto $\mathbf{u}_1$, and hence $\mathbf{w}_2$ is orthogonal to $\mathbf{u}_1$ (see Figure 4-10).

EXAMPLE 4-40: Given

$$\mathbf{v}_1 = \begin{bmatrix} -1 \\ 1 \\ 1 \end{bmatrix}, \qquad \mathbf{v}_2 = \begin{bmatrix} 1 \\ -1 \\ 1 \end{bmatrix}, \qquad \mathbf{v}_3 = \begin{bmatrix} 1 \\ 1 \\ -1 \end{bmatrix}$$

Show that $S = \{\mathbf{v}_1, \mathbf{v}_2, \mathbf{v}_3\}$ is a basis for R^3 and then construct an orthonormal basis $\{\mathbf{u}_1, \mathbf{u}_2, \mathbf{u}_3\}$.

Solution: Since

$$\begin{vmatrix} -1 & 1 & 1 \\ 1 & -1 & 1 \\ 1 & 1 & -1 \end{vmatrix} = 4 \neq 0$$

$S = \{\mathbf{v}_1, \mathbf{v}_2, \mathbf{v}_3\}$ is linearly independent; and since $\dim R^3 = 3$, S forms a basis for R^3. Next, we use the Gram-Schmidt process to obtain an orthonormal basis for R:

Step 1. Let

$$\mathbf{u}_1 = \frac{1}{\|\mathbf{v}_1\|}\mathbf{v}_1 = \frac{1}{\sqrt{3}}\begin{bmatrix} -1 \\ 1 \\ 1 \end{bmatrix} = \begin{bmatrix} -\dfrac{1}{\sqrt{3}} \\ \dfrac{1}{\sqrt{3}} \\ \dfrac{1}{\sqrt{3}} \end{bmatrix}$$

Step 2. Let

$$\mathbf{w}_2 = \mathbf{v}_2 - \langle \mathbf{v}_2, \mathbf{u}_1 \rangle \mathbf{u}_1$$

$$= \begin{bmatrix} 1 \\ -1 \\ 1 \end{bmatrix} - \left(-\frac{1}{\sqrt{3}}\right)\begin{bmatrix} -\dfrac{1}{\sqrt{3}} \\ \dfrac{1}{\sqrt{3}} \\ \dfrac{1}{\sqrt{3}} \end{bmatrix} = \begin{bmatrix} 1 \\ -1 \\ 1 \end{bmatrix} - \begin{bmatrix} \dfrac{1}{3} \\ -\dfrac{1}{3} \\ -\dfrac{1}{3} \end{bmatrix} = \begin{bmatrix} \dfrac{2}{3} \\ -\dfrac{2}{3} \\ \dfrac{4}{3} \end{bmatrix}$$

Step 3. Let

$$\mathbf{u}_2 = \frac{1}{\|\mathbf{w}_2\|}\mathbf{w}_2 = \frac{1}{\frac{2}{3}\sqrt{6}}\begin{bmatrix} \frac{2}{3} \\ -\frac{2}{3} \\ \frac{4}{3} \end{bmatrix} = \begin{bmatrix} \frac{1}{\sqrt{6}} \\ -\frac{1}{\sqrt{6}} \\ \frac{2}{\sqrt{6}} \end{bmatrix}$$

Step 4. Let

$$\mathbf{w}_3 = \mathbf{v}_3 - \langle \mathbf{v}_3, \mathbf{u}_1 \rangle \mathbf{u}_1 - \langle \mathbf{v}_3, \mathbf{u}_2 \rangle \mathbf{u}_2$$

$$= \begin{bmatrix} 1 \\ 1 \\ -1 \end{bmatrix} - \left(-\frac{1}{\sqrt{3}}\right)\begin{bmatrix} -\frac{1}{\sqrt{3}} \\ \frac{1}{\sqrt{3}} \\ \frac{1}{\sqrt{3}} \end{bmatrix} - \left(-\frac{2}{\sqrt{6}}\right)\begin{bmatrix} \frac{1}{\sqrt{6}} \\ -\frac{1}{\sqrt{6}} \\ \frac{2}{\sqrt{6}} \end{bmatrix}$$

$$= \begin{bmatrix} 1 \\ 1 \\ -1 \end{bmatrix} - \begin{bmatrix} \frac{1}{3} \\ -\frac{1}{3} \\ -\frac{1}{3} \end{bmatrix} - \begin{bmatrix} -\frac{1}{3} \\ \frac{1}{3} \\ -\frac{2}{3} \end{bmatrix} = \begin{bmatrix} 1 \\ 1 \\ 0 \end{bmatrix}$$

Step 5. Let

$$\mathbf{u}_3 = \frac{1}{\|\mathbf{w}_3\|}\mathbf{w}_3 = \frac{1}{\sqrt{2}}\begin{bmatrix} 1 \\ 1 \\ 0 \end{bmatrix} = \begin{bmatrix} \frac{1}{\sqrt{2}} \\ \frac{1}{\sqrt{2}} \\ 0 \end{bmatrix}$$

Thus we get an orthonormal basis for R^3, $\{\mathbf{u}_1, \mathbf{u}_2, \mathbf{u}_3\}$ where

$$\mathbf{u}_1 = \begin{bmatrix} -\frac{1}{\sqrt{3}} \\ \frac{1}{\sqrt{3}} \\ \frac{1}{\sqrt{3}} \end{bmatrix} \qquad \mathbf{u}_2 = \begin{bmatrix} \frac{1}{\sqrt{6}} \\ -\frac{1}{\sqrt{6}} \\ \frac{2}{\sqrt{6}} \end{bmatrix} \qquad \mathbf{u}_3 = \begin{bmatrix} \frac{1}{\sqrt{2}} \\ \frac{1}{\sqrt{2}} \\ 0 \end{bmatrix}$$

SUMMARY

1. Let

$$\mathbf{u} = \begin{bmatrix} u_1 \\ u_2 \\ u_3 \end{bmatrix} \quad \text{and} \quad \mathbf{v} = \begin{bmatrix} v_1 \\ v_2 \\ v_3 \end{bmatrix} \in R^3$$

Then the scalar product of **u** and **v**, $\mathbf{u}\cdot\mathbf{v}$ is defined as

$$\mathbf{u}\cdot\mathbf{v} = u_1v_1 + u_2v_2 + u_3v_3$$

2. The Euclidean length of **v** is given by

$$\|\mathbf{v}\| = (\mathbf{v}\cdot\mathbf{v})^{1/2} = \sqrt{v_1^2 + v_2^2 + v_3^2}$$

3. The angle θ between **u** and **v** is given by

$$\theta = \cos^{-1}\left(\frac{\mathbf{u}\cdot\mathbf{v}}{\|\mathbf{u}\|\,\|\mathbf{v}\|}\right)$$

4. The vectors **u** and **v** are orthogonal if $\mathbf{u}\cdot\mathbf{v} = 0$.
5. Let

$$\mathbf{u} = \begin{bmatrix} u_1 \\ u_2 \\ \vdots \\ u_n \end{bmatrix} \quad \text{and} \quad \mathbf{v} = \begin{bmatrix} v_1 \\ v_2 \\ \vdots \\ v_n \end{bmatrix} \in R^n$$

Then the Euclidean inner product of **u** and **v**, $\langle \mathbf{u}, \mathbf{v}\rangle$, is defined as

$$\langle \mathbf{u}, \mathbf{v}\rangle = u_1v_1 + u_2v_2 + \cdots + u_nv_n = \mathbf{v}^T\mathbf{u}$$

6. The Euclidean length (or norm) of **v**, $\|\mathbf{v}\|$ is defined by

$$\|\mathbf{v}\| = \langle \mathbf{v}, \mathbf{v}\rangle^{1/2} = \sqrt{v_1^2 + v_2^2 + \cdots + v_n^2}$$

7. Inner product space is a vector space V with an inner product $\langle \mathbf{u}, \mathbf{v}\rangle$ which is defined by the following axioms:

I1. $\langle \mathbf{u}, \mathbf{v}\rangle = \langle \mathbf{v}, \mathbf{u}\rangle$

I2. $\langle \mathbf{u}, \mathbf{v} + \mathbf{w}\rangle = \langle \mathbf{u}, \mathbf{v}\rangle + \langle \mathbf{u}, \mathbf{w}\rangle$

I3. $\langle \alpha\mathbf{u}, \mathbf{v}\rangle = \alpha\langle \mathbf{u}, \mathbf{v}\rangle$

I4. $\langle \mathbf{v}, \mathbf{v}\rangle \geq 0$ and $\langle \mathbf{v}, \mathbf{v}\rangle = 0$ if and only if $\mathbf{v} = \mathbf{0}$.

8. The norm of $\mathbf{v} \in V$, $\|\mathbf{v}\|$ is defined by

$$\|\mathbf{v}\| = \langle \mathbf{v}, \mathbf{v}\rangle^{1/2}$$

9. The basic properties of the norm of **v** are

$$\|\mathbf{v}\| \geq 0 \quad \text{and} \quad \|\mathbf{v}\| = 0 \quad \text{if and only if } \mathbf{v} = \mathbf{0}$$

$$\|\alpha\mathbf{v}\| = |\alpha|\,\|\mathbf{v}\|$$

$$\|\mathbf{u} + \mathbf{v}\| \leq \|\mathbf{u}\| + \|\mathbf{v}\| \quad \text{(triangle inequality)}$$

10. The distance between **u** and **v** in V, $d(\mathbf{u}, \mathbf{v})$ is defined by $d(\mathbf{u}, \mathbf{v}) = \|\mathbf{v} - \mathbf{u}\|$.
11. The basic properties of distance are

$$d(\mathbf{u}, \mathbf{v}) \geq 0 \text{ and } d(\mathbf{u}, \mathbf{v}) = 0 \text{ if and only if } \mathbf{u} = \mathbf{v}.$$

$$d(\mathbf{u}, \mathbf{v}) = d(\mathbf{v}, \mathbf{u}) \quad \text{(symmetry)}$$

$$d(\mathbf{u}, \mathbf{v}) \leq d(\mathbf{u}, \mathbf{w}) + d(\mathbf{w}, \mathbf{v}) \quad \text{(triangular inequality)}$$

12. The Cauchy-Schwarz inequality is

$$|\langle \mathbf{u}, \mathbf{v}\rangle| \leq \|\mathbf{u}\|\,\|\mathbf{v}\|$$

13. Vectors **u** and **v** are orthogonal if $\langle \mathbf{u}, \mathbf{v}\rangle = 0$.
14. The orthogonal projection of **v** onto **a**, $\text{proj}_{\mathbf{a}}\mathbf{v}$ is defined by

$$\text{proj}_{\mathbf{a}}\mathbf{v} = \frac{\langle \mathbf{v}, \mathbf{a}\rangle}{\|\mathbf{a}\|^2}\mathbf{a}$$

15. The set $\{\mathbf{v}_1, \mathbf{v}_2, \ldots, \mathbf{v}_n\}$ is called orthogonal if $\langle \mathbf{v}_i, \mathbf{v}_j \rangle = 0$ whenever $i \neq j$. In addition, if $\|\mathbf{v}_i\| = 1$ for all i, then the set is called orthonormal.
16. If $\{\mathbf{u}_1, \mathbf{u}_2, \ldots, \mathbf{u}_n\}$ is an orthonormal basis for an inner product space V, then any $\mathbf{v} \in V$ can be expressed as

$$\mathbf{v} = \langle \mathbf{v}, \mathbf{u}_1 \rangle \mathbf{u}_1 + \langle \mathbf{v}, \mathbf{u}_2 \rangle \mathbf{u}_2 + \cdots + \langle \mathbf{v}, \mathbf{u}_n \rangle \mathbf{u}_n$$

17. Gram-Schmidt orthogonalization process is a process to obtain an orthonormal set of vectors from an arbitrary collection of linearly independent vectors.

RAISE YOUR GRADES

Can you explain ...?

☑ how the inner product of two vectors in a vector space V is defined
☑ the basic properties of the inner product
☑ how the norm of a vector is defined
☑ the basic properties of the norm
☑ how the distance between two vectors is defined
☑ how the angle between two vectors is defined
☑ the condition for the orthogonality of two vectors
☑ the orthogonal projection of $\mathbf{v}$ onto $\mathbf{a}$
☑ the basic properties of the distance
☑ the Cauchy-Schwarz inequality
☑ an orthonormal basis
☑ the Gram-Schmidt orthogonalization process

SOLVED PROBLEMS

Scalar Products in R^2 and R^3

PROBLEM 4-1 Let

$$\mathbf{u} = \begin{bmatrix} 1 \\ 2 \\ -3 \end{bmatrix} \quad \text{and} \quad \mathbf{v} = \begin{bmatrix} 0 \\ 1 \\ 1 \end{bmatrix}$$

Compute **(a)** $\|\mathbf{u}\|$, **(b)** $\|\mathbf{v}\|$, and **(c)** $d(\mathbf{u}, \mathbf{v})$.

Solution From (4.2) and (4.4),

(a) $\|\mathbf{u}\| = \sqrt{1^2 + 2^2 + (-3)^2} = \sqrt{12}$

(b) $\|\mathbf{v}\| = \sqrt{0^2 + 1^2 + 1^2} = \sqrt{2}$

(c) $d(\mathbf{u}, \mathbf{v}) = \|\mathbf{v} - \mathbf{u}\| = \sqrt{(0-1)^2 + (1-2)^2 + (1-(-3))^2} = \sqrt{18}$

PROBLEM 4-2 Find all scalars α such that $\|\alpha\mathbf{v}\| = 2$, where

$$\mathbf{v} = \begin{bmatrix} 1 \\ 2 \\ 2 \end{bmatrix}$$

Solution

$$\alpha \mathbf{v} = \alpha \begin{bmatrix} 1 \\ 2 \\ 2 \end{bmatrix} = \begin{bmatrix} \alpha \\ 2\alpha \\ 2\alpha \end{bmatrix}$$

From (4.2),

$$\|\alpha \mathbf{v}\| = \sqrt{\alpha^2 + (2\alpha)^2 + (2\alpha)^2} = |\alpha|\sqrt{9} = 3|\alpha| = 2$$

Thus

$$|\alpha| = \tfrac{2}{3} \quad \text{and} \quad \alpha = \tfrac{2}{3} \quad \text{or} \quad \alpha = -\tfrac{2}{3}$$

PROBLEM 4-3 Prove (4.14) and (4.15) for vectors in R^3.

Solution Let

$$\mathbf{u} = \begin{bmatrix} u_1 \\ u_2 \\ u_3 \end{bmatrix}, \qquad \mathbf{v} = \begin{bmatrix} v_1 \\ v_2 \\ v_3 \end{bmatrix}, \qquad \mathbf{w} = \begin{bmatrix} w_1 \\ w_2 \\ w_3 \end{bmatrix}$$

Then

$$\mathbf{v} + \mathbf{w} = \begin{bmatrix} v_1 + w_1 \\ v_2 + w_2 \\ v_3 + w_3 \end{bmatrix}$$

Now

$$\begin{aligned} \mathbf{u} \cdot \mathbf{v} &= u_1 v_1 + u_2 v_2 + u_3 v_3 \\ &= v_1 u_1 + v_2 u_2 + v_3 u_3 = \mathbf{v} \cdot \mathbf{u} \end{aligned}$$

and

$$\begin{aligned} \mathbf{u} \cdot (\mathbf{v} + \mathbf{w}) &= u_1(v_1 + w_1) + u_2(v_2 + w_2) + u_3(v_3 + w_3) \\ &= (u_1 v_1 + u_2 v_2 + u_3 v_3) + (u_1 w_1 + u_2 w_2 + u_3 w_3) \\ &= \mathbf{u} \cdot \mathbf{v} + \mathbf{u} \cdot \mathbf{w} \end{aligned}$$

PROBLEM 4-4 Let $\mathbf{u} = \begin{bmatrix} 3 \\ 1 \end{bmatrix}$, $\mathbf{v} = \begin{bmatrix} -2 \\ 2 \end{bmatrix}$, and $\mathbf{w} = \begin{bmatrix} 1 \\ 4 \end{bmatrix}$. Calculate $\mathbf{u} \cdot (3\mathbf{v} - 2\mathbf{w})$.

Solution

$$3\mathbf{v} - 2\mathbf{w} = 3\begin{bmatrix} -2 \\ 2 \end{bmatrix} - 2\begin{bmatrix} 1 \\ 4 \end{bmatrix} = \begin{bmatrix} -6 \\ 6 \end{bmatrix} - \begin{bmatrix} 2 \\ 8 \end{bmatrix} = \begin{bmatrix} -8 \\ -2 \end{bmatrix}$$

$$\mathbf{u} \cdot (3\mathbf{v} - 2\mathbf{w}) = (3)(-8) + (1)(-2) = -26$$

Alternate Solution

$$\begin{aligned} \mathbf{u} \cdot (3\mathbf{v} - 2\mathbf{w}) &= 3(\mathbf{u} \cdot \mathbf{v}) - 2(\mathbf{u} \cdot \mathbf{w}) \\ &= 3[(3)(-2) + (1)(2)] - 2[(3)(1) + (1)(4)] \\ &= 3(-4) - 2(7) = -26 \end{aligned}$$

PROBLEM 4-5 Find the angle between

$$\mathbf{u} = \begin{bmatrix} 1 \\ -1 \\ 2 \end{bmatrix} \quad \text{and} \quad \mathbf{v} = \begin{bmatrix} 2 \\ 1 \\ 1 \end{bmatrix}$$

Solution

$$\mathbf{u}\cdot\mathbf{v} = (1)(2) + (-1)(1) + (2)(1) = 3$$
$$\|\mathbf{u}\| = \sqrt{1^2 + (-1)^2 + 2^2} = \sqrt{6}$$
$$\|\mathbf{v}\| = \sqrt{2^2 + 1^2 + 1^2} = \sqrt{6}$$

Then from (4.12)

$$\cos\theta = \frac{\mathbf{u}\cdot\mathbf{v}}{\|\mathbf{u}\|\,\|\mathbf{v}\|} = \frac{3}{\sqrt{6}\sqrt{6}} = \frac{1}{2}$$

Thus

$$\theta = \cos^{-1}(\tfrac{1}{2}) = 60°$$

PROBLEM 4-6 Find the value of k such that

$$\mathbf{u} = \begin{bmatrix} 5 \\ 2 \\ k \end{bmatrix} \quad \text{and} \quad \mathbf{v} = \begin{bmatrix} 3 \\ k \\ 1 \end{bmatrix}$$

are orthogonal.

Solution To be orthogonal, **u** and **v** must satisfy

$$\mathbf{u}\cdot\mathbf{v} = (5)(3) + (2)(k) + (k)(1) = 0$$

Thus $15 + 3k = 0$, so that $k = -5$.

PROBLEM 4-7 Find a vector that is orthogonal to both of the vectors

$$\mathbf{u} = \begin{bmatrix} 1 \\ 2 \\ 3 \end{bmatrix} \quad \text{and} \quad \mathbf{v} = \begin{bmatrix} 1 \\ 1 \\ 1 \end{bmatrix}$$

Solution Let

$$\mathbf{w} = \begin{bmatrix} w_1 \\ w_2 \\ w_3 \end{bmatrix}$$

be the desired vector. Then we must have both

$$\mathbf{u}\cdot\mathbf{w} = w_1 + 2w_2 + 3w_3 = 0$$
$$\mathbf{v}\cdot\mathbf{w} = w_1 + w_2 + w_3 = 0$$

Solving the above system for w_1, w_2, and w_3, we get $w_1 = t$, $w_2 = -2t$, $w_3 = t$, where t is any scalar. Thus,

$$\mathbf{w} = \begin{bmatrix} t \\ -2t \\ t \end{bmatrix}$$

Setting $t = 1$, we have

$$\mathbf{w} = \begin{bmatrix} 1 \\ -2 \\ 1 \end{bmatrix}$$

PROBLEM 4-8 Show that if a vector **w** is orthogonal to both of the vectors **u** and **v**, then **w** is orthogonal to $S\{\mathbf{u}, \mathbf{v}\}$ (span of **u** and **v**), that is, to all linear combination of **u** and **v**.

Solution Since $\mathbf{w} \perp \mathbf{u}$ and $\mathbf{w} \perp \mathbf{v}$, we have

$$\mathbf{w} \cdot \mathbf{u} = 0 \qquad \text{and} \qquad \mathbf{w} \cdot \mathbf{v} = 0$$

If α and β are any scalars, then we have

$$\begin{aligned} \mathbf{w} \cdot (\alpha\mathbf{u} + \beta\mathbf{v}) &= \mathbf{w} \cdot (\alpha\mathbf{u}) + \mathbf{w} \cdot (\beta\mathbf{v}) \\ &= \alpha(\mathbf{w} \cdot \mathbf{u}) + \beta(\mathbf{w} \cdot \mathbf{v}) = (\alpha)(0) + (\beta)(0) = 0 \end{aligned}$$

Hence $\mathbf{w}$ is orthogonal to any linear combination of $\mathbf{u}$ and $\mathbf{v}$.

PROBLEM 4-9 Find the distance from the point (2, 3) to the line expressed as $y = \frac{1}{3}x$ (Figure 4-11).

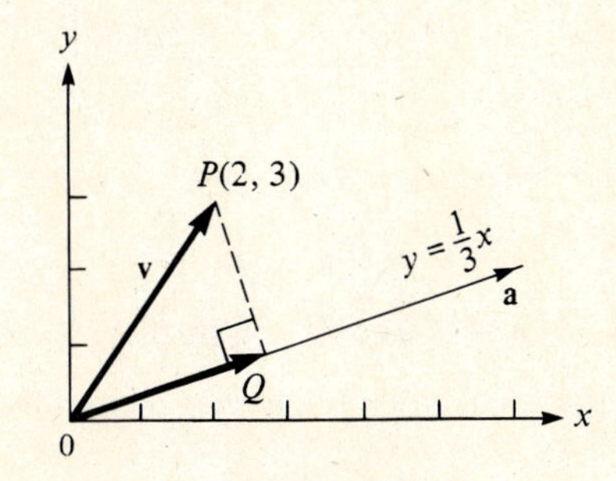

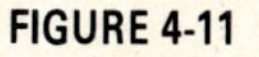

FIGURE 4-11

Solution From geometry we know that the distance from a point to a line is the distance between the given point and the point on the line which is closest to the given point. Let $\mathbf{a} = \begin{bmatrix} 6 \\ 2 \end{bmatrix}$ be the vector in the direction of the line $y = \frac{1}{3}x$. Let $\mathbf{v} = \begin{bmatrix} 2 \\ 3 \end{bmatrix}$, which is represented by $\overrightarrow{OP}$. Then

$$\overrightarrow{OQ} = \text{proj}_{\mathbf{a}}\,\mathbf{v} = \frac{(\mathbf{v} \cdot \mathbf{a})}{\|\mathbf{a}\|^2}\mathbf{a}$$

and the distance from the point (2, 3) to the line $y = \frac{1}{3}x$ is given by $\|\overrightarrow{QP}\| = \|\overrightarrow{OP} - \overrightarrow{OQ}\|$.

Now

$$\mathbf{v} \cdot \mathbf{a} = (2)(6) + (3)(2) = 18$$

$$\|\mathbf{a}\|^2 = \mathbf{a} \cdot \mathbf{a} = 36 + 4 = 40$$

Thus

$$\overrightarrow{OQ} = \frac{18}{40}\begin{bmatrix} 6 \\ 2 \end{bmatrix} = \begin{bmatrix} 2.7 \\ 0.9 \end{bmatrix}$$

and

$$\overrightarrow{OP} - \overrightarrow{OQ} = \begin{bmatrix} 2 \\ 3 \end{bmatrix} - \begin{bmatrix} 2.7 \\ 0.9 \end{bmatrix} = \begin{bmatrix} -0.7 \\ 2.1 \end{bmatrix}$$

Hence the distance from (2, 3) to the line $y = \frac{1}{3}x$ is

$$\|\overrightarrow{QP}\| = \sqrt{(-0.7)^2 + (2.1)^2} = \sqrt{4.9} \simeq 2.21$$

PROBLEM 4-10 If $\mathbf{u}$ and $\mathbf{v}$ are vectors in R^2 or R^3, show that

$$|\mathbf{u} \cdot \mathbf{v}| \le \|\mathbf{u}\|\,\|\mathbf{v}\| \tag{a}$$

where the equality holds if and only if one of the vectors is $\mathbf{0}$ or $\mathbf{u} = \alpha\mathbf{v}$.

Solution From the definition of the scalar product (4.9),

$$|\mathbf{u} \cdot \mathbf{v}| = |\,\|\mathbf{u}\|\,\|\mathbf{v}\| \cos\theta| = \|\mathbf{u}\|\,\|\mathbf{v}\|\,|\cos\theta|$$

Since $|\cos\theta| \le 1$, we get

$$|\mathbf{u}\cdot\mathbf{v}| = \|\mathbf{u}\|\,\|\mathbf{v}\|\,|\cos\theta| \le \|\mathbf{u}\|\,\|\mathbf{v}\|$$

If one of the vectors is $\mathbf{0}$, then both sides of Eq. (a) are 0. If $\mathbf{u} = \alpha\mathbf{v}$, then θ is 0 or π radians and $\cos\theta = \pm 1$ in Eq. (a), so the equality holds in Eq. (a).

PROBLEM 4-11 Show that

$$[a\cos\theta + b\sin\theta]^2 \le a^2 + b^2$$

Solution Let R^2 have the Euclidean inner product and let

$$\mathbf{u} = \begin{bmatrix} a \\ b \end{bmatrix}, \qquad \mathbf{v} = \begin{bmatrix} \cos\theta \\ \sin\theta \end{bmatrix}$$

Then

$$\|\mathbf{u}\|^2 = a^2 + b^2$$

$$\|\mathbf{v}\|^2 = \cos^2\theta + \sin^2\theta = 1$$

$$\mathbf{u}\cdot\mathbf{v} = a\cos\theta + b\sin\theta$$

Applying the result of Problem 4-10a (the Cauchy-Schwarz inequality), we get

$$|\mathbf{u}\cdot\mathbf{v}| \le \|\mathbf{u}\|\,\|\mathbf{v}\|$$

or

$$|\mathbf{u}\cdot\mathbf{v}|^2 \le \|\mathbf{u}\|^2\,\|\mathbf{v}\|^2$$

that is,

$$[a\cos\theta + b\sin\theta]^2 \le a^2 + b^2$$

PROBLEM 4-12 If $\mathbf{u}$ and $\mathbf{v}$ are vectors in R^2 or R^3, show that

$$\|\mathbf{u} + \mathbf{v}\| \le \|\mathbf{u}\| + \|\mathbf{v}\| \tag{a}$$

where the equality holds if and only if $\mathbf{u} = \alpha\mathbf{v}$.

Solution

$$\begin{aligned}\|\mathbf{u} + \mathbf{v}\|^2 &= (\mathbf{u} + \mathbf{v})\cdot(\mathbf{u} + \mathbf{v}) \\ &= \mathbf{u}\cdot\mathbf{u} + 2(\mathbf{u}\cdot\mathbf{v}) + \mathbf{v}\cdot\mathbf{v} \\ &= \|\mathbf{u}\|^2 + 2\|\mathbf{u}\|\,\|\mathbf{v}\|\cos\theta + \|\mathbf{v}\|^2 \\ &\le \|\mathbf{u}\|^2 + 2\|\mathbf{u}\|\,\|\mathbf{v}\| + \|\mathbf{v}\|^2 = (\|\mathbf{u}\| + \|\mathbf{v}\|)^2\end{aligned}$$

since $|\cos\theta| < 1$. Thus

$$\|\mathbf{u} + \mathbf{v}\| \le \|\mathbf{u}\| + \|\mathbf{v}\|$$

The equality holds if and only if $\cos\theta = 1$. This will happen if $\mathbf{u}$ and $\mathbf{v}$ are in the same direction; that is, $\mathbf{u} = \alpha\mathbf{v}$.

Alternate Solution Let $\vec{OP} = \mathbf{u}$ and $\vec{PQ} = \mathbf{v}$, where OPQ is a triangle (Figure 4-12). Then

$$\mathbf{u} + \mathbf{v} = \vec{OQ}$$

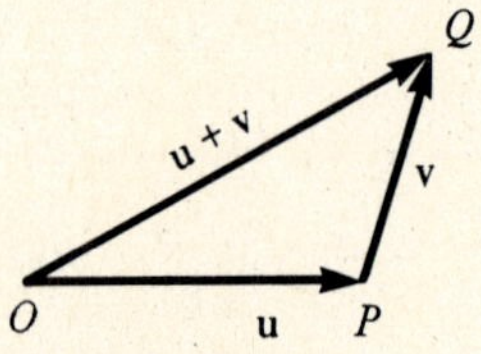

From plane geometry we know that the length of one side of a triangle is less than or equal to the sum of the lengths of the other two sides; that is,

$$\|\vec{OQ}\| \le \|\vec{OP}\| + \|\vec{PQ}\|$$

So we have

$$\|\mathbf{u} + \mathbf{v}\| \le \|\mathbf{u}\| + \|\mathbf{v}\|$$

The equality holds when points O, P, and Q are colinear; that is, when $\mathbf{u} = \alpha\mathbf{v}$.

O u P Q u + v v

FIGURE 4-12

Inner Products in R^n

PROBLEM 4-13 Prove (4.25) and (4.27).

Solution Let α be any scalar and

$$\mathbf{u} = \begin{bmatrix} u_1 \\ u_2 \\ \vdots \\ u_n \end{bmatrix}, \qquad \mathbf{v} = \begin{bmatrix} v_1 \\ v_2 \\ \vdots \\ v_n \end{bmatrix}$$

Then

$$\begin{aligned}
\langle \mathbf{u}, \mathbf{v} \rangle &= u_1 v_1 + u_2 v_2 + \cdots + u_n v_n \\
&= v_1 u_1 + v_2 u_2 + \cdots + v_n u_n \\
&= \langle \mathbf{v}, \mathbf{u} \rangle \\
\alpha \langle \mathbf{u}, \mathbf{v} \rangle &= \alpha(u_1 v_1 + u_2 v_2 + \cdots + u_n v_n) \\
&= (\alpha u_1) v_1 + (\alpha u_2) v_2 + \cdots + (\alpha u_n) v_n = \langle \alpha \mathbf{u}, \mathbf{v} \rangle \\
&= u_1(\alpha v_1) + u_2(\alpha v_2) + \cdots + u_n(\alpha v_n) = \langle \mathbf{u}, \alpha \mathbf{v} \rangle
\end{aligned}$$

PROBLEM 4-14 Find the Euclidean distance between

$$\mathbf{u} = \begin{bmatrix} 5 \\ 0 \\ 1 \\ 3 \\ 0 \end{bmatrix}, \qquad \mathbf{v} = \begin{bmatrix} 2 \\ 1 \\ 0 \\ 2 \\ 1 \end{bmatrix}$$

Solution From (4.30),

$$d(\mathbf{u}, \mathbf{v}) = \|\mathbf{v} - \mathbf{u}\| = \sqrt{(2-5)^2 + (1-0)^2 + (0-1)^2 + (2-3)^2 + (1-0)^2} = \sqrt{13}$$

PROBLEM 4-15 Show that

$$(x_1 y_1 + x_2 y_2 + \cdots + x_n y_n)^2 \le (x_1^2 + x_2^2 + \cdots + x_n^2)(y_1^2 + y_2^2 + \cdots + y_n^2)$$

for any real numbers $x_1 x_2, \ldots, x_n; y_1, y_2, \ldots, y_n$.

Solution Let

$$\mathbf{x} = \begin{bmatrix} x_1 \\ x_2 \\ \vdots \\ x_n \end{bmatrix}, \quad \text{and} \quad \mathbf{y} = \begin{bmatrix} y_1 \\ y_2 \\ \vdots \\ y_n \end{bmatrix}$$

Then

$$\begin{aligned}
\langle \mathbf{x}, \mathbf{y} \rangle &= x_1 y_1 + x_2 y_2 + \cdots + x_n y_n \\
\|\mathbf{x}\|^2 &= x_1^2 + x_2^2 + \cdots + x_n^2 \\
\|\mathbf{y}\|^2 &= y_1^2 + y_2^2 + \cdots + y_n^2
\end{aligned}$$

Applying the Cauchy-Schwarz inequality (4.34) to R^n, that is

$$|\langle \mathbf{x}, \mathbf{y} \rangle| \le \|\mathbf{x}\| \, \|\mathbf{y}\|$$

or

$$|\langle \mathbf{x}, \mathbf{y} \rangle|^2 \le \|\mathbf{x}\|^2 \, \|\mathbf{y}\|^2$$

we obtain the inequality

$$(x_1y_1 + x_2y_2 + \cdots + x_ny_n)^2 \le (x_1^2 + x_2^2 + \cdots + x_n^2)(y_1^2 + y_2^2 + \cdots + y_n^2)$$

PROBLEM 4-16 Let $\mathbf{u}, \mathbf{v} \in R^n$. Show that

$$|\langle \mathbf{u}, \mathbf{v} \rangle| = \|\mathbf{u}\| \, \|\mathbf{v}\|$$

if and only if **u** and **v** are linearly dependent.

Solution First suppose $|\langle \mathbf{u}, \mathbf{v} \rangle| = \|\mathbf{u}\| \, \|\mathbf{v}\|$. If either $\mathbf{u} = \mathbf{0}$ or $\mathbf{v} = \mathbf{0}$, **u** and **v** are clearly linearly dependent, and

$$\langle \mathbf{u}, \mathbf{v} \rangle = \|\mathbf{u}\| \, \|\mathbf{v}\| = 0$$

If $\mathbf{u} \neq \mathbf{0}$ and $\mathbf{v} \neq \mathbf{0}$, we have either

$$\langle \mathbf{u}, \mathbf{v} \rangle = \|\mathbf{u}\| \, \|\mathbf{v}\| \qquad \text{or} \qquad \langle \mathbf{u}, \mathbf{v} \rangle = -\|\mathbf{u}\| \, \|\mathbf{v}\|$$

In the first case we have

$$\begin{aligned}\left\langle \frac{\mathbf{u}}{\|\mathbf{u}\|} - \frac{\mathbf{v}}{\|\mathbf{v}\|}, \frac{\mathbf{u}}{\|\mathbf{u}\|} - \frac{\mathbf{v}}{\|\mathbf{v}\|} \right\rangle &= \frac{\langle \mathbf{u}, \mathbf{u} \rangle}{\|\mathbf{u}\|^2} - \frac{2\langle \mathbf{u}, \mathbf{v} \rangle}{\|\mathbf{u}\| \, \|\mathbf{v}\|} + \frac{\langle \mathbf{v}, \mathbf{v} \rangle}{\|\mathbf{v}\|^2} \\ &= \frac{\|\mathbf{u}\|^2}{\|\mathbf{u}\|^2} - \frac{2\|\mathbf{u}\| \, \|\mathbf{v}\|}{\|\mathbf{u}\| \, \|\mathbf{v}\|} + \frac{\|\mathbf{v}\|^2}{\|\mathbf{v}\|^2} \\ &= 1 - 2 + 1 = 0\end{aligned}$$

Thus by (4.28)

$$\frac{\mathbf{u}}{\|\mathbf{u}\|} - \frac{\mathbf{v}}{\|\mathbf{v}\|} = \mathbf{0} \qquad \text{or} \qquad \mathbf{u} = \frac{\|\mathbf{u}\|}{\|\mathbf{v}\|}\mathbf{v}$$

In the second case, we have

$$\begin{aligned}\left\langle \frac{\mathbf{u}}{\|\mathbf{u}\|} + \frac{\mathbf{v}}{\|\mathbf{v}\|}, \frac{\mathbf{u}}{\|\mathbf{u}\|} + \frac{\mathbf{v}}{\|\mathbf{v}\|} \right\rangle &= \frac{\langle \mathbf{u}, \mathbf{u} \rangle}{\|\mathbf{u}\|^2} + \frac{2\langle \mathbf{u}, \mathbf{v} \rangle}{\|\mathbf{u}\| \, \|\mathbf{v}\|} + \frac{\langle \mathbf{v}, \mathbf{v} \rangle}{\|\mathbf{v}\|^2} \\ &= \frac{\|\mathbf{u}\|^2}{\|\mathbf{u}\|^2} - \frac{2\|\mathbf{u}\| \, \|\mathbf{v}\|}{\|\mathbf{u}\| \, \|\mathbf{v}\|} + \frac{\|\mathbf{v}\|^2}{\|\mathbf{v}\|^2} \\ &= 1 - 2 + 1 = 0\end{aligned}$$

Thus, again by (4.28),

$$\frac{\mathbf{u}}{\|\mathbf{u}\|} + \frac{\mathbf{v}}{\|\mathbf{v}\|} = \mathbf{0} \qquad \text{or} \qquad \mathbf{u} = -\frac{\|\mathbf{u}\|}{\|\mathbf{v}\|}\mathbf{v}$$

Thus in either case **u** and **v** are linearly dependent.

If, on the other hand, **u** and **v** are linearly dependent, say

$$\mathbf{u} = \alpha\mathbf{v}$$

then

$$\langle \mathbf{u}, \mathbf{v} \rangle = \langle \alpha\mathbf{v}, \mathbf{v} \rangle = \alpha\langle \mathbf{v}, \mathbf{v} \rangle = \alpha\|\mathbf{v}\|^2$$

Thus

$$|\langle \mathbf{u}, \mathbf{v} \rangle| = |\alpha\|\mathbf{v}\|^2| = |\alpha| \, \|\mathbf{v}\|^2 = |\alpha| \, \|\mathbf{v}\| \, \|\mathbf{v}\| = \|\mathbf{u}\| \, \|\mathbf{v}\|$$

since $\|\mathbf{u}\| = \|\alpha\mathbf{v}\| = |\alpha| \, \|\mathbf{v}\|$ by (4.32).

PROBLEM 4-17 Prove (4.33); that is, if $\mathbf{u}, \mathbf{v} \in R^n$, then $\|\mathbf{u} + \mathbf{v}\| \le \|\mathbf{u}\| + \|\mathbf{v}\|$

Solution By definition

$$\begin{aligned}\|\mathbf{u} + \mathbf{v}\|^2 &= \langle \mathbf{u} + \mathbf{v}, \mathbf{u} + \mathbf{v} \rangle \\ &= \langle \mathbf{u} + \mathbf{v}, \mathbf{u} \rangle + \langle \mathbf{u} + \mathbf{v}, \mathbf{v} \rangle \qquad &&\text{[by (4.26)]}\end{aligned}$$

$$\begin{aligned}\|\mathbf{u}+\mathbf{v}\|^2 &= \langle \mathbf{u}, \mathbf{u}+\mathbf{v}\rangle + \langle \mathbf{v}, \mathbf{u}+\mathbf{v}\rangle && \text{[by (4.25)]}\\ &= \langle \mathbf{u},\mathbf{u}\rangle + \langle \mathbf{u},\mathbf{v}\rangle + \langle \mathbf{v},\mathbf{u}\rangle + \langle \mathbf{v},\mathbf{v}\rangle && \text{[by (4.26)]}\\ &= \|\mathbf{u}\|^2 + 2\langle \mathbf{u},\mathbf{v}\rangle + \|\mathbf{v}\|^2 && \text{[by (4.25) and (4.28)]}\end{aligned}$$

We recall that $a \le |a|$ for any real number a. Thus,

$$\langle \mathbf{u},\mathbf{v}\rangle < |\langle \mathbf{u},\mathbf{v}\rangle|$$

and hence we get

$$\|\mathbf{u}+\mathbf{v}\|^2 \le \|\mathbf{u}\|^2 + 2|\langle \mathbf{u},\mathbf{v}\rangle| + \|\mathbf{v}\|^2$$

By the Cauchy-Schwarz inequality (4.34),

$$|\langle \mathbf{u},\mathbf{v}\rangle| \le \|\mathbf{u}\|\,\|\mathbf{v}\|$$

Thus

$$\|\mathbf{u}+\mathbf{v}\|^2 \le \|\mathbf{u}\|^2 + 2\|\mathbf{u}\|\,\|\mathbf{v}\| + \|\mathbf{v}\|^2 = (\|\mathbf{u}\| + \|\mathbf{v}\|)^2$$

or

$$\|\mathbf{u}+\mathbf{v}\|^2 \le (\|\mathbf{u}\| + \|\mathbf{v}\|)^2$$

Taking the positive square roots of both sides of this inequality, we get

$$\|\mathbf{u}+\mathbf{v}\| \le \|\mathbf{u}\| + \|\mathbf{v}\|$$

PROBLEM 4-18 Verify the identity

$$\|\mathbf{u}+\mathbf{v}\|^2 + \|\mathbf{u}-\mathbf{v}\|^2 = 2\|\mathbf{u}\|^2 + 2\|\mathbf{v}\|^2$$

for vectors in R^n. Interpret this result geometrically in R^2.

Solution From Problem 4-17, we have

$$\begin{aligned}\|\mathbf{u}+\mathbf{v}\|^2 &= \langle \mathbf{u}+\mathbf{v}, \mathbf{u}+\mathbf{v}\rangle\\ &= \|\mathbf{u}\|^2 + 2\langle \mathbf{u},\mathbf{v}\rangle + \|\mathbf{v}\|^2\end{aligned} \qquad \textbf{(a)}$$

Similarly, by replacing $-\mathbf{v}$ for $\mathbf{v}$ we get

$$\begin{aligned}\|\mathbf{u}-\mathbf{v}\|^2 &= \langle \mathbf{u}-\mathbf{v}, \mathbf{u}-\mathbf{v}\rangle\\ &= \|\mathbf{u}\|^2 - 2\langle \mathbf{u},\mathbf{v}\rangle + \|\mathbf{v}\|^2\end{aligned} \qquad \textbf{(b)}$$

since $\langle \mathbf{u},(-\mathbf{v})\rangle = -\langle \mathbf{u},\mathbf{v}\rangle$, $\|-\mathbf{v}\| = |-1|\,\|\mathbf{v}\| = \|\mathbf{v}\|$. Thus adding Eqs. (a) and (b), we get

$$\|\mathbf{u}+\mathbf{v}\|^2 + \|\mathbf{u}-\mathbf{v}\|^2 = 2\|\mathbf{u}\|^2 + 2\|\mathbf{v}\|^2 \qquad \textbf{(c)}$$

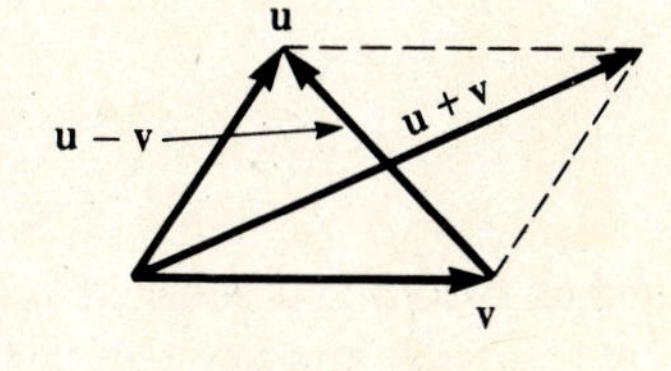

FIGURE 4-13

A geometrical interpretation of this result is as follows: As shown in Figure 4-13, in R^2 $\|\mathbf{u}+\mathbf{v}\|$ and $\|\mathbf{u}-\mathbf{v}\|$ represent the lengths of the diagonals of a parallelogram with sides $\|\mathbf{u}\|$ and $\|\mathbf{v}\|$. Thus the identity (c) shows that the sum of the squares of the lengths of the diagonals of a parallelogram is equal to the sum of the squares of the lengths of its sides.

Inner Product Space

PROBLEM 4-19 Given

$$A = \begin{bmatrix} 1 & -3 \\ 2 & 5 \end{bmatrix}, \qquad B = \begin{bmatrix} 0 & 2 \\ 2 & 1 \end{bmatrix}$$

compute $\langle A, B\rangle$ using the inner product defined by (4.42).

Solution By (4.42),

$$\langle A, B\rangle = (1)(0) + (-3)(2) + (2)(2) + (5)(1) = 3$$

PROBLEM 4-20 Consider the vector space $R_{n \times n}$ of real $n \times n$ matrices. If A and B are two elements of $R_{n \times n}$, then show that

$$\langle A, B \rangle = \text{tr}(AB^T) \tag{a}$$

is an inner product, where the trace of a matrix $C = [c_{ij}]_{n \times n}$ is the sum of the diagonal elements of C;

$$\text{tr}\, C = c_{11} + c_{22} + \cdots + c_{nn} = \sum_{i=1}^{n} c_{ii} \tag{b}$$

Solution Immediate consequences of the definition of the trace are:

$$\text{tr}(\alpha C) = \alpha\, \text{tr}(C) \tag{c}$$

$$\text{tr}(A + B) = \text{tr}(A) + \text{tr}(B) \tag{d}$$

$$\text{tr}\, A = \text{tr}\, A^T \tag{e}$$

Let $A, B \in R_{n \times n}$ and $F = AB^T$. Then $F^T = (AB^T)^T = BA^T$. Now, by definition (a),

$$\begin{aligned} \langle A, B \rangle &= \text{tr}(AB^T) = \text{tr}(F) \\ &= \text{tr}(F^T) \qquad \text{[by (e)]} \\ &= \text{tr}(BA^T) \\ &= \langle B, A \rangle \qquad \text{[by definition (a)]} \end{aligned}$$

Thus $\langle A, B \rangle = \langle B, A \rangle$, and Axiom I1 is satisfied.

Let $C = [c_{ij}]_{n \times n}$. Then

$$\begin{aligned} \langle A, B + C \rangle &= \text{tr}[A(B + C)^T] \\ &= \text{tr}[A(B^T + C^T)] \\ &= \text{tr}(AB^T + AC^T) \\ &= \text{tr}(AB^T) + \text{tr}(AC^T) \qquad \text{[by (d)]} \\ &= \langle A, B \rangle + \langle A, C \rangle \end{aligned}$$

Thus Axiom I2 is satisfied.

Next, if α is any scalar, then

$$\begin{aligned} \langle \alpha A, B \rangle &= \text{tr}[(\alpha A)B^T] \\ &= \text{tr}[\alpha(AB^T)] \\ &= \alpha\, \text{tr}(AB^T) \qquad \text{[by (c)]} \\ &= \alpha \langle A, B \rangle \end{aligned}$$

Thus Axiom I3 is satisfied.

Finally, we note that if $A = [a_{ij}]_{n \times n}$, then

$$\langle A, A \rangle = \text{tr}(AA^T) = \sum_{i,j=1}^{n} a_{ij}^2$$

Thus the trace of AA^T is the sum of the squares of the entries of A. Hence, $\text{tr}(AA^T) \geq 0$. If $\text{tr}(AA^T) = 0$, then all entries of A must be zero, and thus $A = O$. Hence Axiom I4 is satisfied. Thus the function (a) is an inner product.

PROBLEM 4-21 Rework Problem 4-19 using the inner product defined by Problem 4-20a.

Solution

$$A = \begin{bmatrix} 1 & -3 \\ 2 & 5 \end{bmatrix}, \qquad B = \begin{bmatrix} 0 & 2 \\ 2 & 1 \end{bmatrix}$$

Then

$$AB^T = \begin{bmatrix} 1 & -3 \\ 2 & 5 \end{bmatrix} \begin{bmatrix} 0 & 2 \\ 2 & 1 \end{bmatrix} = \begin{bmatrix} -6 & -1 \\ 10 & 9 \end{bmatrix}$$

Hence, by Problem 4-20a,

$$\langle A, B \rangle = \operatorname{tr}(AB^T) = -6 + 9 = 3$$

Note that the results are the same as those for Problem 4-19.

***PROBLEM 4-22** [*For readers who have studied calculus.*] In $C[0, 1]$ with inner product defined by (4.47), compute **(a)** $\langle e^x, e^{-x} \rangle$, **(b)** $\langle x, e^x \rangle$, **(c)** $\langle x^2, x^3 \rangle$.

Solution By (4.47)

(a)
$$\langle e^x, e^{-x} \rangle = \int_0^1 (e^x)(e^{-x})\,dx = \int_0^1 1\,dx = 1$$

(b)
$$\langle x, e^x \rangle = \int_0^1 xe^x\,dx = xe^x \Big|_0^1 - \int_0^1 e^x\,dx = e^1 - e^x \Big|_0^1 = e^0 = 1$$

(c)
$$\langle x^2, x^3 \rangle = \int_0^1 (x^2)(x^3)\,dx = \int_0^1 x^5\,dx = \frac{1}{6}x^6 \Big|_0^1 = \frac{1}{6}$$

PROBLEM 4-23 Verify (4.48), $\langle \mathbf{0}, \mathbf{v} \rangle = 0$.

Solution

$$\begin{aligned} \langle \mathbf{0}, \mathbf{v} \rangle &= \langle \mathbf{v} + (-1)\mathbf{v}, \mathbf{v} \rangle \\ &= \langle \mathbf{v}, \mathbf{v} \rangle + \langle (-1)\mathbf{v}, \mathbf{v} \rangle \\ &= \langle \mathbf{v}, \mathbf{v} \rangle + (-1)\langle \mathbf{v}, \mathbf{v} \rangle \\ &= 0 \end{aligned}$$

PROBLEM 4-24 Verify (4.50), $\langle \mathbf{u}, \alpha\mathbf{v} \rangle = \alpha\langle \mathbf{u}, \mathbf{v} \rangle$.

Solution

$$\begin{aligned} \langle \mathbf{u}, \alpha\mathbf{v} \rangle &= \langle \alpha\mathbf{v}, \mathbf{u} \rangle && \text{[by (4.39)]} \\ &= \alpha\langle \mathbf{v}, \mathbf{u} \rangle && \text{[by (4.41)]} \\ &= \alpha\langle \mathbf{u}, \mathbf{v} \rangle && \text{[by (4.39)]} \end{aligned}$$

PROBLEM 4-25 Let $R_{2\times 2}$ have the inner product of (4.44). Find **(a)** $\|A\|$, $\|B\|$, **(b)** $d(A, B)$, and **(c)** angle θ between A and B, where

$$A = \begin{bmatrix} 1 & -3 \\ 2 & 5 \end{bmatrix}, \qquad B = \begin{bmatrix} 0 & 2 \\ 2 & 1 \end{bmatrix}$$

Solution By (4.44)

(a)
$$\|A\| = \langle A, A \rangle^{1/2} = \sqrt{1^2 + (-3)^2 + 2^2 + 5^2} = \sqrt{39}$$
$$\|B\| = \langle B, B \rangle^{1/2} = \sqrt{0^2 + 2^2 + 2^2 + 1^2} = \sqrt{9} = 3$$

(b)
$$B - A = \begin{bmatrix} 0 & 2 \\ 2 & 1 \end{bmatrix} - \begin{bmatrix} 1 & -3 \\ 2 & 5 \end{bmatrix} = \begin{bmatrix} -1 & 5 \\ 0 & -4 \end{bmatrix}$$

$$d(A, B) = \|\mathbf{B} - \mathbf{A}\| = \sqrt{(-1)^2 + 5^2 + 0^2 + (-4)^2} = \sqrt{42}$$

(c) From Problem 4-21, $\langle A, B \rangle = 3$. Thus, by (4.64),

$$\cos\theta = \frac{\langle A, B \rangle}{\|A\|\,\|B\|} = \frac{3}{\sqrt{39}\cdot 3} = \frac{1}{\sqrt{39}}$$

$$\theta = \cos^{-1}\frac{1}{\sqrt{39}} \simeq 80.79°$$

PROBLEM 4-26 Let $\mathbf{p}, \mathbf{q} \in P_2$ and $\mathbf{p} = 1 + x$, $\mathbf{q} = 1 + x + x^2$. Find **(a)** $\|\mathbf{p}\|$, $\|\mathbf{q}\|$ and **(b)** $d(\mathbf{p}, \mathbf{q})$ with respect to the inner product defined in (4.45).

Solution

(a) By (4.51) and (4.45), we get

$$\|\mathbf{p}\| = \langle \mathbf{p}, \mathbf{p} \rangle^{1/2} = \sqrt{1^2 + 1^2} = \sqrt{2}$$

$$\|\mathbf{q}\| = \langle \mathbf{q}, \mathbf{q} \rangle^{1/2} = \sqrt{1^2 + 1^2 + 1^2} = \sqrt{3}$$

(b)

$$\mathbf{q} - \mathbf{p} = q(x) - p(x) = (1 + x + x^2) - (1 + x) = x^2$$

so by (4.52),

$$d(\mathbf{p}, \mathbf{q}) = \|\mathbf{q} - \mathbf{p}\| = \sqrt{1^2} = 1$$

***PROBLEM 4-27** [*For readers who have studied calculus.*] Rework Example 4-26 with respect to the inner product defined in (4.46) with $a = 0$ and $b = 1$.

Solution

(a) By (4.51) and (4.46) we get

$$\begin{aligned}
\|\mathbf{p}\| = \langle \mathbf{p}, \mathbf{p} \rangle^{1/2} &= \left(\int_0^1 [p(x)]^2\,dx\right)^{1/2} \\
&= \left(\int_0^1 (1 + x)^2\,dx\right)^{1/2} \\
&= \left(\int_0^1 (1 + 2x + x^2)\,dx\right)^{1/2} \\
&= \left[\left(x + x^2 + \frac{x^3}{3}\right)\Bigg|_0^1\right]^{1/2} = \left(1 + 1 + \frac{1}{3}\right)^{1/2} = \sqrt{\frac{7}{3}}
\end{aligned}$$

$$\begin{aligned}
\|\mathbf{q}\| = \langle \mathbf{q}, \mathbf{q} \rangle^{1/2} &= \left(\int_0^1 [q(x)]^2\,dx\right)^{1/2} \\
&= \left(\int_0^1 (1 + x + x^2)^2\,dx\right)^{1/2} \\
&= \left(\int_0^1 (1 + 2x + 3x^2 + 2x^3 + x^4)\,dx\right)^{1/2} \\
&= \left[\left(x + x^2 + x^3 + \frac{x^4}{2} + \frac{x^5}{5}\right)\Bigg|_0^1\right]^{1/2} \\
&= \left(1 + 1 + 1 + \frac{1}{2} + \frac{1}{5}\right)^{1/2} = \sqrt{\frac{37}{10}}
\end{aligned}$$

(b)

$$\mathbf{q} - \mathbf{p} = q(x) - p(x) = x^2$$

so by (4.52),

$$d(\mathbf{p}, \mathbf{q}) = \|\mathbf{q} - \mathbf{p}\| = \left(\int_0^1 (x^2)^2\,dx\right)^{1/2} = \left(\int_0^1 x^4\,dx\right)^{1/2} = \left(\frac{x^5}{5}\bigg|_0^1\right)^{1/2} = \frac{1}{\sqrt{5}}$$

***PROBLEM 4-28** [*For readers who have studied calculus.*] Show that the functions x and x^2 are orthogonal in P_2 with inner product defined by (4.46) with $a = -1$ and $b = 1$, but not orthogonal with $a = 0$ and $b = 1$.

Solution By (4.46), if $a = -1$ and $b = 1$,

$$\langle x, x^2 \rangle = \int_{-1}^1 (x)(x^2)\,dx = \int_{-1}^1 x^3\,dx = \frac{x^4}{4}\bigg|_{-1}^1 = 0$$

but if $a = 0$, $b = 1$,

$$\langle x, x^2 \rangle = \int_0^1 (x)(x^2)\,dx = \int_0^1 x^3\,dx = \frac{x^4}{4}\bigg|_0^1 = \frac{1}{4} \neq 0$$

Thus x and x^2 are orthogonal with inner product

$$\langle \mathbf{p}, \mathbf{q} \rangle = \int_{-1}^{1} p(x)q(x)\,dx$$

but not orthogonal with inner product

$$\langle \mathbf{p}, \mathbf{q} \rangle = \int_{0}^{1} p(x)q(x)\,dx$$

***PROBLEM 4-29** [*For readers who have studied calculus.*] In $C[-\pi, \pi]$ with inner product defined by (4.47), show that $\sin mx$ and $\cos nx$ are orthogonal for any integers m and n.

Solution

$$\begin{aligned}
\langle \sin mx, \cos nx \rangle &= \int_{-\pi}^{\pi} \sin mx \cos nx\,dx \\
&= \int_{-\pi}^{\pi} \tfrac{1}{2}[\sin(m+n)x + \sin(m-n)x]\,dx \\
&= \frac{1}{2}\frac{-1}{m+n}\cos(m+n)x\bigg|_{-\pi}^{\pi} + \frac{1}{2}\frac{-1}{m-n}\cos(m-n)x\bigg|_{-\pi}^{\pi} \\
&= 0 \qquad \text{if } m \neq n
\end{aligned}$$

If $m = n \neq 0$, by using the trigonometric identity $\sin 2\theta = 2\sin\theta\cos\theta$, we get

$$\begin{aligned}
\langle \sin mx, \cos mx \rangle &= \int_{-\pi}^{\pi} \sin mx \cos mx\,dx \\
&= \frac{1}{2}\int_{-\pi}^{\pi} \sin(2mx)\,dx \\
&= -\frac{1}{4m}\cos(2mx)\bigg|_{-\pi}^{\pi} \\
&= 0
\end{aligned}$$

Certainly, for $m = n = 0$, we get

$$\langle \sin 0x, \cos 0x \rangle = \langle 0, 1 \rangle = \int_{-\pi}^{\pi} 0\,dx = 0$$

Hence, $\sin mx$ and $\cos nx$ are orthogonal for any integers m and n.

***PROBLEM 4-30** [*For readers who have studied calculus.*] Show that

$$\left(\int_a^b f(x)g(x)\,dx\right)^2 \le \int_a^b [f(x)]^2\,dx \int_a^b [g(x)]^2\,dx$$

for any continuous functions $f(x)$ and $g(x)$ defined on $[a, b]$.

Solution Let $\mathbf{f} = f(x)$ and $\mathbf{g} = g(x)$ and $\mathbf{f}, \mathbf{g} \in C[a, b]$. Then let $\langle \mathbf{f}, \mathbf{g} \rangle$ be the inner product defined on the vector space $C[a, b]$ defined in (4.47).

Then

$$\|\mathbf{f}\|^2 = \langle \mathbf{f}, \mathbf{f} \rangle = \int_a^b [f(x)]^2\,dx$$

$$\|\mathbf{g}\|^2 = \langle \mathbf{g}, \mathbf{g} \rangle = \int_a^b [g(x)]^2\,dx$$

Applying the Cauchy-Schwarz inequality to $C[a, b]$, that is,

$$|\langle \mathbf{f}, \mathbf{g} \rangle| \le \|\mathbf{f}\|\,\|\mathbf{g}\|$$

or

$$|\langle \mathbf{f}, \mathbf{g} \rangle|^2 \leq \|\mathbf{f}\|^2 \|\mathbf{g}\|^2$$

we obtain

$$\left(\int_a^b f(x)g(x)\,dx \right)^2 \leq \int_a^b [f(x)]^2\,dx \int_a^b [g(x)]^2\,dx$$

PROBLEM 4-31 Let $\mathbf{v}_1, \mathbf{v}_2, \ldots, \mathbf{v}_m$ be nonzero mutually orthogonal vectors of an inner product space V. Let α_i be the component of $\mathbf{v} \in V$ along $\mathbf{v}_i$. Then show that

$$\mathbf{w} = \mathbf{v} - \alpha_1 \mathbf{v}_1 - \alpha_2 \mathbf{v}_2 - \cdots - \alpha_m \mathbf{v}_m$$

is orthogonal to $\mathbf{v}_1, \mathbf{v}_2, \ldots, \mathbf{v}_m$.

Solution Taking the inner product of $\mathbf{w}$ with $\mathbf{v}_j$ for any j,

$$\begin{aligned}
\langle \mathbf{w}, \mathbf{v}_j \rangle &= \langle \mathbf{v} - \alpha_1 \mathbf{v}_1 - \alpha_2 \mathbf{v}_2 - \cdots - \alpha_m \mathbf{v}_m, \mathbf{v}_j \rangle \\
&= \langle \mathbf{v}, \mathbf{v}_j \rangle - \langle \alpha_1 \mathbf{v}_1, \mathbf{v}_j \rangle - \langle \alpha_2 \mathbf{v}_2, \mathbf{v}_j \rangle - \cdots - \langle \alpha_m \mathbf{v}_m, \mathbf{v}_j \rangle \\
&= \langle \mathbf{v}, \mathbf{v}_j \rangle - \alpha_1 \langle \mathbf{v}_1, \mathbf{v}_j \rangle - \alpha_2 \langle \mathbf{v}_2, \mathbf{v}_j \rangle - \cdots - \alpha_m \langle \mathbf{v}_m, \mathbf{v}_j \rangle \\
&= \langle \mathbf{v}, \mathbf{v}_j \rangle - \alpha_j \langle \mathbf{v}_j, \mathbf{v}_j \rangle = \langle \mathbf{v}, \mathbf{v}_j \rangle - \alpha_j \|\mathbf{v}_j\|^2 = 0
\end{aligned}$$

since $\{\mathbf{v}_1, \mathbf{v}_2, \ldots, \mathbf{v}_n\}$ is orthogonal; $\langle \mathbf{v}_i, \mathbf{v}_j \rangle = 0$ if $i \neq j$; α_j is the component of $\mathbf{v}$ along $\mathbf{v}_j$; and by (4.68)

$$\alpha_j = \frac{\langle \mathbf{v}, \mathbf{v}_j \rangle}{\|\mathbf{v}_j\|^2}$$

We conclude that $\mathbf{w}$ is orthogonal to $\mathbf{v}_1, \mathbf{v}_2, \ldots, \mathbf{v}_m$.

PROBLEM 4-32 If $\{\mathbf{u}_1, \mathbf{u}_2, \ldots, \mathbf{u}_n\}$ is mutually orthogonal in an inner product space V and if α_i is the Fourier coefficient of $\mathbf{v}$ with respect to $\mathbf{u}_i$, show that

$$\sum_{i=1}^{n} \alpha_i^2 \leq \|\mathbf{v}\|^2$$

This inequality is known as the **Bessel inequality**.

Solution Now

$$\begin{aligned}
0 \leq \left\| \mathbf{v} - \sum_{i=1}^{n} \alpha_i \mathbf{u}_i \right\|^2 &= \left\langle \mathbf{v} - \sum_{i=1}^{n} \alpha_i \mathbf{u}_i, \mathbf{v} - \sum_{i=1}^{n} \alpha_i \mathbf{u}_i \right\rangle \\
&= \langle \mathbf{v}, \mathbf{v} \rangle - \left\langle \sum_{i=1}^{n} \alpha_i \mathbf{u}_i, \mathbf{v} \right\rangle - \left\langle \mathbf{v}, \sum_{i=1}^{n} \alpha_i \mathbf{u}_i \right\rangle + \left\langle \sum_{i=1}^{n} \alpha_i \mathbf{u}_i, \sum_{i=1}^{n} \alpha_i \mathbf{u}_i \right\rangle \\
&= \langle \mathbf{v}, \mathbf{v} \rangle - \sum_{i=1}^{n} 2\alpha_i \langle \mathbf{u}_i, \mathbf{v} \rangle + \sum_{i=1}^{n} \alpha_i^2 \langle \mathbf{u}_i, \mathbf{u}_i \rangle \\
&= \|\mathbf{v}\|^2 - \sum_{i=1}^{n} 2\alpha_i^2 + \sum_{i=1}^{n} \alpha_i^2 \\
&= \|\mathbf{v}\|^2 - \sum_{i=1}^{n} \alpha_i^2
\end{aligned}$$

since $\langle \mathbf{u}_i, \mathbf{u}_j \rangle = 0$ for $i \neq j$ and $\langle \mathbf{u}_i, \mathbf{u}_i \rangle = \|\mathbf{u}_i\| = 1$ and $\alpha_i = \langle \mathbf{u}_i, \mathbf{v} \rangle$ by (4.69). Thus,

$$\sum_{i=1}^{n} \alpha_i^2 \leq \|\mathbf{v}\|^2$$

***PROBLEM 4-33** [*For readers who have studied calculus.*] Let $C[-\pi, \pi]$ have inner product (4.47) with $a = -\pi$, $b = \pi$; that is,

$$\langle \mathbf{f}, \mathbf{g} \rangle = \int_{-\pi}^{\pi} f(x)g(x)\,dx \tag{a}$$

(a) Show that the set $\{1, \cos x, \sin x\}$ is an orthogonal set of vectors with respect to this inner product.
(b) Construct an orthonormal set from this set.

Solution Since

$$\langle 1, \cos x\rangle = \int_{-\pi}^{\pi} \cos x\, dx = 0$$

$$\langle 1, \sin x\rangle = \int_{-\pi}^{\pi} \sin x\, dx = 0$$

$$\langle \cos x, \sin x\rangle = \int_{-\pi}^{\pi} \cos x \sin x\, dx = \frac{1}{2}\int_{-\pi}^{\pi} \sin 2x\, dx = 0$$

$\{1, \cos x, \sin x\}$ is an orthogonal set with respect to the inner product (a).

To construct an orthonormal set, we calculate the norms of these vectors:

$$\|1\|^2 = \langle 1, 1\rangle = \int_{-\pi}^{\pi} 1\, dx = 2\pi$$

$$\|\cos x\|^2 = \langle \cos x, \cos x\rangle = \int_{-\pi}^{\pi} \cos^2 x\, dx = \frac{1}{2}\int_{-\pi}^{\pi} [1 + \cos 2x]\, dx = \pi$$

$$\|\sin x\|^2 = \langle \sin x, \sin x\rangle = \int_{-\pi}^{\pi} \sin^2 x\, dx = \frac{1}{2}\int_{-\pi}^{\pi} [1 - \cos 2x]\, dx = \pi$$

Thus,

$$\left\{\frac{1}{\sqrt{2\pi}}, \frac{1}{\sqrt{\pi}}\cos x, \frac{1}{\sqrt{\pi}}\sin x\right\}$$

is an orthonormal set with respect to the inner product (a).

PROBLEM 4-34 Let $S = \{\mathbf{u}_1, \mathbf{u}_2, \ldots \mathbf{u}_n\}$ be an orthonormal basis for an inner product space V, and $\mathbf{v} \in V$. If

$$\mathbf{v} = \alpha_1\mathbf{u}_1 + \alpha_2\mathbf{u}_2 + \cdots + \alpha_n\mathbf{u}_n = \sum_{i=1}^{n} \alpha_i\mathbf{u}_i$$

then show that

$$\|\mathbf{v}\|^2 = \sum_{i=1}^{n} \alpha_1^2 \tag{a}$$

Equation (a) is known as **Parseval's identity**.

Solution From Theorem 4-5.1 (4.73), we have

$$\alpha_i = \langle \mathbf{v}, \mathbf{u}_i\rangle$$

Since

$$\begin{aligned}
0 &= \left\|\mathbf{v} - \sum_{i=1}^{n} \alpha_i\mathbf{u}_i\right\| \\
&= \left\langle \mathbf{v} - \sum_{i=1}^{n} \alpha_i\mathbf{u}_i, \mathbf{v} - \sum_{i=1}^{n} \alpha_i\mathbf{u}_i\right\rangle \\
&= \langle \mathbf{v}, \mathbf{v}\rangle - \left\langle \sum_{i=1}^{n} \alpha_i\mathbf{u}_i, \mathbf{v}\right\rangle - \left\langle \mathbf{v}, \sum_{i=1}^{n} \alpha_i\mathbf{u}_i\right\rangle + \left\langle \sum_{i=1}^{n} \alpha_i\mathbf{u}_i, \sum_{i=1}^{n} \alpha_i\mathbf{u}_i\right\rangle \\
&= \langle \mathbf{v}, \mathbf{v}\rangle - \sum_{i=1}^{n} 2\alpha_i\langle \mathbf{v}, \mathbf{u}_i\rangle + \langle \mathbf{v}, \mathbf{v}\rangle \\
&= 2\|\mathbf{v}\|^2 - 2\sum_{i=1}^{n} \alpha_i^2
\end{aligned}$$

Thus we get

$$\|\mathbf{v}\|^2 = \sum_{i=1}^{n} \alpha_i^2$$

PROBLEM 4-35 Let R^3 have the Euclidean inner product. Use the Gram-Schmidt process to obtain an orthonormal basis from

$$\mathbf{v}_1 = \begin{bmatrix} 1 \\ 0 \\ 0 \end{bmatrix}, \qquad \mathbf{v}_2 = \begin{bmatrix} 1 \\ 1 \\ 0 \end{bmatrix}, \qquad \mathbf{v}_3 = \begin{bmatrix} 1 \\ 1 \\ 1 \end{bmatrix}$$

Solution We have shown in Problem 3-29 that $\{\mathbf{v}_1, \mathbf{v}_2, \mathbf{v}_3\}$ forms a basis for R^3. Now:

Step 1. Let $\mathbf{u}_1 = \dfrac{1}{\|\mathbf{v}_1\|}\mathbf{v}_1 = \begin{bmatrix} 1 \\ 0 \\ 0 \end{bmatrix} = \mathbf{i}$

Step 2. Let $\mathbf{w}_2 = \mathbf{v}_2 - \langle \mathbf{v}_2, \mathbf{u}_1 \rangle \mathbf{u}_1 = \begin{bmatrix} 1 \\ 1 \\ 0 \end{bmatrix} - (1)\begin{bmatrix} 1 \\ 0 \\ 0 \end{bmatrix} = \begin{bmatrix} 0 \\ 1 \\ 0 \end{bmatrix}$

Step 3. Let $\mathbf{u}_2 = \dfrac{1}{\|\mathbf{w}_2\|}\mathbf{w}_2 = \begin{bmatrix} 0 \\ 1 \\ 0 \end{bmatrix} = \mathbf{j}$

Step 4. Let $\mathbf{w}_3 = \mathbf{v}_3 - \langle \mathbf{v}_3, \mathbf{u}_1 \rangle - \langle \mathbf{v}_3, \mathbf{u}_2 \rangle \mathbf{u}_2$

$$= \begin{bmatrix} 1 \\ 1 \\ 1 \end{bmatrix} - (1)\begin{bmatrix} 1 \\ 0 \\ 0 \end{bmatrix} - (1)\begin{bmatrix} 0 \\ 1 \\ 0 \end{bmatrix} = \begin{bmatrix} 0 \\ 0 \\ 1 \end{bmatrix}$$

Step 5. Let $\mathbf{u}_3 = \dfrac{1}{\|\mathbf{w}_3\|}\mathbf{w}_3 = \begin{bmatrix} 0 \\ 0 \\ 1 \end{bmatrix} = \mathbf{k}$

Thus we obtain the standard basis for R^3, $\{\mathbf{i}, \mathbf{j}, \mathbf{k}\}$.

PROBLEM 4-36 Let R^3 have the inner product

$$\langle \mathbf{u}, \mathbf{v} \rangle = u_1 v_1 + 2u_2 v_2 + 3u_3 v_3$$

Use the Gram-Schmidt process to obtain an orthonormal basis for R^3 with respect to this inner product from

$$\mathbf{v}_1 = \begin{bmatrix} 1 \\ 0 \\ 0 \end{bmatrix}, \qquad \mathbf{v}_2 = \begin{bmatrix} 1 \\ 1 \\ 0 \end{bmatrix}, \qquad \mathbf{v}_3 = \begin{bmatrix} 1 \\ 1 \\ 1 \end{bmatrix}$$

Solution For the verification of inner product $\langle \mathbf{u}, \mathbf{v} \rangle$, see Example 4-17. Note that with the given inner product,

$$\|\mathbf{v}\| = \sqrt{v_1^2 + 2v_2^2 + 3v_3^2}$$

Now

Step 1. Let $\mathbf{u}_1 = \dfrac{1}{\|\mathbf{v}_1\|}\mathbf{v}_1 = \dfrac{1}{1}\begin{bmatrix} 1 \\ 0 \\ 0 \end{bmatrix} = \begin{bmatrix} 1 \\ 0 \\ 0 \end{bmatrix}$

Step 2. Let $\mathbf{w}_2 = \mathbf{v}_2 - \langle \mathbf{v}_2, \mathbf{u}_1 \rangle \mathbf{u}_1 = \begin{bmatrix} 1 \\ 1 \\ 0 \end{bmatrix} - 1 \begin{bmatrix} 1 \\ 0 \\ 0 \end{bmatrix} = \begin{bmatrix} 0 \\ 1 \\ 0 \end{bmatrix}$

Step 3. Let $\mathbf{u}_2 = \dfrac{1}{\|\mathbf{w}_2\|} \mathbf{w}_2 = \dfrac{1}{\sqrt{2}} \begin{bmatrix} 0 \\ 1 \\ 0 \end{bmatrix} = \begin{bmatrix} 0 \\ \dfrac{1}{\sqrt{2}} \\ 0 \end{bmatrix}$

Step 4. Let $\mathbf{w}_3 = \mathbf{v}_3 - \langle \mathbf{v}_3, \mathbf{u}_1 \rangle \mathbf{u}_1 - \langle \mathbf{v}_3, \mathbf{u}_2 \rangle \mathbf{u}_2$

$$= \begin{bmatrix} 1 \\ 1 \\ 1 \end{bmatrix} - 1 \begin{bmatrix} 1 \\ 0 \\ 0 \end{bmatrix} - \sqrt{2} \begin{bmatrix} 0 \\ \dfrac{1}{\sqrt{2}} \\ 0 \end{bmatrix} = \begin{bmatrix} 0 \\ 0 \\ 1 \end{bmatrix}$$

since $\langle \mathbf{v}_3, \mathbf{u}_1 \rangle = 1, \langle \mathbf{v}_3 \mathbf{u}_2 \rangle = (1)(0) + 2(1)\left(\dfrac{1}{\sqrt{2}}\right) + 3(1)(0) = \sqrt{2}.$

Step 5. Let $\mathbf{u}_3 = \dfrac{1}{\|\mathbf{w}_3\|} \mathbf{w}_3 = \dfrac{1}{\sqrt{3}} \begin{bmatrix} 0 \\ 0 \\ 1 \end{bmatrix} = \begin{bmatrix} 0 \\ 0 \\ \dfrac{1}{\sqrt{3}} \end{bmatrix}$

Thus,

$$\begin{bmatrix} 1 \\ 0 \\ 0 \end{bmatrix}, \quad \begin{bmatrix} 0 \\ \dfrac{1}{\sqrt{2}} \\ 0 \end{bmatrix}, \quad \begin{bmatrix} 0 \\ 0 \\ \dfrac{1}{\sqrt{3}} \end{bmatrix}$$

form an orthonormal basis for R^3 with the given inner product.

PROBLEM 4-37 Let $\mathbf{v}_1, \mathbf{v}_2, \ldots \mathbf{v}_n$ be a set of linearly independent vectors in V. Then show that the set $\{\mathbf{u}_1, \mathbf{u}_2, \ldots, \mathbf{u}_n\}$ constructed by the Gram-Schmidt process of Section 4-5C is an orthonormal set.

Solution It is clear by the linear independence of $\mathbf{v}_1, \mathbf{v}_2, \ldots, \mathbf{v}_n$, that $\mathbf{v}_1 \neq \mathbf{0}$, hence

$$\mathbf{u}_1 = \frac{1}{\|\mathbf{v}_1\|} \mathbf{v}_1 \neq \mathbf{0} \qquad \text{and} \qquad \|\mathbf{u}_1\| = 1$$

If $k \leq n$, consider

$$\mathbf{w}_k = \mathbf{v}_k - \langle \mathbf{v}_k, \mathbf{u}_1 \rangle \mathbf{u}_1 - \cdots - \langle \mathbf{v}_k, \mathbf{u}_{k-1} \rangle \mathbf{u}_{k-1}$$

If

$$\mathbf{v}_k - \langle \mathbf{v}_k, \mathbf{u}_1 \rangle \mathbf{u}_1 - \cdots - \langle \mathbf{v}_k, \mathbf{u}_{k-1} \rangle = \mathbf{0}$$

then

$$\mathbf{v}_k = \langle \mathbf{v}_k, \mathbf{u}_1 \rangle \mathbf{u}_1 + \cdots + \langle \mathbf{v}_k, \mathbf{u}_{k-1} \rangle \mathbf{u}_{k-1}$$

That is, $\mathbf{v}_k$ can be expressed as a linear combination of $\mathbf{u}_1, \mathbf{u}_2, \ldots, \mathbf{u}_{k-1}$ or $\mathbf{v}_1, \mathbf{v}_2, \ldots, \mathbf{v}_{k-1}$. This is inconsistent with the assumption of linear dependence of $\mathbf{v}_1, \mathbf{v}_2, \ldots, \mathbf{v}_n$; therefore $\mathbf{w}_k \neq \mathbf{0}$. Hence

$$\mathbf{u}_k = \frac{1}{\|\mathbf{w}_k\|}\mathbf{w}_k \neq \mathbf{0} \qquad \text{and} \qquad \|\mathbf{u}_k\| = 1$$

Next, in Problem 4-31 we have shown that $\mathbf{w}_k$; hence $\mathbf{u}_k$, is orthogonal to $\mathbf{u}_1, \mathbf{u}_2, \ldots, \mathbf{u}_{k-1}$. Thus $\{\mathbf{u}_1, \mathbf{u}_2, \ldots, \mathbf{u}_n\}$ is an orthonormal set.

Supplementary Exercises

PROBLEM 4-38 Verify the identities:

$$\langle \mathbf{u}, \mathbf{v} \rangle = \tfrac{1}{2}(\|\mathbf{u} + \mathbf{v}\|^2 - \|\mathbf{u}\|^2 - \|\mathbf{v}\|^2)$$

$$\langle \mathbf{u}, \mathbf{v} \rangle = \tfrac{1}{4}\|\mathbf{u} + \mathbf{v}\|^2 - \tfrac{1}{4}\|\mathbf{u} - \mathbf{v}\|^2$$

[*Hint:* Use the results of Problem 4-18.]

PROBLEM 4-39 Let $\mathbf{u}, \mathbf{v} \in R^n$. Show that

$$\|\mathbf{u} + \mathbf{v}\| = \|\mathbf{u} - \mathbf{v}\|$$

if and only if $\mathbf{u}$ and $\mathbf{v}$ are orthogonal. Interpret this result geometrically in R^2. [*Hint:* Use the results of Problem 4-18.]

PROBLEM 4-40 Show that the inner product defined by (4.44) and the inner product of Problem 4-20a are the same. [*Hint:* Use matrices in $R_{2\times 2}$ to show the same results.]

***PROBLEM 4-41** [*For readers who have studied calculus.*] Find the norm of $\cos x$, which is an element of $C[0, \pi]$ with respect to the inner product defined in (4.47).

Answer $\sqrt{\dfrac{\pi}{2}}$

PROBLEM 4-42 Let V be an inner product space. If $\mathbf{w}$ is orthogonal to each of the vectors $\mathbf{v}_1$, $\mathbf{v}_2, \ldots, \mathbf{v}_m$, is it orthogonal to any linear combination of $\mathbf{v}_1, \mathbf{v}_2, \ldots, \mathbf{v}_m$?

Answer Yes

PROBLEM 4-43 Let $\{\mathbf{v}_1, \mathbf{v}_2, \ldots, \mathbf{v}_n\}$ be an orthonormal set of an inner product space with $\|\mathbf{v}_i\| \neq 0$ for all i. Let $\mathbf{v} \in V$ and let α_i be the component of $\mathbf{v}$ along $\mathbf{v}_i$. Let $\beta_1, \beta_2, \ldots, \beta_n$ be real numbers. Then show that

$$\left\|\mathbf{v} - \sum_{i=1}^{n} \alpha_i \mathbf{v}_i\right\| \leq \left\|\mathbf{v} - \sum_{i=1}^{n} \beta_i \mathbf{v}_i\right\|$$

[*Hint:* Use the result of Problem 4-31 and the generalized Pythagorean theorem (4.65).]

PROBLEM 4-44 Let R^4 have the Euclidean inner product. Use the Gram-Schmidt process to obtain an orthonormal basis for the subspace W of R^4 spanned by

$$\mathbf{v}_1 = \begin{bmatrix} 1 \\ 0 \\ 0 \\ 1 \end{bmatrix}, \qquad \mathbf{v}_2 = \begin{bmatrix} 1 \\ 1 \\ 0 \\ 1 \end{bmatrix}, \qquad \mathbf{v}_3 = \begin{bmatrix} 0 \\ 1 \\ 1 \\ 1 \end{bmatrix}$$

Answer

$$\mathbf{u}_1 = \begin{bmatrix} \frac{1}{\sqrt{2}} \\ 0 \\ 0 \\ \frac{1}{\sqrt{2}} \end{bmatrix}, \qquad \mathbf{u}_2 = \begin{bmatrix} 0 \\ 1 \\ 0 \\ 0 \end{bmatrix}, \qquad \mathbf{u}_3 = \begin{bmatrix} -\frac{1}{\sqrt{6}} \\ 0 \\ \frac{2}{\sqrt{6}} \\ \frac{1}{\sqrt{6}} \end{bmatrix}$$

***PROBLEM 4-45** [*For readers who have studied calculus.*] Let P_2 have the inner product of (4.46) with $a = 0$, $b = 1$; that is,

$$\langle \mathbf{p}, \mathbf{q} \rangle = \int_0^1 p(x)q(x)\,dx$$

Apply the Gram-Schmidt process to obtain an orthonormal basis from the standard basis $\mathscr{B} = \{1, x, x^2\}$.

Answer $\{1, \sqrt{12}(x - \frac{1}{2}), 6\sqrt{5}(x^2 - x + \frac{1}{6})\}$
These polynomials are known as the first three normalized **Legendre polynomials**.

***PROBLEM 4-46** [*For readers who have studied calculus.*] Express $p(x) = 1 + x + 3x^2$ as a linear combination of the three normalized Legendre polynomials obtained in Problem 4-45.

Answer

$$\begin{aligned} 1 + x + 3x^2 &= \frac{5}{2} + \frac{2}{\sqrt{3}}\sqrt{12}\left(x - \frac{1}{2}\right) + \frac{1}{2\sqrt{5}}6\sqrt{5}\left(x^2 - x + \frac{1}{6}\right) \\ &= \frac{5}{2} + 4\left(x - \frac{1}{2}\right) + 3\left(x^2 - x + \frac{1}{6}\right) \end{aligned}$$

5 LINEAR TRANSFORMATION

THIS CHAPTER IS ABOUT

☑ Mappings
☑ Linear Transformations
☑ Kernel and Range of Linear Transformations
☑ Matrix Representation of a Linear Transformation
☑ Changes of Basis and Similarity
☑ Orthogonal Linear Transformations

5-1. Mappings

A. Definitions

If U and V are two vector spaces, a **mapping** (or **transformation**) T associates a unique vector in V with each vector in U. When T maps U into V, we write

$$T: U \to V \tag{5.1}$$

where U is the **domain** of T, V is the **codomain** of T, and T is a rule of correspondence (Figure 5-1).

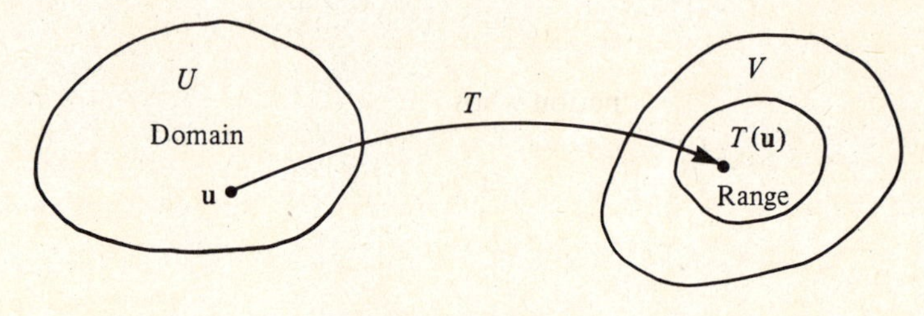

FIGURE 5-1. Mapping T.

If T associates the vector $\mathbf{v} \in V$ with the vector $\mathbf{u} \in U$, we write

$$\mathbf{v} = T(\mathbf{u}) \tag{5.2}$$

Here we say that $\mathbf{v}$ is the **image** of $\mathbf{u}$ under T, or we can say that $\mathbf{v}$ is the **value** of T at $\mathbf{u}$. (The symbol $T(\mathbf{u})$ is read "T of $\mathbf{u}$.") The set of all vectors $T(\mathbf{u})$, when $\mathbf{u}$ ranges over all vectors of U, is called the **range** of T. Observe that the range of T is always contained in the codomain of T, V (see Figure 5-1).

When the range of T is exactly equal to V (the codomain of T), we say that the mapping T is **onto**. And if $T(\mathbf{u}_1) = T(\mathbf{u}_2)$ always implies that $\mathbf{u}_1 = \mathbf{u}_2$, then a mapping $T: U \to V$ is said to be *one-to-one*. In other words, T is one-to-one if distinct vectors $\mathbf{u}_1, \mathbf{u}_2$ in U have distinct images $T(\mathbf{u}_1)$, $T(\mathbf{u}_2)$ in V.

note: The words *mapping* and *transformation* are often used interchangeably.

B. Examples

EXAMPLE 5-1: Let

$$\mathbf{a} = \begin{bmatrix} 1 \\ 2 \\ 3 \end{bmatrix} \in R^3$$

and let $T: R^3 \to R^1$ be the mapping such that if $\mathbf{u} \in R^3$

$$T\langle \mathbf{u} \rangle = \langle \mathbf{u}, \mathbf{a} \rangle \tag{5.3}$$

where $\langle \mathbf{u}, \mathbf{a} \rangle$ denotes the Euclidean inner product. Find the image of

$$\mathbf{u} = \begin{bmatrix} 1 \\ 1 \\ -1 \end{bmatrix}$$

under T.

Solution

$$\begin{aligned} T\langle \mathbf{u} \rangle &= \langle \mathbf{u}, \mathbf{a} \rangle \\ &= u_1 a_1 + u_2 a_2 + u_3 a_3 \\ &= (1)(1) + (1)(2) + (-1)(3) = 0 \end{aligned}$$

EXAMPLE 5-2: Let $f: R^1 \to R^1$ be the function

$$f(x) = x^2 \tag{5.4}$$

Find the range of f and show that f is neither onto nor one-to-one.

Solution: Since the mapping f is the function whose image (or value) at a number x is x^2, the range of f is the set of numbers ≥ 0. Thus the range of f is not equal to total R^1, and therefore f is not onto. Furthermore, $f(1) = f(-1) = 1$, so f is not one-to-one either.

5-2. Linear Transformations

A. Definition

Let U and V be vector spaces. A **linear transformation**

$$T: U \to V$$

is a mapping from U into V that satisfies the following two properties:

LT1. $T(\mathbf{u}_1 + \mathbf{u}_2) = T(\mathbf{u}_1) + T(\mathbf{u}_2)$ for any vectors $\mathbf{u}_1, \mathbf{u}_2 \in U$ additivity **(5.5)**

LT2. $T(\alpha\mathbf{u}) = \alpha T(\mathbf{u})$ for all $\mathbf{u} \in U$ and any scalar α homogeneity **(5.6)**

If $T: V \to V$ is a linear transformation from a vector space V into itself, then T may be called a **linear operator** on V.

B. Examples

EXAMPLE 5-3: Show that mapping T defined in Example 5-1, that is,

$$T: R^3 \to R^1 \qquad \text{and} \qquad T\langle \mathbf{u} \rangle = \langle \mathbf{u}, \mathbf{a} \rangle$$

where $\mathbf{a} \in R^3$ is a fixed vector and $\mathbf{u} \in R^3$, is a linear transformation.

Solution: Let $\mathbf{u}_1, \mathbf{u}_2 \in R^3$ and α be any scalar. Then from the properties of inner product,

$$\begin{aligned} T(\mathbf{u}_1 + \mathbf{u}_2) &= \langle \mathbf{u}_1 + \mathbf{u}_2, \mathbf{a} \rangle \\ &= \langle \mathbf{u}_1, \mathbf{a} \rangle + \langle \mathbf{u}_2, \mathbf{a} \rangle \qquad \text{[by (4.49)]} \\ &= T(\mathbf{u}_1) + T(\mathbf{u}_2) \end{aligned}$$

and

$$\begin{aligned} T(\alpha \mathbf{u}) &= \langle \alpha \mathbf{u}, \mathbf{a} \rangle \\ &= \alpha \langle \mathbf{u}, \mathbf{a} \rangle \qquad \text{[by (4.27)]} \\ &= \alpha T(\mathbf{u}) \end{aligned}$$

Thus properties LT1 and LT2 are satisfied and the mapping T is a linear transformation.

EXAMPLE 5-4: Show that the mapping defined in Example 5-2, that is

$$f: R^1 \to R^1$$
$$f(x) = x^2$$

is not a linear transformation.

Solution: Let $x, y \in R^1$. Then

$$f(x + y) = (x + y)^2 = x^2 + y^2 + 2xy = f(x) + f(y) + 2xy \neq f(x) + f(y)$$

and

$$f(\alpha x) = (\alpha x)^2 = \alpha^2 x^2 \neq \alpha f(x)$$

Thus neither property LT1 nor LT2 is satisfied, so f is not a linear transformation.

EXAMPLE 5-5: Let A be a fixed $m \times n$ matrix. Let the mapping $T: R^n \to R^m$ be defined by

$$T(\mathbf{x}) = A\mathbf{x} \tag{5.7}$$

Show that mapping T is a linear transformation.

Solution: Observe that if

$$\mathbf{x} = \begin{bmatrix} x_1 \\ x_2 \\ \vdots \\ x_n \end{bmatrix} \in R^n \qquad \text{(i.e., an } n \times 1 \text{ matrix)}$$

and if $A = [a_{ij}]_{m \times n}$, then

$$A\mathbf{x} = \begin{bmatrix} a_{11} & a_{12} & \cdots & a_{1n} \\ a_{21} & a_{22} & \cdots & a_{2n} \\ \vdots & \vdots & & \vdots \\ a_{m1} & a_{m2} & \cdots & a_{mn} \end{bmatrix} \begin{bmatrix} x_1 \\ x_2 \\ \vdots \\ x_n \end{bmatrix} = \begin{bmatrix} b_1 \\ b_2 \\ \vdots \\ b_m \end{bmatrix} \in R^m \qquad \text{(i.e., an } m \times 1 \text{ matrix)}$$

Let $\mathbf{x}$ and $\mathbf{y} \in R^n$. Using properties of matrix multiplication, we get

$$T(\mathbf{x} + \mathbf{y}) = A(\mathbf{x} + \mathbf{y}) = A\mathbf{x} + A\mathbf{y} = T(\mathbf{x}) + T(\mathbf{y})$$

and

$$T(\alpha \mathbf{x}) = A(\alpha \mathbf{x}) = \alpha A\mathbf{x} = \alpha T(\mathbf{x})$$

Thus properties LT1 and LT2 are satisfied and the mapping T is a linear transformation.

The linear transformation (5.7) shown in Example 5-5 is called a multiplication by matrix A. Linear transformations of this kind are called **matrix transformations**.

EXAMPLE 5-6: Let U and V be any two vector spaces. Show that the mapping $T: U \to V$ defined by

$$T(\mathbf{u}) = \mathbf{0}, \qquad \text{for all } \mathbf{u} \in U \tag{5.8}$$

is a linear transformation.

Solution: Let $\mathbf{u},\mathbf{u}_1,\mathbf{u}_2 \in U$ and α be any scalar; then $\mathbf{u}_1 + \mathbf{u}_2 \in U$ and $\alpha\mathbf{u} \in U$. Now by definition,

$$T(\mathbf{u}_1) = \mathbf{0}, \qquad T(\mathbf{u}_2) = \mathbf{0}, \qquad T(\mathbf{u}_1 + \mathbf{u}_2) = \mathbf{0}$$

and

$$T(\mathbf{u}) = \mathbf{0}, \qquad T(\alpha\mathbf{u}) = \mathbf{0}$$

Thus

$$T(\mathbf{u}_1 + \mathbf{u}_2) = T(\mathbf{u}_1) + T(\mathbf{u}_2)$$

and

$$T(\alpha\mathbf{u}) = \alpha T(\mathbf{u})$$

That is, T satisfies the properties LT1 and LT2, and therefore T is a linear transformation.

The transformation T shown in Example 5-6 is called the **zero transformation.**

EXAMPLE 5-7: Let V be any vector space. Show that the mapping $T: V \to V$ defined by

$$T(\mathbf{v}) = \mathbf{v} \tag{5.9}$$

is a linear operator.

Solution: By definition,

$$T(\mathbf{v}_1) = \mathbf{v}_1, \qquad T(\mathbf{v}_2) = \mathbf{v}_2, \qquad T(\mathbf{v}_1 + \mathbf{v}_2) = \mathbf{v}_1 + \mathbf{v}_2 = T(\mathbf{v}_1) + T(\mathbf{v}_2)$$

and

$$T(\alpha\mathbf{v}) = \alpha\mathbf{v} = \alpha T(\mathbf{v})$$

Hence T is a linear operator.

The transformation T shown in Example 5-7 is called the **identity operator** and is usually denoted by I.

C. Properties of linear transformations

If $T: U \to V$ is a linear transformation, then

$$T(\mathbf{0}_U) = \mathbf{0}_V \qquad \text{where } \mathbf{0}_U \text{ and } \mathbf{0}_V \text{ are the zero vectors in } U \text{ and } V\text{, respectively.} \tag{5.10}$$

$$T(-\mathbf{u}) = -T(\mathbf{u}) \qquad \text{for all } \mathbf{u} \in U \tag{5.11}$$

$$T(\alpha_1\mathbf{u}_1 + \alpha_2\mathbf{u}_2 + \cdots + \alpha_n\mathbf{u}_n) = \alpha_1 T(\mathbf{u}_1) + \alpha_2 T(\mathbf{u}_2) + \cdots + \alpha_n T(\mathbf{u}_n) \tag{5.12}$$

where $\mathbf{u}_1, \mathbf{u}_2, \ldots, \mathbf{u}_n \in U$ and $\alpha_1, \alpha_2, \ldots, \alpha_n$ are scalars.

EXAMPLE 5-8: Verify properties (5.10) and (5.11).

Solution: Let $\mathbf{u}$ be any vector in U. Since $T(\alpha\mathbf{u}) = \alpha T(\mathbf{u})$ and $0\mathbf{u} = \mathbf{0}$, we have

$$T(\mathbf{0}_U) = T(0\mathbf{u}) = 0T(\mathbf{u}) = \mathbf{0}$$

Also, since $-\mathbf{u} = (-1)\mathbf{u}$, we have

$$T(-\mathbf{u}) = T((-1)\mathbf{u}) = (-1)T(\mathbf{u}) = -T(\mathbf{u})$$

EXAMPLE 5-9: If $T: U \to V$ is a linear transformation, show that

$$T(\alpha_1 \mathbf{u}_1 + \alpha_2 \mathbf{u}_2) = \alpha_1 T(\mathbf{u}_1) + \alpha_2 T(\mathbf{u}_2) \quad \textbf{(5.13)}$$

where $\mathbf{u}_1, \mathbf{u}_2 \in U$ and α_1, α_2 are scalars.

Solution: Using the properties of linear transformation LT1 and LT2, we have

$$\begin{aligned} T(\alpha_1 \mathbf{u}_1 + \alpha_2 \mathbf{u}_2) &= T(\alpha_1 \mathbf{u}_1) + T(\alpha_2 \mathbf{u}_2) \\ &= \alpha_1 T(\mathbf{u}_1) + \alpha_2 T(\mathbf{u}_2) \end{aligned}$$

note: For the proof of property (5.12), see Problem 5-12.

EXAMPLE 5-10: Let $\mathscr{B} = \{\mathbf{u}_1, \mathbf{u}_2, \ldots, \mathbf{u}_n\}$ be a basis of a vector space U and $T: U \to V$ a linear transformation. Show that T is completely determined by its values at a basis (i.e., the images of the basis vectors).

Solution: Suppose we know the images of the basis vectors, that is,

$$T(\mathbf{u}_1) = \mathbf{v}_1,\ T(\mathbf{u}_2) = \mathbf{v}_2, \ldots,\ T(\mathbf{u}_n) = \mathbf{v}_n \quad \textbf{(5.14)}$$

If $\mathbf{u}$ is any vector in U, then $\mathbf{u}$ can be expressed as a linear combination of the basis vectors

$$\mathbf{u} = \alpha_1 \mathbf{u}_1 + \alpha_2 \mathbf{u}_2 + \cdots + \alpha_n \mathbf{u}_n \quad \textbf{(5.15)}$$

Then using property (5.12),

$$\begin{aligned} T(\mathbf{u}) &= T(\alpha_1 \mathbf{u}_1 + \alpha_2 \mathbf{u}_2 + \cdots + \alpha_n \mathbf{u}_n) \quad \textbf{(5.16)} \\ &= \alpha_1 T(\mathbf{u}_1) + \alpha_2 T(\mathbf{u}_2) + \cdots + \alpha_n T(\mathbf{u}_n) \\ &= \alpha_1 \mathbf{v}_1 + \alpha_2 \mathbf{v}_2 + \cdots + \alpha_n \mathbf{v}_n \end{aligned}$$

Therefore, the action of T on U is completely determined by its values at a basis of U.

EXAMPLE 5-11: Consider the basis $\mathscr{B} = \{\mathbf{u}_1, \mathbf{u}_2, \mathbf{u}_3\}$ for R^3, where

$$\mathbf{u}_1 = \begin{bmatrix} 1 \\ 0 \\ 0 \end{bmatrix}, \qquad \mathbf{u}_2 = \begin{bmatrix} 1 \\ 1 \\ 0 \end{bmatrix}, \qquad \mathbf{u}_3 = \begin{bmatrix} 1 \\ 1 \\ 1 \end{bmatrix}$$

and let a linear transformation $T: R^3 \to R^2$ be defined as

$$T(\mathbf{u}_1) = \begin{bmatrix} 3 \\ 2 \end{bmatrix}, \qquad T(\mathbf{u}_2) = \begin{bmatrix} -1 \\ 2 \end{bmatrix}, \qquad T(\mathbf{u}_3) = \begin{bmatrix} 0 \\ 1 \end{bmatrix}$$

Find $T(\mathbf{u})$ where

$$\mathbf{u} = \begin{bmatrix} 5 \\ 3 \\ 1 \end{bmatrix}$$

Solution: We first express $\mathbf{u}$ as a linear combination of $\mathbf{u}_1, \mathbf{u}_2$, and $\mathbf{u}_3$. From the result of Problem 3-29 we have

$$\begin{bmatrix} 5 \\ 3 \\ 1 \end{bmatrix} = 2 \begin{bmatrix} 1 \\ 0 \\ 0 \end{bmatrix} + 2 \begin{bmatrix} 1 \\ 1 \\ 0 \end{bmatrix} + \begin{bmatrix} 1 \\ 1 \\ 1 \end{bmatrix}$$

or $\mathbf{u} = 2\mathbf{u}_1 + 2\mathbf{u}_2 + \mathbf{u}_3$. Thus, by (5.16) we have

$$T(\mathbf{u}) = 2T(\mathbf{u}_1) + 2T(\mathbf{u}_2) + T(\mathbf{u}_3)$$

or

$$T\left(\begin{bmatrix}5\\3\\1\end{bmatrix}\right) = 2\begin{bmatrix}3\\2\end{bmatrix} + 2\begin{bmatrix}-1\\2\end{bmatrix} + \begin{bmatrix}0\\1\end{bmatrix} = \begin{bmatrix}4\\9\end{bmatrix}$$

5-3. Kernel and Range of Linear Transformations

A. Kernel and range

DEFINITIONS Let $T\colon U \to V$ be a linear transformation. The set of all vectors $\mathbf{u}$ in U such that $T(\mathbf{u}) = \mathbf{0}$ is called the **kernel** (or **null space**) of T, denoted by $\ker(T)$. Thus

$$\ker(T) = \{\mathbf{u}\colon T(\mathbf{u}) = \mathbf{0}\} \tag{5.17}$$

The set of all vectors $\mathbf{v} = T(\mathbf{u})$ in V (where $\mathbf{u} \in U$) is called the *range* of T, denoted by $R(T)$; thus

$$R(T) = \{\mathbf{v}\colon \mathbf{v} = T(\mathbf{u}) \text{ for } \mathbf{u} \in U\} \tag{5.18}$$

The kernel of T is a subspace of the domain U and the range of T is a subspace of the codomain V (see Figure 5-2).

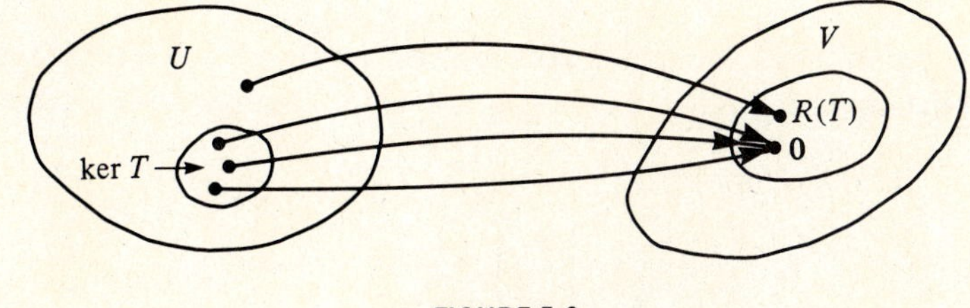

FIGURE 5-2

EXAMPLE 5-12: Let $T\colon R^3 \to R^1$ be the linear transformation defined by

$$T(\mathbf{u}) = \langle \mathbf{u},\mathbf{a} \rangle$$

where $\mathbf{a} \in R^3$ is a fixed vector (see Example 5-3). Find the kernel and range of T.

Solution: Let $\mathbf{u} \in R^3$ such that

$$T(\mathbf{u}) = \langle \mathbf{u},\mathbf{a} \rangle = 0$$

Then from (4.38) we see that $\mathbf{u}$ is orthogonal to $\mathbf{a}$. Thus the kernel of T is the set of all $\mathbf{u}$ that are orthogonal to $\mathbf{a}$. Since the image of $\mathbf{u}$ under T is a scalar, the range of T is R^1.

THEOREM 5-3.1 If $T\colon U \to V$ is a linear transformation, then

(1) The kernel of T is a subspace of U.
(2) The range of T is a subspace of V.

For the proof of Theorem 5-3.1, see Problem 5-17.

THEOREM 5-3.2 If $T\colon U \to V$ is a linear transformation, then T is one-to-one if and only if $\ker(T) = \{\mathbf{0}\}$.

EXAMPLE 5-13: Verify Theorem 5-3.2.

Solution: First, we assume T is one-to-one. Let $\mathbf{u} \in \ker(T)$. Then $T(\mathbf{u}) = \mathbf{0}$. Since $T(\mathbf{0}) = \mathbf{0}$ we have $T(\mathbf{u}) = T(\mathbf{0})$. Since T is one-to-one $\mathbf{u} = \mathbf{0}$. Hence, we have shown that any vector in $\ker(T)$ is $\mathbf{0}$ and it follows that $\ker(T)$ is the zero subspace.

On the other hand, if $\ker(T)$ is the zero subspace, suppose $T(\mathbf{u}_1) = T(\mathbf{u}_2)$. Then $T(\mathbf{u}_1 - \mathbf{u}_2) = T(\mathbf{u}_1) - T(\mathbf{u}_2) = \mathbf{0}$. But since $\ker(T) = \mathbf{0}$, we have $\mathbf{u}_1 - \mathbf{u}_2 = \mathbf{0}$, or $\mathbf{u}_1 = \mathbf{u}_2$. And since $T(\mathbf{u}_1) = T(\mathbf{u}_2)$ implies $\mathbf{u}_1 = \mathbf{u}_2$, we know that T is one-to-one.

B. Nullity and rank

DEFINITIONS Let $T: U \to V$ be a linear transformation.

- The dimension of the kernel of T is called the **nullity** of T, i.e.,

$$\text{nullity of } T = \dim \ker(T) \tag{5.19}$$

- The dimension of the range of T is called the **rank** of T, i.e.

$$\text{rank of } T = \dim R(T) \tag{5.20}$$

THEOREM 5-3.3 (Dimension Theorem) If $T: U \to V$ is a linear transformation, then

$$\text{nullity of } T + \text{rank of } T = \text{dimension of } U \tag{5.21}$$

Theorem 5-3.3 is proved in Problem 5-18.

EXAMPLE 5-14: Let $T: R^3 \to R^2$ be the transformation defined by

$$T\left(\begin{bmatrix} x_1 \\ x_2 \\ x_3 \end{bmatrix}\right) = \begin{bmatrix} x_1 - x_2 \\ x_1 - x_3 \end{bmatrix}$$

Find the nullity and the rank of T.

Solution: If

$$\mathbf{x} = \begin{bmatrix} x_1 \\ x_2 \\ x_3 \end{bmatrix} \in \ker(T)$$

then $T(\mathbf{x}) = \mathbf{0}$; that is,

$$\begin{bmatrix} x_1 - x_2 \\ x_1 - x_3 \end{bmatrix} = \begin{bmatrix} 0 \\ 0 \end{bmatrix}$$

Therefore,

$$x_1 - x_2 = 0$$
$$x_1 - x_3 = 0$$

Hence $x_1 = x_2 = x_3$, so that each vector in $\ker(T)$ is a scalar multiple of the vector

$$\mathbf{u} = \begin{bmatrix} 1 \\ 1 \\ 1 \end{bmatrix}$$

This vector forms a basis for $\ker(T)$. Thus $\dim \ker(T) = 1$ and the nullity of $T = 1$.

Next, if $\mathbf{v} = \begin{bmatrix} a \\ b \end{bmatrix} \in R(T)$, then

$$T(\mathbf{u}) = \mathbf{v}$$

or

$$T\left(\begin{bmatrix} x_1 \\ x_2 \\ x_3 \end{bmatrix}\right) = \begin{bmatrix} x_1 - x_2 \\ x_1 - x_3 \end{bmatrix} = \begin{bmatrix} a \\ b \end{bmatrix}$$

Therefore,

$$x_1 - x_2 = a$$
$$x_1 - x_3 = b$$

Solving this system, we obtain

$$x_1 = b + t, \qquad x_2 = b - a + t, \qquad x_3 = t$$

Therefore, every $\mathbf{v} = \begin{bmatrix} a \\ b \end{bmatrix} \in R^2$ is the image $T(\mathbf{u})$ of a vector

$$\mathbf{u} = \begin{bmatrix} b + t \\ b - a + t \\ t \end{bmatrix}$$

in R^3 for all t. Thus $R(T) = R^2$ and $\dim R^2 = 2$. Therefore, the rank of $T = 2$. Note that

$$\text{nullity of } T + \text{rank of } T = 1 + 2 = 3 = \dim R^3 \text{ (domain of } T\text{)}$$

C. Isomorphism

DEFINITION Let $T: U \to V$ be a linear transformation from a vector space U to a vector space V. If T is both one-to-one and onto, then T is said to be an **isomorphism**.

THEOREM 5-3.4 Let $T: U \to V$ be an isomorphism of two vector spaces. Then:

(a) If $\mathbf{u}_1, \mathbf{u}_2, \ldots, \mathbf{u}_n$ are linearly independent in U, then $T(\mathbf{u}_1), T(\mathbf{u}_2), \ldots, T(\mathbf{u}_n)$ are linearly independent in V.
(b) If $\mathbf{u}_1, \mathbf{u}_2, \ldots, \mathbf{u}_n$ span U, then $T(\mathbf{u}_1), T(\mathbf{u}_2), \ldots, T(\mathbf{u}_n)$ span V.
(c) If $\{\mathbf{u}_1, \mathbf{u}_2, \ldots, \mathbf{u}_n\}$ is a basis for U, then $\{T(\mathbf{u}_1), T(\mathbf{u}_2), \ldots, T(\mathbf{u}_n)\}$ is a basis for V.
(d) $\dim U = \dim V$.

For the proof of Theorem 5-3.4, see Problem 5-19.

5-4. Matrix Representation of a Linear Transformation

A. Linear transformation from R^n to R^m

THEOREM 5-4.1 If $T: R^n \to R^m$ is a linear transformation, then there is an $m \times n$ matrix A such that

$$T(\mathbf{x}) = A\mathbf{x} \tag{5.22}$$

for each $\mathbf{x} \in R^n$.

EXAMPLE 5-15: Verify Theorem 5-4.1.

Solution: Let $\{\mathbf{e}_1, \mathbf{e}_2, \ldots, \mathbf{e}_n\}$ be the standard basis for R^n (see Example 3-32).
Define

$$T(\mathbf{e}_1) = \mathbf{a}_1 = \begin{bmatrix} a_{11} \\ a_{21} \\ \vdots \\ a_{m1} \end{bmatrix}, T(\mathbf{e}_2) = \mathbf{a}_2 = \begin{bmatrix} a_{12} \\ a_{22} \\ \vdots \\ a_{m2} \end{bmatrix}, \ldots, T(\mathbf{e}_n) = \mathbf{a}_n = \begin{bmatrix} a_{1n} \\ a_{2n} \\ \vdots \\ a_{mn} \end{bmatrix} \tag{5.23}$$

Let

$$A = [a_{ij}]_{m \times n} = [\mathbf{a}_1, \mathbf{a}_2, \ldots, \mathbf{a}_n] = [T(\mathbf{e}_1), T(\mathbf{e}_2), \ldots, T(\mathbf{e}_n)] \tag{5.24}$$

Now if $\mathbf{x} \in R^n$, then

$$\mathbf{x} = \begin{bmatrix} x_1 \\ x_2 \\ \vdots \\ x_n \end{bmatrix} = x_1\mathbf{e}_1 + x_2\mathbf{e}_2 + \cdots + x_n\mathbf{e}_n \tag{5.25}$$

Then applying T, we get

$$\begin{aligned} T\mathbf{x} &= x_1 T(\mathbf{e}_1) + x_2 T(\mathbf{e}_2) + \cdots + x_n T(\mathbf{e}_n) \\ &= x_1 \mathbf{a}_1 + x_2 \mathbf{a}_2 + \cdots + x_n \mathbf{a}_n \\ &= [\mathbf{a}_1, \mathbf{a}_2, \ldots, \mathbf{a}_n] \begin{bmatrix} x_1 \\ x_2 \\ \vdots \\ x_n \end{bmatrix} \\ &= A\mathbf{x} \end{aligned} \quad \textbf{(5.26)}$$

Thus we have shown that each linear transformation T from R^n into R^m can be represented by multiplication by an $m \times n$ matrix A. We refer to the matrix A in (5.24) as the **standard matrix** of T.

EXAMPLE 5-16: Find the standard matrix of the transformation $T: R^3 \to R^2$ defined by

$$T\left(\begin{bmatrix} x_1 \\ x_2 \\ x_3 \end{bmatrix}\right) = \begin{bmatrix} x_1 - x_2 \\ x_1 - x_3 \end{bmatrix}$$

(see Example 5-14).

Solution

$$T(\mathbf{e}_1) = T\left(\begin{bmatrix} 1 \\ 0 \\ 0 \end{bmatrix}\right) = \begin{bmatrix} 1 \\ 1 \end{bmatrix}, \qquad T(\mathbf{e}_2) = T\left(\begin{bmatrix} 0 \\ 1 \\ 0 \end{bmatrix}\right) = \begin{bmatrix} -1 \\ 0 \end{bmatrix}, \qquad T(\mathbf{e}_3) = T\left(\begin{bmatrix} 0 \\ 0 \\ 1 \end{bmatrix}\right) = \begin{bmatrix} 0 \\ -1 \end{bmatrix}$$

Thus by (5.24) we have

$$A = \begin{bmatrix} 1 & -1 & 0 \\ 1 & 0 & -1 \end{bmatrix}$$

Check:

$$A\mathbf{x} = \begin{bmatrix} 1 & -1 & 0 \\ 1 & 0 & -1 \end{bmatrix} \begin{bmatrix} x_1 \\ x_2 \\ x_3 \end{bmatrix} = \begin{bmatrix} x_1 - x_2 \\ x_1 - x_3 \end{bmatrix}$$

which agrees with the given transformation.

B. Matrix representation of a linear transformation

In Theorem 5-4.1 we show that any $T: R^n \to R^m$ can be represented by matrix multiplication. We can show that the same is true for any $T: U \to V$ for finite-dimensional vector spaces.

Recall from Section 3-8 that we associated with each vector $\mathbf{v} \in V$ having ordered basis $\mathscr{B} = \{\mathbf{v}_1, \mathbf{v}_2, \ldots, \mathbf{v}_n\}$ the coordinate vector

$$[\mathbf{v}]_{\mathscr{B}} = \begin{bmatrix} x_1 \\ x_2 \\ \vdots \\ x_n \end{bmatrix}_{\mathscr{B}} \quad \textbf{(5.27)}$$

where $x_1, x_2, \ldots, x_n$ are numbers such that

$$\mathbf{v} = x_1\mathbf{v}_1 + x_2\mathbf{v}_2 + \cdots + x_n\mathbf{v}_n \quad \textbf{(5.28)}$$

note: The subscript $\mathscr{B}$ in (5.27) denotes which basis for V we are using in our calculation.

THEOREM 5-4.2 Let $T: U \to V$ be a linear transformation from a finite-dimensional vector space U into a finite-dimensional vector space V. Let $\mathscr{B} = \{\mathbf{u}_1, \mathbf{u}_2, \ldots, \mathbf{u}_n\}$ be an ordered basis for U, $\mathscr{B}' = \{\mathbf{v}_1, \mathbf{v}_2, \ldots, \mathbf{v}_m\}$ be an ordered basis of V, and A the $m \times n$ matrix having $[T(\mathbf{u}_1)]_{\mathscr{B}'}$, $[T(\mathbf{u}_2)]_{\mathscr{B}'}, \ldots, [T(\mathbf{u}_n)]_{\mathscr{B}'}$ as its columns. That is,

$$A = [[T(\mathbf{u}_1)]_{\mathscr{B}'}, [T(\mathbf{u}_2)]_{\mathscr{B}'}, \ldots, [T(\mathbf{u}_n)]_{\mathscr{B}'}] \tag{5.29}$$

Then

$$[T(\mathbf{u})]_{\mathscr{B}'} = A[\mathbf{u}]_{\mathscr{B}} \tag{5.30}$$

for every $\mathbf{u} \in U$.

EXAMPLE 5-17: Verify Theorem 5-4.2

Solution: Let $\mathbf{u} \in U$, and

$$[\mathbf{u}]_{\mathscr{B}} = \begin{bmatrix} x_1 \\ x_2 \\ \vdots \\ x_n \end{bmatrix}_{\mathscr{B}}$$

Then

$$\mathbf{u} = x_1\mathbf{u}_1 + x_2\mathbf{u}_2 + \cdots + \mathbf{x}_n\mathbf{u}_n$$

Now,

$$\begin{aligned} A[\mathbf{u}]_{\mathscr{B}} &= [[T(\mathbf{u}_1)]_{\mathscr{B}'}, [T(\mathbf{u}_2)]_{\mathscr{B}'}, \ldots, [T(\mathbf{u}_n)]_{\mathscr{B}'}] \begin{bmatrix} x_1 \\ x_2 \\ \vdots \\ x_n \end{bmatrix}_{\mathscr{B}} \\ &= x_1[T(\mathbf{u}_1)]_{\mathscr{B}'} + x_2[T(\mathbf{u}_2)]_{\mathscr{B}'} + \cdots + x_n[T(\mathbf{u}_n)]_{\mathscr{B}'} \\ &= [x_1T(\mathbf{u}_1) + x_2T(\mathbf{u}_2) + \cdots + x_nT(\mathbf{u}_n)]_{\mathscr{B}'} \\ &= [T(x_1\mathbf{u}_1 + x_2\mathbf{u}_2 + \cdots + x_n\mathbf{u}_n)]_{\mathscr{B}'} \\ &= [T(\mathbf{u})]_{\mathscr{B}'} \end{aligned}$$

DEFINITIONS We call matrix A obtained by (5.29) the **matrix of T relative to the bases $\mathscr{B}$ and $\mathscr{B}'$**. In the special case $U = V$ (so that $T: V \to V$ is a linear operator), we usually take $\mathscr{B} = \mathscr{B}'$ when constructing matrix A of T. In this case we call matrix A the **matrix of T relative to the basis $\mathscr{B}$**.

EXAMPLE 5-18: Let $T: R^3 \to R^2$ be a linear transformation defined by

$$T\left(\begin{bmatrix} x_1 \\ x_2 \\ x_3 \end{bmatrix}\right) = x_1\mathbf{u}_1 + (x_2 + x_3)\mathbf{u}_2$$

where

$$\mathbf{u}_1 = \begin{bmatrix} 1 \\ 1 \end{bmatrix} \quad \text{and} \quad \mathbf{u}_2 = \begin{bmatrix} 1 \\ -1 \end{bmatrix}$$

Find the matrix of T relative to the ordered basis $\mathscr{B} = \{\mathbf{i},\mathbf{j},\mathbf{k}\}$ and $\mathscr{B}' = \{\mathbf{u}_1,\mathbf{u}_2\}$, where $\{\mathbf{i},\mathbf{j},\mathbf{k}\}$ is the standard basis for R^3.

Solution

$$T(\mathbf{i}) = T\left(\begin{bmatrix}1\\0\\0\end{bmatrix}\right) = 1\mathbf{u}_1 + 0\mathbf{u}_2, \qquad [T(\mathbf{i})]_{\mathscr{B}'} = \begin{bmatrix}1\\0\end{bmatrix}$$

$$T(\mathbf{j}) = T\left(\begin{bmatrix}0\\1\\0\end{bmatrix}\right) = 0\mathbf{u}_1 + 1\mathbf{u}_2, \qquad [T(\mathbf{j})]_{\mathscr{B}'} = \begin{bmatrix}0\\1\end{bmatrix}$$

$$T(\mathbf{k}) = T\left(\begin{bmatrix}0\\0\\1\end{bmatrix}\right) = 0\mathbf{u}_1 + 1\mathbf{u}_2, \qquad [T(\mathbf{k})]_{\mathscr{B}'} = \begin{bmatrix}0\\1\end{bmatrix}$$

Thus, by (5.29) the matrix of T relative to $\mathscr{B}$ and $\mathscr{B}'$ is

$$A = \begin{bmatrix}1 & 0 & 0\\0 & 1 & 1\end{bmatrix}$$

EXAMPLE 5-19: Let $T\colon R^2 \to R^2$ be the linear operator defined by

$$T\left(\begin{bmatrix}x_1\\x_2\end{bmatrix}\right) = \begin{bmatrix}3x_1 + 2x_2\\-5x_1 + 4x_2\end{bmatrix}$$

Find the matrix of T relative to the basis $\mathscr{B} = \{\mathbf{u}_1, \mathbf{u}_2\}$, where

$$\mathbf{u}_1 = \begin{bmatrix}1\\0\end{bmatrix}, \qquad \mathbf{u}_2 = \begin{bmatrix}1\\1\end{bmatrix}$$

Solution: From the formula of T

$$T(\mathbf{u}_1) = T\left(\begin{bmatrix}1\\0\end{bmatrix}\right) = \begin{bmatrix}3\\-5\end{bmatrix} \quad \text{and} \quad T(\mathbf{u}_2) = T\left(\begin{bmatrix}1\\1\end{bmatrix}\right) = \begin{bmatrix}5\\-1\end{bmatrix}$$

Now, let

$$\begin{bmatrix}3\\-5\end{bmatrix} = x_1\begin{bmatrix}1\\0\end{bmatrix} + x_2\begin{bmatrix}1\\1\end{bmatrix} = \begin{bmatrix}x_1 + x_2\\x_2\end{bmatrix}$$

$$\begin{bmatrix}5\\-1\end{bmatrix} = y_1\begin{bmatrix}1\\0\end{bmatrix} + y_2\begin{bmatrix}1\\1\end{bmatrix} = \begin{bmatrix}y_1 + y_2\\y_2\end{bmatrix}$$

Thus, we obtain

$$x_1 = 8, \qquad x_2 = -5, \qquad y_1 = 6, \qquad y_2 = -1$$

Therefore

$$[T(\mathbf{u}_1)]_{\mathscr{B}} = \begin{bmatrix}8\\-5\end{bmatrix}, \qquad [T(\mathbf{u}_2)]_{\mathscr{B}} = \begin{bmatrix}6\\-1\end{bmatrix}$$

Hence, the matrix of T relative to $\mathscr{B}$ is

$$A = \begin{bmatrix}8 & 6\\-5 & -1\end{bmatrix}$$

5-5. Changes of Basis and Similarity

A. Change of basis

Let $T\colon V \to V$ be a linear operator. If $\mathscr{B}$ is any basis for V, then from Theorem 5-3.4 (with $U = V$, $\mathscr{B}' = \mathscr{B}$), we know that there is a matrix A such that

$$[T(\mathbf{v})]_{\mathscr{B}} = A[\mathbf{v}]_{\mathscr{B}} \tag{5.31}$$

Similarly, if $\mathscr{B}'$ is another basis for V, then there is a matrix A' such that

$$[T(\mathbf{v})]_{\mathscr{B}'} = A'[\mathbf{v}]_{\mathscr{B}'} \tag{5.32}$$

The following theorem shows how a change of basis affects the matrix of a linear operator.

THEOREM 5-5.1 Let $T: V \to V$ be a linear operator on a finite-dimensional vector space V. If A is the matrix of T relative to a basis $\mathscr{B}$ and A' is the matrix of T relative to a basis $\mathscr{B}'$, then

$$A' = P^{-1}AP \tag{5.33}$$

where P is the transition matrix from $\mathscr{B}'$ to $\mathscr{B}$.

EXAMPLE 5-20: Verify Theorem 5-5.1.

Solution: Let $\mathbf{v} \in V$. If P is the transition matrix from $\mathscr{B}'$ to $\mathscr{B}$, then from Theorem 3-8.2 of Section 3-8, P^{-1} is the transition matrix from $\mathscr{B}$ to $\mathscr{B}'$. Thus, by (3.68) and (3.69), we have

$$[\mathbf{v}]_{\mathscr{B}} = P[\mathbf{v}]_{\mathscr{B}'} \tag{5.34}$$

$$[\mathbf{v}]_{\mathscr{B}'} = P^{-1}[\mathbf{v}]_{\mathscr{B}} \tag{5.35}$$

Since $T(\mathbf{v})$ is a vector of V for every $\mathbf{v} \in V$, we have

$$\begin{aligned} [T(\mathbf{v})]_{\mathscr{B}'} &= P^{-1}[T(\mathbf{v})]_{\mathscr{B}} && \text{[by (5.35)]} \\ &= P^{-1}A[\mathbf{v}]_{\mathscr{B}} && \text{[by (5.31)]} \\ &= P^{-1}AP[\mathbf{v}]_{\mathscr{B}'} && \text{[by (5.34)]} \\ &= A'[\mathbf{v}]_{\mathscr{B}'} && \text{[by (5.32)]} \end{aligned}$$

Hence it follows that

$$A' = P^{-1}AP$$

The above relationships can be depicted diagrammatically as shown in Figure 5-3.

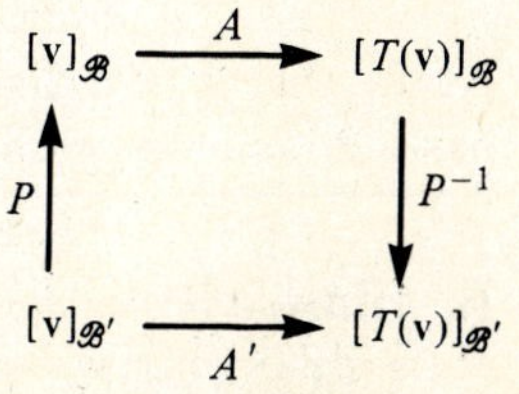

$$\begin{array}{ccc} [\mathbf{v}]_{\mathscr{B}} & \xrightarrow{\;A\;} & [T(\mathbf{v})]_{\mathscr{B}} \\ P \uparrow & & \downarrow P^{-1} \\ [\mathbf{v}]_{\mathscr{B}'} & \xrightarrow[\;A'\;]{} & [T(\mathbf{v})]_{\mathscr{B}'} \end{array}$$

FIGURE 5-3

EXAMPLE 5-21: Let $T: R^2 \to R^2$ be defined by

$$T\left(\begin{bmatrix} x_1 \\ x_2 \end{bmatrix}\right) = \begin{bmatrix} 2x_1 + x_2 \\ x_1 + 2x_2 \end{bmatrix}$$

Find the standard matrix of T, that is, the matrix of T relative to the standard basis $\mathscr{B} = \{\mathbf{e}_1, \mathbf{e}_2\}$ where

$$\mathbf{e}_1 = \begin{bmatrix} 1 \\ 0 \end{bmatrix}, \qquad \mathbf{e}_2 = \begin{bmatrix} 0 \\ 1 \end{bmatrix}$$

Then use (5.33) to find the matrix of T relative to the basis $\mathscr{B}' = \{\mathbf{u}_1, \mathbf{u}_2\}$ where

$$\mathbf{u}_1 = \begin{bmatrix} 1 \\ 1 \end{bmatrix}, \qquad \mathbf{u}_2 = \begin{bmatrix} -1 \\ 1 \end{bmatrix}$$

Solution: From the formula of T,

$$T(\mathbf{e}_1) = T\left(\begin{bmatrix} 1 \\ 0 \end{bmatrix}\right) = \begin{bmatrix} 2 \\ 1 \end{bmatrix} = 2\begin{bmatrix} 1 \\ 0 \end{bmatrix} + 1\begin{bmatrix} 0 \\ 1 \end{bmatrix}$$

$$T(\mathbf{e}_2) = T\left(\begin{bmatrix} 0 \\ 1 \end{bmatrix}\right) = \begin{bmatrix} 1 \\ 2 \end{bmatrix} = 1\begin{bmatrix} 1 \\ 0 \end{bmatrix} + 2\begin{bmatrix} 0 \\ 1 \end{bmatrix}$$

so that

$$[T(\mathbf{e}_1)]_{\mathscr{B}} = \begin{bmatrix} 2 \\ 1 \end{bmatrix}, \qquad [T(\mathbf{e}_2)]_{\mathscr{B}} = \begin{bmatrix} 1 \\ 2 \end{bmatrix}$$

Therefore

$$A = \begin{bmatrix} 2 & 1 \\ 1 & 2 \end{bmatrix}$$

Next, we need to find the transition matrix P from $\mathscr{B}'$ to $\mathscr{B}$. Notice that

$$\mathbf{u}_1 = \begin{bmatrix} 1 \\ 1 \end{bmatrix} = 1\begin{bmatrix} 1 \\ 0 \end{bmatrix} + 1\begin{bmatrix} 0 \\ 1 \end{bmatrix}, \qquad \mathbf{u}_2 = \begin{bmatrix} -1 \\ 1 \end{bmatrix} = -1\begin{bmatrix} 1 \\ 0 \end{bmatrix} + 1\begin{bmatrix} 0 \\ 1 \end{bmatrix}$$

$$[\mathbf{u}_1]_{\mathscr{B}} = \begin{bmatrix} 1 \\ 1 \end{bmatrix}, \qquad [\mathbf{u}_2]_{\mathscr{B}} = \begin{bmatrix} -1 \\ 1 \end{bmatrix}$$

Therefore, by (3.59)

$$P = \begin{bmatrix} 1 & -1 \\ 1 & 1 \end{bmatrix}$$

Taking the inverse, we have

$$P^{-1} = \frac{1}{2}\begin{bmatrix} 1 & 1 \\ -1 & 1 \end{bmatrix} = \begin{bmatrix} \frac{1}{2} & \frac{1}{2} \\ -\frac{1}{2} & \frac{1}{2} \end{bmatrix}$$

Hence by (5.33), the matrix of T relative to the basis $\mathscr{B}'$ is

$$A' = P^{-1}AP = \begin{bmatrix} \frac{1}{2} & \frac{1}{2} \\ -\frac{1}{2} & \frac{1}{2} \end{bmatrix}\begin{bmatrix} 2 & 1 \\ 1 & 2 \end{bmatrix}\begin{bmatrix} 1 & -1 \\ 1 & 1 \end{bmatrix} = \begin{bmatrix} 3 & 0 \\ 0 & 1 \end{bmatrix}$$

Notice that A' is a diagonal matrix. Thus in this example, using the basis $\mathscr{B}'$ instead of the standard basis $\mathscr{B}$, we obtained the more simply structured matrix of T.

B. Similarity

DEFINITION If A and B are square matrices, we say that B is *similar* to A if there is a nonsingular matrix P such that

SIMILARITY $$B = P^{-1}AP \qquad \textbf{(5.36)}$$

The relation of similarity has the following properties:

(1) Reflexivity: A is similar to A
(2) Symmetry: If B is similar to A, then A is similar to B
(3) Transitivity: If A is similar to B and B is similar to C, then A is similar to C

EXAMPLE 5-22: Verify the properties of reflexivity, symmetry, and transitivity for the relation of similarity.

Solution

(1) Since $A = I^{-1}AI$, where I is the identity matrix, A is similar to A.
(2) Suppose that B is similar to A. Then there exists a nonsingular P such that

$$B = P^{-1}AP$$

Solving for A, we get

$$A = (P^{-1})^{-1}BP^{-1}$$

since $(P^{-1})^{-1}BP^{-1} = (P^{-1})^{-1}P^{-1}APP^{-1} = IAI = A$. Then letting $Q = P^{-1}$ yields

$$A = Q^{-1}BQ$$

which says that A is similar to B.
(3) Suppose A is similar to B and B is similar to C. Then for some nonsingular matrices P and Q, we have

$$A = P^{-1}BP \qquad \text{and} \qquad B = Q^{-1}CQ$$

Hence

$$A = P^{-1}(Q^{-1}CQ)P = (P^{-1}Q^{-1})C(QP) = (QP)^{-1}C(QP)$$

since $(QP)^{-1} = P^{-1}Q^{-1}$ [see (1.38)]. Then letting $S = QP$ yields

$$A = S^{-1}CS$$

which says that A is similar to C.

note: Because of the symmetry property of the similarity relation (that is, if B is similar to A, then A is similar to B), we usually say simply that A and B are "similar." In this terminology, Theorem 5-5.1 asserts that two matrices representing the same linear operator $T: V \to V$ relative to different bases are similar.

5-6. Orthogonal Linear Transformations

In some applications we are interested only in those linear transformations that preserve the norm (or length) of a vector and the angle between two vectors, when these are defined in the vector space. Recall that these ideas are meaningful only in real inner product space (see Section 4-4).

A. Orthogonal linear operator

DEFINITION If $T: V \to V$ is a linear operator defined on a real inner product space V, we say that T is an **orthogonal linear operator** if, for each $\mathbf{u}$, $\mathbf{v}$ in V, we have

$$\langle T(\mathbf{u}), T(\mathbf{v}) \rangle = \langle \mathbf{u}, \mathbf{v} \rangle \qquad \textbf{(5.37)}$$

That is, T preserves inner products.

EXAMPLE 5-23: Let $T: R^2 \to R^2$ be defined by

$$T\left(\begin{bmatrix} x_1 \\ x_2 \end{bmatrix}\right) = \begin{bmatrix} -x_2 \\ x_1 \end{bmatrix}$$

Show that T is orthogonal.

Solution: Let $\mathbf{u} = \begin{bmatrix} u_1 \\ u_2 \end{bmatrix}$ and $\mathbf{v} = \begin{bmatrix} v_1 \\ v_2 \end{bmatrix}$. Then, by the formula of T,

$$T(\mathbf{u}) = \begin{bmatrix} -u_2 \\ u_1 \end{bmatrix}, \qquad T(\mathbf{v}) = \begin{bmatrix} -v_2 \\ v_1 \end{bmatrix}$$

Thus

$$\begin{aligned} \langle T(\mathbf{u}), T(\mathbf{v}) \rangle &= (-u_2)(-v_2) + u_1v_1 \\ &= u_2v_2 + u_1v_1 = u_1v_1 + u_2v_2 = \langle \mathbf{u}, \mathbf{v} \rangle \end{aligned}$$

Therefore T is orthogonal.

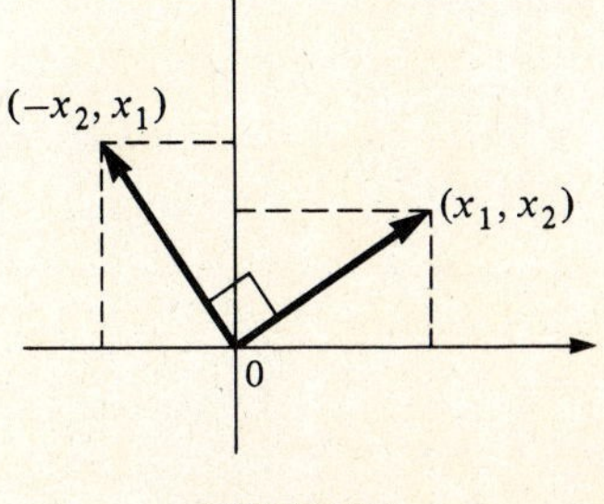

FIGURE 5-4

note: Here T is equal to 90° counterclockwise rotation (see Figure 5-4).

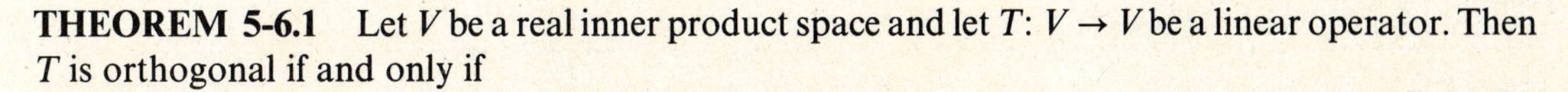
THEOREM 5-6.1 Let V be a real inner product space and let $T: V \to V$ be a linear operator. Then T is orthogonal if and only if

$$\|T(\mathbf{v})\| = \|\mathbf{v}\| \qquad \textbf{(5.38)}$$

for all $\mathbf{v} \in V$.

Thus, Theorem 5-6.1 says that if T preserves the norms (or lengths), it preserves inner products and vice versa. Because of (5.38), an orthogonal linear operator T is said to be *isometry*.

For the proof of Theorem 5-6.1, see Problem 5-32.

EXAMPLE 5-24: Show that an orthogonal linear operator preserves angles too.

Solution: In Section 4-4 we introduced the angle θ between two nonzero vectors **u**, **v** in an inner product space as

$$\cos\theta = \frac{\langle \mathbf{u},\mathbf{v}\rangle}{\|\mathbf{u}\|\,\|\mathbf{v}\|}$$

[see (4.64)]. Thus, if T is orthogonal, we have

$$\cos\theta = \frac{\langle \mathbf{u},\mathbf{v}\rangle}{\|\mathbf{u}\|\,\|\mathbf{v}\|} = \frac{\langle T(\mathbf{u}),T(\mathbf{v})\rangle}{\|T(\mathbf{u})\|\,\|T(\mathbf{v})\|}$$

Thus T preserves angles too.

B. Other properties of an orthogonal operator

THEOREM 5-6.2 Let T: $V \to V$ be an orthogonal operator on an inner product space V. Then

(1) T is one-to-one.
(2) If **u** and **v** are orthogonal vectors in V, then $T(\mathbf{u})$ and $T(\mathbf{v})$ are also orthogonal.
(3) If $\{\mathbf{v}_1, \mathbf{v}_2, \ldots, \mathbf{v}_n\}$ is an orthonormal basis for V, then $\{T(\mathbf{v}_1), T(\mathbf{v}_2), \ldots, T(\mathbf{v}_n)\}$ is also an orthonormal basis.

For the proof of Theorem 5-6.2 see Problem 5-33.

C. Orthogonal matrix

DEFINITION A square matrix A is called an **orthogonal matrix** if

ORTHOGONAL MATRIX

$$AA^T = A^TA = I_n \tag{5.39}$$

EXAMPLE 5-25: Show that

$$A = \begin{bmatrix} \cos\theta & -\sin\theta \\ \sin\theta & \cos\theta \end{bmatrix}$$

is an orthogonal matrix.

Solution

$$\begin{aligned} AA^T &= \begin{bmatrix} \cos\theta & -\sin\theta \\ \sin\theta & \cos\theta \end{bmatrix}\begin{bmatrix} \cos\theta & \sin\theta \\ -\sin\theta & \cos\theta \end{bmatrix} \\ &= \begin{bmatrix} \cos^2\theta + \sin^2\theta & \cos\theta\sin\theta - \sin\theta\cos\theta \\ \sin\theta\cos\theta - \cos\theta\sin\theta & \sin^2\theta + \cos^2\theta \end{bmatrix} \\ &= \begin{bmatrix} 1 & 0 \\ 0 & 1 \end{bmatrix} = I_2 \end{aligned}$$

$$A^TA = \begin{bmatrix} \cos\theta & \sin\theta \\ -\sin\theta & \cos\theta \end{bmatrix}\begin{bmatrix} \cos\theta & -\sin\theta \\ \sin\theta & \cos\theta \end{bmatrix} = \begin{bmatrix} 1 & 0 \\ 0 & 1 \end{bmatrix} = I_2$$

Thus A is an orthogonal matrix.

EXAMPLE 5-26: Let T: $R^n \to R^n$ be a linear operator defined by the matrix transformation, i.e.,

$$T(\mathbf{x}) = A\mathbf{x}$$

where $\mathbf{x} \in R^n$ and A is an $n \times n$ orthogonal matrix. Show that T is orthogonal.

Solution: Let $\mathbf{u}, \mathbf{v} \in R^n$. Then

$$T(\mathbf{u}) = A\mathbf{u}, \qquad T(\mathbf{v}) = A\mathbf{v}$$

Recall from Section 4-2 (4.23) that $\langle \mathbf{u},\mathbf{v} \rangle = \mathbf{u}^T\mathbf{v}$ Thus

$$\begin{aligned}\langle T(\mathbf{u}),T(\mathbf{v})\rangle &= \langle A\mathbf{u},A\mathbf{v}\rangle \\ &= (A\mathbf{u})^T(A\mathbf{v}) \\ &= (\mathbf{u}^T A^T)(A\mathbf{v}) && \text{[by (1.43)]} \\ &= \mathbf{u}^T(A^T A)\mathbf{v} \\ &= \langle \mathbf{u},A^T A\mathbf{v}\rangle && \text{[by (4.23)]} \\ &= \langle \mathbf{u},I\mathbf{v}\rangle = \langle \mathbf{u},\mathbf{v}\rangle && \text{[by (5.39)]}\end{aligned}$$

Therefore, T is orthogonal.

THEOREM 5-6.3 If A is an orthogonal matrix, then:

(1) $A^T = A^{-1}$ **(5.40)**

(2) $\det A = \pm 1$ **(5.41)**

(3) $\langle A\mathbf{x},A\mathbf{y}\rangle = \langle \mathbf{x},\mathbf{y}\rangle$ for $\mathbf{x}, \mathbf{y} \in R^n$ **(5.42)**

(4) $\|A\mathbf{x}\|^2 = \|\mathbf{x}\|^2$ for $\mathbf{x} \in R^n$ **(5.43)**

(5) The column vectors of A are mutually orthonormal

(6) The row vectors of A are mutually orthonormal.

Properties 5 and 6 are proved in Problems 5-36 and 5-46.

EXAMPLE 5-27: Verify Properties 1–4 of Theorem 5-6.3.

Solution

(1) If A is an orthogonal matrix, then by definition,

$$AA^T = A^T A = I$$

Thus, by definition (1.36) of the inverse of A it follows that $A^T = A^{-1}$.

(2) $\det(AA^T) = (\det A)(\det A^T) = (\det A)^2 = \det I = 1$. Thus $\det A = \pm\sqrt{1} = \pm 1$.

(3) From (4.23), we have $\langle \mathbf{u},\mathbf{v}\rangle = \mathbf{u}^T\mathbf{v}$. Thus

$$\begin{aligned}\langle A\mathbf{x},A\mathbf{y}\rangle &= (A\mathbf{x})^T(A\mathbf{y}) \\ &= (\mathbf{x}^T A^T)(A\mathbf{y}) \\ &= \mathbf{x}^T(A^T A\mathbf{y}) \\ &= \mathbf{x}^T(I\mathbf{y}) = \mathbf{x}^T\mathbf{y} = \langle \mathbf{x},\mathbf{y}\rangle\end{aligned}$$

(4)
$$\begin{aligned}\|A\mathbf{x}\|^2 &= \langle A\mathbf{x},A\mathbf{x}\rangle \\ &= \langle \mathbf{x},\mathbf{x}\rangle && \text{(by Property 3)} \\ &= \|\mathbf{x}\|^2\end{aligned}$$

SUMMARY

1. Let U and V be two vector spaces. A mapping $T: U \to V$ is the process of association of a unique vector in V with each vector in U, denoted by $\mathbf{v} = T(\mathbf{u})$, $\mathbf{u} \in U$, $\mathbf{v} \in V$.
2. The domain of T is U, and the range $R(T)$ of T is the set of all vectors $T(\mathbf{u})$ when $\mathbf{u}$ ranges over all vectors of U.
3. When $R(T) = V$, we say that T is onto.
4. A mapping $T: U \to V$ is called one-to-one if $T(\mathbf{u}_1) = T(\mathbf{u}_2)$ implies $\mathbf{u}_1 = \mathbf{u}_2$.
5. A mapping $T: U \to V$ is called a linear transformation if

$$T(\mathbf{u}_1 + \mathbf{u}_2) = T(\mathbf{u}_1) + T(\mathbf{u}_2)$$

$$T(\alpha\mathbf{u}) = \alpha T(\mathbf{u})$$

If $V = U$, then T is called a linear operator.

6. The kernel (or null space) ker T of a linear transformation T is the set of $\mathbf{u} \in U$ such that $T(\mathbf{u}) = \mathbf{0}$:

$$\ker(T) = \{\mathbf{u}\colon T(\mathbf{u}) = \mathbf{0}\}$$

7. The range $R(T)$ of T, is the set of all vectors $\mathbf{v} = T(\mathbf{u}) \in V$:

$$R(T) = \{\mathbf{v}, \mathbf{v} = T(\mathbf{u})\}$$

8. If $T\colon U \to V$ is a linear transformation,

$$\text{nullity of } T + \text{rank of } T = \text{dimension of } U$$

where the nullity of T is the dimension of ker T and the rank of T is the dimension of $R(T)$.

9. If $T\colon U \to V$ is both one-to-one and onto, then T is said to be an isomorphism.
10. If $T\colon U \to V$ is an isomorphism and if $\{\mathbf{u}_1, \ldots, \mathbf{u}_n\}$ is a basis for U, then $\{T(\mathbf{u}_1), \ldots, T(\mathbf{u}_n)\}$ is a basis for V.
11. If $T\colon R^n \to R^m$ is a linear transformation, there is an $m \times n$ matrix A such that $T(\mathbf{x}) = A\mathbf{x}$ for each $\mathbf{x} \in R^n$.
12. Let $T\colon U \to V$ be a linear transformation. Let $\mathscr{B} = \{\mathbf{u}_1, \ldots, \mathbf{u}_n\}$ be an ordered basis for U and $\mathscr{B}' = \{\mathbf{v}_1, \ldots, \mathbf{v}_m\}$ be an ordered basis for V. Then $[T(\mathbf{u})]_{\mathscr{B}'} = A[\mathbf{u}]_{\mathscr{B}}$ where matrix A is given by

$$A = [[T(\mathbf{u}_1)]_{\mathscr{B}'}, \ldots, [T(\mathbf{u}_n)]_{\mathscr{B}'}]$$

where $[T(\mathbf{u}_i)]_{\mathscr{B}'}$ is the coordinate vector of $T(\mathbf{u}_i)$ relative to $\mathscr{B}'$ and $[\mathbf{u}]_{\mathscr{B}}$ is the coordinate vector of $\mathbf{u}$ relative to $\mathscr{B}$.

13. Let $T\colon V \to V$ be a linear operator. If A is the matrix of T relative to a basis $\mathscr{B}$ and A' is the matrix of T relative to a basis $\mathscr{B}'$, then $A' = P^{-1}AP$, where P is the transition matrix from $\mathscr{B}'$ to $\mathscr{B}$.
14. If A and B are square matrices, then A is similar to B if there exists a nonsingular matrix P such that $B = P^{-1}AP$.
15. If $T\colon V \to V$ is a linear operator on a real inner product space V, then T is orthogonal if $\langle T(\mathbf{u}), T(\mathbf{v})\rangle = \langle \mathbf{u}, \mathbf{v}\rangle$ or $\|T(\mathbf{v})\| = \|\mathbf{v}\|$.
16. If $T\colon V \to V$ is an orthogonal linear operator, then

    (a) T is one-to-one.
    (b) If $\mathbf{u} \perp \mathbf{v}$; then $T(\mathbf{u}) \perp T(\mathbf{v})$.
    (c) If $\{\mathbf{v}_1, \ldots, \mathbf{v}_n\}$ is an orthonormal basis for V, then $\{T(\mathbf{v}_1), \ldots, T(\mathbf{v}_n)\}$ is also an orthonormal basis.

17. A square matrix A is orthogonal if $AA^T = A^TA = I$. Then

    (a) $A^T = A^{-1}$.
    (b) $\det A = \pm 1$.
    (c) $\langle A\mathbf{x}, A\mathbf{y}\rangle = \langle \mathbf{x}, \mathbf{y}\rangle$ for $\mathbf{x}, \mathbf{y} \in R^n$.
    (d) $\|A\mathbf{x}\| = \|\mathbf{x}\|$.
    (e) The column vectors of A are orthonormal.
    (f) The row vectors of A are orthonormal.

RAISE YOUR GRADES

Can you explain ...?

- ☑ what conditions must be satisfied for a mapping T to be linear
- ☑ the kernel of T
- ☑ the range of T
- ☑ one-to-one and onto mapping
- ☑ how to find a matrix representation of T
- ☑ the relationship between two matrix representations of T relative to two different bases

- ☑ an orthogonal linear operator
- ☑ the properties of an orthogonal matrix

SOLVED PROBLEMS

Linear Transformation

PROBLEM 5-1 Let the transformation $T: V \to V$ be defined by $T(\mathbf{v}) = \mathbf{v} + \mathbf{a}$, where $\mathbf{a} \in V$ is a fixed vector (see Figure 5-5). Show that T is not a linear transformation.

Solution Let $\mathbf{u}, \mathbf{v} \in V$, and α any scalar. Then

$$T(\mathbf{u}) = \mathbf{u} + \mathbf{a}, \qquad T(\mathbf{v}) = \mathbf{v} + \mathbf{a}$$

and

$$T(\mathbf{u} + \mathbf{v}) = \mathbf{u} + \mathbf{v} + \mathbf{a} \neq \mathbf{u} + \mathbf{a} + \mathbf{v} + \mathbf{a} = T(\mathbf{u}) + T(\mathbf{v})$$

$$T(\alpha\mathbf{v}) = \alpha\mathbf{v} + \mathbf{a} \neq \alpha(\mathbf{v} + \mathbf{a}) = \alpha T(\mathbf{v})$$

Neither property LT1 nor property LT2 is satisfied; therefore, T is not a linear transformation. Here T is called the **translation** by $\mathbf{a}$.

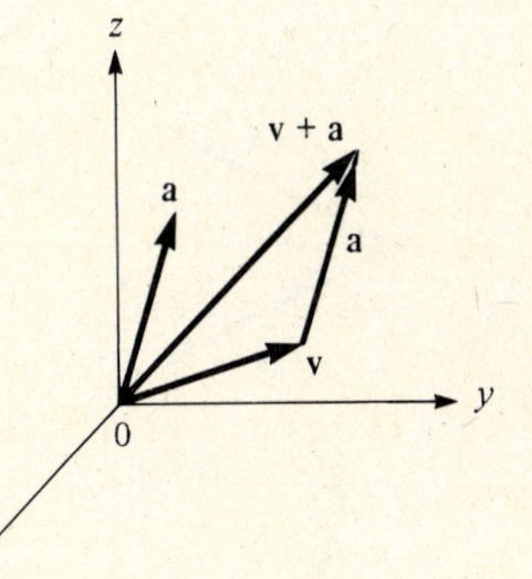

FIGURE 5-5

PROBLEM 5-2 Let V be any vector space and k any fixed scalar. Show that the transformation $T: V \to V$ defined by $T(\mathbf{v}) = k\mathbf{v}$ is a linear operator.

Solution By definition,

$$T(\mathbf{v}_1) = k\mathbf{v}_1, \qquad T(\mathbf{v}_2) = k\mathbf{v}_2$$

$$T(\mathbf{v}_1 + \mathbf{v}_2) = k(\mathbf{v}_1 + \mathbf{v}_2) = k\mathbf{v}_1 + k\mathbf{v}_2 = T(\mathbf{v}_1) + T(\mathbf{v}_2)$$

and

$$T(\mathbf{v}) = k\mathbf{v}, \qquad T(\alpha\mathbf{v}) = k(\alpha\mathbf{v}) = \alpha(k\mathbf{v}) = \alpha T(\mathbf{v})$$

Hence T is a linear operator.

If $k > 1$, the linear operator T is called a **dilation** of V (see Figure 5-6a). And if $0 < k < 1$, then T is called a **contraction** of V (see Figure 5-6b).

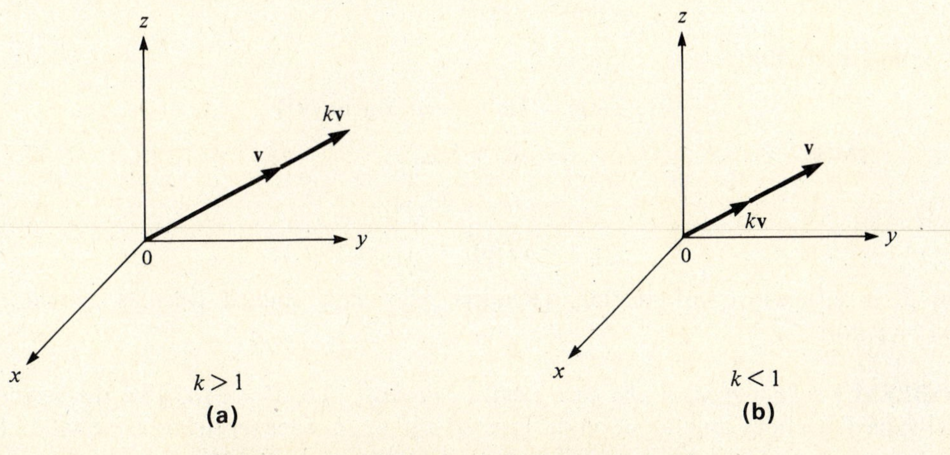

FIGURE 5-6. (a) Dilation $k > 1$. (b) Contraction $k < 1$.

PROBLEM 5-3 Let $T: R^2 \to R^2$ be a transformation defined by

$$T\left(\begin{bmatrix} x_1 \\ x_2 \end{bmatrix}\right) = \begin{bmatrix} x_1 \\ -x_2 \end{bmatrix}$$

Show that T is a linear transformation.

Solution Let $\mathbf{u} = \begin{bmatrix} u_1 \\ u_2 \end{bmatrix}$ and $\mathbf{v} = \begin{bmatrix} v_1 \\ v_2 \end{bmatrix} \in R^2$ and α be any scalar. Then

$$T(\mathbf{u}) = \begin{bmatrix} u_1 \\ -u_2 \end{bmatrix}, \qquad T(\mathbf{v}) = \begin{bmatrix} v_1 \\ -v_2 \end{bmatrix}$$

and

$$T(\mathbf{u} + \mathbf{v}) = T\left(\begin{bmatrix} u_1 + v_1 \\ u_2 + v_2 \end{bmatrix}\right) = \begin{bmatrix} u_1 + v_1 \\ -(u_2 + v_2) \end{bmatrix}$$

$$= \begin{bmatrix} u_1 \\ -u_2 \end{bmatrix} + \begin{bmatrix} v_1 \\ -v_2 \end{bmatrix} = T(\mathbf{u}) + T(\mathbf{v})$$

$$T(\alpha\mathbf{u}) = T\left(\begin{bmatrix} \alpha u_1 \\ \alpha u_2 \end{bmatrix}\right) = \begin{bmatrix} \alpha u_1 \\ -\alpha u_2 \end{bmatrix} = \alpha \begin{bmatrix} u_1 \\ -u_2 \end{bmatrix} = \alpha T(\mathbf{u})$$

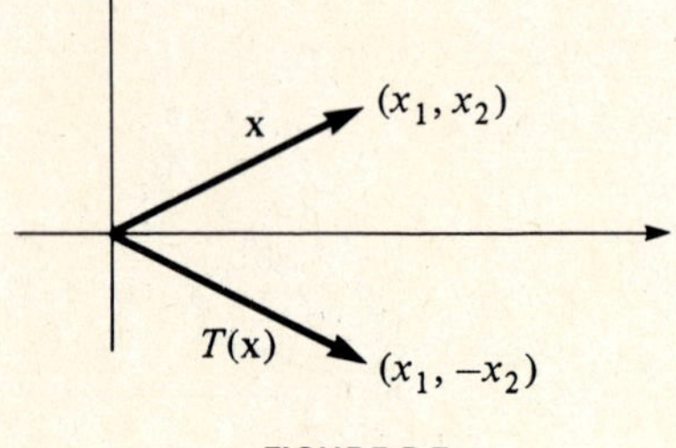

FIGURE 5-7

Thus T is a linear operator.

The operator T has the effect of reflecting a vector $\mathbf{u}$ about the x-axis (Figure 5-7), so T is called a **reflection operator** about the x-axis.

PROBLEM 5-4 Consider the transformation $T: P_2 \to P_1$ defined by $T(a_0 + a_1x + a_2x^2) = a_1 + 2a_2x$. Show that T is a linear transformation.

Solution Let

$$\mathbf{p} = p(x) = a_0 + a_1x + a_2x^2$$
$$\mathbf{q} = q(x) = b_0 + b_1x + b_2x^2$$

be any vectors in P_2. Then

$$\begin{aligned} T(\mathbf{p} + \mathbf{q}) &= T[(a_0 + b_0) + (a_1 + b_1)x + (a_2 + b_2)x^2] \\ &= (a_1 + b_1) + 2(a_2 + b_2)x \\ &= (a_1 + 2a_2x) + (b_1 + 2b_2)x \\ &= T(\mathbf{p}) + T(\mathbf{q}) \end{aligned}$$

Let α be any scalar, then

$$\begin{aligned} T(\alpha\mathbf{p}) &= T(\alpha a_0 + \alpha a_1x + \alpha a_2x^2) \\ &= \alpha a_1 + 2\alpha a_2x \\ &= \alpha(a_1 + 2a_2x) \\ &= \alpha T(\mathbf{p}) \end{aligned}$$

Hence T is a linear transformation. [Readers who have studied calculus should note that $T(p(x)) = p'(x)$.]

***PROBLEM 5-5** [*For readers who have studied calculus.*] Let $V = C[a,b]$ be the vector space of all real-valued functions continuous on the interval $a \leq x \leq b$. Then let the transformation $D: V \to W$

be defined by $D(\mathbf{f}) = \mathbf{f}'$, where $\mathbf{f}' = f'(x)$ is the derivative of $\mathbf{f} = f(x) \in V$ with respect to x. Show that D is a linear transformation.

Solution From the properties of differentiation, we have

$$D(\mathbf{f} + \mathbf{g}) = [f(x) + g(x)]' = f'(x) + g'(x) = D(\mathbf{f}) + D(\mathbf{g})$$

and

$$D(\alpha\mathbf{f}) = [\alpha f(x)]' = \alpha f'(x) = \alpha D(\mathbf{f})$$

Hence D is a linear transformation.

***PROBLEM 5-6** [*For readers who have studied calculus.*] Let $V = C[0, 1]$ and let the transformation $T\colon V \to R^1$ be defined by $T(\mathbf{f}) = \int_0^1 f(x)\,dx$. Show that T is a linear transformation.

Solution Since

$$T(\mathbf{f} + \mathbf{g}) = \int_0^1 (f(x) + g(x))\,dx = \int_0^1 f(x)\,dx + \int_0^1 g(x)\,dx = T(\mathbf{f}) + T(\mathbf{g})$$

and

$$T(\alpha\mathbf{f}) = \int_0^1 (\alpha f(x))\,dx = \alpha \int_0^1 f(x)\,dx = \alpha T(\mathbf{f})$$

for any constant α, it follows that T is a linear transformation.

PROBLEM 5-7 Let T be a transformation $T\colon R^3 \to R^2$ defined by projection, that is,

$$T\left(\begin{bmatrix} x_1 \\ x_2 \\ x_3 \end{bmatrix}\right) = \begin{bmatrix} x_1 \\ x_2 \end{bmatrix}$$

Show that T is a linear transformation.

Solution Note that the mapping T is essentially an orthogonal projection of $\mathbf{u} \in R^3$ onto R^2 considered as a subspace of R^3 (see Figure 5-8). Hence if $\mathbf{u} = \begin{bmatrix} u_1 \\ u_2 \\ u_3 \end{bmatrix}$, $\mathbf{v} = \begin{bmatrix} v_1 \\ v_2 \\ v_3 \end{bmatrix} \in R^3$ and α is any scalar, then

$$T(\mathbf{u}) = \begin{bmatrix} u_1 \\ u_2 \end{bmatrix}, \qquad T(\mathbf{v}) = \begin{bmatrix} v_1 \\ v_2 \end{bmatrix}$$

and

$$T(\mathbf{u} + \mathbf{v}) = T\left(\begin{bmatrix} u_1 + v_1 \\ u_2 + v_2 \\ u_3 + v_3 \end{bmatrix}\right) = \begin{bmatrix} u_1 + v_1 \\ u_2 + v_2 \end{bmatrix}$$

$$= \begin{bmatrix} u_1 \\ u_2 \end{bmatrix} + \begin{bmatrix} v_1 \\ v_2 \end{bmatrix} = T(\mathbf{u}) + T(\mathbf{v})$$

$$T(\alpha\mathbf{u}) = T\left(\begin{bmatrix} \alpha u_1 \\ \alpha u_2 \\ \alpha u_3 \end{bmatrix}\right) = \begin{bmatrix} \alpha u_1 \\ \alpha u_2 \end{bmatrix} = \alpha \begin{bmatrix} u_1 \\ u_2 \end{bmatrix} = \alpha T(\mathbf{u})$$

Hence T is a linear transformation.

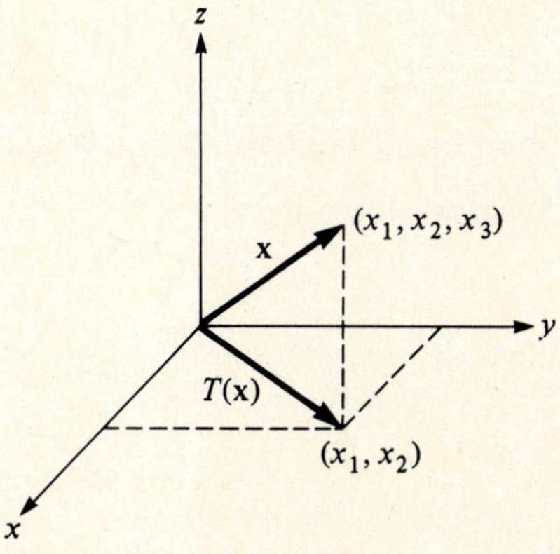

FIGURE 5-8

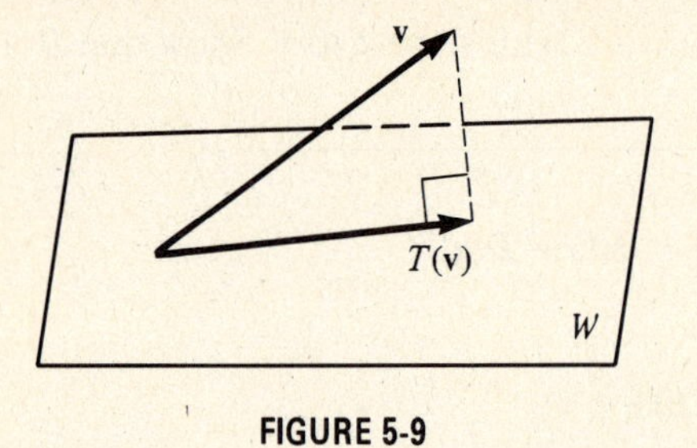

FIGURE 5-9

PROBLEM 5-8 Given that V is an inner product space, suppose W is a finite-dimensional subspace of V having $\mathscr{B} = \{\mathbf{w}_1, \mathbf{w}_2, \ldots, \mathbf{w}_m\}$ as an orthonormal basis. Let $T: V \to W$ be the mapping that maps $\mathbf{v} \in V$ into its orthogonal projection on W (see Figure 5-9); that is,

$$T(\mathbf{v}) = \langle \mathbf{v},\mathbf{w}_1 \rangle \mathbf{w}_1 + \langle \mathbf{v},\mathbf{w}_2 \rangle \mathbf{w}_2 + \cdots + \langle \mathbf{v},\mathbf{w}_m \rangle \mathbf{w}_m$$

The mapping T is called the **orthogonal projection** of V onto W. Show that T is a linear transformation.

Solution Let $\mathbf{u}, \mathbf{v} \in V$ and α any scalar. Then from the properties of inner product

$$\begin{aligned}
T(\mathbf{u} + \mathbf{v}) &= \langle \mathbf{u} + \mathbf{v},\mathbf{w}_1 \rangle \mathbf{w}_1 + \langle \mathbf{u} + \mathbf{v},\mathbf{w}_2 \rangle \mathbf{w}_2 + \cdots + (\mathbf{u} + \mathbf{v},\mathbf{w}_m \rangle \mathbf{w}_m \\
&= \langle \mathbf{u},\mathbf{w}_1 \rangle \mathbf{w}_1 + \langle \mathbf{u},\mathbf{w}_2 \rangle \mathbf{w}_2 + \cdots + \langle \mathbf{u},\mathbf{w}_m \rangle \mathbf{w}_m \\
&\quad + \langle \mathbf{v},\mathbf{w}_1 \rangle \mathbf{w}_1 + \langle \mathbf{v},\mathbf{w}_2 \rangle \mathbf{w}_2 + \cdots + \langle \mathbf{v},\mathbf{w}_m \rangle \mathbf{w}_m \\
&= T(\mathbf{u}) + T(\mathbf{v})
\end{aligned}$$

and

$$\begin{aligned}
T(\alpha\mathbf{v}) &= \langle \alpha\mathbf{v},\mathbf{w}_1 \rangle \mathbf{w}_1 + \langle \alpha\mathbf{v},\mathbf{w}_2 \rangle \mathbf{w}_2 + \cdots + \langle \alpha\mathbf{v},\mathbf{w}_m \rangle \mathbf{w}_m \\
&= \alpha[\langle \mathbf{v},\mathbf{w}_1 \rangle \mathbf{w}_1 + \langle \mathbf{v},\mathbf{w}_2 \rangle \mathbf{w}_2 + \cdots + \langle \mathbf{v},\mathbf{w}_m \rangle \mathbf{w}_m] \\
&= \alpha T(\mathbf{v})
\end{aligned}$$

Thus T is a linear transformation.

PROBLEM 5-9 Let V be a finite-dimensional vector space and let $\mathscr{B} = \{\mathbf{v}_1, \mathbf{v}_2, \ldots, \mathbf{v}_n\}$ be a basis of V. We define a transformation $T: V \to R^n$ by associating to each $\mathbf{v} \in V$ its coordinate vector $[\mathbf{v}]_{\mathscr{B}}$ with respect to the basis $\mathscr{B}$ (see Section 3-8). Thus, if

$$\mathbf{v} = x_1\mathbf{v}_1 + x_2\mathbf{v}_2 + \cdots + x_n\mathbf{v}_n$$

then

$$T(\mathbf{v}) = \begin{bmatrix} x_1 \\ x_2 \\ \vdots \\ x_n \end{bmatrix}_{\mathscr{B}} = [\mathbf{v}]_{\mathscr{B}}$$

Show that T is a linear transformation.

Solution If $\mathbf{u} \in V$ and

$$\mathbf{u} = y_1\mathbf{v}_1 + y_2\mathbf{v}_2 + \cdots + y_n\mathbf{v}_n$$

then

$$T(\mathbf{u}) = \begin{bmatrix} y_1 \\ y_2 \\ \vdots \\ y_n \end{bmatrix}_{\mathscr{B}} = [\mathbf{u}]_{\mathscr{B}}$$

Now

$$\mathbf{v} + \mathbf{u} = (x_1 + y_1)\mathbf{v}_1 + (x_2 + y_2)\mathbf{v}_2 + \cdots + (x_n + y_n)\mathbf{v}_n$$
$$\alpha\mathbf{v} = (\alpha x_1)\mathbf{v}_1 + (\alpha x_2)\mathbf{v}_2 + \cdots + (\alpha x_n)\mathbf{v}_n$$

Thus

$$T(\mathbf{v} + \mathbf{u}) = \begin{bmatrix} x_1 + y_1 \\ x_2 + y_2 \\ \vdots \\ x_n + y_n \end{bmatrix}_{\mathscr{B}} = \begin{bmatrix} x_1 \\ x_2 \\ \vdots \\ x_n \end{bmatrix}_{\mathscr{B}} + \begin{bmatrix} y_1 \\ y_2 \\ \vdots \\ y_n \end{bmatrix}_{\mathscr{B}} = [\mathbf{v}]_{\mathscr{B}} + [\mathbf{u}]_{\mathscr{B}} = T(\mathbf{v}) + T(\mathbf{u})$$

$$T(\alpha\mathbf{v}) = \begin{bmatrix} \alpha x_1 \\ \alpha x_2 \\ \vdots \\ \alpha x_n \end{bmatrix}_{\mathscr{B}} = \alpha \begin{bmatrix} x_1 \\ x_2 \\ \vdots \\ x_n \end{bmatrix}_{\mathscr{B}} = \alpha[\mathbf{v}]_{\mathscr{B}} = \alpha T(\mathbf{v})$$

Therefore, T is a linear transformation.

PROBLEM 5-10 Let $T: R^2 \to R^1$ be a linear transformation for which

$$T\left(\begin{bmatrix} 1 \\ 0 \end{bmatrix}\right) = 2 \quad \text{and} \quad T\left(\begin{bmatrix} 1 \\ 1 \end{bmatrix}\right) = -3$$

(a) Find $T\left(\begin{bmatrix} x_1 \\ x_2 \end{bmatrix}\right)$. **(b)** Find $T\left(\begin{bmatrix} 2 \\ 4 \end{bmatrix}\right)$.

Solution

(a) Since $\left\{\begin{bmatrix} 1 \\ 0 \end{bmatrix}\begin{bmatrix} 1 \\ 1 \end{bmatrix}\right\}$ is a basis for R^2, we can express

$$\begin{bmatrix} x_1 \\ x_2 \end{bmatrix} = \alpha_1 \begin{bmatrix} 1 \\ 0 \end{bmatrix} + \alpha_2 \begin{bmatrix} 1 \\ 1 \end{bmatrix}$$

where α_1 and α_2 are scalars. Then

$$\begin{bmatrix} x_1 \\ x_2 \end{bmatrix} = \begin{bmatrix} \alpha_1 + \alpha_2 \\ \alpha_2 \end{bmatrix}$$

and so $x_1 = \alpha_1 + \alpha_2$, $x_2 = \alpha_2$. Solving for α_1 and α_2, we obtain $\alpha_1 = x_1 - x_2$, $\alpha_2 = x_2$. Thus

$$\begin{bmatrix} x_1 \\ x_2 \end{bmatrix} = (x_1 - x_2)\begin{bmatrix} 1 \\ 0 \end{bmatrix} + x_2 \begin{bmatrix} 1 \\ 1 \end{bmatrix}$$

Now applying T, we get

$$\begin{aligned} T\left(\begin{bmatrix} x_1 \\ x_2 \end{bmatrix}\right) &= T\left((x_1 - x_2)\begin{bmatrix} 1 \\ 0 \end{bmatrix} + x_2 \begin{bmatrix} 1 \\ 1 \end{bmatrix}\right) \\ &= (x_1 - x_2)T\left(\begin{bmatrix} 1 \\ 0 \end{bmatrix}\right) + x_2 T\left(\begin{bmatrix} 1 \\ 1 \end{bmatrix}\right) \\ &= (x_1 - x_2)(2) + x_2(-3) = 2x_1 - 5x_2 \end{aligned}$$

(**b**) Using the above result, we get

$$T\left(\begin{bmatrix} 2 \\ 4 \end{bmatrix}\right) = 2(2) - 5(4) = -16$$

PROBLEM 5-11 Let $T: R^3 \to R^2$ be a matrix transformation, and suppose that

$$T(\mathbf{i}) = T\left(\begin{bmatrix} 1 \\ 0 \\ 0 \end{bmatrix}\right) = \begin{bmatrix} 1 \\ 1 \end{bmatrix}, \qquad T(\mathbf{j}) = T\left(\begin{bmatrix} 0 \\ 1 \\ 0 \end{bmatrix}\right) = \begin{bmatrix} 2 \\ 0 \end{bmatrix}, \qquad T(\mathbf{k}) = T\left(\begin{bmatrix} 0 \\ 0 \\ 1 \end{bmatrix}\right) = \begin{bmatrix} 3 \\ -1 \end{bmatrix}$$

Find (**a**) the matrix A such that $T(\mathbf{x}) = A\mathbf{x}$, and (**b**) $T\left(\begin{bmatrix} 1 \\ 1 \\ 1 \end{bmatrix}\right)$.

Solution

(**a**) Let $T(\mathbf{x}) = A\mathbf{x}$ and

$$A = [a_{ij}]_{2 \times 3} = \begin{bmatrix} a_{11} & a_{12} & a_{13} \\ a_{21} & a_{22} & a_{23} \end{bmatrix}$$

Let

$$\mathbf{x} = \begin{bmatrix} x_1 \\ x_2 \\ x_3 \end{bmatrix}$$

then

$$T(\mathbf{x}) = A\mathbf{x} = \begin{bmatrix} a_{11} & a_{12} & a_{13} \\ a_{21} & a_{22} & a_{23} \end{bmatrix}\begin{bmatrix} x_1 \\ x_2 \\ x_3 \end{bmatrix} = \begin{bmatrix} a_{11}x_1 + a_{12}x_2 + a_{13}x_3 \\ a_{21}x_1 + a_{22}x_2 + a_{23}x_3 \end{bmatrix}$$

Now from

$$T\left(\begin{bmatrix} 1 \\ 0 \\ 0 \end{bmatrix}\right) = \begin{bmatrix} 1 \\ 1 \end{bmatrix}$$

we have

$$a_{11}(1) + a_{12}(0) + a_{13}(0) = 1$$
$$a_{21}(1) + a_{22}(0) + a_{23}(0) = 1$$

Thus $a_{11} = 1$ and $a_{21} = 1$. From

$$T\left(\begin{bmatrix} 0 \\ 1 \\ 0 \end{bmatrix}\right) = \begin{bmatrix} 2 \\ 0 \end{bmatrix}$$

we have

$$a_{11}(0) + a_{12}(1) + a_{13}(0) = 2$$
$$a_{21}(0) + a_{22}(1) + a_{23}(0) = 0$$

Thus $a_{12} = 2$, $a_{22} = 0$. From

$$T\left(\begin{bmatrix} 0 \\ 0 \\ 1 \end{bmatrix}\right) = \begin{bmatrix} 3 \\ -1 \end{bmatrix}$$

we have

$$a_{11}(0) + a_{12}(0) + a_{13}(1) = 3$$
$$a_{21}(0) + a_{22}(0) + a_{23}(1) = -1$$

Thus $a_{13} = 3$ and $a_{23} = -1$. Hence we obtain

$$A = \begin{bmatrix} 1 & 2 & 3 \\ 1 & 0 & -1 \end{bmatrix}$$

(b) Let

$$\mathbf{x} = \begin{bmatrix} 1 \\ 1 \\ 1 \end{bmatrix}$$

Then

$$A\mathbf{x} = \begin{bmatrix} 1 & 2 & 3 \\ 1 & 0 & -1 \end{bmatrix}\begin{bmatrix} 1 \\ 1 \\ 1 \end{bmatrix} = \begin{bmatrix} 6 \\ 0 \end{bmatrix}$$

Thus

$$T\left(\begin{bmatrix} 1 \\ 1 \\ 1 \end{bmatrix}\right) = \begin{bmatrix} 6 \\ 0 \end{bmatrix}$$

PROBLEM 5-12 Verify property (5.12), that is,

$$T(\alpha_1\mathbf{u}_1 + \alpha_2\mathbf{u}_2 + \cdots + \alpha_n\mathbf{u}_n) = \alpha_1 T(\mathbf{u}_1) + \alpha_2 T(\mathbf{u}_2) + \cdots + \alpha_n T(\mathbf{u}_n)$$

by induction.

Solution If $n = 1$, property (5.12) clearly holds by property LT2 of a linear transformation, that is,

$$T(\alpha_1\mathbf{u}_1) = \alpha_1 T(\mathbf{u}_1)$$

If $n > 1$, we assume that

$$T(\alpha_1\mathbf{u}_1 + \cdots + \alpha_{n-1}\mathbf{u}_{n-1}) = \alpha_1 T(\mathbf{u}_1) + \cdots + \alpha_{n-1} T(\mathbf{u}_{n-1})$$

or

$$T\left(\sum_{i=1}^{n-1} \alpha_i\mathbf{u}_i\right) = \sum_{i=1}^{n-1} \alpha_i T(\mathbf{u}_i) \qquad \textbf{(a)}$$

Now

$$T\left(\sum_{i=1}^{n} \alpha_i\mathbf{u}_i\right) = T\left(\sum_{i=1}^{n-1} \alpha_i\mathbf{u}_i + \alpha_n\mathbf{u}_n\right) \qquad \textbf{(b)}$$
$$= T\left(\sum_{i=1}^{n-1} \alpha_i\mathbf{u}_i\right) + T(\alpha_n\mathbf{u}_n)$$

by property LT1 of a linear transformation. Substituting (a) into (b), we obtain

$$T\left(\sum_{i=1}^{n} \alpha_i\mathbf{u}_i\right) = \sum_{i=1}^{n-1} \alpha_i T(\mathbf{u}_i) + \alpha_n T(\mathbf{u}_n) = \sum_{i=1}^{n} \alpha_i T(\mathbf{u}_i)$$

that is,

$$T(\alpha_1\mathbf{u}_1 + \alpha_2\mathbf{u}_2 + \cdots + \alpha_n\mathbf{u}_n) = \alpha_1 T(\mathbf{u}_1) + \alpha_2 T(\mathbf{u}_2) + \cdots + \alpha_n T(\mathbf{u}_n)$$

Kernel and Range of Linear Transformations

PROBLEM 5-13 Let $0: U \to V$ be the zero transformation of Example 5-6. Find the kernel and the range of 0.

Solution Since 0 maps every vector in U into $\mathbf{0}$,

$$\ker(0) = U$$

Since $\mathbf{0}$ is the only possible image under T,

$$R(0) = \{\mathbf{0}\}$$

PROBLEM 5-14 Let $I: V \to V$ be the identity operator of Example 5-7. Find the kernel and range of I.

Solution Since I maps every vector in V into itself, $R(I) = V$ and $\ker(I) = \{\mathbf{0}\}$.

PROBLEM 5-15 Let $T_A: R^n \to R^m$ be the linear transformation of Example 5-5, defined by multiplication by A:

$$A = \begin{bmatrix} a_{11} & a_{12} & \cdots & a_{1n} \\ a_{21} & a_{22} & \cdots & a_{2n} \\ \vdots & \vdots & & \vdots \\ a_{m1} & a_{m2} & \cdots & a_{mn} \end{bmatrix}$$

Find the kernel and range of T_A.

Solution The kernel of T_A consists of all

$$\mathbf{x} = \begin{bmatrix} x_1 \\ x_2 \\ \vdots \\ x_n \end{bmatrix} \in R^n$$

which are solution vectors of the homogeneous system

$$A\mathbf{x} = \mathbf{0} \qquad \textbf{(a)}$$

Thus the kernel of T_A is the solution space of system (a).

The range of T_A consists of vectors

$$\mathbf{b} = \begin{bmatrix} b_1 \\ b_2 \\ \vdots \\ b_m \end{bmatrix} \in R^m$$

such that the system $A\mathbf{x} = \mathbf{b}$ is consistent. Thus the range of T_A is the column space of A (Example 3-57).

PROBLEM 5-16 Let $T: R^n \to R^m$ be the linear transformation of Example 5-5, defined by multiplication by a matrix A. Find the nullity and the rank of T.

Solution In Problem 5-15, we observed that the kernel of T is the solution space of the system $A\mathbf{x} = \mathbf{0}$. The nullity of T is the dimension of this solution space, that is, the dimension of the null space of A (see Section 3-8D). Hence the nullity of T is equal to the nullity of A.

Also in Problem 5-15 we saw that the range of T is the column space of A. The rank of T is the dimension of the column space (see Section 3-8D). Therefore, the rank of T is equal to the rank of A.

PROBLEM 5-17 Verify Theorem 5-3.1.

Solution Recall that a subset of a vector space is a subspace if it is nonempty and if it is closed under addition and scalar multiplication (Theorem 3-4.1).

Notice that both ker T and $R(T)$ are nonempty, since

$$T(\mathbf{0}_u) = \mathbf{0}_v$$

Now let $\mathbf{u}_1, \mathbf{u}_2 \in \ker(T)$ and α be any scalar. Then

$$T(\mathbf{u}_1) = \mathbf{0} \qquad \text{and} \qquad T(\mathbf{u}_2) = \mathbf{0}$$

Since T is a linear transformation, we get

$$T(\mathbf{u}_1 + \mathbf{u}_2) = T(\mathbf{u}_1) + T(\mathbf{u}_2) = \mathbf{0} + \mathbf{0} = \mathbf{0}$$

$$T(\alpha\mathbf{u}_1) = \alpha T(\mathbf{u}_1) = \alpha\mathbf{0} = \mathbf{0}$$

Thus ker T is closed under addition and scalar multiplication. Therefore the kernel of T is a subspace of U.

Now let $\mathbf{v}_1, \mathbf{v}_2 \in R(T)$. Then there are vectors $\mathbf{u}_1, \mathbf{u}_2 \in U$ such that

$$\mathbf{v}_1 = T(\mathbf{u}_1) \qquad \text{and} \qquad \mathbf{v}_2 = T(\mathbf{u}_2)$$

Then

$$\mathbf{v}_1 + \mathbf{v}_2 = T(\mathbf{u}_1) + T(\mathbf{u}_2) = T(\mathbf{u}_1 + \mathbf{u}_2)$$

$$\alpha\mathbf{v}_1 = \alpha T(\mathbf{u}_1) = T(\alpha\mathbf{u}_1)$$

for any scalar α. The above identities show that $\mathbf{v}_1 + \mathbf{v}_2$ and $\alpha\mathbf{v}_1$ are images of $\mathbf{u}_1 + \mathbf{u}_2$ and $\alpha\mathbf{u}_1$ under T, respectively. Thus $\mathbf{v}_1 + \mathbf{v}_2 \in R(T)$, $\alpha\mathbf{v}_1 \in R(T)$ and hence $R(T)$ is closed under addition and scalar multiplication. Therefore, the range of T is a subspace of V.

PROBLEM 5-18 Verify Theorem 5-3.3, that is, if $T: U \to V$ is a linear transformation, then

$$\text{nullity of } T + \text{rank of } T = \dim U$$

Solution We must show that

$$\dim \ker(T) + \dim R(T) = \dim U \tag{a}$$

Let $\{\mathbf{u}_1, \mathbf{u}_2, \ldots, \mathbf{u}_m\}$ be a basis for ker T and $\{\mathbf{v}_1, \mathbf{v}_2, \ldots, \mathbf{v}_r\}$ be a basis for $R(T)$. Then

$$\dim \ker(T) = m, \qquad \dim R(T) = r \tag{b}$$

Since $\mathbf{v}_1, \mathbf{v}_2, \ldots, \mathbf{v}_r \in R(T)$, there are vectors $\mathbf{w}_1, \mathbf{w}_2, \ldots, \mathbf{w}_r \in U$ such that

$$T(\mathbf{w}_1) = \mathbf{v}_1,\ T(\mathbf{w}_2) = \mathbf{v}_2, \ldots,\ T(\mathbf{w}_r) = \mathbf{v}_r \tag{c}$$

We want to show that $\{\mathbf{u}_1, \mathbf{u}_2, \ldots, \mathbf{u}_m, \mathbf{w}_1, \mathbf{w}_2, \ldots, \mathbf{w}_r\}$ forms a basis for U. First we show that $\mathbf{u}_1, \mathbf{u}_2, \ldots, \mathbf{u}_m, \mathbf{w}_1, \mathbf{w}_2, \ldots, \mathbf{w}_r$ are linearly independent. Suppose we have

$$\alpha_1\mathbf{u}_1 + \alpha_2\mathbf{u}_2 + \cdots + \alpha_m\mathbf{u}_m + \beta_1\mathbf{w}_1 + \beta_2\mathbf{w}_2 + \cdots + \beta_r\mathbf{w}_r = \mathbf{0} \tag{d}$$

Applying T, we get

$$\alpha_1 T(\mathbf{u}_1) + \alpha_2 T(\mathbf{u}_2) + \cdots + \alpha_m T(\mathbf{u}_m) + \beta_1 T(\mathbf{w}_1) + \beta_2 T(\mathbf{w}_2) + \cdots + \beta_r T(\mathbf{w}_r) = \mathbf{0} \tag{e}$$

Since $\mathbf{u}_1, \mathbf{u}_2, \ldots, \mathbf{u}_m \in \ker(T)$,

$$T(\mathbf{u}_1) = T(\mathbf{u}_2) = \cdots = T(\mathbf{u}_m) = \mathbf{0} \tag{f}$$

Hence

$$\beta_1 T(\mathbf{w}_1) + \beta_2 T(\mathbf{w}_2) + \cdots + \beta_r T(\mathbf{w}_r) = \mathbf{0} \tag{g}$$

or

$$\beta_1\mathbf{v}_1 + \beta_2\mathbf{v}_2 + \cdots + \beta_r\mathbf{v}_r = \mathbf{0} \tag{h}$$

Since $\mathbf{v}_1, \mathbf{v}_2, \ldots, \mathbf{v}_r$ form a basis for $R(T)$, they are linearly independent and we have

$$\beta_1 = \beta_2 = \cdots = \beta_r = 0$$

and so

$$\alpha_1\mathbf{u}_1 + \alpha_2\mathbf{u}_2 + \cdots + \alpha_m\mathbf{u}_m = \mathbf{0} \tag{i}$$

Again, since $\mathbf{u}_1, \mathbf{u}_2, \ldots, \mathbf{u}_m$ form a basis for $\ker(T)$, they are linearly independent, so $\alpha_1 = \alpha_2 = \cdots = \alpha_m = 0$. Therefore it follows that

$$\mathbf{u}_1, \mathbf{u}_2, \ldots, \mathbf{u}_m, \mathbf{w}_1, \mathbf{w}_2, \ldots, \mathbf{w}_r$$

are linearly independent.

Next, we show that $\{\mathbf{u}_1, \mathbf{u}_2, \ldots, \mathbf{u}_m, \mathbf{w}_1, \mathbf{w}_2, \ldots, \mathbf{w}_r\}$ spans U. Suppose $\mathbf{u} \in U$. Then $T(\mathbf{u}) \in R(T)$. Since $\{\mathbf{v}_1, \mathbf{v}_2, \ldots, \mathbf{v}_r\}$ is a basis for $R(T)$, we can express $T(\mathbf{u})$ as

$$T(\mathbf{u}) = \beta_1\mathbf{v}_1 + \beta_2\mathbf{v}_2 + \cdots + \beta_r\mathbf{v}_r \tag{j}$$

Consider the vector

$$\mathbf{w} = \mathbf{u} - (\beta_1\mathbf{w}_1 + \beta_2\mathbf{w}_2 + \cdots + \beta_r\mathbf{w}_r) \tag{k}$$

Then

$$\begin{aligned} T(\mathbf{w}) &= T(\mathbf{u}) - \beta_1 T(\mathbf{w}_1) - \beta_2 T(\mathbf{w}_2) - \cdots - \beta_r T(\mathbf{w}_r) \\ &= \beta_1\mathbf{v}_1 + \beta_2\mathbf{v}_2 + \cdots + \beta_r\mathbf{v}_r - \beta_1\mathbf{v}_1 - \beta_2\mathbf{v}_2 - \cdots - \beta_r\mathbf{v}_r = \mathbf{0} \end{aligned} \tag{l}$$

Thus $\mathbf{w} \in \ker(T)$.

Since $\{\mathbf{u}_1, \mathbf{u}_2, \ldots, \mathbf{u}_m\}$ is a basis for $\ker(T)$, we can express $\mathbf{w}$ as

$$\mathbf{w} = \alpha_1\mathbf{u}_1 + \alpha_2\mathbf{u}_2 + \cdots + \alpha_m\mathbf{u}_m \tag{m}$$

Then from (k), we get

$$\begin{aligned} \mathbf{u} &= \mathbf{w} + (\beta_1\mathbf{w}_1 + \beta_2\mathbf{w}_2 + \cdots + \beta_r\mathbf{w}_r) \\ &= \alpha_1\mathbf{u}_1 + \alpha_2\mathbf{u}_2 + \cdots + \alpha_m\mathbf{u}_m + \beta_1\mathbf{w}_1 + \beta_2\mathbf{w}_2 + \cdots + \beta_r\mathbf{w}_r \end{aligned} \tag{n}$$

Thus $\{\mathbf{u}_1, \mathbf{u}_2, \ldots, \mathbf{u}_m, \mathbf{w}_1, \mathbf{w}_2, \ldots, \mathbf{w}_r\}$ spans U, and hence we conclude that $\{\mathbf{u}_1, \mathbf{u}_2, \ldots, \mathbf{u}_m, \mathbf{w}_1, \mathbf{w}_2, \ldots, \mathbf{w}_r\}$ forms a basis for U. Hence

$$\dim U = m + r = \dim\ker(T) + \dim R(T).$$

PROBLEM 5-19 Verify Theorem 5-3.4.

Solution

(a) Suppose $\mathbf{u}_1, \mathbf{u}_2, \ldots, \mathbf{u}_n$ are linearly independent in U. If

$$\alpha_1 T(\mathbf{u}_1) + \alpha_2 T(\mathbf{u}_2) + \cdots + \alpha_n T(\mathbf{u}_n) = \mathbf{0}$$

then by linearity of T,

$$T(\alpha_1\mathbf{u}_1 + \alpha_2\mathbf{u}_2 + \cdots + \alpha_n\mathbf{u}_n) = \mathbf{0}$$

Because T is one-to-one, we conclude that

$$\alpha_1\mathbf{u}_1 + \alpha_2\mathbf{u}_2 + \cdots + \alpha_n\mathbf{u}_n = \mathbf{0}$$

Since $\mathbf{u}_1, \mathbf{u}_2, \ldots, \mathbf{u}_n$ are linearly independent in U, $\alpha_1 = \alpha_2 = \cdots = \alpha_n = 0$. Thus $\alpha_1 T(\mathbf{u}_1) + \alpha_2 T(\mathbf{u}_2) + \cdots + \alpha_n T(\mathbf{u}_n) = \mathbf{0}$ implies that $\alpha_1 = \alpha_2 = \cdots = \alpha_n = 0$. So $T(\mathbf{u}_1)$, $T(\mathbf{u}_2)$, $\ldots$, $T(\mathbf{u}_n)$ are linearly independent in V.

(b) Suppose $\mathbf{u}_1, \mathbf{u}_2, \ldots, \mathbf{u}_n$ span U and $\mathbf{v} \in V$. Since T is onto, there is a vector $\mathbf{u} \in U$ such that $T(\mathbf{u}) = \mathbf{v}$. If $\mathbf{u} = \alpha_1\mathbf{u}_1 + \alpha_2\mathbf{u}_2 + \cdots + \alpha_n\mathbf{u}_n$, then

$$\begin{aligned} \mathbf{v} = T(\mathbf{u}) &= T(\alpha_1\mathbf{u}_1 + \alpha_2\mathbf{u}_2 + \cdots + \alpha_n\mathbf{u}_n) \\ &= \alpha_1 T(\mathbf{u}_1) + \alpha_2 T(\mathbf{u}_2) + \cdots + \alpha_n T(\mathbf{u}_n) \end{aligned}$$

Thus $T(\mathbf{u}_1)$, $T(\mathbf{u}_2)$, $\ldots$, $T(\mathbf{u}_n)$ span V.

(c) Suppose $\{\mathbf{u}_1, \mathbf{u}_2, \ldots, \mathbf{u}_n\}$ is a basis for U. Then by (a) and (b), $T(\mathbf{u}_1)$, $T(\mathbf{u}_2)$, $\ldots$, $T(\mathbf{u}_n)$ are linearly independent and span V. Thus $\{T(\mathbf{u}_1), T(\mathbf{u}_2), \ldots, T(\mathbf{u}_n)\}$ forms a basis for V.

(d) If $\dim U = n$, then some basis, say $\{\mathbf{u}_1, \mathbf{u}_2, \ldots, \mathbf{u}_n\}$, is a basis for U with n vectors. Since $\{T(\mathbf{u}_1), T(\mathbf{u}_2), \ldots, T(\mathbf{u}_n)\}$ is a basis for V with n vectors, we see that $\dim U = \dim V$.

PROBLEM 5-20 Let $T\colon U \to V$ be a linear transformation between two finite-dimensional vector spaces with $\dim U = \dim V$. Show that the following are equivalent:

(1) T is an isomorphism.
(2) T is one-to-one (or $\ker(T) = \{\mathbf{0}\}$).
(3) T is onto (or $R(T) = V$).

Solution If T is an isomorphism, then T is one-to-one and onto. So suppose that T is one-to-one; then by Theorem 5-3.2 $\ker(T) = \{\mathbf{0}\}$. Since by Theorem 5-3.3 $\dim \ker(T) + \dim R(T) = \dim U$, it follows that $\dim R(T) = \dim U$. Since $\dim U = \dim V$, $\dim R(T) = \dim V$. Since $R(T)$ is a subspace of V by Theorem 5-3.1, $R(T) = V$ and T is onto. Similarly, if T is onto, $\dim R(T) = \dim V = \dim U$. Since $\dim \ker(T) + \dim R(T) = \dim U$, it follows that $\dim \ker(T) = 0$. Thus $\ker(T) = \{\mathbf{0}\}$, and T is one-to-one by Theorem 5-3.2.

Matrix Representation of a Linear Transformation

PROBLEM 5-21 Find the standard matrix of $T: R^3 \to R^2$ of Problem 5-11.

Solution The standard basis for R^3 is

$$\left\{ \begin{bmatrix} 1 \\ 0 \\ 0 \end{bmatrix}, \begin{bmatrix} 0 \\ 1 \\ 0 \end{bmatrix}, \begin{bmatrix} 0 \\ 0 \\ 1 \end{bmatrix} \right\}$$

Thus from Problem 5-11,

$$T\left(\begin{bmatrix} 1 \\ 0 \\ 0 \end{bmatrix} \right) = \begin{bmatrix} 1 \\ 1 \end{bmatrix}, \qquad T\left(\begin{bmatrix} 0 \\ 1 \\ 0 \end{bmatrix} \right) = \begin{bmatrix} 2 \\ 0 \end{bmatrix}, \qquad T\left(\begin{bmatrix} 0 \\ 0 \\ 1 \end{bmatrix} \right) = \begin{bmatrix} 3 \\ -1 \end{bmatrix}$$

Thus by (5.24) the standard matrix of T is

$$A = \begin{bmatrix} 1 & 2 & 3 \\ 1 & 0 & -1 \end{bmatrix}$$

PROBLEM 5-22 Let $T: R^2 \to R^2$ be the linear operator that rotates each vector counterclockwise by an angle θ. Find the standard matrix of T.

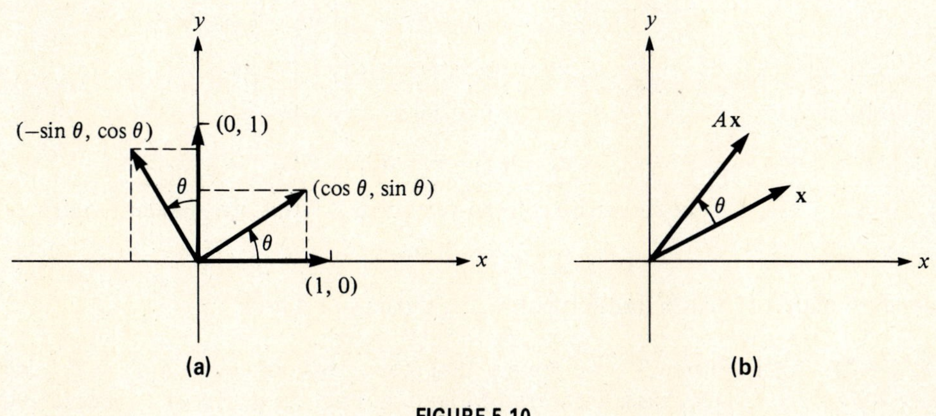

FIGURE 5-10

Solution From Figure 5-10a we see that

$$T(\mathbf{e}_1) = T\left(\begin{bmatrix} 1 \\ 0 \end{bmatrix} \right) = \begin{bmatrix} \cos\theta \\ \sin\theta \end{bmatrix}$$

$$T(\mathbf{e}_2) = T\left(\begin{bmatrix} 0 \\ 1 \end{bmatrix} \right) = \begin{bmatrix} \cos\left(\dfrac{\pi}{2} + \theta\right) \\ \sin\left(\dfrac{\pi}{2} + \theta\right) \end{bmatrix} = \begin{bmatrix} -\sin\theta \\ \cos\theta \end{bmatrix}$$

Thus, by (5.24) we get

$$A = \begin{bmatrix} \cos\theta & -\sin\theta \\ \sin\theta & \cos\theta \end{bmatrix}$$

If $\mathbf{x} = \begin{bmatrix} x_1 \\ x_2 \end{bmatrix}$, then

$$T\mathbf{x} = A\mathbf{x} = \begin{bmatrix} \cos\theta & -\sin\theta \\ \sin\theta & \cos\theta \end{bmatrix}\begin{bmatrix} x_1 \\ x_2 \end{bmatrix} = \begin{bmatrix} x_1\cos\theta - x_2\sin\theta \\ x_1\sin\theta + x_2\cos\theta \end{bmatrix}$$

represents the vector obtained by rotating $\mathbf{x}$ counterclockwise by an angle θ (see Figure 5-10b). The operator T is often referred to as the **rotation operator**. For example, let $\theta = \pi/2$. Then

$$A = \begin{bmatrix} \cos\frac{\pi}{2} & -\sin\frac{\pi}{2} \\ \sin\frac{\pi}{2} & \cos\frac{\pi}{2} \end{bmatrix} = \begin{bmatrix} 0 & -1 \\ 1 & 0 \end{bmatrix}$$

is the matrix representation of the counterclockwise rotation by an angle of $\pi/2$ radians ($=90°$; see Example 5-23).

PROBLEM 5-23 Let $T: R^n \to R^m$ be multiplication by

$$A = \begin{bmatrix} a_{11} & a_{12} & \cdots & a_{1n} \\ a_{21} & a_{22} & \cdots & a_{2n} \\ \vdots & & & \vdots \\ a_{m1} & a_{m2} & \cdots & a_{mn} \end{bmatrix}$$

Find the standard matrix for T.

Solution Let $\{\mathbf{e}_1, \mathbf{e}_2, \ldots, \mathbf{e}_n\}$ be the standard basis for R^n. Then

$$T(\mathbf{e}_1) = A\mathbf{e}_1 = \begin{bmatrix} a_{11} & a_{12} & \cdots & a_{1n} \\ a_{21} & a_{22} & \cdots & a_{2n} \\ \vdots & & & \\ a_{m1} & a_{m2} & \cdots & a_{mn} \end{bmatrix}\begin{bmatrix} 1 \\ 0 \\ \vdots \\ 0 \end{bmatrix} = \begin{bmatrix} a_{11} \\ a_{21} \\ \vdots \\ a_{m1} \end{bmatrix}$$

$$T(\mathbf{e}_2) = A\mathbf{e}_2 = \begin{bmatrix} a_{11} & a_{12} & \cdots & a_{1n} \\ a_{21} & a_{22} & \cdots & a_{2n} \\ \vdots & \vdots & & \vdots \\ a_{m1} & a_{m2} & \cdots & a_{mn} \end{bmatrix}\begin{bmatrix} 0 \\ 1 \\ \vdots \\ 0 \end{bmatrix} = \begin{bmatrix} a_{12} \\ a_{22} \\ \vdots \\ a_{m2} \end{bmatrix}$$

So $T(\mathbf{e}_1), T(\mathbf{e}_2), \ldots, T(\mathbf{e}_n)$ are the successive column vectors of A. Thus, the standard matrix of T is

$$[T(\mathbf{e}_1) \vdots T(\mathbf{e}_2) \vdots \cdots \vdots T(\mathbf{e}_n)] = A$$

i.e., the standard matrix of matrix transformation T is the matrix A itself.

PROBLEM 5-24 In Problem 5-4 we found that the transformation $T: P_2 \to P_1$ defined by $T(a_0 + a_1x + a_2x^2) = a_1 + 2a_2x$ is a linear transformation. Find the matrix of T relative to the bases

$$\mathscr{B} = \{\mathbf{u}_1, \mathbf{u}_2, \mathbf{u}_3\} \quad \text{and} \quad \mathscr{B}' = \{\mathbf{v}_1, \mathbf{v}_2\}$$

where $\mathbf{u}_1 = 1$, $\mathbf{u}_2 = x$, $\mathbf{u}_3 = x^2$, and $\mathbf{v}_1 = 1$, $\mathbf{v}_2 = x$.

Solution From the formula for T we have

$$T(\mathbf{u}_1) = T(1) = 0 = 0\mathbf{v}_1 + 0\mathbf{v}_2$$

$$T(\mathbf{u}_2) = T(x) = 1 = 1\mathbf{v}_1 + 0\mathbf{v}_2$$

$$T(\mathbf{u}_3) = T(x^2) = 2x = 0\mathbf{v}_1 + 2\mathbf{v}_2$$

Thus

$$[T(\mathbf{u}_1)]_{\mathscr{B}'} = \begin{bmatrix} 0 \\ 0 \end{bmatrix}, \qquad [T(\mathbf{u}_2)]_{\mathscr{B}'} = \begin{bmatrix} 1 \\ 0 \end{bmatrix}, \qquad [T(\mathbf{u}_3)]_{\mathscr{B}'} = \begin{bmatrix} 0 \\ 2 \end{bmatrix}$$

Thus, by (5.29) the matrix of T relative to $\mathscr{B}$ and $\mathscr{B}'$ is

$$A = [[T(\mathbf{u}_1)]_{\mathscr{B}'}, [T(\mathbf{u}_2)]_{\mathscr{B}'}, [T(\mathbf{u}_3)]_{\mathscr{B}'}] = \begin{bmatrix} 0 & 1 & 0 \\ 0 & 0 & 2 \end{bmatrix}$$

PROBLEM 5-25 Let $T: P_2 \to P_1$, $\mathscr{B}$ and $\mathscr{B}'$ be as in Problem 5-24 and let $\mathbf{p} = 2 + x + 5x^2$. Use the matrix obtained in Problem 5-24 to compute $T(\mathbf{p})$.

Solution By inspection, the coordinate vector of $\mathbf{p}$ relative to $\mathscr{B}$ is

$$[\mathbf{p}]_{\mathscr{B}} = \begin{bmatrix} 2 \\ 1 \\ 5 \end{bmatrix}$$

Therefore, by (5.30) and the result of Problem 5-24, we get

$$[T(\mathbf{p})]_{\mathscr{B}'} = A[\mathbf{p}]_{\mathscr{B}} = \begin{bmatrix} 0 & 1 & 0 \\ 0 & 0 & 2 \end{bmatrix} \begin{bmatrix} 2 \\ 1 \\ 5 \end{bmatrix} = \begin{bmatrix} 1 \\ 10 \end{bmatrix}$$

Thus

$$T(\mathbf{p}) = 1\mathbf{v}_1 + 10\mathbf{v}_2 = 1(1) + 10(x) = 1 + 10x$$

Check: By the transformation formula,

$$T(2 + x + 5x^2) = 1 + 2(5)x = 1 + 10x.$$

Change of Basis and Similarity

PROBLEM 5-26 Let $T: R^2 \to R^2$ be defined by

$$T\left(\begin{bmatrix} x_1 \\ x_2 \end{bmatrix}\right) = \begin{bmatrix} 3x_1 + 2x_2 \\ -5x_1 + 4x_2 \end{bmatrix}$$

Find the standard matrix of T, that is, the matrix of T relative to the standard basis $\mathscr{B} = \{\mathbf{e}_1, \mathbf{e}_2\}$; then use (5.33) to find the matrix of T relative to the basis $\mathscr{B}' = \{\mathbf{u}_1, \mathbf{u}_2\}$ where

$$\mathbf{e}_1 = \begin{bmatrix} 1 \\ 0 \end{bmatrix}, \qquad \mathbf{e}_2 = \begin{bmatrix} 0 \\ 1 \end{bmatrix}, \qquad \mathbf{u}_1 = \begin{bmatrix} 1 \\ 0 \end{bmatrix}, \qquad \mathbf{u}_2 = \begin{bmatrix} 1 \\ 1 \end{bmatrix}$$

Solution From the formula for T,

$$T(\mathbf{e}_1) = T\left(\begin{bmatrix} 1 \\ 0 \end{bmatrix}\right) = \begin{bmatrix} 3 \\ -5 \end{bmatrix}, \qquad T(\mathbf{e}_2) = T\left(\begin{bmatrix} 0 \\ 1 \end{bmatrix}\right) = \begin{bmatrix} 2 \\ 4 \end{bmatrix}$$

Hence the standard matrix of T is

$$A = \begin{bmatrix} 3 & 2 \\ -5 & 4 \end{bmatrix}$$

Next we need to find the transition matrix P from $\mathscr{B}'$ to $\mathscr{B}$. Notice that

$$[\mathbf{u}_1]_{\mathscr{B}} = \begin{bmatrix} 1 \\ 0 \end{bmatrix}, \qquad [\mathbf{u}_2]_{\mathscr{B}} = \begin{bmatrix} 1 \\ 1 \end{bmatrix}$$

Therefore

$$P = \begin{bmatrix} 1 & 1 \\ 0 & 1 \end{bmatrix}$$

Taking the inverse, we have

$$P^{-1} = \begin{bmatrix} 1 & -1 \\ 0 & 1 \end{bmatrix}$$

Hence by (5.33) the matrix of T relative to the basis $\mathscr{B}'$ is

$$A' = P^{-1}AP = \begin{bmatrix} 1 & -1 \\ 0 & 1 \end{bmatrix}\begin{bmatrix} 3 & 2 \\ -5 & 4 \end{bmatrix}\begin{bmatrix} 1 & 1 \\ 0 & 1 \end{bmatrix} = \begin{bmatrix} 8 & 6 \\ -5 & -1 \end{bmatrix}$$

which is the result obtained in Example 5-19.

PROBLEM 5-27 Let $T: R^3 \to R^3$ be the linear operator defined by $T(\mathbf{x}) = A\mathbf{x}$, where

$$A = \begin{bmatrix} 2 & 0 & 2 \\ 1 & 2 & 1 \\ 1 & 2 & 1 \end{bmatrix}$$

(a) Find the matrix A' of T relative to the basis $\mathscr{B}' = \{\mathbf{u}_1, \mathbf{u}_2, \mathbf{u}_3\}$ where

$$\mathbf{u}_1 = \begin{bmatrix} -2 \\ 1 \\ 1 \end{bmatrix}, \qquad \mathbf{u}_2 = \begin{bmatrix} 1 \\ 1 \\ 1 \end{bmatrix}, \qquad \mathbf{u}_3 = \begin{bmatrix} 1 \\ -1 \\ 0 \end{bmatrix}$$

(b) Find the transition matrix P from $\mathscr{B}' = \{\mathbf{u}_1, \mathbf{u}_2, \mathbf{u}_3\}$ to $\mathscr{B} = \{\mathbf{i}, \mathbf{j}, \mathbf{k}\}$ and verify that $A' = P^{-1}AP$.

Solution

(a)

$$T(\mathbf{u}_1) = A\mathbf{u}_1 = \begin{bmatrix} 2 & 0 & 2 \\ 1 & 2 & 1 \\ 1 & 2 & 1 \end{bmatrix}\begin{bmatrix} -2 \\ 1 \\ 1 \end{bmatrix} = \begin{bmatrix} -2 \\ 1 \\ 1 \end{bmatrix} = 1\mathbf{u}_1 + 0\mathbf{u}_2 + 0\mathbf{u}_3$$

$$T(\mathbf{u}_2) = A\mathbf{u}_2 = \begin{bmatrix} 2 & 0 & 2 \\ 1 & 2 & 1 \\ 1 & 2 & 1 \end{bmatrix}\begin{bmatrix} 1 \\ 1 \\ 1 \end{bmatrix} = \begin{bmatrix} 4 \\ 4 \\ 4 \end{bmatrix} = 0\mathbf{u}_1 + 4\mathbf{u}_2 + 0\mathbf{u}_3$$

$$T(\mathbf{u}_3) = A\mathbf{u}_3 = \begin{bmatrix} 2 & 0 & 2 \\ 1 & 2 & 1 \\ 1 & 2 & 1 \end{bmatrix}\begin{bmatrix} 1 \\ -1 \\ 0 \end{bmatrix} = \begin{bmatrix} 2 \\ -1 \\ -1 \end{bmatrix} = -1\mathbf{u}_1 + 0\mathbf{u}_2 + 0\mathbf{u}_3$$

Thus

$$[T(\mathbf{u}_1)]_{\mathscr{B}'} = \begin{bmatrix} 1 \\ 0 \\ 0 \end{bmatrix}, \qquad [T(\mathbf{u}_2)]_{\mathscr{B}'} = \begin{bmatrix} 0 \\ 4 \\ 0 \end{bmatrix}, \qquad [T(\mathbf{u}_3)]_{\mathscr{B}'} = \begin{bmatrix} -1 \\ 0 \\ 0 \end{bmatrix}$$

Therefore, by (5.29) we get

$$A' = \begin{bmatrix} 1 & 0 & -1 \\ 0 & 4 & 0 \\ 0 & 0 & 0 \end{bmatrix}$$

(b) From (3.59) the transition matrix P from $\mathscr{B}'$ to $\mathscr{B} = \{\mathbf{i},\mathbf{j},\mathbf{k}\}$ is

$$P = \begin{bmatrix} -2 & 1 & 1 \\ 1 & 1 & -1 \\ 1 & 1 & 0 \end{bmatrix}$$

Taking the inverse, we get

$$P^{-1} = \begin{bmatrix} -\frac{1}{3} & -\frac{1}{3} & \frac{2}{3} \\ \frac{1}{3} & \frac{1}{3} & \frac{1}{3} \\ 0 & -1 & 1 \end{bmatrix}$$

and

$$P^{-1}AP = \begin{bmatrix} -\frac{1}{3} & -\frac{1}{3} & \frac{2}{3} \\ \frac{1}{3} & \frac{1}{3} & \frac{1}{3} \\ 0 & -1 & 1 \end{bmatrix}\begin{bmatrix} 2 & 0 & 2 \\ 1 & 2 & 1 \\ 1 & 2 & 1 \end{bmatrix}\begin{bmatrix} -2 & 1 & 1 \\ 1 & 1 & -1 \\ 1 & 1 & 0 \end{bmatrix} = \begin{bmatrix} 1 & 0 & -1 \\ 0 & 4 & 0 \\ 0 & 0 & 0 \end{bmatrix} = A'$$

***PROBLEM 5-28** [*For readers who have studied calculus.*] Let $T: P_2 \to P_2$ be the differential operator on P_2 defined by $T(p(x)) = p'(x)$ where $p(x) \in P_2$.

(a) Find the standard matrix A of T relative to the standard basis $\mathscr{B} = \{1, x, x^2\}$.
(b) Find the matrix A' of T relative to the basis $\mathscr{B}' = \{1, 1 + x, 1 + x + x^2\}$.
(c) Show that A and A' are similar.

Solution

(a)

$$T(1) = 0 = (0)1 + (0)x + (0)x^2$$
$$T(x) = 1 = (1)1 + (0)x + (0)x^2$$
$$T(x^2) = 2x = (0)1 + (2)x + (0)x^2$$

Thus

$$[T(1)]_{\mathscr{B}} = \begin{bmatrix} 0 \\ 0 \\ 0 \end{bmatrix}, \qquad [T(x)]_{\mathscr{B}} = \begin{bmatrix} 1 \\ 0 \\ 0 \end{bmatrix}, \qquad [T(x^2)]_{\mathscr{B}} = \begin{bmatrix} 0 \\ 2 \\ 0 \end{bmatrix}$$

Hence

$$A = \begin{bmatrix} 0 & 1 & 0 \\ 0 & 0 & 2 \\ 0 & 0 & 0 \end{bmatrix}$$

(b)

$$T(1) = 0 = (0)(1) + (0)(1 + x) + (0)(1 + x + x^2)$$
$$T(1 + x) = 1 = (1)(1) + (0)(1 + x) + (0)(1 + x + x^2)$$
$$T(1 + x + x^2) = 1 + 2x = \alpha_1(1) + \alpha_2(1 + x) + \alpha_3(1 + x + x^2)$$
$$= (\alpha_1 + \alpha_2 + \alpha_3) + (\alpha_2 + \alpha_3)x + \alpha_3 x^2$$

Thus $\alpha_1 + \alpha_2 + \alpha_3 = 1$, $\alpha_2 + \alpha_3 = 2$, $\alpha_3 = 0$. Solving for $\alpha_1, \alpha_2, \alpha_3$, we get $\alpha_1 = -1$, $\alpha_2 = 2$, $\alpha_3 = 0$. Hence $T(1 + x + x^2) = (-1)1 + 2(1 + x) + 0(1 + x + x^2)$. Thus

$$[T(1)]_{\mathscr{B}'} = \begin{bmatrix} 0 \\ 0 \\ 0 \end{bmatrix}, \qquad [T(1 + x)]_{\mathscr{B}'} = \begin{bmatrix} 1 \\ 0 \\ 0 \end{bmatrix}, \qquad [T(1 + x + x^2)]_{\mathscr{B}'} = \begin{bmatrix} -1 \\ 2 \\ 0 \end{bmatrix}$$

Therefore

$$A' = \begin{bmatrix} 0 & 1 & -1 \\ 0 & 0 & 2 \\ 0 & 0 & 0 \end{bmatrix}$$

(c) By inspection, the coordinate vectors of the elements of $\mathscr{B}'$ with respect to $\mathscr{B}$ are

$$[1]_{\mathscr{B}} = \begin{bmatrix} 1 \\ 0 \\ 0 \end{bmatrix}, \qquad [1+x]_{\mathscr{B}} = \begin{bmatrix} 1 \\ 1 \\ 0 \end{bmatrix}, \qquad [1+x+x^2]_{\mathscr{B}} = \begin{bmatrix} 1 \\ 1 \\ 1 \end{bmatrix}$$

Hence from (3.59) the transition matrix P from $\mathscr{B}'$ to $\mathscr{B}$ is

$$P = \begin{bmatrix} 1 & 1 & 1 \\ 0 & 1 & 1 \\ 0 & 0 & 1 \end{bmatrix}$$

Taking the inverse, we get

$$P^{-1} = \begin{bmatrix} 1 & -1 & 0 \\ 0 & 1 & -1 \\ 0 & 0 & 1 \end{bmatrix}$$

and

$$P^{-1}AP = \begin{bmatrix} 1 & -1 & 0 \\ 0 & 1 & -1 \\ 0 & 0 & 1 \end{bmatrix} \begin{bmatrix} 0 & 1 & 0 \\ 0 & 0 & 2 \\ 0 & 0 & 0 \end{bmatrix} \begin{bmatrix} 1 & 1 & 1 \\ 0 & 1 & 1 \\ 0 & 0 & 1 \end{bmatrix} = \begin{bmatrix} 0 & 1 & -1 \\ 0 & 0 & 2 \\ 0 & 0 & 0 \end{bmatrix} = A'$$

Thus A and A' are similar.

PROBLEM 5-29 Let

$$A = \begin{bmatrix} 1 & 0 \\ 0 & -1 \end{bmatrix} \quad \text{and} \quad B = \begin{bmatrix} -1 & 0 \\ 0 & 1 \end{bmatrix}$$

Is A similar to B?

Solution In order for A to be similar to B, there must exist a 2×2 nonsingular matrix P such that

$$B = P^{-1}AP \tag{a}$$

or

$$PB = AP \tag{b}$$

Let

$$P = \begin{bmatrix} a & b \\ c & d \end{bmatrix} \tag{c}$$

where a, b, c, and d are unknown scalars. Then substituting (c) into (b), we get

$$\begin{bmatrix} a & b \\ c & d \end{bmatrix} \begin{bmatrix} -1 & 0 \\ 0 & 1 \end{bmatrix} = \begin{bmatrix} 1 & 0 \\ 0 & -1 \end{bmatrix} \begin{bmatrix} a & b \\ c & d \end{bmatrix}$$

Simplifying, we get

$$\begin{bmatrix} -a & b \\ -c & d \end{bmatrix} = \begin{bmatrix} a & b \\ -c & -d \end{bmatrix}$$

Equating each entry, we must have $-a = a$, $d = -d$; that is, $a = d = 0$. Thus

$$P = \begin{bmatrix} 0 & b \\ c & 0 \end{bmatrix}$$

Since P is nonsingular, that is, $\det P = -(bc) \neq 0$, we must have $b \neq 0$ and $c \neq 0$. Let $b = c = 1$; then

$$P = \begin{bmatrix} 0 & 1 \\ 1 & 0 \end{bmatrix} \quad \text{and} \quad P^{-1} = \begin{bmatrix} 0 & 1 \\ 1 & 0 \end{bmatrix}$$

Now

$$P^{-1}AP = \begin{bmatrix} 0 & 1 \\ 1 & 0 \end{bmatrix}\begin{bmatrix} 1 & 0 \\ 0 & -1 \end{bmatrix}\begin{bmatrix} 0 & 1 \\ 1 & 0 \end{bmatrix} = \begin{bmatrix} -1 & 0 \\ 0 & 1 \end{bmatrix} = B$$

Thus A is similar to B.

PROBLEM 5-30 Show that if A and B are similar matrices, then $\det(A) = \det(B)$.

Solution If A and B are similar, then there exists a nonsingular matrix P such that

$$A = P^{-1}BP$$

Thus, using (2.7) and (2.9), we have

$$\begin{aligned} \det(A) &= \det(P^{-1}BP) \\ &= \det(P^{-1})\det(B)\det(P) \\ &= \det(P^{-1})\det(P)\det(B) \\ &= \det(B) \end{aligned}$$

Orthogonal Linear Transformations

PROBLEM 5-31 Show that the linear operator $T: R^2 \to R^2$ of Problem 5-22 is an orthogonal linear operator.

Solution Let

$$\mathbf{u} = \begin{bmatrix} u_1 \\ u_2 \end{bmatrix} \quad \text{and} \quad \mathbf{v} = \begin{bmatrix} v_1 \\ v_2 \end{bmatrix}$$

be any two vectors in R^2. (Here R^2 is Euclidean 2-space with the Euclidean inner product.) From Problem 5-22, we have

$$T\mathbf{u} = A\mathbf{u} = \begin{bmatrix} \cos\theta & -\sin\theta \\ \sin\theta & \cos\theta \end{bmatrix}\begin{bmatrix} u_1 \\ u_2 \end{bmatrix} = \begin{bmatrix} u_1\cos\theta - u_2\sin\theta \\ u_1\sin\theta + u_2\cos\theta \end{bmatrix}$$

$$T\mathbf{v} = A\mathbf{v} = \begin{bmatrix} \cos\theta & -\sin\theta \\ \sin\theta & \cos\theta \end{bmatrix}\begin{bmatrix} v_1 \\ v_2 \end{bmatrix} = \begin{bmatrix} v_1\cos\theta - v_2\sin\theta \\ v_1\sin\theta + v_2\cos\theta \end{bmatrix}$$

Thus, by the definition of inner product (4.22),

$$\begin{aligned} \langle T\mathbf{u}, T\mathbf{v}\rangle &= (u_1\cos\theta - u_2\sin\theta)(v_1\cos\theta - v_2\sin\theta) + (u_1\sin\theta + u_2\cos\theta)(v_1\sin\theta + v_2\cos\theta) \\ &= u_1v_1\cos^2\theta + u_2v_2\sin^2\theta - u_2v_1\sin\theta\cos\theta - u_1v_2\cos\theta\sin\theta \\ &\quad + u_1v_1\sin^2\theta + u_2v_2\cos^2\theta + u_2v_1\cos\theta\sin\theta + u_1v_2\sin\theta\cos\theta \\ &= u_1v_1(\cos^2\theta + \sin^2\theta) + u_2v_2(\sin^2\theta + \cos^2\theta) \\ &= u_1v_1 + u_2v_2 = \langle \mathbf{u}, \mathbf{v}\rangle \end{aligned}$$

since $\cos^2\theta + \sin^2\theta = 1$. Therefore T is an orthogonal linear operator.

PROBLEM 5-32 Verify Theorem 5-6.1.

Solution If T is orthogonal, then

$$\|\mathbf{v}\|^2 = \langle \mathbf{v},\mathbf{v} \rangle = \langle T(\mathbf{v}),T(\mathbf{v}) \rangle = \|T(\mathbf{v})\|^2$$

for all $\mathbf{v} \in V$. Therefore $\|T(\mathbf{v})\| = \|\mathbf{v}\|$.

Next, suppose $\|T(\mathbf{v})\| = \|\mathbf{v}\|$ for every $\mathbf{v} \in V$. Then $\|T(\mathbf{v})\|^2 = \|\mathbf{v}\|^2$. Now using the identity of Problem 4-38 and the fact that T is linear, we get

$$\begin{aligned}
\langle \mathbf{u},\mathbf{v} \rangle &= \tfrac{1}{4}\|\mathbf{u} + \mathbf{v}\|^2 - \tfrac{1}{4}\|\mathbf{u} - \mathbf{v}\|^2 \\
&= \tfrac{1}{4}\|T(\mathbf{u} + \mathbf{v})\|^2 - \tfrac{1}{4}\|T(\mathbf{u} - \mathbf{v})\|^2 \\
&= \tfrac{1}{4}\langle T(\mathbf{u} + \mathbf{v}),T(\mathbf{u} + \mathbf{v}) \rangle - \tfrac{1}{4}\langle T(\mathbf{u} - \mathbf{v}),T(\mathbf{u} - \mathbf{v}) \rangle \\
&= \tfrac{1}{4}\langle T(\mathbf{u}) + T(\mathbf{v}),T(\mathbf{u}) + T(\mathbf{v}) \rangle - \tfrac{1}{4}\langle T(\mathbf{u}) - T(\mathbf{v}),T(\mathbf{u}) - T(\mathbf{v}) \rangle \\
&= \tfrac{1}{4}\{\langle T(\mathbf{u}),T(\mathbf{u}) \rangle + 2\langle T(\mathbf{u}),T(\mathbf{v}) \rangle + \langle T(\mathbf{v}),T(\mathbf{v}) \rangle\} \\
&\quad -\tfrac{1}{4}\{\langle T(\mathbf{u}),T(\mathbf{u}) \rangle - 2\langle T(\mathbf{u}),T(\mathbf{v}) \rangle + \langle T(\mathbf{v}),T(\mathbf{v}) \rangle\} \\
&= \langle T(\mathbf{u}),T(\mathbf{v}) \rangle
\end{aligned}$$

for all $\mathbf{u}, \mathbf{v} \in V$.

PROBLEM 5-33 Verify the Theorem 5-6.2.

Solution

(1) Since $\|T(\mathbf{v})\| = \|\mathbf{v}\|$, if $T(\mathbf{v}) = \mathbf{0}$, then $\mathbf{v} = \mathbf{0}$. Thus, by Theorem 5-3.2, T is one-to-one.
(2) If $\mathbf{u}$ and $\mathbf{v}$ are orthogonal, then $\langle \mathbf{u},\mathbf{v} \rangle = 0$. Since $\langle T(\mathbf{u}),T(\mathbf{v}) \rangle = \langle \mathbf{u},\mathbf{v} \rangle = 0$, $T(\mathbf{u})$ and $T(\mathbf{v})$ are also orthogonal.
(3) If $(\mathbf{v}_1, \mathbf{v}_2, \ldots, \mathbf{v}_n)$ is an orthonormal basis, then

$$\|T(\mathbf{v}_i)\| = \|\mathbf{v}_i\| = 1 \qquad (i = 1, 2, \ldots, n)$$

Thus $T(\mathbf{v}_1)$, $T(\mathbf{v}_2)$, $\ldots$, $T(\mathbf{v}_n)$ are each unit vectors.

Next, by property (2), if $i \neq j$, then $T(\mathbf{v}_i)$ and $T(\mathbf{v}_j)$ are orthogonal, since $\mathbf{v}_i$ and $\mathbf{v}_j$ are orthogonal. Again, by property (1), if $\{\mathbf{v}_1, \mathbf{v}_2, \ldots, \mathbf{v}_n\}$ forms a basis, so does $\{T(\mathbf{v}_1), T(\mathbf{v}_2), \ldots, T(\mathbf{v}_n)\}$. Hence, $\{T(\mathbf{v}_1), T(\mathbf{v}_2), \ldots, T(\mathbf{v}_n)\}$ forms an orthonormal basis.

PROBLEM 5-34 Find the matrix of the linear operator T of Example 5-23 relative to the standard (orthonormal) basis for R^2 and show that the matrix is an orthogonal matrix.

Solution Let $\mathscr{B} = \{\mathbf{e}_1,\mathbf{e}_2\}$ be the standard basis for R^2:

$$\mathbf{e}_1 = \begin{bmatrix} 1 \\ 0 \end{bmatrix}, \qquad \mathbf{e}_2 = \begin{bmatrix} 0 \\ 1 \end{bmatrix}$$

Then

$$T(\mathbf{e}_1) = T\left(\begin{bmatrix} 1 \\ 0 \end{bmatrix}\right) = \begin{bmatrix} 0 \\ 1 \end{bmatrix}, \qquad T(\mathbf{e}_2) = T\left(\begin{bmatrix} 0 \\ 1 \end{bmatrix}\right) = \begin{bmatrix} -1 \\ 0 \end{bmatrix}$$

Hence the matrix of T relative to $\mathscr{B}$ is

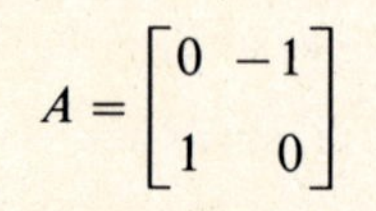

$$A = \begin{bmatrix} 0 & -1 \\ 1 & 0 \end{bmatrix}$$

Now

$$A^{-1} = \begin{bmatrix} 0 & 1 \\ -1 & 0 \end{bmatrix} = A^T$$

Thus A is an orthogonal matrix (see Problem 5-22).

PROBLEM 5-35 Show that the linear operator T of Example 5-23 preserves lengths.

Solution Let

$$\mathbf{v} = \begin{bmatrix} v_1 \\ v_2 \end{bmatrix}$$

Then

$$\|\mathbf{v}\| = \sqrt{\langle \mathbf{v},\mathbf{v} \rangle} = \sqrt{v_1^2 + v_2^2}$$

On the other hand,

$$T(\mathbf{v}) = \begin{bmatrix} -v_2 \\ v_1 \end{bmatrix}$$

and

$$\|T(\mathbf{v})\| = \sqrt{\langle T(\mathbf{v}),T(\mathbf{v}) \rangle} = \sqrt{(-v_2)^2 + v_1^2} = \sqrt{v_1^2 + v_2^2}$$

Hence $\|T(\mathbf{v})\| = \|\mathbf{v}\|$.

PROBLEM 5-36 Show that the column vectors of an orthogonal matrix are mutually orthogonal unit vectors.

Solution Let

$$A = \begin{bmatrix} a_{11} & a_{12} & \cdots & a_{1n} \\ a_{21} & a_{22} & \cdots & a_{2n} \\ \vdots & \vdots & & \vdots \\ a_{n1} & a_{n2} & \cdots & a_{nn} \end{bmatrix} = [\mathbf{a}_1, \mathbf{a}_2, \ldots, \mathbf{a}_n]$$

where $\mathbf{a}_1, \mathbf{a}_2, \ldots, \mathbf{a}_n$ are column vectors of A. Then

$$A^T = \begin{bmatrix} a_{11} & a_{21} & \cdots & a_{n1} \\ a_{12} & a_{22} & \cdots & a_{n2} \\ \vdots & \vdots & & \vdots \\ a_{1n} & a_{2n} & \cdots & a_{nn} \end{bmatrix}$$

Now, if A is an orthogonal matrix, then $A^TA = I_n$ or

$$\begin{bmatrix} a_{11} & a_{21} & \cdots & a_{n1} \\ a_{12} & a_{22} & \cdots & a_{n2} \\ \vdots & \vdots & & \vdots \\ a_{1n} & a_{2n} & \cdots & a_{nn} \end{bmatrix} \begin{bmatrix} a_{11} & a_{12} & \cdots & a_{1n} \\ a_{21} & a_{22} & \cdots & a_{2n} \\ \vdots & \vdots & & \vdots \\ a_{n1} & a_{n2} & \cdots & a_{nn} \end{bmatrix} = \begin{bmatrix} 1 & 0 & \cdots & 0 \\ 0 & 1 & \cdots & 0 \\ \vdots & \vdots & & \vdots \\ 0 & 0 & \cdots & 1 \end{bmatrix}$$

Thus we have

$$\sum_{i=1}^{n} a_{ij}a_{ik} = \delta_{jk} = \begin{cases} 1 & j = k \\ 0 & j \neq k \end{cases}$$

or equivalently, $\langle \mathbf{a}_j,\mathbf{a}_k \rangle = 0$ if $j \neq k$ and $\langle \mathbf{a}_j,\mathbf{a}_j \rangle = 1$ for all j, which indicates that the column vectors of A are mutually orthogonal unit vectors.

Supplementary Exercises

PROBLEM 5-37 Is the mapping $T: R^2 \to R^1$ defined by $T(\mathbf{u}) = \|\mathbf{u}\|$ a linear transformation? [*Hint:* Use (4.32).]

Answer No

PROBLEM 5-38 Let $T: R_{n \times n} \to R_{n \times n}$ be the mapping defined by $T(A) = AB - BA$, where B is a fixed $n \times n$ matrix and $A \in R_{n \times n}$. Is T a linear operator?

Answer Yes

PROBLEM 5-39 Let U and V be vector spaces. Let $\{\mathbf{u}_1, \mathbf{u}_2, \ldots, \mathbf{u}_n\}$ be a basis of U and let $\mathbf{v}_1, \mathbf{v}_2, \ldots, \mathbf{v}_n$ be arbitrary vectors of V. Show that there exists a unique linear transformation $T: U \to V$ such that

$$T(\mathbf{u}_1) = \mathbf{v}_1, T(\mathbf{u}_2) = \mathbf{v}_2, \ldots, T(\mathbf{u}_n) = \mathbf{v}_n$$

PROBLEM 5-40 Let $T: R^3 \to R^4$ be the linear transformation defined by

$$T\left(\begin{bmatrix} x_1 \\ x_2 \\ x_3 \end{bmatrix}\right) = \begin{bmatrix} x_1 \\ x_1 \\ x_2 \\ x_1 \end{bmatrix}$$

Find a basis and the dimension of (**a**) the ker(T) and (**b**) the $R(T)$.

Answer (**a**) $= \begin{bmatrix} 0 \\ 0 \\ 1 \end{bmatrix}$, $\dim \ker(T) = 1$ (**b**) $\left\{\begin{bmatrix} 1 \\ 1 \\ 0 \\ 1 \end{bmatrix}, \begin{bmatrix} 0 \\ 0 \\ 1 \\ 0 \end{bmatrix}\right\}$, $\dim R(T) = 2$

PROBLEM 5-41 If $I: V \to V$ is the identity operator on a finite-dimensional vector space V, show that the matrix of I relative to any basis $\mathscr{B}$ for V is given by the identity matrix of nth order I_n if $\dim V = n$.

PROBLEM 5-42 Let $T: R^2 \to R^2$ be a linear operator defined by

$$T\left(\begin{bmatrix} x_1 \\ x_2 \end{bmatrix}\right) = \begin{bmatrix} x_1 + x_2 \\ 0 \end{bmatrix}$$

(**a**) Find the matrix of T relative to

$$\mathscr{B} = \left\{\begin{bmatrix} 1 \\ 0 \end{bmatrix}, \begin{bmatrix} 0 \\ 1 \end{bmatrix}\right\} \quad \text{and} \quad \mathscr{B}' = \left\{\begin{bmatrix} 1 \\ 0 \end{bmatrix}, \begin{bmatrix} 0 \\ 1 \end{bmatrix}\right\}$$

(**b**) Find the matrix of T relative to

$$\mathscr{B} = \left\{\begin{bmatrix} 1 \\ 0 \end{bmatrix}, \begin{bmatrix} 0 \\ 1 \end{bmatrix}\right\} \quad \text{and} \quad \mathscr{B}' = \left\{\begin{bmatrix} 0 \\ 1 \end{bmatrix}, \begin{bmatrix} 1 \\ 0 \end{bmatrix}\right\}$$

where $\mathscr{B}$ is the basis for the domain of T and $\mathscr{B}'$ is the basis for the codomain of T.

Answer (**a**) $A = \begin{bmatrix} 1 & 1 \\ 0 & 0 \end{bmatrix}$ (**b**) $A = \begin{bmatrix} 0 & 0 \\ 1 & 1 \end{bmatrix}$

PROBLEM 5-43 Let $T: P_2 \to P_1$ be the linear transformation defined by $T(a_0 + a_1x + a_2x^2) = a_1 + 2a_2x$ (Problem 5-24). Find the matrix of T relative to the bases $\mathscr{B} = \{\mathbf{u}_1, \mathbf{u}_2, \mathbf{u}_3\}$ and $\mathscr{B}' = \{\mathbf{v}_1, \mathbf{v}_2\}$ where $\mathbf{u}_1 = 1, \mathbf{u}_2 = x, \mathbf{u}_3 = x^2, \mathbf{v}_1 = 1, \mathbf{v}_2 = 1 + x$.

Answer $A = \begin{bmatrix} 0 & 1 & -2 \\ 0 & 0 & 2 \end{bmatrix}$

PROBLEM 5-44 Let A and B be $n \times n$ square matrices. Show that if A and B are similar, then $\text{tr}(A) = \text{tr}(B)$. [*Hint:* See Problem 4-20.]

PROBLEM 5-45 Let $T: V \to V$ be a linear operator and let $\mathscr{B} = \{\mathbf{v}_1, \mathbf{v}_2, \ldots, \mathbf{v}_n\}$ be an orthonormal basis for the real inner product space V. Suppose A is the matrix of T relative to $\mathscr{B}$. Then show that T is orthogonal if and only if A is an orthogonal matrix.

PROBLEM 5-46 Show that the row vectors of an orthogonal matrix are mutually orthogonal. [*Hint:* Use the result of Problem 5-36.]

6 COMPLEX VECTOR SPACES

THIS CHAPTER IS ABOUT

☑ Complex Numbers
☑ Complex Vector Spaces
☑ Complex Inner Product Spaces
☑ Unitary Transformation and Unitary Matrix

6-1. Complex Numbers

A. Definitions

A **complex number** z can be defined as an ordered pair of real numbers x, y that satisfies certain laws of operation. It is written in either of two forms

$$z = x + iy \qquad \text{or} \qquad z = x + yi \tag{6.1}$$

where i is called the **imaginary unit**.

When $y = 0$, z becomes the real number x; that is, complex numbers include all the real numbers:

$$x + 0i = x \tag{6.2}$$

When $x = 0$, z is called the pure imaginary number:

$$0 + yi = yi \tag{6.3}$$

The real number x is called the *real part* of z, denoted by $\mathcal{Re}(z)$; and the real number y is called the *imaginary part* of z, denoted by $\mathcal{Im}(z)$:

$$\mathcal{Re}(z) = x, \qquad \mathcal{Im}(z) = y \tag{6.4}$$

B. Fundamental operations

Let z_1 and z_2 be any two complex numbers:

$$z_1 = x_1 + iy_1, \quad z_2 = x_2 + iy_2$$

(1) $z_1 = z_2$ implies $x_1 = x_2$, $y_1 = y_2$ **(6.5)**
In particular,
$z = x + iy = 0$ implies $x = y = 0$ **(6.6)**

(2) $z_1 + z_2 = (x_1 + x_2) + i(y_1 + y_2)$ **(6.7)**

(3) $z_1 - z_2 = (x_1 - x_2) + i(y_1 - y_2)$ **(6.8)**

(4) $kz = k(x + iy) = kx + i(ky)$, k is a real number **(6.9)**
In particular, when $k = -1$, then $(-1)z = -z = -(x + iy)$ is called the *negative* of z.

(5) $z_1 z_2 = (x_1 x_2 - y_1 y_2) + i(x_1 y_2 + x_2 y_1)$ **(6.10)**
In particular, when $z_1 = z_2 = i$, operation (6.10) becomes
$i^2 = -1$ **(6.11)**

C. Properties of the fundamental operations

(1) $z_1 + z_2 = z_2 + z_1$

(2) $z_1 + (z_2 + z_3) = (z_1 + z_2) + z_3$

(3) $z + 0 = z$

(4) $z + (-z) = 0$

(5) $z_1 z_2 = z_2 z_1$

(6) $z_1(z_2 z_3) = (z_1 z_2)z_3$

(7) $z_1(z_2 + z_3) = z_1 z_2 + z_1 z_3$

(8) $1z = z$

EXAMPLE 6-1: Show that the law of multiplication (6.10) can be obtained by simply expanding the product $(x_1 + iy_1)(x_2 + iy_2)$ by the formal use of the operation on real numbers and replacing i^2 by -1.

Solution

$$\begin{aligned}(x_1 + iy_1)(x_2 + iy_2) &= x_1 x_2 + (iy_1)(iy_2) + x_1(iy_2) + (iy_1)x_2 \\ &= x_1 x_2 + i^2 y_1 y_2 + i(x_1 y_2 + x_2 y_1) \\ &= (x_1 x_2 - y_1 y_2) + i(x_1 y_2 + x_2 y_1)\end{aligned}$$

EXAMPLE 6-2: Given that $z_1 = 1 - i2$ and $z_2 = 3 + i4$, find

(a) $z_1 + z_2$, **(b)** $z_1 - z_2$, **(c)** $2z_1$, **(d)** $-z_2$, **(e)** $z_1 z_2$.

Solution

(a)

$$\begin{aligned}z_1 + z_2 &= (1 - i2) + (3 + i4) \\ &= (1 + 3) + i(-2 + 4) \\ &= 4 + i2\end{aligned}$$

(b)

$$\begin{aligned}z_1 - z_2 &= (1 - i2) - (3 + i4) \\ &= (1 - 3) + i(-2 - 4) \\ &= -2 + i(-6) \\ &= -2 - i6\end{aligned}$$

(c)

$$\begin{aligned}2z_1 &= 2(1 - i2) \\ &= 2 - i4\end{aligned}$$

(d)

$$\begin{aligned}-z_2 = (-1)z_2 &= (-1)(3 + i4) \\ &= -3 - i4\end{aligned}$$

(e) Using operation (6.10), we have

$$\begin{aligned}z_1 z_2 &= (1 - i2)(3 + i4) \\ &= [(1)(3) - (-2)(4)] + i[(1)(4) + (-2)(3)] \\ &= 11 - i2\end{aligned}$$

or, as in Example 6-1, we get

$$
\begin{aligned}
z_1 z_2 &= (1 - i2)(3 + i4) \\
&= (1)(3) + (-i2)(3) + (1)(i4) + (-i2)(i4) \\
&= 3 - i6 + i4 - i^2 8 \\
&= 3 - i2 - (-1)8 \\
&= 11 - i2
\end{aligned}
$$

D. Geometric representation of complex numbers

Since a complex number is an ordered pair of real numbers, we can represent it as a point in a plane (2-space) like the xy-plane (see Section 3-1). Corresponding to each point (x, y) there is a single complex number $z = x + iy$, and corresponding to each complex number z there is a definite point (Figure 6-1). In this case we refer to the xy-plane as the **complex plane** or the **z-plane**. The x-axis will be called the *real axis* and the y-axis, the *imaginary axis*.

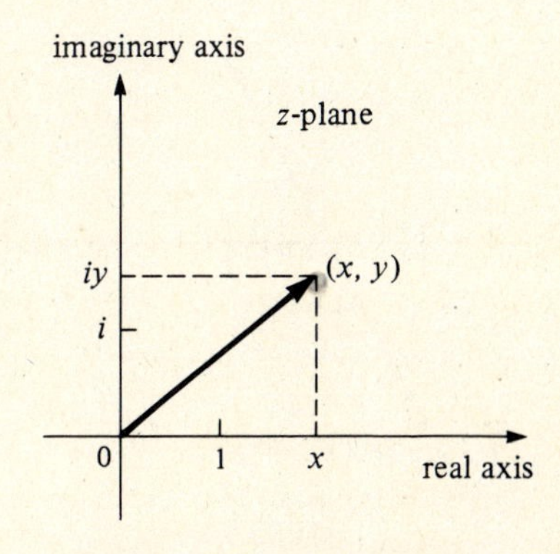

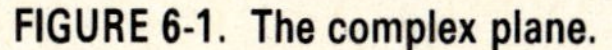
FIGURE 6-1. The complex plane.

E. Polar representation, absolute value, and argument

Let $z = x + iy$ be a nonzero complex number represented by a point with rectangular coordinates (x, y). Next let (r, θ) be the polar coordinates of the point P (Figure 6-2). Then

$$x = r\cos\theta, \qquad y = r\sin\theta \tag{6.12}$$

and the complex number $z = x + iy$ can be written in its **polar form**:

$$z = r(\cos\theta + i\sin\theta) \tag{6.13}$$

We shall assume that $r \geq 0$. Then we call r the **absolute value**, or **modulus**, of the complex number z, denoted by $|z|$. Thus if $z = x + iy$,

$$r = |z| = \sqrt{x^2 + y^2} \tag{6.14}$$

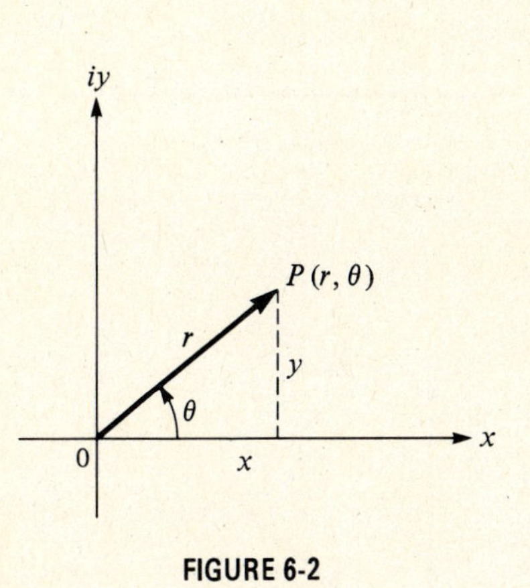

FIGURE 6-2

We call θ the **argument** of the complex number $z \neq 0$, which can be obtained from

$$\theta = \arg z = \tan^{-1}\frac{y}{x} \tag{6.15}$$

note: θ is multivalued. Thus we shall agree that θ will always be such that $-\pi < \theta < \pi$. Such a value of θ is called the **principal argument** of z.

Finally, using **Euler's identity**

$$e^{i\theta} = \cos\theta + i\sin\theta \tag{6.16}$$

we can rewrite (6.13) as

$$z = re^{i\theta} \tag{6.17}$$

EXAMPLE 6-3: Determine $|z|$ and $\arg z$ if **(a)** $z = 1 + i\sqrt{3}$, **(b)** $z = -1 - i\sqrt{3}$.

Solution

(a)

$$z = 1 + i\sqrt{3}$$

$$|z| = \sqrt{1^2 + (\sqrt{3})^2} = \sqrt{1+3} = \sqrt{4} = 2$$

Now

$$\tan^{-1}\left(\frac{\sqrt{3}}{1}\right) = \frac{\pi}{3} + k\pi, \qquad k = 0, \pm 1, \ldots$$

Since $1 + i\sqrt{3}$ is in the first quadrant (see Figure 6-3),

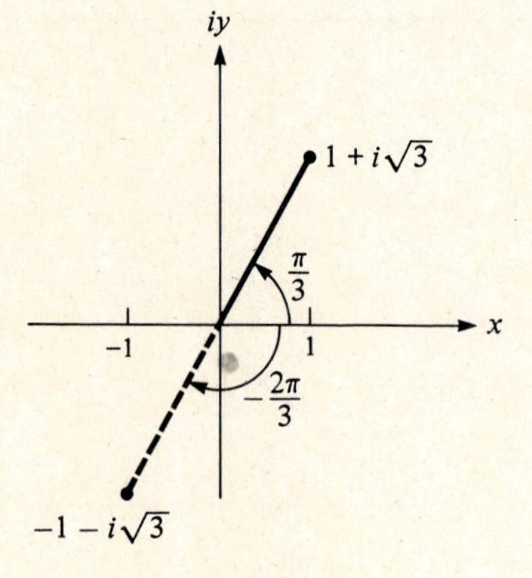

FIGURE 6-3

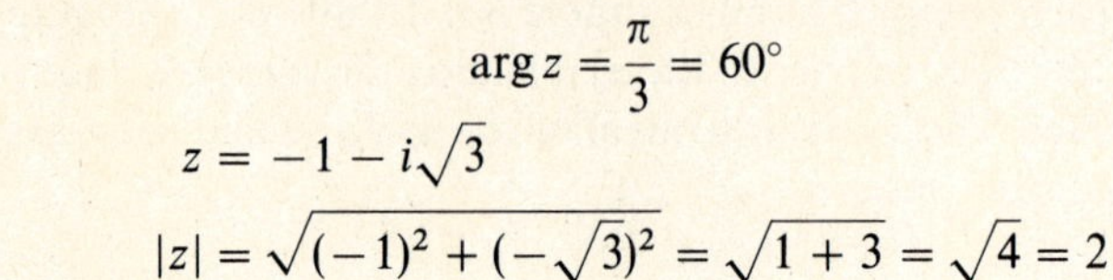

$$\arg z = \frac{\pi}{3} = 60°$$

(b)

$$z = -1 - i\sqrt{3}$$

$$|z| = \sqrt{(-1)^2 + (-\sqrt{3})^2} = \sqrt{1+3} = \sqrt{4} = 2$$

Now

$$\tan^{-1}\left(\frac{-\sqrt{3}}{-1}\right) = \tan^{-1}(\sqrt{3}) = \frac{\pi}{3} + k\pi, \qquad k = 0, \pm 1, \ldots$$

Since $-1 - i\sqrt{3}$ is in the third quadrant (see Figure 6-3),

$$\arg z = -\frac{2}{3}\pi = -120°$$

F. Conjugate of a complex number

We define the **conjugate** of $z = x + iy$ as the complex number $x - iy$ and denote it by $\bar{z}$:

$$\bar{z} = x - iy \tag{6.18}$$

In its polar form (see Figure 6-4),

$$\bar{z} = r(\cos\theta - i\sin\theta) = r[\cos(-\theta) + i\sin(-\theta)] \tag{6.19}$$

or

$$\bar{z} = re^{-i\theta} \tag{6.20}$$

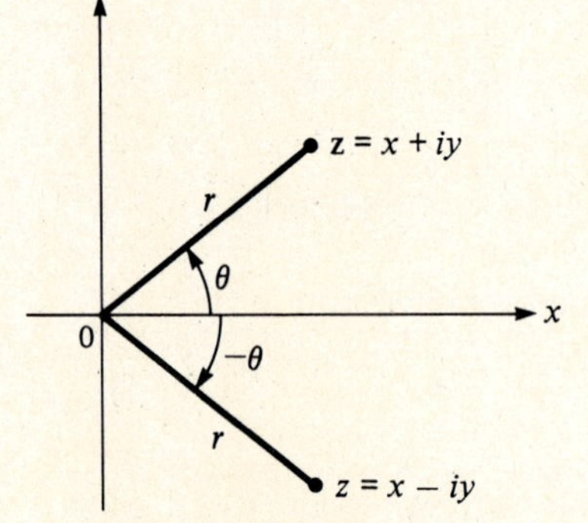

FIGURE 6-4. Conjugate of *z*.

We have the following properties involving conjugates:

(1) $\overline{z_1 + z_2} = \bar{z}_1 + \bar{z}_2$ **(6.21)**

(2) $\overline{z_1 z_2} = \bar{z}_1 \bar{z}_2$ **(6.22)**

(3) $z + \bar{z} = 2\mathcal{Re}(z)$ **(6.23)**

(4) $z - \bar{z} = 2i\,\mathcal{Im}(z)$ **(6.24)**

(5) $\bar{z} = z$ if and only if z is real **(6.25)**

(6) $z\bar{z} = |z|^2$ **(6.26)**

(7) $\bar{\bar{z}} = z$ **(6.27)**

With the properties of fundamental operations on complex numbers, (6.21) can be extended to n terms and (6.22) to n factors; that is,

(8) $\overline{z_1 + z_2 + \cdots + z_n} = \bar{z}_1 + \bar{z}_2 + \cdots + \bar{z}_n$ **(6.28)**

(9) $\overline{z_1 z_2 \ldots z_n} = \bar{z}_1 \bar{z}_2 \ldots . \bar{z}_n$ **(6.29)**

EXAMPLE 6-4: Verify (**a**) property (6.21), (**b**) property (6.23), and (**c**) property (6.26).

Solution

(**a**) If $z_1 = x_1 + iy_1$, $z_2 = x_2 + iy_2$, then

$$z_1 + z_2 = (x_1 + x_2) + i(y_1 + y_2)$$

and

$$\overline{z_1 + z_2} = (x_1 + x_2) - i(y_1 + y_2) = x_1 - iy_1 + x_2 - iy_2 = \bar{z}_1 + \bar{z}_2$$

Thus property (6.21) is verified.

(**b**) If $z = x + iy$, then $\bar{z} = x - iy$. Thus

$$z + \bar{z} = x + iy + x - iy = 2x = 2\mathcal{R}e(z)$$

Thus property (6.23) is verified.

(**c**)
$$\begin{aligned} z\bar{z} &= (x + iy)(x - iy) \\ &= x^2 + iyx - ixy - i^2y = x^2 + y^2 = |z|^2 \end{aligned}$$

Thus property (6.26) is verified.

EXAMPLE 6-5: Show that if z is a root of

$$a_n z^n + a_{n-1} z^{n-1} + \cdots + a_1 z + a_0 = 0$$

where $a_n, a_{n-1}, \ldots, a_0$ are real, then $\bar{z}$ is also a root of this equation.

Solution

$$\begin{aligned} a_n\bar{z}^n + a_{n-1}\bar{z}^{n-1} + \cdots + a_1\bar{z} + a_0 &= \bar{a}_n\bar{z}^n + \bar{a}_{n-1}\bar{z}^{n-1} + \cdots + \bar{a}_1\bar{z} + \bar{a}_0 && \text{[by (6.25)]} \\ &= \overline{a_n z^n} + \overline{a_{n-1} z^{n-1}} + \cdots + \overline{a_1 z} + \bar{a}_0 && \text{[by (6.22)]} \\ &= \overline{a_n z^n + a_{n-1} z^{n-1} + \cdots + a_1 z + a_0} && \text{[by (6.28)]} \\ &= \bar{0} = 0 && \text{[by (6.25)]} \end{aligned}$$

Thus $\bar{z}$ is also a root of the given equation.

6-2. Complex Vector Spaces

In the definition of a real vector space in Chapter 3, by a *scalar* we mean a real number. When the set R of real numbers (scalars) is replaced by the set C of complex numbers in the axioms for a vector space, we say that V is a **complex vector space** or a **vector space over the complex numbers.**

A. Examples of complex vector spaces

1. The vector space of *n*-tuples of complex numbers: C^n

A vector $\mathbf{v} \in C^n$ can be written as

$$\mathbf{v} = \begin{bmatrix} v_1 \\ v_2 \\ \vdots \\ v_n \end{bmatrix} \qquad \text{where } v_1 = x_1 + iy_1, v_2 = x_2 + iy_2, \ldots, v_n = x_n + iy_n$$

with addition and scalar multiplication performed coordinatewise; that is,

$$\begin{bmatrix} v_1 \\ v_2 \\ \vdots \\ v_n \end{bmatrix} + \begin{bmatrix} u_1 \\ u_2 \\ \vdots \\ u_n \end{bmatrix} = \begin{bmatrix} v_1 + u_1 \\ v_2 + u_2 \\ \vdots \\ v_n + u_n \end{bmatrix} \qquad \textbf{(6.30)}$$

and

$$\alpha \begin{bmatrix} v_1 \\ v_2 \\ \vdots \\ v_n \end{bmatrix} = \begin{bmatrix} \alpha v_1 \\ \alpha v_2 \\ \vdots \\ \alpha v_n \end{bmatrix} \qquad \textbf{(6.31)}$$

where v_j, u_j are complex numbers and α is a complex scalar.

In C^n, as in R^n, the vectors

$$\mathbf{e}_1 = \begin{bmatrix} 1 \\ 0 \\ \vdots \\ 0 \end{bmatrix}, \mathbf{e}_2 = \begin{bmatrix} 0 \\ 1 \\ \vdots \\ 0 \end{bmatrix}, \ldots, \mathbf{e}_n = \begin{bmatrix} 0 \\ 0 \\ \vdots \\ 1 \end{bmatrix}$$

form a basis that is called the **standard basis** for C^n. Since there are n vectors in this basis, C^n is an n-dimensional vector space.

*2. **Complex $C[a,b]$** [*For readers who have studied calculus.*]

This is the vector space of all complex-valued functions that are continuous on the closed interval $[a,b]$. Complex $C[a,b]$ is the complex analog of the vector space $C[a,b]$ discussed in Example 3-17.

If $f_1(x)$ and $f_2(x)$ are real-valued functions of the real variable x, then

$$f(x) = f_1(x) + if_2(x) \qquad \textbf{(6.32)}$$

is called a complex-valued function of the real variable x. And if $f_1(x)$ and $f_2(x)$ are continuous, then $f(x)$ is said to be continuous.

We also define

$$f'(x) = f_1'(x) + if_2'(x) \qquad \textbf{(6.33)}$$

where the prime denotes the derivative with respect to x, and

$$\int_a^b f(x)\,dx = \int_a^b [f_1(x) + if_2(x)]\,dx = \int_a^b f_1(x)\,dx + i\int_a^b f_2(x)\,dx \qquad \textbf{(6.34)}$$

6-3. Complex Inner Product Spaces

A. Definition

An inner product on a complex vector space V is a function that assigns a complex number $\langle \mathbf{u},\mathbf{v} \rangle$ to each pair $\mathbf{u}$ and $\mathbf{v}$ of vectors in V such that the following axioms are satisfied for all $\mathbf{u}$, $\mathbf{v}$, $\mathbf{w}$ in V and for a complex scalar (number) α.

CI1. $\langle \mathbf{u},\mathbf{v} \rangle = \overline{\langle \mathbf{v},\mathbf{u} \rangle}$ **(6.35)**

CI2. $\langle \mathbf{u},\mathbf{v} + \mathbf{w} \rangle = \langle \mathbf{u},\mathbf{v} \rangle + \langle \mathbf{u},\mathbf{w} \rangle$ **(6.36)**

CI3. $\langle \alpha\mathbf{u},\mathbf{v} \rangle = \alpha\langle \mathbf{u},\mathbf{v} \rangle$ **(6.37)**

CI4. $\langle \mathbf{v},\mathbf{v} \rangle \geq 0$ and $\langle \mathbf{v},\mathbf{v} \rangle = 0$ if and only if $\mathbf{v} = \mathbf{0}$ **(6.38)**

note: The bar stands, as usual, for complex conjugation.

A complex vector space V with an inner product is called the **complex inner product space** or a **unitary space**.

B. Additional properties of an inner product

(1) $\langle \mathbf{0},\mathbf{v} \rangle = 0$ **(6.39)**

(2) $\langle \mathbf{u} + \mathbf{v},\mathbf{w} \rangle = \langle \mathbf{u},\mathbf{w} \rangle + \langle \mathbf{v},\mathbf{w} \rangle$ **(6.40)**

(3) $\langle \mathbf{u},\alpha\mathbf{v} \rangle = \bar{\alpha}\langle \mathbf{u},\mathbf{v} \rangle$ **(6.41)**

It should be observed that it is Axiom CI1 (6.35) and property (6.41) that distinguish a complex inner product from the familiar real inner product.

EXAMPLE 6-6: Verify property (6.41).

Solution

$$\begin{aligned} \langle \mathbf{u},\alpha\mathbf{v}\rangle &= \overline{\langle \alpha\mathbf{v},\mathbf{u}\rangle} && \text{[by (6.35)]} \\ &= \overline{\alpha\langle \mathbf{v},\mathbf{u}\rangle} && \text{[by (6.37)]} \\ &= \bar{\alpha}\overline{\langle \mathbf{v},\mathbf{u}\rangle} && \text{[by (6.22)]} \\ &= \bar{\alpha}\langle \mathbf{u},\mathbf{v}\rangle && \text{[by (6.35)]} \end{aligned}$$

EXAMPLE 6-7: Let

$$\mathbf{u} = \begin{bmatrix} u_1 \\ u_2 \\ \vdots \\ u_n \end{bmatrix} \quad \text{and} \quad \mathbf{v} = \begin{bmatrix} v_1 \\ v_2 \\ \vdots \\ v_n \end{bmatrix}$$

be vectors in C^n. Show that the inner product defined by

$$\langle \mathbf{u},\mathbf{v}\rangle = u_1\bar{v}_1 + u_2\bar{v}_2 + \cdots + u_n\bar{v}_n \tag{6.42}$$

satisfies all the complex inner product axioms.

Solution: Let

$$\mathbf{u} = \begin{bmatrix} u_1 \\ u_2 \\ \vdots \\ u_n \end{bmatrix}, \quad \mathbf{v} = \begin{bmatrix} v_1 \\ v_2 \\ \vdots \\ v_n \end{bmatrix}, \quad \mathbf{w} = \begin{bmatrix} w_1 \\ w_2 \\ \vdots \\ w_n \end{bmatrix}$$

(1) By definition,

$$\langle \mathbf{u},\mathbf{v}\rangle = u_1\bar{v}_1 + u_2\bar{v}_2 + \cdots + u_n\bar{v}_n$$
$$\langle \mathbf{v},\mathbf{u}\rangle = v_1\bar{u}_1 + v_2\bar{u}_2 + \cdots + v_n\bar{u}_n$$

so

$$\begin{aligned} \overline{\langle \mathbf{v},\mathbf{u}\rangle} &= \overline{v_1\bar{u}_1 + v_2\bar{u}_2 + \cdots + v_n\bar{u}_n} \\ &= \bar{v}_1\bar{\bar{u}}_1 + \bar{v}_2\bar{\bar{u}}_2 + \cdots + \bar{v}_n\bar{\bar{u}}_n && \text{[by (6.28) and (6.22)]} \\ &= \bar{v}_1 u_1 + \bar{v}_2 u_2 + \cdots + \bar{v}_n u_n && \text{[by (6.27)]} \\ &= u_1\bar{v}_1 + u_2\bar{v}_2 + \cdots + u_n\bar{v}_n \\ &= \langle \mathbf{u},\mathbf{v}\rangle \end{aligned}$$

Thus Axiom CI1 is satisfied.

(2) Next

$$\begin{aligned} \langle \mathbf{u},\mathbf{v}+\mathbf{w}\rangle &= u_1(\overline{v_1 + w_1}) + u_2(\overline{v_2 + w_2}) + \cdots + u_n(\overline{v_n + w_n}) \\ &= u_1(\bar{v}_1 + \bar{w}_1) + u_2(\bar{v}_2 + \bar{w}_2) + \cdots + u_n(\bar{v}_n + \bar{w}_n) && \text{[by (6.21)]} \\ &= u_1\bar{v}_1 + u_2\bar{v}_2 + \cdots + u_n\bar{v}_n + u_1\bar{w}_1 + u_2\bar{w}_2 + \cdots + u_n\bar{w}_n \\ &= \langle \mathbf{u},\mathbf{v}\rangle + \langle \mathbf{u},\mathbf{w}\rangle \end{aligned}$$

Thus Axiom CI2 is satisfied.

(3) Let α be any complex number; then

$$\begin{aligned}\langle \alpha\mathbf{u},\mathbf{v}\rangle &= \alpha u_1\bar{v}_1 + \alpha u_2\bar{v}_2 + \cdots + \alpha u_n\bar{v}_n\\ &= \alpha\{u_1\bar{v}_1 + u_2\bar{v}_2 + \cdots + u_n\bar{v}_n)\\ &= \alpha\langle \mathbf{u},\mathbf{v}\rangle\end{aligned}$$

Thus Axiom CI3 is satisfied.

(4) Now

$$\begin{aligned}\langle \mathbf{v},\mathbf{v}\rangle &= v_1\bar{v}_1 + v_2\bar{v}_2 + \cdots + v_n\bar{v}_n\\ &= |v_1|^2 + |v_2|^2 + \cdots + |v_n|^2 \geq 0 \qquad \text{[by (6.26)]}\end{aligned}$$

Moreover, the equality holds if and only if $|v_1| = |v_2| = \cdots = |v_n| = 0$. But this is true if and only if $v_1 = v_2 = \cdots = v_n = 0$, that is, if and only if $\mathbf{v} = \mathbf{0}$. So Axiom CI4 is satisfied. The inner product defined by (6.42) is called the **standard inner product** on C^n. The inner product (6.42) can also be written in the form

$$\langle \mathbf{u},\mathbf{v}\rangle = [\bar{v}_1, \bar{v}_2, \ldots, \bar{v}_n]\begin{bmatrix}u_1\\u_2\\\vdots\\u_n\end{bmatrix} = \bar{\mathbf{v}}^T\mathbf{u} = \mathbf{v}^*\mathbf{u} \tag{6.43}$$

where $\mathbf{v}^* = \bar{\mathbf{v}}^T$ represents the **conjugate transpose** of $\mathbf{v}$.

***EXAMPLE 6-8** [*For readers who have studied calculus.*]: Let $\mathbf{f} = f(x) = f_1(x) + if_2(x)$ and $\mathbf{g} = g(x) = g_1(x) + ig_2(x)$ be vectors in complex $C[a, b]$. Show that the formula defined by

$$\langle \mathbf{f},\mathbf{g}\rangle = \int_a^b f(x)\overline{g(x)}\,dx \tag{6.44}$$

satisfies all the complex inner product axioms.

Solution

(1)

$$\begin{aligned}\langle \mathbf{f},\mathbf{g}\rangle &= \int_a^b f(x)\overline{g(x)}\,dx\\ &= \int_a^b [f_1(x) + if_2(x)]\overline{[g_1(x) + ig_2(x)]}\,dx\\ &= \int_a^b [f_1(x) + if_2(x)][g_1(x) - ig_2(x)]\,dx\\ &= \int_a^b [f_1(x)g_1(x) + f_2(x)g_2(x)]\,dx + i\int_a^b [f_2(x)g_1(x) - f_1(x)g_2(x)]\,dx\\ \langle \mathbf{g},\mathbf{f}\rangle &= \int_a^b g(x)\overline{f(x)}\,dx\\ &= \int_a^b [g_1(x) + ig_2(x)]\overline{[f_1(x) + if_2(x)]}\,dx\\ &= \int_a^b [g_1(x) + ig_2(x)][f_1(x) - if_2(x)]\,dx\\ &= \int_a^b [g_1(x)f_1(x) + g_2(x)f_2(x)]\,dx - i\int_a^b [g_1(x)f_2(x) - g_2(x)f_1(x)]\,dx\\ &= \overline{\langle \mathbf{f},\mathbf{g}\rangle}\end{aligned}$$

Thus Axiom CI1 is satisfied.

(2) Let $\mathbf{h} = h(x) = h_1(x) + ih_2(x)$.

$$\langle \mathbf{f},\mathbf{g} + \mathbf{h}\rangle = \int_a^b f(x)\overline{[g(x) + h(x)]}\,dx$$
$$= \int_a^b f(x)[\overline{g(x)} + \overline{h(x)}]\,dx$$
$$= \int_a^b f(x)\overline{g(x)}\,dx + \int_a^b f(x)\overline{h(x)}\,dx$$
$$= \langle \mathbf{f},\mathbf{g}\rangle + \langle \mathbf{f},\mathbf{h}\rangle$$

Thus Axiom CI2 is satisfied.

(3) If α is any scalar,

$$\langle \alpha\mathbf{f},\mathbf{g}\rangle = \int_a^b \alpha f(x)\overline{g(x)}\,dx$$
$$= \alpha\int_a^b f(x)\overline{g(x)}\,dx = \alpha\langle \mathbf{f},\mathbf{g}\rangle$$

Thus Axiom CI3 is satisfied.

(4)
$$\langle \mathbf{f},\mathbf{f}\rangle = \int_a^b f(x)\overline{f(x)}\,dx$$
$$= \int_a^b |f(x)|^2\,dx$$
$$= \int_a^b [f_1^2(x) + f_2^2(x)]\,dx \geq 0$$

The equality holds if and only if the integrand $[f_1^2(x) + f_2^2(x)] = 0$, but this is true if and only if $f_1(x) = f_2(x) = 0$, that is, $\mathbf{f} = f(x) = \mathbf{0}$. Thus Axiom CI4 is satisfied.

C. Norm and distance

In complex inner product spaces, as in real inner product spaces, the *norm* (or *length*) of a vector $\mathbf{v}$ is defined by

$$\|\mathbf{v}\| = \langle \mathbf{v},\mathbf{v}\rangle^{1/2} \tag{6.45}$$

and the distance between two vectors $\mathbf{u}$ and $\mathbf{v}$ is defined by

$$d(\mathbf{u},\mathbf{v}) = \|\mathbf{u} - \mathbf{v}\| \tag{6.46}$$

EXAMPLE 6-9: Let C^2 have the Euclidean inner product of (6.42), that is,

$$\langle \mathbf{u},\mathbf{v}\rangle = u_1\bar{v}_1 + u_2\bar{v}_2 \tag{6.47}$$

Find **(a)** $\|\mathbf{u}\|$, **(b)** $\|\mathbf{v}\|$, and **(c)** $d(\mathbf{u},\mathbf{v})$ if

$$\mathbf{u} = \begin{bmatrix} 1 \\ 1 \end{bmatrix}, \qquad \mathbf{v} = \begin{bmatrix} i \\ -i \end{bmatrix}$$

Solution

(a)
$$\|\mathbf{u}\| = \langle \mathbf{u},\mathbf{u}\rangle^{1/2}$$
$$= (u_1\bar{u}_1 + u_2\bar{u}_2)^{1/2}$$
$$= \sqrt{(1)(1) + (1)(1)}$$
$$= \sqrt{2}$$

(b)

$$\begin{aligned}\|\mathbf{v}\| &= \langle \mathbf{v},\mathbf{v}\rangle^{1/2} \\ &= (v_1\bar{v}_1 + v_2\bar{v}_2)^{1/2} = \sqrt{i(-i)+(-i)(i)} \\ &= \sqrt{-i^2 - i^2} = \sqrt{1+1} = \sqrt{2}\end{aligned}$$

(c) Since

$$\mathbf{u} - \mathbf{v} = \begin{bmatrix} 1 \\ 1 \end{bmatrix} - \begin{bmatrix} i \\ -i \end{bmatrix} = \begin{bmatrix} 1-i \\ 1+i \end{bmatrix}$$

then

$$\begin{aligned} d(\mathbf{u},\mathbf{v}) &= \|\mathbf{u}-\mathbf{v}\| \\ &= \langle \mathbf{u}-\mathbf{v}, \mathbf{u}-\mathbf{v}\rangle^{1/2} \\ &= [(1-i)(\overline{1-i}) + (1+i)(\overline{1+i})]^{1/2} \\ &= \sqrt{(1-i)(1+i)+(1+i)(1-i)} \\ &= \sqrt{(1-i^2)+(1-i^2)} = \sqrt{2+2} = \sqrt{4} = 2 \end{aligned}$$

***EXAMPLE 6-10** [*For readers who have studied calculus.*]: Let complex $C[0,2\pi]$ have the inner product of (6.44), that is,

$$\langle \mathbf{f},\mathbf{g}\rangle = \int_0^{2\pi} f(x)\overline{g(x)}\,dx \tag{6.48}$$

Find $\|\mathbf{f}\|$ if $\mathbf{f} = e^{ikx}$, where k is any integer.

Solution

$$\begin{aligned}\|\mathbf{f}\| = \langle \mathbf{f},\mathbf{f}\rangle^{1/2} &= \left[\int_0^{2\pi} e^{ikx}\overline{e^{ikx}}\,dx\right]^{1/2} \\ &= \left[\int_0^{2\pi} e^{ikx}e^{-ikx}\,dx\right]^{1/2} = \left[\int_0^{2\pi} dx\right]^{1/2} = \sqrt{2\pi}\end{aligned}$$

D. Orthogonality

The concepts of orthogonal vectors, orthogonal set, and orthonormal basis carry over to complex inner product spaces without alteration.

EXAMPLE 6-11: Show that the vectors

$$\mathbf{u} = \begin{bmatrix} 1 \\ i \end{bmatrix} \quad \text{and} \quad \mathbf{v} = \begin{bmatrix} i \\ 1 \end{bmatrix}$$

in C^2 are orthogonal with respect to the inner product of (6.47).

Solution

$$\begin{aligned}\langle \mathbf{u},\mathbf{v}\rangle &= u_1\bar{v}_1 + u_2\bar{v}_2 \\ &= (1)(\bar{i}) + (i)(\bar{1}) = (1)(-i) + (i)(1) \\ &= -i + i = 0\end{aligned}$$

Thus $\mathbf{u}$ and $\mathbf{v}$ are orthogonal.

***EXAMPLE 6-12** [*For readers who have studied calculus.*]: Show that the set

$$\left\{\frac{1}{\sqrt{2\pi}}e^{ikx}, \qquad k = 0, \pm 1, \pm 2, \ldots\right\}$$

forms an orthonormal set in complex $C[0, 2\pi]$.

Solution: Let $\mathbf{f} = e^{ikx}$ and $\mathbf{g} = e^{imx}$, where $k \neq m$. Then

$$\begin{aligned}
\langle \mathbf{f},\mathbf{g} \rangle &= \int_0^{2\pi} e^{ikx}\overline{e^{imx}}\, dx \\
&= \int_0^{2\pi} e^{ikx}e^{-imx}\, dx \\
&= \int_0^{2\pi} e^{i(k-m)x}\, dx \\
&= \int_0^{2\pi} [\cos(k-m)x + i\sin(k-m)x]\, dx \\
&= \int_0^{2\pi} \cos(k-m)x\, dx + i\int_0^{2\pi} \sin(k-m)x\, dx \\
&= \frac{1}{k-m}\sin(k-m)x\Big|_0^{2\pi} - i\frac{1}{k-m}\cos(k-m)x\Big|_0^{2\pi} \\
&= 0 - i0 = 0
\end{aligned}$$

Thus the set $\{e^{ikx}, k = 0, \pm 1, \pm 2, \ldots\}$ is an orthogonal set. But in Example 6-10 we showed that each vector in this set has norm $\sqrt{2\pi}$, so the set

$$\left\{\frac{1}{\sqrt{2\pi}}e^{ikx}, k = 0, \pm 1, \pm 2, \ldots\right\}$$

forms an orthonormal set in complex $C[0,2\pi]$.

6-4. Unitary Transformation and Unitary Matrix

The complex analog of an *orthogonal operator* on a real inner product space is a *unitary operator* on a complex inner product space.

A. Unitary operator

DEFINITION If $T: V \to V$ is a linear operator defined on the complex inner product space V, we say that T is a **unitary operator** if for each $\mathbf{u}$, $\mathbf{v}$ in V, we have

$$\langle T(\mathbf{u}), T(\mathbf{v}) \rangle = \langle \mathbf{u},\mathbf{v} \rangle \qquad \textbf{(6.49)}$$

that is, T preserves inner products.

As shown in Theorem 5-6.1, the equivalent condition of (6.49) is

$$\|T(\mathbf{u})\| = \|\mathbf{u}\| \qquad \textbf{(6.50)}$$

EXAMPLE 6-13: Let $T: C^n \to C^n$ be a unitary operator on C^n. If A is the matrix of T relative to the standard basis of C^n, show that $A^*A = I_n$ where $A^* = \bar{A}^T$ is the conjugate transpose of A.

Solution: If A is the matrix of T relative to the standard basis of C^n, then

$$T\mathbf{u} = A\mathbf{u} \quad \text{and} \quad T\mathbf{v} = A\mathbf{v} \quad \text{for } \mathbf{u}, \mathbf{v} \in C^n$$

and

$$\begin{aligned} \langle T\mathbf{u}, T\mathbf{v} \rangle &= \langle A\mathbf{u}, A\mathbf{v} \rangle \\ &= (A\mathbf{v})^* A\mathbf{u} \qquad \text{[by (6.43)]} \\ &= \mathbf{v}^* A^* A\mathbf{u} \qquad \text{[by (1.43) and (6.22)]} \end{aligned}$$

Since T is unitary, we get

$$\langle T\mathbf{u}, T\mathbf{v} \rangle = \langle \mathbf{u}, \mathbf{v} \rangle = \mathbf{v}^*\mathbf{u} = \mathbf{v}^* A^* A\mathbf{u} \tag{6.51}$$

for all $\mathbf{u}$ and $\mathbf{v}$. Thus it follows that $A^*A = I_n$.

note: The matrix A^*, the complex conjugate transpose of a matrix A, is called the **adjoint** of A.

B. Unitary matrix

DEFINITION *A square matrix A is called a* **unitary matrix** *if*

UNITARY MATRIX

$$AA^* = A^*A = I_n \tag{6.52}$$

where $A^* = \bar{A}^T$ is the conjugate transpose of A.

Observe that a unitary matrix with real entries is an orthogonal matrix. From (6.52) it follows that if A is unitary, then

(1) $A^* = A^{-1}$ **(6.53)**

(2) $\det(A^*) = \pm 1$ **(6.54)**

(3) $\langle A\mathbf{x}, A\mathbf{y} \rangle = \langle \mathbf{x}, \mathbf{y} \rangle, \quad \mathbf{x}, \mathbf{y} \in C^n$ **(6.55)**

(4) $\|A\mathbf{x}\|^2 = \|\mathbf{x}\|^2, \quad \mathbf{x} \in C^n$ **(6.56)**

(5) The column vectors of A are orthonormal.

(6) The row vectors of A are orthonormal.

EXAMPLE 6-14: If $|a|^2 + |b|^2 = 1$, show that the matrix

$$A = \begin{bmatrix} a & b \\ -\bar{b} & \bar{a} \end{bmatrix}$$

is unitary.

Solution

$$A^* = \bar{A}^T = \begin{bmatrix} \bar{a} & \bar{b} \\ -\bar{\bar{b}} & \bar{\bar{a}} \end{bmatrix}^T = \begin{bmatrix} \bar{a} & \bar{b} \\ -b & a \end{bmatrix}^T = \begin{bmatrix} \bar{a} & -b \\ \bar{b} & a \end{bmatrix}$$

$$A^*A = \begin{bmatrix} \bar{a} & -b \\ \bar{b} & a \end{bmatrix}\begin{bmatrix} a & b \\ -\bar{b} & \bar{a} \end{bmatrix} = \begin{bmatrix} \bar{a}a + b\bar{b} & \bar{a}b - b\bar{a} \\ \bar{b}a - a\bar{b} & \bar{b}b + a\bar{a} \end{bmatrix}$$

$$= \begin{bmatrix} |a|^2 + |b|^2 & 0 \\ 0 & |b|^2 + |a|^2 \end{bmatrix} = \begin{bmatrix} 1 & 0 \\ 0 & 1 \end{bmatrix} = I_2$$

$$AA^* = \begin{bmatrix} a & b \\ -\bar{b} & \bar{a} \end{bmatrix}\begin{bmatrix} \bar{a} & -b \\ \bar{b} & a \end{bmatrix} = \begin{bmatrix} a\bar{a} + b\bar{b} & -ab + ba \\ -\bar{b}\bar{a} + \bar{a}\bar{b} & \bar{b}b + \bar{a}a \end{bmatrix}$$

$$= \begin{bmatrix} |a|^2 + |b|^2 & 0 \\ 0 & |b|^2 + |a|^2 \end{bmatrix} = \begin{bmatrix} 1 & 0 \\ 0 & 1 \end{bmatrix} = I_2$$

Thus, $AA^* = A^*A = I_2$, and by (6.52) A is unitary.

SUMMARY

1. A complex number z is an ordered pair of real numbers x,y denoted by $z = x + iy$, that satisfies the following laws of operation:

   Let $z_1 = x_1 + iy_1, \qquad z_2 = x_2 + iy_2$

   (1) $z_1 = z_2 \Rightarrow x_1 = x_2, \quad y_1 = y_2$

   (2) $z_1 + z_2 = (x_1 + x_2) + i(y_1 + y_2)$

   (3) $z_1 - z_2 = (x_1 - x_2) + i(y_1 - y_2)$

   (4) $kz = k(x + iy) = kx + i(ky)$, k is a real number

   (5) $z_1 z_2 = (x_1 x_2 - y_1 y_2) + i(x_1 y_2 + x_2 y_1)$

2. The polar form of $z = x + iy$ is

$$z = r(\cos\theta + i\sin\theta) = re^{i\theta}$$

   where

$$r = |z| = \sqrt{x^2 + y^2}, \qquad \theta = \arg z = \tan^{-1}\frac{y}{x}$$

3. The conjugate of $z = x + iy$, denoted by $\bar{z}$, is

$$\bar{z} = x - iy$$
$$\bar{z} = re^{-i\theta}$$

4. The properties involving conjugates are

   (1) $\overline{z_1 + z_2} = \bar{z}_1 + \bar{z}_2$

   (2) $\overline{z_1 z_2} = \bar{z}_1 \bar{z}_2$

   (3) $z + \bar{z} = 2\mathcal{Re}(z)$

   (4) $z - \bar{z} = 2i\,\mathcal{Im}(z)$

   (5) $\bar{z} = z$ if and only if z is real

   (6) $z\bar{z} = |z|^2$

   (7) $\bar{\bar{z}} = z$

5. A complex vector space is a vector space defined over complex numbers.
6. A complex inner product space is a vector space V over complex numbers with an inner product $\langle \mathbf{u},\mathbf{v} \rangle$ defined by the following axioms:

   CI1. $\langle \mathbf{u},\mathbf{v} \rangle = \overline{\langle \mathbf{v},\mathbf{u} \rangle}$

   CI2. $\langle \mathbf{u},\mathbf{v} + \mathbf{w} \rangle = \langle \mathbf{u},\mathbf{v} \rangle + \langle \mathbf{u},\mathbf{w} \rangle$

   CI3. $\langle \alpha\mathbf{u},\mathbf{v} \rangle = \alpha\langle \mathbf{u},\mathbf{v} \rangle$

   CI4. $\langle \mathbf{v},\mathbf{v} \rangle \geq 0$ and $\langle \mathbf{v},\mathbf{v} \rangle = 0$ if and only if $\mathbf{v} = \mathbf{0}$

7. If

$$\mathbf{u} = \begin{bmatrix} u_1 \\ u_2 \\ \vdots \\ u_n \end{bmatrix}, \qquad \mathbf{v} = \begin{bmatrix} v_1 \\ v_2 \\ \vdots \\ v_n \end{bmatrix} \in C^n$$

   then the standard inner product is defined by

$$\langle \mathbf{u},\mathbf{v} \rangle = u_1\bar{v}_1 + u_2\bar{v}_2 + \cdots + u_n\bar{v}_n$$

8. If complex functions $f(x), g(x) \in C[a,b]$, then the standard inner product is defined by

$$\langle \mathbf{f},\mathbf{g} \rangle = \int_a^b f(x)\overline{g(x)}\,dx$$

9. The norm of $\mathbf{v} \in V$, $\|\mathbf{v}\|$ is defined by

$$\|\mathbf{v}\| = \langle \mathbf{v},\mathbf{v} \rangle^{1/2}$$

10. The distance between $\mathbf{u}$ and $\mathbf{v} \in V$, $d(\mathbf{u},\mathbf{v})$, is defined by

$$d(\mathbf{u},\mathbf{v}) = \|\mathbf{v} - \mathbf{u}\|$$

11. If $\langle \mathbf{u},\mathbf{v} \rangle = 0$, then $\mathbf{u} \perp \mathbf{v}$.

12. The set $\mathbf{u}_1, \mathbf{u}_2, \ldots, \mathbf{u}_n$ is called orthonormal if

$$\langle \mathbf{u}_i,\mathbf{u}_j \rangle = \delta_{ij} = \begin{Bmatrix} 1 & i = j \\ 0 & i \neq j \end{Bmatrix}$$

13. If $T\colon V \to V$ is a linear operator on a complex inner product space V, T is unitary if

$$\langle T(\mathbf{u}),T(\mathbf{v}) \rangle = \langle \mathbf{u},\mathbf{v} \rangle$$

or

$$\|T(\mathbf{u})\| = \|\mathbf{u}\|$$

The unitary operator is the complex analog of the orthogonal operator on a real inner product space.

14. A square matrix A is called unitary if

$$AA^* = A^*A = I_n$$

where $A^* = \bar{A}^T$ is the conjugate transpose of A. Then

(1) $A^* = A^{-1}$

(2) $\det A^* = \pm 1$

(3) $\langle A\mathbf{x},A\mathbf{y} \rangle = \langle \mathbf{x},\mathbf{y} \rangle$, $\mathbf{x}, \mathbf{y} \in C^n$

(4) $\|A\mathbf{x}\|^2 = \|\mathbf{x}\|^2$, $\mathbf{x} \in C^n$

(5) The column vectors of A are orthonormal

(6) The vectors of A are orthonormal

RAISE YOUR GRADES

Can you explain ...?

☑ the imaginary unit of a complex number
☑ how to add two complex numbers
☑ how to multiply two complex numbers
☑ the properties of complex conjugates
☑ the main difference between the real inner product and the complex inner product
☑ a unitary operator
☑ a unitary matrix

SOLVED PROBLEMS

Complex Numbers

PROBLEM 6-1 Let $z_1 = x_1 + iy_1$ and $z_2 = x_2 + iy_2$. Show that

$$\frac{z_1}{z_2} = \frac{x_1x_2 + y_1y_2}{x_2^2 + y_2^2} + i\frac{x_2y_1 - x_1y_2}{x_2^2 + y_2^2}, \qquad z_2 \neq 0$$

Solution

$$\frac{z_1}{z_2} = \frac{x_1 + iy_1}{x_2 + iy_2} = \frac{(x_1 + iy_1)(x_2 - iy_2)}{(x_2 + iy_2)(x_2 - iy_2)}$$

Now

$$\begin{aligned}
(x_1 + iy_1)(x_2 - iy_2) &= x_1x_2 + (iy_1)(-iy_2) + x_1(-iy_2) + (iy_1)x_2 \\
&= x_1x_2 - i^2y_1y_2 - ix_1y_2 + ix_2y_1 \\
&= x_1x_2 - (-1)y_1y_2 + i(x_2y_1 - x_1y_2) \\
&= x_1x_2 + y_1y_2 + i(x_2y_1 - x_1y_2) \\
(x_2 + iy_2)(x_2 - iy_2) &= x_2^2 + (iy_2)(-iy_2) + x_2(-iy_2) + (iy_2)x_2 \\
&= x_2^2 - (i)^2y_2^2 + i(y_2x_2 - x_2y_2) \\
&= x_2^2 - (-1)y_2^2 + i0 \\
&= x_2^2 + y_2^2
\end{aligned}$$

Thus

$$\begin{aligned}
\frac{x_1 + iy_1}{x_2 + iy_2} &= \frac{x_1x_2 + y_1y_2 + i(x_2y_1 - x_1y_2)}{x_2^2 + y_2^2} \\
&= \frac{x_1x_2 + y_1y_2}{x_2^2 + y_2^2} + i\frac{(x_2y_1 - x_1y_2)}{x_2^2 + y_2^2}
\end{aligned}$$

PROBLEM 6-2 Given

$$z_1 = 2\left(\cos\frac{\pi}{3} + i\sin\frac{\pi}{3}\right), \qquad z_2 = \sqrt{2}\left(\cos\frac{\pi}{4} - i\sin\frac{\pi}{4}\right)$$

Find **(a)** $z_1 + z_2$ and **(b)** z_1z_2.

Solution

(a)

$$\begin{aligned}
z_1 &= 2\cos\frac{\pi}{3} + i2\sin\frac{\pi}{3} \\
&= 2\left(\frac{1}{2}\right) + i2\left(\frac{\sqrt{3}}{2}\right) \\
&= 1 + i\sqrt{3}
\end{aligned}$$

and

$$\begin{aligned} z_2 &= \sqrt{2}\cos\frac{\pi}{4} + i\sqrt{2}\sin\frac{\pi}{4} \\ &= \sqrt{2}\left(\frac{1}{\sqrt{2}}\right) + i\sqrt{2}\left(\frac{1}{\sqrt{2}}\right) \\ &= 1 + i \end{aligned}$$

Thus,

$$\begin{aligned} z_1 + z_2 &= 1 + i\sqrt{3} + 1 + i \\ &= 2 + i(1 + \sqrt{3}) \end{aligned}$$

(b)

$$\begin{aligned} z_1 z_2 &= (1 + i\sqrt{3})(1 + i) \\ &= 1 + i\sqrt{3} + i + i^2\sqrt{3} \\ &= 1 + i(\sqrt{3} + 1) - \sqrt{3} \\ &= (1 - \sqrt{3}) + i(1 + \sqrt{3}) \end{aligned}$$

PROBLEM 6-3 Verify **DeMoivre's formula**, that is,

$$(\cos\theta + i\sin\theta)^n = \cos(n\theta) + i\sin(n\theta)$$

Solution Using (6.16), we have

$$(\cos\theta + i\sin\theta)^n = (e^{i\theta})^n = e^{in\theta} = \cos(n\theta) + i\sin(n\theta)$$

PROBLEM 6-4 Verify properties **(a)** (6.22), **(b)** (6.24), **(c)** (6.25), and **(d)** (6.27).

Solution

(a) Let $z_1 = x_1 + iy_1$ and $z_2 = x_2 + iy_2$. Then by (6.10)

$$z_1 z_2 = (x_1 x_2 - y_1 y_2) + i(x_1 y_2 + x_2 y_1)$$

Thus

$$\overline{z_1 z_2} = (x_1 x_2 - y_1 y_2) - i(x_1 y_2 + x_2 y_1)$$

Next,

$$\begin{aligned} \bar{z}_1 \bar{z}_2 &= (x_1 - iy_1)(x_2 - iy_2) \\ &= x_1 x_2 - iy_1 x_2 - ix_1 y_2 + i^2 y_1 y_2 \\ &= x_1 x_2 - y_1 y_2 - i(x_1 y_2 + x_2 y_1) \\ &= \overline{z_1 z_2} \end{aligned}$$

Thus property (6.22) is verified.

(b) If $z = x + iy$, then $\bar{z} = x - iy$. Thus

$$\begin{aligned} z - \bar{z} &= (x + iy) - (x - iy) \\ &= 2iy = 2i\,\mathscr{Im}(z) \end{aligned}$$

Thus property (6.24) is verified.

(c) If $\bar{z} = z$, then $x + iy = x' - iy$. By (6.5) we get $x = x$, $y = -y$, which is true if and only if $y = 0$. Thus if $\bar{z} = z$, then $z = x + i0$, that is, z is real. So property (6.25) is verified.

(d) If $\bar{z} = x - iy$, then $\bar{\bar{z}} = \overline{x - iy} = x + iy = z$. So property (6.27) is verified.

PROBLEM 6-5 Let A and B be complex matrices whose entries are complex numbers and let α be a complex number. Show that

(a) $\overline{A + B} = \bar{A} + \bar{B}$ **(b)** $\overline{\alpha A} = \bar{\alpha}\bar{A}$ **(c)** $\overline{AB} = \bar{A}\bar{B}$

where $\bar{A}$ is the complex conjugate of A (that is, if $A = [a_{ij}]$, then $\bar{A} = [\bar{a}_{ij}]$), and $\bar{B}$ and $\overline{A + B}$ have similar definitions.

Solution

(a) Let $A = [a_{ij}]$ and $B = [b_{ij}]$. Then $A + B = [a_{ij} + b_{ij}]$. Thus

$$\begin{aligned}\overline{A + B} &= [\overline{a_{ij} + b_{ij}}] \\ &= [\bar{a}_{ij} + \bar{b}_{ij}] \qquad \text{[by (6.21)]} \\ &= [\bar{a}_{ij}] + [\bar{b}_{ij}] = \bar{A} + \bar{B}\end{aligned}$$

(b) $\alpha A = [\alpha a_{ij}]$. Thus

$$\begin{aligned}\overline{\alpha A} &= [\overline{\alpha a_{ij}}] \\ &= [\bar{\alpha}\bar{a}_{ij}] \qquad \text{[by (6.22)]} \\ &= \bar{\alpha}[\bar{a}_{ij}] = \bar{\alpha}\bar{A}\end{aligned}$$

(c) Let $A = [a_{ij}]_{n \times m}$ and $B = [b_{ij}]_{m \times p}$. Then $\bar{A} = [\bar{a}_{ij}]_{n \times m}$ and $\bar{B} = [\bar{b}_{ij}]_{m \times p}$. Let $AB = C = [c_{ik}]_{n \times p}$. Then

$$c_{ik} = \sum_{j=1}^{m} a_{ij} b_{jk} \qquad \text{[by (1.18)]}$$

and

$$\begin{aligned}\bar{c}_{ik} &= \overline{\sum_{j=1}^{m} a_{ij} b_{jk}} = \sum_{j=1}^{m} \overline{a_{ij} b_{jk}} \qquad \text{[by (6.28)]} \\ &= \sum_{j=1}^{m} \bar{a}_{ij} \bar{b}_{jk} \qquad \text{[by (6.22)]}\end{aligned}$$

Thus it follows that $\overline{AB} = \bar{A}\bar{B}$.

Complex Vector Spaces

PROBLEM 6-6 Let

$$\mathbf{v}_1 = \begin{bmatrix} 1 \\ 1 + i \end{bmatrix}, \qquad \mathbf{v}_2 = \begin{bmatrix} 1 - i \\ 0 \end{bmatrix}$$

Find α_1 and α_2 such that

$$\alpha_1 \mathbf{v}_1 + \alpha_2 \mathbf{v}_2 = \begin{bmatrix} -1 - i \\ 3 + i3 \end{bmatrix}$$

Solution

$$\begin{aligned}\alpha_1 \mathbf{v}_1 + \alpha_2 \mathbf{v}_2 &= \alpha_1 \begin{bmatrix} 1 \\ 1 + i \end{bmatrix} + \alpha_2 \begin{bmatrix} 1 - i \\ 0 \end{bmatrix} \\ &= \begin{bmatrix} \alpha_1 \\ \alpha_1 + i\alpha_1 \end{bmatrix} + \begin{bmatrix} \alpha_2 - i\alpha_2 \\ 0 \end{bmatrix} = \begin{bmatrix} \alpha_1 + \alpha_2 - i\alpha_2 \\ \alpha_1 + i\alpha_1 \end{bmatrix} = \begin{bmatrix} -1 - i \\ 3 + i3 \end{bmatrix}\end{aligned}$$

Thus

$$\begin{aligned}\alpha_1 + \alpha_2 - i\alpha_2 &= -1 - i \\ \alpha_1 + i\alpha_1 &= 3 + i3\end{aligned}$$

From the last equation, $\alpha_1 = 3$.

Substituting $\alpha_1 = 3$ into the first equation, we get

$$3 + \alpha_2 - i\alpha_2 = -1 - i$$

or

$$\alpha_2 - i\alpha_2 = -4 - i$$

Let $\alpha_2 = x_2 + iy_2$. Then

$$(x_2 + iy_2) - i(x_2 + iy_2) = -4 - i$$

or

$$x_2 + y_2 - i(x_2 - y_2) = -4 - i$$

Thus

$$x_2 + y_2 = -4$$

$$x_2 - y_2 = 1$$

Solving for x_2 and y_2, we get $x_2 = -\frac{3}{2}$ and $y_2 = -\frac{5}{2}$. Hence $\alpha_2 = -\frac{3}{2} - i\frac{5}{2}$.

PROBLEM 6-7 Show that $\mathbf{v}_1$ and $\mathbf{v}_2$ of Problem 6-6 are linearly independent.

Solution Let α_1, α_2 be any scalars and suppose

$$\alpha_1\mathbf{v}_1 + \alpha_2\mathbf{v}_2 = \mathbf{0}$$

or

$$\alpha_1\begin{bmatrix} 1 \\ 1+i \end{bmatrix} + \alpha_2\begin{bmatrix} 1-i \\ 0 \end{bmatrix} = \begin{bmatrix} 0 \\ 0 \end{bmatrix}$$

or

$$\begin{bmatrix} \alpha_1 + \alpha_2 - i\alpha_2 \\ (1+i)\alpha_1 \end{bmatrix} = \begin{bmatrix} 0 \\ 0 \end{bmatrix}$$

Equating each component, we get

$$\alpha_1 + \alpha_2 - i\alpha_2 = 0$$

$$(1+i)\alpha_1 = 0$$

From the last equation $\alpha_1 = 0$ since $(1 + i) \neq 0$. Substituting $\alpha_1 = 0$ into the first equation, we get $(1 - i)\alpha_2 = 0$ and $\alpha_2 = 0$ since $(1 - i) \neq 0$. Since $\alpha_1\mathbf{v}_1 + \alpha_2\mathbf{v}_2 = \mathbf{0}$ implies $\alpha_1 = \alpha_2 = 0$, hence $\mathbf{v}_1$ and $\mathbf{v}_2$ are linearly independent.

PROBLEM 6-8 Show that the following A and B are linearly dependent vectors in complex $C_{2\times 2}$, where $C_{2\times 2}$ denotes the set of all complex 2×2 matrices:

$$A = \begin{bmatrix} i & i2 \\ -i & 0 \end{bmatrix} \quad \text{and} \quad B = \begin{bmatrix} 1 & 2 \\ -1 & 0 \end{bmatrix}$$

Solution By inspection, we see that

$$A = iB$$

Thus A and B are linearly dependent.

Complex Inner Product Spaces

PROBLEM 6-9 Let

$$\mathbf{u} = \begin{bmatrix} 1 \\ i \\ 1-i \end{bmatrix}, \qquad \mathbf{v} = \begin{bmatrix} -i \\ 2i \\ 1+i \end{bmatrix}$$

Find $\langle \mathbf{u}, \mathbf{v} \rangle$.

Solution Using (6.43), we get

$$\langle \mathbf{u},\mathbf{v}\rangle = \mathbf{v}^*\mathbf{u} = [i,-2i,1-i]\begin{bmatrix} 1 \\ i \\ 1-i \end{bmatrix}$$

$$= (i)(1) + (-2i)(i) + (1-i)(1-i)$$

$$= i - 2i^2 + 1 - 2i + i^2$$

$$= i + 2 + 1 - 2i - 1 = 2 - i$$

PROBLEM 6-10 Let

$$A = \begin{bmatrix} a_1 & a_2 \\ a_3 & a_4 \end{bmatrix} \quad \text{and} \quad B = \begin{bmatrix} b_1 & b_2 \\ b_3 & b_4 \end{bmatrix}$$

be any two complex matrices in $C_{2\times 2}$. Show that

$$\langle A,B\rangle = a_1\bar{b}_1 + a_2\bar{b}_2 + a_3\bar{b}_3 + a_4\bar{b}_4 \qquad \textbf{(a)}$$

defines an inner product on $C_{2\times 2}$.

Solution

(1)

$$\overline{\langle B,A\rangle} = \overline{b_1\bar{a}_1 + b_2\bar{a}_2 + b_3\bar{a}_3 + b_4\bar{a}_4}$$

$$= \overline{b_1\bar{a}_1} + \overline{b_2\bar{a}_2} + \overline{b_3\bar{a}_3} + \overline{b_4\bar{a}_4}$$

$$= \bar{b}_1\bar{\bar{a}}_1 + \bar{b}_2\bar{\bar{a}}_2 + \bar{b}_3\bar{\bar{a}}_3 + \bar{b}_4\bar{\bar{a}}_4$$

$$= \bar{b}_1 a_1 + \bar{b}_2 a_2 + \bar{b}_3 a_3 + \bar{b}_4 a_4$$

$$= a_1\bar{b}_1 + a_2\bar{b}_2 + a_3\bar{b}_3 + a_4\bar{b}_4 = \langle A,B\rangle$$

So Axiom CI1 is satisfied.

(2) If

$$C = \begin{bmatrix} c_1 & c_2 \\ c_3 & c_4 \end{bmatrix} \in C_{2\times 2}$$

then

$$\langle A,B+C\rangle = a_1\overline{(b_1+c_1)} + a_2\overline{(b_2+c_2)} + a_3\overline{(b_3+c_3)} + a_4\overline{(b_4+c_4)}$$

$$= a_1(\bar{b}_1+\bar{c}_1) + a_2(\bar{b}_2+\bar{c}_2) + a_3(\bar{b}_3+\bar{c}_3) + a_4(\bar{b}_4+\bar{c}_4)$$

$$= (a_1\bar{b}_1 + a_2\bar{b}_2 + a_3\bar{b}_3 + a_4\bar{b}_4) + (a_1\bar{c}_1 + a_2\bar{c}_2 + a_3\bar{c}_3 + a_4\bar{c}_4)$$

$$= \langle A,B\rangle + \langle A,C\rangle$$

So Axiom CI2 is satisfied.

(3) If α is any scalar, then

$$\langle \alpha A,B\rangle = \alpha a_1\bar{b}_1 + \alpha a_2\bar{b}_2 + \alpha a_3\bar{b}_3 + \alpha a_4\bar{b}_4$$

$$= \alpha(a_1\bar{b}_1 + a_2\bar{b}_2 + a_3\bar{b}_3 + a_4\bar{b}_4)$$

$$= \alpha\langle A,B\rangle$$

So Axiom CI3 is satisfied.

(4)

$$\langle A,A\rangle = a_1\bar{a}_1 + a_2\bar{a}_2 + a_3\bar{a}_3 + a_4\bar{a}_4$$

$$= |a_1|^2 + |a_2|^2 + |a_3|^2 + |a_4|^2 \geq 0$$

and $\langle A,A\rangle = 0$ if and only if $a_1 = a_2 = a_3 = a_4 = 0$, that is, if and only if $A = O$. So Axiom CI4 is satisfied.

Thus (a) defines an inner product on $C_{2\times 2}$.

PROBLEM 6-11 Use the inner product of Problem 6-10 to find $\|A\|$ if

$$A = \begin{bmatrix} -1 & 1-i \\ 1+i & 2 \end{bmatrix}$$

Solution

$$\begin{aligned}\langle A,A\rangle &= (-1)(\overline{-1}) + (1-i)(\overline{1-i}) + (1+i)(\overline{1+i}) + (2)(\overline{2}) \\ &= (-1)(-1) + (1-i)(1+i) + (1+i)(1-i) + (2)(2) \\ &= 1 + 1 - i^2 + 1 - i^2 + 4 \\ &= 1 + 1 + 1 + 1 + 1 + 4 = 9\end{aligned}$$

Thus, by (6.45),

$$\|A\| = \langle A,A\rangle^{1/2} = \sqrt{9} = 3$$

PROBLEM 6-12 Show that if α is a complex number and $\langle \mathbf{u},\mathbf{v}\rangle$ is an inner product on a complex vector space, then

$$\|\mathbf{u} - \alpha\mathbf{v}\|^2 = \|\mathbf{u}\|^2 - \bar{\alpha}\langle \mathbf{u},\mathbf{v}\rangle - \alpha\langle \overline{\mathbf{u},\mathbf{v}}\rangle + |\alpha|^2\|\mathbf{v}\|^2$$

Solution

$$\begin{aligned}\|\mathbf{u} - \alpha\mathbf{v}\|^2 &= \langle \mathbf{u} - \alpha\mathbf{v},\mathbf{u} - \alpha\mathbf{v}\rangle \\ &= \langle \mathbf{u},\mathbf{u}\rangle - \langle \mathbf{u},\alpha\mathbf{v}\rangle - \langle \alpha\mathbf{v},\mathbf{u}\rangle + \langle \alpha\mathbf{v},\alpha\mathbf{v}\rangle \\ &= \langle \mathbf{u},\mathbf{u}\rangle - \bar{\alpha}\langle \mathbf{u},\mathbf{v}\rangle - \alpha\langle \mathbf{v},\mathbf{u}\rangle + \alpha\bar{\alpha}\langle \mathbf{v},\mathbf{v}\rangle \\ &= \|\mathbf{u}\|^2 - \bar{\alpha}\langle \mathbf{u},\mathbf{v}\rangle - \alpha\langle \overline{\mathbf{u},\mathbf{v}}\rangle + |\alpha|^2\|\mathbf{v}\|^2\end{aligned}$$

PROBLEM 6-13 Prove the following **Cauchy-Schwarz inequality** for complex inner product spaces. If $\mathbf{u}$ and $\mathbf{v}$ are vectors in a complex inner product space, then

$$|\langle \mathbf{u},\mathbf{v}\rangle| \leq \|\mathbf{u}\|\,\|\mathbf{v}\|$$

Solution If $\mathbf{u} = \mathbf{0}$ or $\mathbf{v} = \mathbf{0}$ then $\langle \mathbf{u},\mathbf{v}\rangle = 0$ and then

$$|\langle \mathbf{u},\mathbf{v}\rangle| = \|\mathbf{u}\|\,\|\mathbf{v}\| = 0.$$

Next, from the result of Problem 6-12, we have for any complex α

$$0 \leq \|\mathbf{u} - \alpha\mathbf{v}\|^2 = \|\mathbf{u}\|^2 - \bar{\alpha}\langle \mathbf{u},\mathbf{v}\rangle - \alpha\langle \overline{\mathbf{u},\mathbf{v}}\rangle + |\alpha|^2\|\mathbf{v}\|$$

If $\mathbf{v} \neq \mathbf{0}$ and if $\alpha = \langle \mathbf{u},\mathbf{v}\rangle/\|\mathbf{v}\|^2$ in the above inequality, we get

$$0 \leq \|\mathbf{u}\|^2 - \frac{\langle \overline{\mathbf{u},\mathbf{v}}\rangle}{\|\mathbf{v}\|^2}\langle \mathbf{u},\mathbf{v}\rangle - \frac{\langle \mathbf{u},\mathbf{v}\rangle}{\|\mathbf{v}\|^2}\langle \overline{\mathbf{u},\mathbf{v}}\rangle + \frac{|\langle \mathbf{u},\mathbf{v}\rangle|^2}{\|\mathbf{v}\|^4}\|\mathbf{v}\|^2$$

or

$$0 \leq \|\mathbf{u}\|^2 - \frac{|\langle \mathbf{u},\mathbf{v}\rangle|^2}{\|\mathbf{v}\|^2} - \frac{|\langle \mathbf{u},\mathbf{v}\rangle|^2}{\|\mathbf{v}\|^2} + \frac{|\langle \mathbf{u},\mathbf{v}\rangle|^2}{\|\mathbf{v}\|^2}$$

or

$$0 \leq \|\mathbf{u}\|^2 - \frac{|\langle \mathbf{u},\mathbf{v}\rangle|^2}{\|\mathbf{v}\|^2}$$

Multiplying both sides by $\|\mathbf{v}\|^2$, we get $0 \leq \|\mathbf{u}\|^2\|\mathbf{v}\|^2 - |\langle \mathbf{u},\mathbf{v}\rangle|^2$. Thus $|\langle \mathbf{u},\mathbf{v}\rangle|^2 \leq \|\mathbf{u}\|^2\|\mathbf{v}\|^2$. Taking the square root of both sides, we get $|\langle \mathbf{u},\mathbf{v}\rangle| \leq \|\mathbf{u}\|\,\|\mathbf{v}\|$.

PROBLEM 6-14 Let C^3 have the Euclidean inner product. Use the Gram-Schmidt process to construct an orthonormal basis from the basis

$$\mathbf{v}_1 = \begin{bmatrix} i \\ 0 \\ 0 \end{bmatrix}, \qquad \mathbf{v}_2 = \begin{bmatrix} i \\ i \\ 0 \end{bmatrix}, \qquad \mathbf{v}_3 = \begin{bmatrix} i \\ i \\ i \end{bmatrix}$$

Solution

Step 1:

$$\|\mathbf{v}_1\| = \sqrt{i(-i) + 0 + 0} = \sqrt{-i^2} = \sqrt{1} = 1$$

$$\mathbf{u}_1 = \frac{1}{\|\mathbf{v}_1\|}\mathbf{v}_1 = \begin{bmatrix} i \\ 0 \\ 0 \end{bmatrix}$$

Step 2:

$$\langle \mathbf{v}_2,\mathbf{u}_1 \rangle = \sqrt{i(-i) + i(0) + (0)(0)} = 1$$

$$\mathbf{w}_2 = \mathbf{v}_2 - \langle \mathbf{v}_2,\mathbf{u}_1 \rangle \mathbf{u}_1$$

$$= \begin{bmatrix} i \\ i \\ 0 \end{bmatrix} - \begin{bmatrix} i \\ 0 \\ 0 \end{bmatrix} = \begin{bmatrix} 0 \\ i \\ 0 \end{bmatrix}$$

Step 3:

$$\|\mathbf{w}_2\| = \sqrt{0 + i(-i) + 0} = 1$$

$$\mathbf{u}_2 = \frac{1}{\|w_2\|}\mathbf{w}_2 = \begin{bmatrix} 0 \\ i \\ 0 \end{bmatrix}$$

Step 4:

$$\langle \mathbf{v}_3,\mathbf{u}_1 \rangle = \sqrt{i(-i) + 0 + 0} = 1$$

$$\langle \mathbf{v}_3,\mathbf{u}_2 \rangle = \sqrt{0 + i(-i) + 0} = 1$$

$$\mathbf{w}_3 = \mathbf{v}_3 - \langle \mathbf{v}_3,\mathbf{u}_1 \rangle \mathbf{u}_1 - \langle \mathbf{v}_3,\mathbf{u}_2 \rangle \mathbf{u}_2$$

$$= \begin{bmatrix} i \\ i \\ i \end{bmatrix} - \begin{bmatrix} i \\ 0 \\ 0 \end{bmatrix} - \begin{bmatrix} 0 \\ i \\ 0 \end{bmatrix} = \begin{bmatrix} 0 \\ 0 \\ i \end{bmatrix}$$

Step 5:

$$\|\mathbf{w}_3\| = \sqrt{0 + 0 + i(-i)} = 1$$

$$\mathbf{u}_3 = \frac{1}{\|w_3\|}\mathbf{w}_3 = \begin{bmatrix} 0 \\ 0 \\ i \end{bmatrix}$$

PROBLEM 6-15 Let

$$\mathbf{u} = \begin{bmatrix} 1 \\ i \\ 1 - i \end{bmatrix}, \qquad \mathbf{v} = \begin{bmatrix} -i \\ 2i \\ 1 + i \end{bmatrix}$$

Find the coordinate vectors of **u** and **v** relative to the orthonormal basis of $\mathscr{B} = \{\mathbf{u}_1,\mathbf{u}_2,\mathbf{u}_3\}$ of Problem 6-14; that is

$$\mathbf{u}_1 = \begin{bmatrix} i \\ 0 \\ 0 \end{bmatrix}, \qquad \mathbf{u}_2 = \begin{bmatrix} 0 \\ i \\ 0 \end{bmatrix}, \qquad \mathbf{u}_3 = \begin{bmatrix} 0 \\ 0 \\ i \end{bmatrix}$$

Solution By Theorem 4-5.1 of Section 4-5 we can express

$$\mathbf{u} = \langle \mathbf{u},\mathbf{u}_1 \rangle \mathbf{u}_1 + \langle \mathbf{u},\mathbf{u}_2 \rangle \mathbf{u}_2 + \langle \mathbf{u},\mathbf{u}_3 \rangle \mathbf{u}_3 = \alpha_1 \mathbf{u}_1 + \alpha_2 \mathbf{u}_2 + \alpha_3 \mathbf{u}_3$$

$$\mathbf{v} = \langle \mathbf{v},\mathbf{u}_1 \rangle \mathbf{u}_1 + \langle \mathbf{v},\mathbf{u}_2 \rangle \mathbf{u}_2 + \langle \mathbf{v},\mathbf{u}_3 \rangle \mathbf{u}_3 = \beta_1 \mathbf{u}_1 + \beta_2 \mathbf{u}_2 + \beta_3 \mathbf{u}_3$$

Then the coordinate vectors of **u** and **v** relative to $\mathscr{B}$ are

$$[\mathbf{u}]_{\mathscr{B}} = \begin{bmatrix} \alpha_1 \\ \alpha_2 \\ \alpha_3 \end{bmatrix} \quad \text{and} \quad [\mathbf{v}]_{\mathscr{B}} = \begin{bmatrix} \beta_1 \\ \beta_2 \\ \beta_3 \end{bmatrix}$$

Now

$$\alpha_1 = \langle \mathbf{u},\mathbf{u}_1 \rangle = \mathbf{u}_1^* \mathbf{u} = [-i, 0, 0]\begin{bmatrix} 1 \\ i \\ 1-i \end{bmatrix} = -i$$

$$\alpha_2 = \langle \mathbf{u},\mathbf{u}_2 \rangle = \mathbf{u}_2^* \mathbf{u} = [0, -i, 0]\begin{bmatrix} 1 \\ i \\ 1-i \end{bmatrix} = -i^2 = 1$$

$$\alpha_3 = \langle \mathbf{u},\mathbf{u}_3 \rangle = \mathbf{u}_3^* \mathbf{u} = [0, 0, -i]\begin{bmatrix} 1 \\ i \\ 1-i \end{bmatrix} = -i + i^2 = -1 - i$$

$$\beta_1 = \langle \mathbf{v},\mathbf{u}_1 \rangle = \mathbf{u}_1^* \mathbf{v} = [-i, 0, 0]\begin{bmatrix} -i \\ 2i \\ 1+i \end{bmatrix} = +i^2 = -1$$

$$\beta_2 = \langle \mathbf{v},\mathbf{u}_2 \rangle = \mathbf{u}_2^* \mathbf{v} = [0, -i, 0]\begin{bmatrix} -i \\ 2i \\ 1+i \end{bmatrix} = -2i^2 = 2$$

$$\beta_3 = \langle \mathbf{v},\mathbf{u}_3 \rangle = \mathbf{u}_3^* \mathbf{v} = [0, 0, -i]\begin{bmatrix} -i \\ 2i \\ 1+i \end{bmatrix} = -i - i^2 = 1 - i$$

Thus

$$[\mathbf{u}]_{\mathscr{B}} = \begin{bmatrix} -i \\ 1 \\ -1-i \end{bmatrix}, \qquad [\mathbf{v}]_{\mathscr{B}} = \begin{bmatrix} -1 \\ 2 \\ 1-i \end{bmatrix}$$

PROBLEM 6-16 Let $\mathscr{B} = \{\mathbf{u}_1, \mathbf{u}_2, \ldots, \mathbf{u}_n\}$ be an orthonormal basis for a complex inner product V. If $\mathbf{u}$ and $\mathbf{v}$ are any two vectors in V such that

$$\mathbf{u} = \alpha_1 \mathbf{u}_1 + \alpha_2 \mathbf{u}_2 + \cdots + \alpha_n \mathbf{u}_n$$
$$\mathbf{v} = \beta_1 \mathbf{u}_1 + \beta_2 \mathbf{u}_2 + \cdots + \beta_n \mathbf{u}_n$$

show that

$$\langle \mathbf{u},\mathbf{v} \rangle = \alpha_1 \bar{\beta}_1 + \alpha_2 \bar{\beta}_2 + \cdots + \alpha_n \bar{\beta}_n$$

(cf. Example 4-39).

Solution Since $\mathscr{B}$ is an orthonormal basis

$$\langle \mathbf{u}_i,\mathbf{u}_j \rangle = \delta_{ij} = \begin{cases} 1 & i = j \\ 0 & i \neq j \end{cases}$$

we have

$$\begin{aligned} \langle \mathbf{u},\mathbf{u} \rangle &= \left\langle \sum_{i=1}^{n} \alpha_i \mathbf{u}_i, \sum_{j=1}^{n} \beta_j \mathbf{u}_j \right\rangle \\ &= \sum_{i=1}^{n} \sum_{j=1}^{n} \alpha_i \bar{\beta}_j \langle \mathbf{u}_i,\mathbf{u}_j \rangle \\ &= \sum_{i=1}^{n} \sum_{j=1}^{n} \alpha_i \bar{\beta}_j \delta_{ij} \\ &= \sum_{i=1}^{n} \alpha_i \bar{\beta}_i = \alpha_1 \bar{\beta}_1 + \alpha_2 \bar{\beta}_2 + \cdots + \alpha_n \bar{\beta}_n \end{aligned}$$

PROBLEM 6-17 Rework Problem 6-9 using the results of Problems 6-15 and 6-16.

Solution The result of Problem 6-16 can be rewritten as

$$\langle \mathbf{u},\mathbf{v}\rangle = [\bar{\beta}_1, \bar{\beta}_2, \ldots, \bar{\beta}_n]\begin{bmatrix}\alpha_1\\ \alpha_2\\ \vdots\\ \alpha_n\end{bmatrix} = [\mathbf{v}]^*_{\mathscr{B}}[\mathbf{u}]_{\mathscr{B}}$$

where $[\mathbf{u}]_{\mathscr{B}}$ and $[\mathbf{v}]_{\mathscr{B}}$ are the coordinate vectors of **u** and **v**, respectively, relative to the orthonormal basis $\mathscr{B}$, and $[\mathbf{v}]^*_{\mathscr{B}} = [\bar{\mathbf{v}}]^T_{\mathscr{B}}$. Now, from Problem 6-15, we have

$$\begin{aligned}\langle \mathbf{u},\mathbf{v}\rangle &= [\mathbf{v}]^*_{\mathscr{B}}[\mathbf{u}]_{\mathscr{B}}\\ &= [-1, 2, 1+i]\begin{bmatrix}-i\\ 1\\ -1-i\end{bmatrix}\\ &= (-1)(-i) + (2)(1) + (1+i)(-1-i)\\ &= i + 2 - 1 - i - i - i^2\\ &= 2 - i\end{aligned}$$

Unitary Transformation and Unitary Matrix

PROBLEM 6-18 If A and B are complex matrices and α is a complex number, show that

(a) $(A^*)^* = A$

(b) $(A+B)^* = A^* + B^*$

(c) $(\alpha A)^* = \bar{\alpha}A^*$

(d) $(AB)^* = B^*A^*$

Solution

(a) By definition,

$$(A^*)^* = [\overline{(\bar{A})^T}]^T = [(\bar{\bar{A}})^T]^T = [(A)^T]^T = A$$

(b)

$$\begin{aligned}(A+B)^* &= (\overline{A+B})^T\\ &= (\bar{A}+\bar{B})^T && \text{[Problem 6-5(a)]}\\ &= \bar{A}^T + \bar{B}^T\\ &= A^* + B^*\end{aligned}$$

(c)

$$\begin{aligned}(\alpha A)^* &= (\overline{\alpha A})^T\\ &= (\bar{\alpha}\bar{A})^T && \text{[Problem 6-5(b)]}\\ &= \bar{\alpha}(\bar{A})^T\\ &= \bar{\alpha}A^*\end{aligned}$$

(d)

$$\begin{aligned}(AB)^* &= (\overline{AB})^T\\ &= (\bar{A}\bar{B})^T && \text{[Problem 6-5(c)]}\\ &= \bar{B}^T\bar{A}^T && \text{[by (1.43)]}\\ &= B^*A^*\end{aligned}$$

PROBLEM 6-19 Let $\mathbf{u}, \mathbf{v} \in C^n$ and A be an $n \times n$ complex matrix. Show that

$$\langle A\mathbf{u},\mathbf{v}\rangle = \langle \mathbf{u}, A^*\mathbf{v}\rangle$$

Solution

$$\langle A\mathbf{u},\mathbf{v}\rangle = \mathbf{v}^*A\mathbf{u}$$

$$\langle \mathbf{u},A^*\mathbf{v}\rangle = (A^*\mathbf{v})^*\mathbf{u}$$

$$= \mathbf{v}^*(A^*)^*\mathbf{u}$$

$$= \mathbf{v}^*A\mathbf{u}$$

Hence

$$\langle A\mathbf{u},\mathbf{v}\rangle = \langle \mathbf{u},A^*\mathbf{v}\rangle.$$

PROBLEM 6-20 Using the result of Problem 6-19, verify (6.55); that is, if $\mathbf{u}, \mathbf{v} \in C^n$ and A is a unitary matrix, then

$$\langle A\mathbf{u},A\mathbf{v}\rangle = \langle \mathbf{u},\mathbf{v}\rangle$$

Solution Applying the result of Problem 6-19, we have

$$\langle A\mathbf{u},A\mathbf{v}\rangle = \langle \mathbf{u},A^*A\mathbf{v}\rangle$$

$$= \langle \mathbf{u},I\mathbf{v}\rangle \qquad [\text{since } A \text{ is unitary and } A^*A = I]$$

$$= \langle \mathbf{u},\mathbf{v}\rangle$$

PROBLEM 6-21 Given

$$A = \begin{bmatrix} \frac{1}{2}+i\frac{1}{2} & \frac{1}{2}+i\frac{1}{2} \\ \frac{1}{2}-i\frac{1}{2} & -\frac{1}{2}+i\frac{1}{2} \end{bmatrix} = \frac{1}{2}\begin{bmatrix} 1+i & 1+i \\ 1-i & -1+i \end{bmatrix}$$

Show **(a)** that A is unitary and **(b)** that the column vectors of A form an orthonormal set.

Solution

(a) $$A^* = (\bar{A})^T = \frac{1}{2}\begin{bmatrix} 1-i & 1-i \\ 1+i & -1-i \end{bmatrix}^T = \frac{1}{2}\begin{bmatrix} 1-i & 1+i \\ 1-i & -1-i \end{bmatrix}$$

$$A^*A = \frac{1}{4}\begin{bmatrix} 1-i & 1+i \\ 1-i & -1-i \end{bmatrix}\begin{bmatrix} 1+i & 1+i \\ 1-i & -1+i \end{bmatrix}$$

$$= \frac{1}{4}\begin{bmatrix} (1-i)(1+i)+(1+i)(1-i) & (1-i)(1+i)+(1+i)(-1+i) \\ (1-i)(1+i)+(-1-i)(1-i) & (1-i)(1+i)+(-1-i)(-1+i) \end{bmatrix}$$

$$= \frac{1}{4}\begin{bmatrix} 1-i^2+1-i^2 & 0 \\ 0 & 1-i^2+1-i^2 \end{bmatrix}$$

$$= \frac{1}{4}\begin{bmatrix} 4 & 0 \\ 0 & 4 \end{bmatrix} = \begin{bmatrix} 1 & 0 \\ 0 & 1 \end{bmatrix} = I_2$$

Thus A is unitary.

(b) Let $\mathbf{c}_1$ and $\mathbf{c}_2$ be the column vectors of A; then

$$\mathbf{c}_1 = \begin{bmatrix} \frac{1}{2}+i\frac{1}{2} \\ \frac{1}{2}-i\frac{1}{2} \end{bmatrix}, \qquad \mathbf{c}_2 = \begin{bmatrix} \frac{1}{2}+i\frac{1}{2} \\ -\frac{1}{2}+i\frac{1}{2} \end{bmatrix}$$

Relative to the Euclidean inner product on C^n, and by (6.45) and (6.47), we have

$$\|\mathbf{c}_1\| = \sqrt{(\tfrac{1}{2}+i\tfrac{1}{2})(\tfrac{1}{2}-i\tfrac{1}{2}) + (\tfrac{1}{2}-i\tfrac{1}{2})(\tfrac{1}{2}+i\tfrac{1}{2})} = \sqrt{1} = 1$$

$$\|\mathbf{c}_2\| = \sqrt{(\tfrac{1}{2}+i\tfrac{1}{2})(\tfrac{1}{2}-i\tfrac{1}{2}) + (-\tfrac{1}{2}+i\tfrac{1}{2})(-\tfrac{1}{2}-i\tfrac{1}{2})} = \sqrt{1} = 1$$

$$\langle \mathbf{c}_1,\mathbf{c}_2\rangle = (\tfrac{1}{2}+i\tfrac{1}{2})(\tfrac{1}{2}-i\tfrac{1}{2}) + (\tfrac{1}{2}-i\tfrac{1}{2})(-\tfrac{1}{2}-i\tfrac{1}{2}) = 0$$

Thus $\{\mathbf{c}_1,\mathbf{c}_2\}$ forms an orthonormal set.

PROBLEM 6-22 Let

$$\mathbf{u} = \begin{bmatrix} 1 \\ i \end{bmatrix} \quad \text{and} \quad \mathbf{v} = \begin{bmatrix} -i \\ 2 \end{bmatrix}$$

Using A of Problem 6-21, show that

$$\langle A\mathbf{u}, A\mathbf{v}\rangle = \langle \mathbf{u}, \mathbf{v}\rangle$$

Solution

$$A\mathbf{u} = \frac{1}{2}\begin{bmatrix} 1+i & 1+i \\ 1-i & -1+i \end{bmatrix}\begin{bmatrix} 1 \\ i \end{bmatrix} = \frac{1}{2}\begin{bmatrix} 1+i+i+i^2 \\ 1-i-i+i^2 \end{bmatrix} = \frac{1}{2}\begin{bmatrix} i2 \\ -i2 \end{bmatrix} = \begin{bmatrix} i \\ -i \end{bmatrix}$$

$$A\mathbf{v} = \frac{1}{2}\begin{bmatrix} 1+i & 1+i \\ 1-i & -1+i \end{bmatrix}\begin{bmatrix} -i \\ 2 \end{bmatrix} = \frac{1}{2}\begin{bmatrix} -i-i^2+2+i2 \\ -i+i^2-2+i2 \end{bmatrix} = \frac{1}{2}\begin{bmatrix} 3+i \\ -3+i \end{bmatrix} = \begin{bmatrix} \frac{3}{2}+i\frac{1}{2} \\ -\frac{3}{2}+i\frac{1}{2} \end{bmatrix}$$

$$\langle A\mathbf{u}, A\mathbf{v}\rangle = \langle A\mathbf{v}\rangle^* A\mathbf{u}$$

$$= \frac{1}{2}[3-i, -3-i]\begin{bmatrix} i \\ -i \end{bmatrix} = \frac{1}{2}(i3 - i^2 + i3 + i^2) = i3$$

$$\langle \mathbf{u}, \mathbf{v}\rangle = \mathbf{v}^*\mathbf{u}$$

$$= [i, 2]\begin{bmatrix} 1 \\ i \end{bmatrix} = (i + i2) = i3$$

Hence

$$\langle A\mathbf{u}, A\mathbf{v}\rangle = \langle \mathbf{u}, \mathbf{v}\rangle$$

Supplementary Exercises

PROBLEM 6-23 Use the definition of the product of two complex numbers (6.10) to show that if k is a real number, then

$$kz = kx + iky$$

PROBLEM 6-24 Determine $\mathscr{Re}(z)$ and $\mathscr{Im}(z)$ if

$$z = \frac{(1+i2)(1-i) - (2+i3)}{1+i}$$

Answer $\mathscr{Re}(z) = -\frac{1}{2}$ $\mathscr{Im}(z) = -\frac{3}{2}$.

PROBLEM 6-25 If $z_1 = r_1(\cos\theta_1 + i\sin\theta_1)$ and $z_2 = r_2(\cos\theta_2 + i\sin\theta_2)$, show that

$$z_1 z_2 = r_1 r_2[\cos(\theta_1 + \theta_2) + i\sin(\theta_1 + \theta_2)]$$

PROBLEM 6-26 Let

$$\mathbf{u} = \begin{bmatrix} -i \\ 2 \\ 1+i3 \end{bmatrix}, \quad \mathbf{v} = \begin{bmatrix} 1+i \\ 0 \\ 1-i2 \end{bmatrix}$$

Find **(a)** $\mathbf{u} + \mathbf{v}$, **(b)** $2\mathbf{u} + i\mathbf{v}$, **(c)** $\mathbf{u} - 2i\mathbf{v}$.

Answer **(a)** $\begin{bmatrix} 1 \\ 2 \\ 2+i \end{bmatrix}$ **(b)** $\begin{bmatrix} -1-i \\ 4 \\ 4+i7 \end{bmatrix}$ **(c)** $\begin{bmatrix} 2-i3 \\ 2 \\ -3+i \end{bmatrix}$.

PROBLEM 6-27 Prove that if **u** and **v** are vectors in a complex inner product space, then

$$\langle \mathbf{u},\mathbf{v}\rangle = \frac{1}{4}\|\mathbf{u}+\mathbf{v}\|^2 - \frac{1}{4}\|\mathbf{u}-\mathbf{v}\|^2 + \frac{i}{4}\|\mathbf{u}+i\mathbf{v}\|^2 - \frac{i}{4}\|\mathbf{u}-i\mathbf{v}\|^2$$

(cf. Problem 4-38).

PROBLEM 6-28 Use the inner product of Problem 6-10 to show that

$$A = \begin{bmatrix} i2 & i \\ -i & i3 \end{bmatrix} \quad \text{and} \quad B = \begin{bmatrix} -3 & 1-i \\ 1-i & 2 \end{bmatrix}$$

are orthogonal.

PROBLEM 6-29 Let

$$A = \begin{bmatrix} 2 & 1-i \\ 1+i & 1 \end{bmatrix}, \qquad B = \begin{bmatrix} i & 0 \\ 0 & i \end{bmatrix}$$

Show that $(AB)^* = B^*A^*$.

7 EIGENVALUES AND EIGENVECTORS

THIS CHAPTER IS ABOUT

- ☑ Eigenvalues and Eigenvectors
- ☑ Eigenspaces
- ☑ Diagonalization
- ☑ Symmetric Matrices and Hermitian Matrices
- ☑ The Cayley-Hamilton Theorem

7-1. Eigenvalues and Eigenvectors

A. Eigenvalues and eigenvectors for a matrix A

1. **Definition**

Let A be an $n \times n$ matrix. If

$$A\mathbf{x} = \lambda\mathbf{x} \tag{7.1}$$

for some nonzero vector $\mathbf{x}$ and some scalar λ, then λ is called an **eigenvalue** (or *characteristic value*) of A and $\mathbf{x}$ is said to be an **eigenvector** (or *characteristic vector*) of A corresponding to λ.

It is important to note that, if $\mathbf{x} \in R^n$, then λ must be real; and if $\mathbf{x} \in C^n$, then λ may be complex. Thus A may have no eigenvalues if $\mathbf{x} \in R^n$.

2. **Characteristic equation for A**

Equation (7.1) can be rewritten as

$$(\lambda I_n - A)\mathbf{x} = \mathbf{0} \tag{7.2}$$

where I_n is the nth-order identity matrix. Equation (7.2) will have a nontrivial solution if and only if $\lambda I_n - A$ is singular, or equivalently,

$$\det(\lambda I_n - A) = 0 \tag{7.3}$$

This is called the **characteristic equation** of A. If the determinant in (7.3) is expanded, we obtain an nth-degree polynomial in the variable λ. That is, if

$$A = \begin{bmatrix} a_{11} & a_{12} & \cdots & a_{1n} \\ a_{21} & a_{22} & \cdots & a_{2n} \\ \vdots & \vdots & & \vdots \\ a_{n1} & a_{n2} & \cdots & a_{nn} \end{bmatrix}$$

then

$$c(\lambda) = \det(\lambda I_n - A) = \begin{vmatrix} \lambda - a_{11} & -a_{12} & \cdots & -a_{1n} \\ -a_{21} & \lambda - a_{22} & \cdots & -a_{2n} \\ \vdots & \vdots & & \vdots \\ -a_{n1} & -a_{n2} & \cdots & \lambda - a_{nn} \end{vmatrix} \tag{7.4}$$

$$= \lambda^n + c_1\lambda^{n-1} + \cdots + c_n$$

This polynomial is called the **characteristic polynomial** of A. Now if $\lambda_1, \lambda_2, \ldots, \lambda_k$ are distinct eigenvalues of A, then we have

$$c(\lambda) = (\lambda - \lambda_1)^{m_1}(\lambda - \lambda_2)^{m_2} \cdots (\lambda - \lambda_k)^{m_k} \tag{7.5}$$

where $m_1 + m_2 + \cdots + m_k = n$ and m_k is called the **algebraic multiplicity** of λ_k.

EXAMPLE 7-1: Find (**a**) the eigenvalues and (**b**) the corresponding eigenvectors of the matrix

$$A = \begin{bmatrix} 1 & 2 \\ 2 & 1 \end{bmatrix}$$

Solution

(**a**) Since

$$\lambda I_2 - A = \lambda \begin{bmatrix} 1 & 0 \\ 0 & 1 \end{bmatrix} - \begin{bmatrix} 1 & 2 \\ 2 & 1 \end{bmatrix} = \begin{bmatrix} \lambda - 1 & -2 \\ -2 & \lambda - 1 \end{bmatrix}$$

the characteristic polynomial of A is

$$c(\lambda) = \det(\lambda I_2 - A) = \begin{vmatrix} \lambda - 1 & -2 \\ -2 & \lambda - 1 \end{vmatrix} = (\lambda - 1)^2 - 4 = \lambda^2 - 2\lambda - 3$$

and the characteristic equation of A is

$$\lambda^2 - 2\lambda - 3 = (\lambda + 1)(\lambda - 3) = 0$$

Therefore the eigenvalues of A are -1 and 3.

(**b**) To find the eigenvectors of A corresponding to $\lambda = -1$, we solve the equation

$$(\lambda I_2 - A)\mathbf{x} = \mathbf{0} \quad \text{with } \lambda = -1$$

That is,

$$[(-1)I_2 - A]\mathbf{x} = \mathbf{0}$$

or

$$\begin{bmatrix} (-1) - 1 & -2 \\ -2 & (-1) - 1 \end{bmatrix} \begin{bmatrix} x_1 \\ x_2 \end{bmatrix} = \begin{bmatrix} 0 \\ 0 \end{bmatrix}$$

or

$$\begin{bmatrix} -2 & -2 \\ -2 & -2 \end{bmatrix} \begin{bmatrix} x_1 \\ x_2 \end{bmatrix} = \begin{bmatrix} 0 \\ 0 \end{bmatrix}$$

or

$$\begin{bmatrix} 1 & 1 \\ 1 & 1 \end{bmatrix} \begin{bmatrix} x_1 \\ x_2 \end{bmatrix} = \begin{bmatrix} 0 \\ 0 \end{bmatrix}$$

which corresponds to the single equation

$$x_1 + x_2 = 0$$

Since $x_2 = -x_1$, $\mathbf{x}$ has the form

$$\mathbf{x} = \begin{bmatrix} x_1 \\ -x_1 \end{bmatrix} = x_1 \begin{bmatrix} 1 \\ -1 \end{bmatrix}$$

Therefore the eigenvectors of A corresponding to $\lambda = -1$ are all the nonzero scalar multiples of

$$\mathbf{u} = \begin{bmatrix} 1 \\ -1 \end{bmatrix}$$

To find the eigenvectors of A corresponding to $\lambda = 3$, we solve the equation

$$(\lambda I_2 - A)\mathbf{x} = \mathbf{0} \quad \text{with } \lambda = 3.$$

That is,

$$\begin{bmatrix} 3-1 & -2 \\ -2 & 3-1 \end{bmatrix}\begin{bmatrix} x_1 \\ x_2 \end{bmatrix} = \begin{bmatrix} 0 \\ 0 \end{bmatrix}$$

or

$$\begin{bmatrix} 2 & -2 \\ -2 & 2 \end{bmatrix}\begin{bmatrix} x_1 \\ x_2 \end{bmatrix} = \begin{bmatrix} 0 \\ 0 \end{bmatrix}$$

or

$$\begin{bmatrix} 1 & -1 \\ -1 & 1 \end{bmatrix}\begin{bmatrix} x_1 \\ x_2 \end{bmatrix} = \begin{bmatrix} 0 \\ 0 \end{bmatrix}$$

which corresponds to the single equation

$$x_1 - x_2 = 0$$

Since $x_1 = x_2$, $\mathbf{x}$ has the form

$$\mathbf{x} = \begin{bmatrix} x_1 \\ x_2 \end{bmatrix} = x_1\begin{bmatrix} 1 \\ 1 \end{bmatrix}$$

Therefore the eigenvectors of A corresponding to $\lambda = 3$ are all the nonzero real scalar multiples of

$$\mathbf{v} = \begin{bmatrix} 1 \\ 1 \end{bmatrix}$$

B. Eigenvalues and eigenvectors for a linear operator T

DEFINITION Let T: $V \to V$ be a linear operator on a finite-dimensional vector space V. If

$$T\mathbf{v} = \lambda\mathbf{v} \tag{7.6}$$

for some nonzero vector $\mathbf{v}$ in V and some scalar λ, then λ is called an *eigenvalue* of T and $\mathbf{v}$ is said to be an *eigenvector* of T corresponding to λ.

As in the case of matrices, if V is a real vector space (vector space over R), then only real numbers λ are allowed. If V is a complex vector space (vector space over C), then complex numbers λ are allowed.

The important connection between the eigenvalues of linear operators and those of matrices is given in Theorem 7-1.1.

THEOREM 7-1.1 Let A be the matrix of T relative to a basis $\mathscr{B}$. Then T and A have the same eigenvalues.

EXAMPLE 7-2: Verify Theorem 7-1.1.

Solution: Suppose that T: $V \to V$ is represented by the matrix A relative to the basis $\mathscr{B}$ of V. Then let λ be an eigenvalue of T and

$$\mathbf{x} = \begin{bmatrix} x_1 \\ x_2 \\ \vdots \\ x_n \end{bmatrix}_{\mathscr{B}}$$

be the coordinate vector relative to $\mathscr{B}$ of a corresponding eigenvector $\mathbf{v}$. Then

$$T(\mathbf{v}) = \lambda \mathbf{v}$$

and hence

$$A\mathbf{x} = \lambda \mathbf{x} \tag{7.7}$$

since A represents T (see Theorem 5-4.2).

Thus (7.7) indicates that λ is an eigenvalue of A.

From Theorem 7-1.1 and Theorem 5-5.1 we can conclude that similar matrices have the same eigenvalues (see Problem 7-9).

EXAMPLE 7-3: Consider the linear operator $T: R^2 \to R^2$ of Example 5-21 given by

$$T\left(\begin{bmatrix} x_1 \\ x_2 \end{bmatrix}\right) = \begin{bmatrix} 2x_1 + x_2 \\ x_1 + 2x_2 \end{bmatrix}$$

Find the eigenvalues of T.

Solution: The standard matrix of T is given by

$$A = \begin{bmatrix} 2 & 1 \\ 1 & 2 \end{bmatrix}$$

(see Example 5-21). The characteristic polynomial of A is

$$c(\lambda) = |\lambda I_2 - A| = \begin{vmatrix} \lambda - 2 & -1 \\ -1 & \lambda - 2 \end{vmatrix} = (\lambda - 2)^2 - 1 = \lambda^2 - 4\lambda + 3 = (\lambda - 1)(\lambda - 3)$$

Thus the eigenvalues of T are $\lambda_1 = 1, \lambda_2 = 3$.

C. Other properties of eigenvalues

Let A be an $n \times n$ matrix with eigenvalues $\lambda_1, \lambda_2, \ldots, \lambda_n$ (counting multiplicities). Then

(1) $\det A = \lambda_1 \lambda_2 \ldots \lambda_n$ **(7.8)**

(2) $\text{tr}(A) = \lambda_1 + \lambda_2 + \cdots + \lambda_n$ **(7.9)**

(3) A is singular if and only if 0 is an eigenvalue of A.

(4) A and A^T have the same eigenvalues.

(5) If A is nonsingular, then $\lambda_1^{-1}, \lambda_2^{-1}, \ldots, \lambda_n^{-1}$ are the eigenvalues of A^{-1}.

The above properties are proved in Problems 7-5 to 7-8.

EXAMPLE 7-4: Verify properties (7.8) and (7.9) for the matrix A of Example 7-1.

Solution: In Example 7-1 we have $A = \begin{bmatrix} 1 & 2 \\ 2 & 1 \end{bmatrix}$ and $\lambda_1 = -1, \lambda_2 = 3$. Now

$$\det A = \begin{vmatrix} 1 & 2 \\ 2 & 1 \end{vmatrix} = 1 - 4 = -3$$

and

$$\lambda_1 \lambda_2 = (-1)(3) = -3 = \det A$$

$$\text{tr}(A) = 1 + 1 = 2$$

and

$$\lambda_1 + \lambda_2 = -1 + 3 = 2 = \text{tr}(A)$$

7-2. Eigenspaces

THEOREM 7-2.1 Let $T: V \to V$ be a linear operator on a vector space V. For each eigenvalue λ of T, let V_λ be the set consisting of the zero vector together with all eigenvectors corresponding to λ. Then each V_λ is a subspace of V.

For the proof of Theorem 7-2.1, see Problem 7-13.

DEFINITION The subspace V_λ of Theorem 7-2.1 is called the **eigenspace** of T corresponding to λ. The dimension of V_λ is called the **geometric multiplicity** of λ.

THEOREM 7-2.2 If $\lambda_1, \lambda_2, \ldots, \lambda_k$ are distinct eigenvalues of an $n \times n$ matrix A with corresponding eigenvectors $\mathbf{x}_1, \mathbf{x}_2, \ldots, \mathbf{x}_k$, then $\{\mathbf{x}_1, \mathbf{x}_2, \ldots, \mathbf{x}_k\}$ is a linearly independent set.

For the proof of Theorem 7-2.2, see Problem 7-14.

EXAMPLE 7-5: Find the eigenvalues of A and the bases of the corresponding eigenspaces for

$$A = \begin{bmatrix} -1 & 1 & 1 \\ -2 & 2 & 1 \\ -1 & 1 & 1 \end{bmatrix}$$

Solution

$$c(\lambda) = |\lambda I_3 - A| = \begin{vmatrix} \lambda + 1 & -1 & -1 \\ 2 & \lambda - 2 & -1 \\ 1 & -1 & \lambda - 1 \end{vmatrix} = \lambda^3 - 2\lambda^2 + \lambda = \lambda(\lambda - 1)^2$$

Hence the eigenvalues of A are $\lambda_1 = 0$ and $\lambda_2 = 1$ with multiplicity 2.

Let

$$\mathbf{x} = \begin{bmatrix} x_1 \\ x_2 \\ x_3 \end{bmatrix}$$

be an eigenvector of A. For $\lambda = 0$,

$$\begin{bmatrix} 1 & -1 & -1 \\ 2 & -2 & -1 \\ 1 & -1 & -1 \end{bmatrix} \begin{bmatrix} x_1 \\ x_2 \\ x_3 \end{bmatrix} = \begin{bmatrix} 0 \\ 0 \\ 0 \end{bmatrix}$$

or

$$x_1 - x_2 - x_3 = 0$$
$$2x_1 - 2x_2 - x_3 = 0$$

Solving for x_1, x_2, x_3, we get $x_1 = t$, $x_2 = t$, $x_3 = 0$, where t is a parameter. Hence the eigenvector corresponding to $\lambda = 0$ is of the form

$$\mathbf{x} = \begin{bmatrix} t \\ t \\ 0 \end{bmatrix} = t \begin{bmatrix} 1 \\ 1 \\ 0 \end{bmatrix}$$

and the basis for the eigenspace corresponding to $\lambda = 0$ is $\begin{bmatrix} 1 \\ 1 \\ 0 \end{bmatrix}$.

For $\lambda = 1$,

$$\begin{bmatrix} 2 & -1 & -1 \\ 2 & -1 & -1 \\ 1 & -1 & 0 \end{bmatrix} \begin{bmatrix} x_1 \\ x_2 \\ x_3 \end{bmatrix} = \begin{bmatrix} 0 \\ 0 \\ 0 \end{bmatrix}$$

or

$$\begin{aligned} 2x_1 - x_2 - x_3 &= 0 \\ x_1 - x_2 \qquad &= 0 \end{aligned}$$

Solving for x_1, x_2, x_3, we get $x_1 = x_2 = x_3 = s$, where s is a parameter. Hence the eigenvector corresponding to $\lambda = 1$ is of the form

$$\mathbf{x} = \begin{bmatrix} s \\ s \\ s \end{bmatrix} = s \begin{bmatrix} 1 \\ 1 \\ 1 \end{bmatrix}$$

and the basis for the eigenspace corresponding to $\lambda = 1$ is $\begin{bmatrix} 1 \\ 1 \\ 1 \end{bmatrix}$. Note that the geometric multiplicity of $\lambda = 1$ is 1.

7-3. Diagonalization

A. Definition

An $n \times n$ matrix A is said to be *diagonalizable* if there exists a nonsingular matrix P and a diagonal matrix D such that

$$P^{-1}AP = D \tag{7.10}$$

We say that P *diagonalizes* A.

THEOREM 7-3.1 An $n \times n$ matrix A is diagonalizable if and only if A has n linearly independent eigenvectors.

Combining Theorems 7-2.2 and 7-3.1, we have the following theorem.

THEOREM 7-3.2 If an $n \times n$ matrix A has n distinct eigenvalues, then A is diagonalizable.

For the proof of Theorem 7-3.1 see Problem 7-17.

EXAMPLE 7-6: Determine whether matrix A of Example 7-1 is diagonalizable. If it is, show that matrix $P = [\mathbf{x}_1, \mathbf{x}_2]$, where $\mathbf{x}_1$, $\mathbf{x}_2$ are the eigenvectors of A, will diagonalize A.

Solution: From Example 7-1,

$$A = \begin{bmatrix} 1 & 2 \\ 2 & 1 \end{bmatrix}$$

and the eigenvalues of A are $\lambda_1 = -1$ and $\lambda_2 = 3$. Thus, by Theorem 7-3.2, A is diagonalizable. From Example 7-1, the eigenvectors corresponding to λ_1 and λ_2 are given by

$$\mathbf{x}_1 = \begin{bmatrix} 1 \\ -1 \end{bmatrix} \quad \text{and} \quad \mathbf{x}_2 = \begin{bmatrix} 1 \\ 1 \end{bmatrix}$$

Thus let

$$P = [\mathbf{x}_1, \mathbf{x}_2] = \begin{bmatrix} 1 & 1 \\ -1 & 1 \end{bmatrix}$$

Then

$$P^{-1} = \begin{bmatrix} 1 & 1 \\ -1 & 1 \end{bmatrix}^{-1} = \frac{1}{2}\begin{bmatrix} 1 & -1 \\ 1 & 1 \end{bmatrix} = \begin{bmatrix} \frac{1}{2} & -\frac{1}{2} \\ \frac{1}{2} & \frac{1}{2} \end{bmatrix}$$

and we get

$$P^{-1}AP = \begin{bmatrix} \frac{1}{2} & -\frac{1}{2} \\ \frac{1}{2} & \frac{1}{2} \end{bmatrix}\begin{bmatrix} 1 & 2 \\ 2 & 1 \end{bmatrix}\begin{bmatrix} 1 & 1 \\ -1 & 1 \end{bmatrix} = \begin{bmatrix} -1 & 0 \\ 0 & 3 \end{bmatrix} = \begin{bmatrix} \lambda_1 & 0 \\ 0 & \lambda_2 \end{bmatrix}$$

B. Representation of a linear operator by a diagonal matrix

THEOREM 7-3.3 Let $T: V \to V$ be a linear operator on a finite-dimensional vector space V. If T has linearly independent eigenvectors $\mathbf{v}_1, \mathbf{v}_2, \ldots, \mathbf{v}_n$ corresponding to the eigenvalues $\lambda_1, \lambda_2, \ldots, \lambda_n$, respectively, then the matrix of T relative to the ordered basis $\mathscr{B} = \{\mathbf{v}_1, \mathbf{v}_2, \ldots, \mathbf{v}_n\}$ is

$$\begin{bmatrix} \lambda_1 & 0 & \cdots & 0 \\ 0 & \lambda_2 & \cdots & 0 \\ \vdots & \vdots & & \vdots \\ 0 & 0 & \cdots & \lambda_n \end{bmatrix} \tag{7.11}$$

EXAMPLE 7-7: Verify Theorem 7-3.3.

Solution: Let $\mathscr{B} = \{\mathbf{v}_1, \mathbf{v}_2, \ldots, \mathbf{v}_n\}$ be an ordered basis of V and let $\mathbf{v} \in V$. Then the coordinate vector of $\mathbf{v}$ relative to $\mathscr{B}$ is [see (5.27)]

$$[\mathbf{v}]_{\mathscr{B}} = \begin{bmatrix} x_1 \\ x_2 \\ \vdots \\ x_n \end{bmatrix}_{\mathscr{B}} \tag{7.12}$$

such that

$$\mathbf{v} = x_1\mathbf{v}_1 + x_2\mathbf{v}_2 + \cdots + x_n\mathbf{v}_n \tag{7.13}$$

Now

$$T\mathbf{v}_i = \lambda_i\mathbf{v}_i \qquad i = 1, 2, \ldots, n \tag{7.14}$$

Thus

$$[T\mathbf{v}_1]_{\mathscr{B}} = \begin{bmatrix} \lambda_1 \\ 0 \\ \vdots \\ 0 \end{bmatrix}, [T\mathbf{v}_2]_{\mathscr{B}} = \begin{bmatrix} 0 \\ \lambda_2 \\ \vdots \\ 0 \end{bmatrix}, \ldots, [T\mathbf{v}_n]_{\mathscr{B}} = \begin{bmatrix} 0 \\ 0 \\ \vdots \\ \lambda_n \end{bmatrix} \tag{7.15}$$

Hence, by (5.29), the matrix of T relative to $\mathscr{B}$ is

$$A = \begin{bmatrix} \lambda_1 & 0 & \cdots & 0 \\ 0 & \lambda_2 & \cdots & 0 \\ \vdots & \vdots & & \vdots \\ 0 & 0 & \cdots & \lambda_n \end{bmatrix} \tag{7.16}$$

EXAMPLE 7-8: Let $T: R^2 \to R^2$ be the linear operator of Example 7-3. Find a basis for R^2 relative to which the matrix of T is diagonal.

Solution: From Example 7-3 or Example 5-21, the standard matrix of T is

$$A = \begin{bmatrix} 2 & 1 \\ 1 & 2 \end{bmatrix}$$

Now

$$c(\lambda) = |\lambda I_2 - A| = \begin{vmatrix} \lambda - 2 & -1 \\ -1 & \lambda - 2 \end{vmatrix} = (\lambda - 2)^2 - 1 = \lambda^2 - 4\lambda + 3 = (\lambda - 1)(\lambda - 3)$$

so the eigenvalues of A (and also T) are $\lambda_1 = 1, \lambda_2 = 3$.

Let $\mathbf{x} = \begin{bmatrix} x_1 \\ x_2 \end{bmatrix}$ be the eigenvector of A. Then for $\lambda = 1$,

$$\begin{bmatrix} -1 & -1 \\ -1 & -1 \end{bmatrix}\begin{bmatrix} x_1 \\ x_2 \end{bmatrix} = \begin{bmatrix} 0 \\ 0 \end{bmatrix}$$

or $-x_1 - x_2 = 0$. Thus $x_2 = -x_1$, and

$$\mathbf{x} = \begin{bmatrix} x_1 \\ -x_1 \end{bmatrix} = x_1 \begin{bmatrix} 1 \\ -1 \end{bmatrix}$$

For $\lambda = 3$,

$$\begin{bmatrix} 1 & -1 \\ -1 & 1 \end{bmatrix}\begin{bmatrix} x_1 \\ x_2 \end{bmatrix} = \begin{bmatrix} 0 \\ 0 \end{bmatrix}$$

or $x_1 - x_2 = 0$. Thus $x_1 = x_2$, and

$$\mathbf{x} = \begin{bmatrix} x_1 \\ x_1 \end{bmatrix} = x_1 \begin{bmatrix} 1 \\ 1 \end{bmatrix}.$$

Thus the basis for R^2 relative to which the matrix of T is diagonalizable is $\mathscr{B}' = \{\mathbf{u}_1, \mathbf{u}_2\}$ where

$$\mathbf{u}_1 = \begin{bmatrix} 1 \\ -1 \end{bmatrix} \quad \text{and} \quad \mathbf{u}_2 = \begin{bmatrix} 1 \\ 1 \end{bmatrix}$$

Check:

$$T(\mathbf{u}_1) = T\left(\begin{bmatrix} 1 \\ -1 \end{bmatrix}\right) = \begin{bmatrix} 1 \\ -1 \end{bmatrix} = 1\mathbf{u}_1$$

and

$$T(\mathbf{u}_2) = T\left(\begin{bmatrix} 1 \\ 1 \end{bmatrix}\right) = \begin{bmatrix} 3 \\ 3 \end{bmatrix} = 3\mathbf{u}_2$$

Thus

$$[T(\mathbf{u}_1)]_{\mathscr{B}'} = \begin{bmatrix} 1 \\ 0 \end{bmatrix}, \qquad [T(\mathbf{u}_2)]_{\mathscr{B}'} = \begin{bmatrix} 0 \\ 3 \end{bmatrix}$$

and by (5.29), the matrix of T relative to $\mathscr{B}'$ is

$$A' = \begin{bmatrix} 1 & 0 \\ 0 & 3 \end{bmatrix} = \begin{bmatrix} \lambda_1 & 0 \\ 0 & \lambda_2 \end{bmatrix}$$

(cf. Example 5-21).

7-4. Symmetric Matrices and Hermitian Matrices

A. Symmetric matrices

In Section 1-6D we defined an $n \times n$ real matrix A to be symmetric if

$$A^T = A \tag{7.17}$$

EXAMPLE 7-9: Let A be a symmetric matrix and let $\langle \mathbf{u}, \mathbf{v} \rangle$ denote the Euclidean inner product on R^n. Then show that

$$\langle A\mathbf{u},\mathbf{v}\rangle = \langle \mathbf{u},A\mathbf{v}\rangle \tag{7.18}$$

Solution: Recall from Section 4-2 [(4.23) and (4.24)] that the Euclidean inner product $\langle \mathbf{u},\mathbf{v}\rangle$ can be expressed as

$$\langle \mathbf{u},\mathbf{v}\rangle = \mathbf{u}^T\mathbf{v} \tag{7.19}$$

$$\langle \mathbf{u},\mathbf{v}\rangle = \mathbf{v}^T\mathbf{u} \tag{7.20}$$

Then

$$\begin{aligned} \langle A\mathbf{u},\mathbf{v}\rangle &= (A\mathbf{u})^T\mathbf{v} \\ &= (\mathbf{u}^T A^T)\mathbf{v} \\ &= \mathbf{u}^T(A^T\mathbf{v}) \\ &= \langle \mathbf{u},A^T\mathbf{v}\rangle \end{aligned} \tag{7.21}$$

Since A is symmetric—that is, $A^T = A$—we get

$$\langle A\mathbf{u},\mathbf{v}\rangle = \langle \mathbf{u},A^T\mathbf{v}\rangle = \langle \mathbf{u},A\mathbf{v}\rangle$$

THEOREM 7-4.1 An $n \times n$ real symmetric matrix has only real eigenvalues.

EXAMPLE 7-10: Verify Theorem 7-4.1.

Solution: In general, the roots of the characteristic equation of a real matrix may be complex numbers (see Problem 7-1). Thus we assume that $\lambda \in C$ where λ is an eigenvalue of A and we shall use the inner product in C^n such that if $\mathbf{u}, \mathbf{v} \in C^n$, then

$$\langle \mathbf{u},\mathbf{v}\rangle = \overline{\langle \mathbf{v},\mathbf{u}\rangle} \qquad \text{[see (6.35)]} \tag{7.22}$$

Let $\mathbf{x}$ be an eigenvector of A corresponding to an eigenvalue λ so that $\mathbf{x} \neq \mathbf{0}$, and let

$$A\mathbf{x} = \lambda\mathbf{x}$$

Then

$$\langle A\mathbf{x},\mathbf{x}\rangle = \langle \lambda\mathbf{x},\mathbf{x}\rangle = \lambda\langle \mathbf{x},\mathbf{x}\rangle \tag{7.23}$$

On the other hand, since A is symmetric and real, by (7.18) and (6.41), we have

$$\langle A\mathbf{x},\mathbf{x}\rangle = \langle \mathbf{x},A\mathbf{x}\rangle = \langle \mathbf{x},\lambda\mathbf{x}\rangle = \bar{\lambda}\langle \mathbf{x},\mathbf{x}\rangle \tag{7.24}$$

Hence

$$\lambda\langle \mathbf{x},\mathbf{x}\rangle = \bar{\lambda}\langle \mathbf{x},\mathbf{x}\rangle \tag{7.25}$$

Since $\langle \mathbf{x},\mathbf{x}\rangle$ is real and nonzero, it follows that

$$\lambda = \bar{\lambda} \tag{7.26}$$

that is, λ is equal to its complex conjugate, and by (6.25) we conclude that λ is real.

---

note: The matrices A of Examples 7-1 and 7-8 are symmetric and their eigenvalues are real.

THEOREM 7-4.2 Eigenvectors corresponding to distinct eigenvalues of a symmetric matrix are orthogonal with respect to the Euclidean inner product.

EXAMPLE 7-11: Verify Theorem 7-4.2.

Solution: Let $\mathbf{x}$ and $\mathbf{y}$ be eigenvectors of a symmetric matrix A corresponding to distinct eigenvalues λ and μ. Then

$$A\mathbf{x} = \lambda\mathbf{x}$$

and

$$A\mathbf{y} = \mu\mathbf{y}$$

Since A is symmetric, by (7.18) we have

$$\langle A\mathbf{x},\mathbf{y}\rangle = \langle \mathbf{x},A\mathbf{y}\rangle \tag{7.27}$$

or

$$\langle \lambda\mathbf{x},\mathbf{y}\rangle = \langle \mathbf{x},\mu\mathbf{y}\rangle$$

or

$$\lambda\langle \mathbf{x},\mathbf{y}\rangle = \mu\langle \mathbf{x},\mathbf{y}\rangle \tag{7.28}$$

Hence

$$(\lambda - \mu)\langle \mathbf{x},\mathbf{y}\rangle = 0 \tag{7.29}$$

Since $\lambda \neq \mu$, we must have

$$\langle \mathbf{x},\mathbf{y}\rangle = 0 \tag{7.30}$$

so that **x** and **y** are orthogonal.

EXAMPLE 7-12: Verify Theorem 7-4.2 for the matrix of Example 7-1.

Solution: From Example 7-1, we have

$$A = \begin{bmatrix} 1 & 2 \\ 2 & 1 \end{bmatrix}$$

which is symmetric. The eigenvalues of A are -1 and 3 and the corresponding eigenvectors are

$$\mathbf{u} = \begin{bmatrix} 1 \\ -1 \end{bmatrix}, \qquad \mathbf{v} = \begin{bmatrix} 1 \\ 1 \end{bmatrix}$$

hence

$$\langle \mathbf{u},\mathbf{v}\rangle = (1)(1) + (-1)(1) = 0$$

Thus **u** and **v** are orthogonal.

B. Orthogonal diagonalization

DEFINITION An $n \times n$ real matrix A is *orthogonally diagonalizable* if there is an orthogonal matrix P such that

$$P^{-1}AP = P^TAP = D \tag{7.31}$$

where D is a diagonal matrix of nth order, and the matrix P is said to *orthogonally diagonalize A*.

If A is a real symmetric matrix, then A is orthogonally diagonalizable. (The proof of this statement is beyond the scope of this text.)

EXAMPLE 7-13: Show that the matrix $A = \begin{bmatrix} 1 & 2 \\ 2 & 1 \end{bmatrix}$ of Example 7-1 is orthogonally diagonalizable.

Solution: The eigenvalues of A from Example 7-1 are -1 and 3, and the corresponding eigenvectors are

$$\mathbf{u} = \begin{bmatrix} 1 \\ -1 \end{bmatrix} \quad \text{and} \quad \mathbf{v} = \begin{bmatrix} 1 \\ 1 \end{bmatrix}$$

which are orthogonal (see Example 7-12). Now, let

$$\mathbf{u}_1 = \frac{1}{\|\mathbf{u}\|}\mathbf{u} = \frac{1}{\sqrt{2}}\begin{bmatrix} 1 \\ -1 \end{bmatrix} = \begin{bmatrix} \dfrac{1}{\sqrt{2}} \\ -\dfrac{1}{\sqrt{2}} \end{bmatrix}$$

and

$$\mathbf{u}_2 = \frac{1}{\|\mathbf{v}\|}\mathbf{v} = \frac{1}{\sqrt{2}}\begin{bmatrix} 1 \\ 1 \end{bmatrix} = \begin{bmatrix} \frac{1}{\sqrt{2}} \\ \frac{1}{\sqrt{2}} \end{bmatrix}$$

Thus

$$P = [\mathbf{u}_1, \mathbf{u}_2] = \begin{bmatrix} \frac{1}{\sqrt{2}} & \frac{1}{\sqrt{2}} \\ -\frac{1}{\sqrt{2}} & \frac{1}{\sqrt{2}} \end{bmatrix}$$

and

$$P^{-1} = \begin{bmatrix} \frac{1}{\sqrt{2}} & -\frac{1}{\sqrt{2}} \\ \frac{1}{\sqrt{2}} & \frac{1}{\sqrt{2}} \end{bmatrix} = P^T$$

Hence P is an orthogonal matrix. Then

$$P^{-1}AP = P^TAP = \begin{bmatrix} \frac{1}{\sqrt{2}} & -\frac{1}{\sqrt{2}} \\ \frac{1}{\sqrt{2}} & \frac{1}{\sqrt{2}} \end{bmatrix}\begin{bmatrix} 1 & 2 \\ 2 & 1 \end{bmatrix}\begin{bmatrix} \frac{1}{\sqrt{2}} & \frac{1}{\sqrt{2}} \\ -\frac{1}{\sqrt{2}} & \frac{1}{\sqrt{2}} \end{bmatrix}$$

$$= \begin{bmatrix} -1 & 0 \\ 0 & 3 \end{bmatrix}$$

Thus A is orthogonally diagonalizable.

C. Hermitian matrices

DEFINITION An $n \times n$ matrix A is called **Hermitian** if

HERMITIAN MATRIX

$$A^* = (\bar{A})^T = A \tag{7.32}$$

That is, if $A = [a_{ij}]$, then A is Hermitian if

$$a_{ji} = \bar{a}_{ij} \tag{7.33}$$

EXAMPLE 7-14: Let A be a Hermitian matrix and let $\langle \mathbf{u},\mathbf{v} \rangle$ denote the standard inner product on C^n. Then show that

$$\langle A\mathbf{u},\mathbf{v} \rangle = \langle \mathbf{u},A\mathbf{v} \rangle \tag{7.34}$$

Solution: Recall from Section 6-3 [see (6.43)] that the standard inner product $\langle \mathbf{u},\mathbf{v} \rangle$ can be expressed as

$$\langle \mathbf{u},\mathbf{v} \rangle = (\bar{\mathbf{v}})^T\mathbf{u} = \mathbf{v}^*\mathbf{u} \tag{7.35}$$

Then

$$\begin{aligned} \langle A\mathbf{u},\mathbf{v} \rangle &= \mathbf{v}^*A\mathbf{u} && \text{[by (7.35)]} \\ &= \mathbf{v}^*A^*\mathbf{u} && \text{[by (7.32)]} \\ &= (A\mathbf{v})^*\mathbf{u} && \text{[by Prob. 6-18(d)]} \\ &= \langle \mathbf{u},A\mathbf{v} \rangle && \text{[by (7.35)]} \end{aligned}$$

THEOREM 7-4.3 The eigenvalues of a Hermitian matrix are all real. Furthermore, eigenvectors corresponding to distinct eigenvalues are orthogonal.

In view of property (7.34), the proof of Theorem 7-4.3 follows exactly as Examples 7-10 and 7-11.

EXAMPLE 7-15: Given

$$A = \begin{bmatrix} 2 & 1-i \\ 1+i & 1 \end{bmatrix}$$

(a) show that A is Hermitian; **(b)** find the eigenvalues and eigenvectors of A; and **(c)** show that eigenvectors of A are orthogonal.

Solution

(a)

$$A^* = (\bar{A})^T = \begin{bmatrix} 2 & 1+i \\ 1-i & 1 \end{bmatrix}^T = \begin{bmatrix} 2 & 1-i \\ 1+i & 1 \end{bmatrix} = A$$

Thus A is Hermitian.

(b)

$$|\lambda I_2 - A| = \begin{vmatrix} \lambda-2 & -1+i \\ -1-i & \lambda-1 \end{vmatrix}$$

$$= (\lambda-2)(\lambda-1) - (-1+i)(-1-i)$$

$$= \lambda^2 - 3\lambda + 2 - 2 = \lambda^2 - 3\lambda = \lambda(\lambda - 3)$$

Thus the eigenvalues of A are $\lambda_1 = 0$, $\lambda_2 = 3$. Note that they are real.

Let $\mathbf{x} = \begin{bmatrix} x_1 \\ x_2 \end{bmatrix}$ be an eigenvector of A. When $\lambda = 0$,

$$\begin{bmatrix} -2 & -1+i \\ -1-i & -1 \end{bmatrix}\begin{bmatrix} x_1 \\ x_2 \end{bmatrix} = \begin{bmatrix} 0 \\ 0 \end{bmatrix}$$

which corresponds to

$$\begin{aligned} -2x_1 + (-1+i)x_2 &= 0 \\ (-1-i)x_1 - \quad x_2 &= 0 \end{aligned}$$

Solving for x_1 and x_2, we get

$$x_2 = -(1+i)x_1$$

Hence the eigenvector $\mathbf{x}$ corresponding to $\lambda = 0$ is proportional to

$$\mathbf{u} = \begin{bmatrix} -1 \\ 1+i \end{bmatrix}$$

When $\lambda = 3$, we have

$$\begin{bmatrix} 1 & -1+i \\ -1-i & 2 \end{bmatrix}\begin{bmatrix} x_1 \\ x_2 \end{bmatrix} = \begin{bmatrix} 0 \\ 0 \end{bmatrix}$$

which corresponds to

$$\begin{aligned} x_1 + (-1+i)x_2 &= 0 \\ (-1-i)x_1 + \quad 2x_2 &= 0 \end{aligned}$$

Solving for x_1 and x_2, we get

$$x_1 = (1-i)x_2$$

Hence the eigenvector $\mathbf{x}$ corresponding to $\lambda = 3$ is proportional to

$$\mathbf{v} = \begin{bmatrix} 1 - i \\ 1 \end{bmatrix}$$

(c)
$$\begin{aligned} \langle \mathbf{u}, \mathbf{v} \rangle &= \mathbf{v}^*\mathbf{u} \\ &= [1 + i, 1]\begin{bmatrix} -1 \\ 1 + i \end{bmatrix} = -(1 + i) + (1 + i) = 0 \end{aligned}$$

Hence **u** and **v** are orthogonal.

D. Unitary diagonalization

DEFINITION An $n \times n$ complex matrix A is called *unitarily diagonalizable* if there is a unitary matrix P such that

$$P^{-1}AP = P^*AP = D \tag{7.36}$$

where D is a diagonal matrix of nth order and the matrix P is said to *unitarily diagonalize A.*

If A is a Hermitian matrix, then A is unitarily diagonalizable. (The proof of this statement is also beyond the scope of this text.)

EXAMPLE 7-16: Show that

$$A = \begin{bmatrix} 2 & 1 - i \\ 1 + i & 1 \end{bmatrix}$$

is unitarily diagonalizable.

Solution: From Example 7-15, we have $\lambda_1 = 0$, $\lambda_2 = 3$ and the corresponding eigenvectors

$$\mathbf{u} = \begin{bmatrix} -1 \\ 1 + i \end{bmatrix}, \qquad \mathbf{v} = \begin{bmatrix} 1 - i \\ 1 \end{bmatrix}$$

and **u** and **v** are orthogonal. Now

$$\|\mathbf{u}\|^2 = \langle \mathbf{u}, \mathbf{u} \rangle = \mathbf{u}^*\mathbf{u} = [-1, 1 - i]\begin{bmatrix} -1 \\ 1 + i \end{bmatrix} = 1 + (1 - i)(1 + i) = 3$$

$$\|\mathbf{v}\|^2 = \langle \mathbf{v}, \mathbf{v} \rangle = \mathbf{v}^*\mathbf{v} = [1 + i, 1]\begin{bmatrix} 1 - i \\ 1 \end{bmatrix} = (1 + i)(1 - i) + 1 = 3$$

Thus $\|\mathbf{u}\| = \|\mathbf{v}\| = \sqrt{3}$.

Let

$$\mathbf{u}_1 = \frac{1}{\|\mathbf{u}\|}\mathbf{u} = \frac{1}{\sqrt{3}}\begin{bmatrix} -1 \\ 1 + i \end{bmatrix}$$

and

$$\mathbf{u}_2 = \frac{1}{\|\mathbf{v}\|}\mathbf{v} = \frac{1}{\sqrt{3}}\begin{bmatrix} 1 - i \\ 1 \end{bmatrix}$$

and let

$$P = [\mathbf{u}_1, \mathbf{u}_2] = \frac{1}{\sqrt{3}}\begin{bmatrix} -1 & 1 - i \\ 1 + i & 1 \end{bmatrix}$$

then

$$P^{-1} = \frac{1}{\sqrt{3}}\begin{bmatrix} -1 & 1 - i \\ 1 + i & 1 \end{bmatrix} = (\bar{P})^T = P^*$$

and

$$P^{-1}AP = P^*AP = \frac{1}{3}\begin{bmatrix} -1 & 1-i \\ 1+i & 1 \end{bmatrix}\begin{bmatrix} 2 & 1-i \\ 1+i & 1 \end{bmatrix}\begin{bmatrix} -1 & 1-i \\ 1+i & 1 \end{bmatrix}$$

$$= \begin{bmatrix} 0 & 0 \\ 0 & 3 \end{bmatrix} = \begin{bmatrix} \lambda_1 & 0 \\ 0 & \lambda_2 \end{bmatrix}$$

Thus A is unitarily diagonalizable.

THEOREM 7-4.4 Let A be a Hermitian matrix with eigenvalues $\lambda_1, \lambda_2, \ldots, \lambda_n$ and orthonormal eigenvectors $\mathbf{u}_1, \mathbf{u}_2, \ldots, \mathbf{u}_n$. Then A can be decomposed into

$$A = \lambda_1 \mathbf{u}_1 \mathbf{u}_1^* + \lambda_2 \mathbf{u}_2 \mathbf{u}_2^* + \cdots + \lambda_n \mathbf{u}_n \mathbf{u}_n^* \qquad \textbf{(7.37)}$$

This decomposition is known as the **spectral theorem.**

EXAMPLE 7-17: Verify (7.37).

Solution: If A is Hermitian, then there is a unitary matrix P that diagonalizes A. Thus from (7.36) we get

$$A = PDP^*$$

$$= [\mathbf{u}_1, \mathbf{u}_2, \ldots, \mathbf{u}_n]\begin{bmatrix} \lambda_1 & 0 & \cdots & 0 \\ 0 & \lambda_2 & \cdots & 0 \\ \vdots & \vdots & & \vdots \\ 0 & 0 & \cdots & \lambda_n \end{bmatrix}\begin{bmatrix} \mathbf{u}_1^* \\ \mathbf{u}_2^* \\ \vdots \\ \mathbf{u}_n^* \end{bmatrix}$$

$$= [\lambda_1 \mathbf{u}_1, \lambda_2 \mathbf{u}_2, \ldots, \lambda \mathbf{u}_n)\begin{bmatrix} \mathbf{u}_1^* \\ \mathbf{u}_2^* \\ \vdots \\ \mathbf{u}_n^* \end{bmatrix}$$

$$= \lambda_1 \mathbf{u}_1 \mathbf{u}_1^* + \lambda_2 \mathbf{u}_2 \mathbf{u}_2^* + \cdots + \lambda_n \mathbf{u}_n \mathbf{u}_n^*$$

7-5. The Cayley-Hamilton Theorem

A. Matrix polynomial

Let

$$p(\lambda) = \lambda^m + a_{m-1}\lambda^{m-1} + \cdots + a_1\lambda + a_0 \qquad \textbf{(7.38)}$$

be a polynomial of λ with scalar coefficients a_i. If A is an $n \times n$ matrix, the **matrix polynomial** $p(A)$ is defined by

$$p(A) = A^m + a_{m-1}A^{m-1} + \cdots + a_1 A + a_0 I_n \qquad \textbf{(7.39)}$$

Notice particularly that the constant term a_0 must be replaced by $a_0 I_n$ so that each term of $p(A)$ will be an $n \times n$ matrix.

Expression (7.39) is a polynomial with scalar coefficients defined for a matrix variable. We can also define a polynomial of λ with $n \times n$ matrix coefficients by

$$Q(\lambda) = B_m\lambda^m + B_{m-1}\lambda^{m-1} + \cdots + B_1\lambda + B_0 \qquad \textbf{(7.40)}$$

If A is an $n \times n$ matrix, then we define

$$Q(A) = B_m A^m + B_{m-1} A^{m-1} + \cdots + B_1 A + B_0 \tag{7.41}$$

note: We must be careful in writing expression (7.41), since matrices do not commute under multiplication.

EXAMPLE 7-18: Consider

$$A = \begin{bmatrix} 1 & 2 \\ 2 & 1 \end{bmatrix}$$

If $p(\lambda) = \lambda^2 + 2\lambda - 3$, find $p(A)$.

Solution

$$\begin{aligned} p(A) &= A^2 + 2A - 3I_2 \\ &= \begin{bmatrix} 1 & 2 \\ 2 & 1 \end{bmatrix}^2 + 2\begin{bmatrix} 1 & 2 \\ 2 & 1 \end{bmatrix} - 3\begin{bmatrix} 1 & 0 \\ 0 & 1 \end{bmatrix} \\ &= \begin{bmatrix} 5 & 4 \\ 4 & 5 \end{bmatrix} + \begin{bmatrix} 2 & 4 \\ 4 & 2 \end{bmatrix} - \begin{bmatrix} 3 & 0 \\ 0 & 3 \end{bmatrix} \\ &= \begin{bmatrix} 4 & 8 \\ 8 & 4 \end{bmatrix} \end{aligned}$$

EXAMPLE 7-19: Consider a diagonal matrix

$$D = \begin{bmatrix} \lambda_1 & 0 & \cdots & 0 \\ 0 & \lambda_2 & \cdots & 0 \\ \vdots & \vdots & & \vdots \\ 0 & 0 & \cdots & \lambda_n \end{bmatrix} \tag{7.42}$$

If $p(\lambda)$ is a polynomial of λ, then show that

$$p(D) = \begin{bmatrix} p(\lambda_1) & 0 & \cdots & 0 \\ 0 & p(\lambda_2) & \cdots & 0 \\ \vdots & \vdots & & \vdots \\ 0 & 0 & \cdots & p(\lambda_n) \end{bmatrix} \tag{7.43}$$

Solution: Let

$$p(\lambda) = \lambda^m + a_{m-1}\lambda^{m-1} + \cdots + a_1\lambda + a_0$$

Then

$$p(D) = D^m + a_{m-1}D^{m-1} + \cdots + a_1 D + a_0 I_n$$

Now

$$D^k = \begin{bmatrix} \lambda_1^k & 0 & \cdots & 0 \\ 0 & \lambda_1^k & \cdots & 0 \\ \vdots & \vdots & & \vdots \\ 0 & 0 & \cdots & \lambda_n^k \end{bmatrix}$$

for every positive integer k. Thus

$$p(D) = \begin{bmatrix} \lambda_1^m & 0 & \cdots & 0 \\ 0 & \lambda_2^m & \cdots & 0 \\ \vdots & \vdots & & \vdots \\ 0 & 0 & \cdots & \lambda_n^m \end{bmatrix} + a_{m-1} \begin{bmatrix} \lambda_1^{m-1} & 0 & \cdots & 0 \\ 0 & \lambda_2^{m-1} & \cdots & 0 \\ \vdots & \vdots & & \vdots \\ 0 & 0 & \cdots & \lambda_n^{m-1} \end{bmatrix}$$

$$+ \cdots + a_1 \begin{bmatrix} \lambda_1 & 0 & \cdots & 0 \\ 0 & \lambda_2 & \cdots & 0 \\ \vdots & \vdots & & \vdots \\ 0 & 0 & \cdots & \lambda_n \end{bmatrix} + a_0 \begin{bmatrix} 1 & 0 & \cdots & 0 \\ 0 & 1 & \cdots & 0 \\ \vdots & \vdots & & \vdots \\ 0 & 0 & \cdots & 1 \end{bmatrix}$$

$$= \begin{bmatrix} p(\lambda_1) & 0 & \cdots & 0 \\ 0 & p(\lambda_2) & \cdots & 0 \\ \vdots & \vdots & & \vdots \\ 0 & 0 & \cdots & p(\lambda_n) \end{bmatrix}$$

EXAMPLE 7-20: Express

$$\begin{bmatrix} \lambda^2 - 2\lambda + 1 & 2\lambda^2 - 3\lambda + 4 \\ -\lambda^2 + 3\lambda + 5 & 3\lambda^2 + \lambda - 3 \end{bmatrix}$$

as a polynomial of λ with matrix coefficients.

Solution

$$\begin{bmatrix} \lambda^2 - 2\lambda + 1 & 2\lambda^2 - 3\lambda + 4 \\ -\lambda^2 + 3\lambda + 5 & 3\lambda^2 + \lambda - 3 \end{bmatrix} = \begin{bmatrix} \lambda^2 & 2\lambda^2 \\ -\lambda^2 & 3\lambda^2 \end{bmatrix} + \begin{bmatrix} -2\lambda & -3\lambda \\ 3\lambda & \lambda \end{bmatrix} + \begin{bmatrix} 1 & 4 \\ 5 & -3 \end{bmatrix}$$

$$= \begin{bmatrix} 1 & 2 \\ -1 & 3 \end{bmatrix} \lambda^2 + \begin{bmatrix} -2 & -3 \\ 3 & 1 \end{bmatrix} \lambda + \begin{bmatrix} 1 & 4 \\ 5 & -3 \end{bmatrix}$$

B. The Cayley-Hamilton theorem

Let the square matrix A of nth order have the characteristic equation

$$c(\lambda) = |\lambda I_n - A| = \lambda^n + c_{n-1}\lambda^{n-1} + \cdots + c_1\lambda + c_0 = 0 \tag{7.44}$$

Then the **Cayley-Hamilton theorem** states that

$$c(A) = A^n + c_{n-1}A^{n-1} + \cdots + c_1 A + c_0 I_n = O \tag{7.45}$$

That is, every square matrix satisfies its own characteristic equation. (For the proof of the Cayley-Hamilton theorem, see Problem 7-28.)

EXAMPLE 7-21: Verify (7.45) for the matrix

$$A = \begin{bmatrix} 1 & 2 \\ 2 & 1 \end{bmatrix}$$

Solution: From Example 7-1 the characteristic equation of A is

$$c(\lambda) = \lambda^2 - 2\lambda - 3 = 0$$

Now

$$c(A) = A^2 - 2A - 3I_2$$
$$= \begin{bmatrix} 1 & 2 \\ -2 & 1 \end{bmatrix}^2 - 2\begin{bmatrix} 1 & 2 \\ 2 & 1 \end{bmatrix} - 3\begin{bmatrix} 1 & 0 \\ 0 & 1 \end{bmatrix}$$
$$= \begin{bmatrix} 5 & 4 \\ 4 & 5 \end{bmatrix} - \begin{bmatrix} 2 & 4 \\ 4 & 2 \end{bmatrix} - \begin{bmatrix} 3 & 0 \\ 0 & 3 \end{bmatrix} = \begin{bmatrix} 0 & 0 \\ 0 & 0 \end{bmatrix} = O$$

C. Function of matrix A

By means of the Cayley-Hamilton theorem it is possible to reduce a polynomial of nth order A to a polynomial whose highest degree in A is $n - 1$. This follows from (7.45), which is rewritten as

$$A^n = -c_{n-1}A^{n-1} - \cdots - c_1 A - c_0 I_n \qquad \textbf{(7.46)}$$

Multiplying this expression by A yields

$$A^{n+1} = -c_{n-1}A^n - \cdots - c_1 A^2 - c_0 A \qquad \textbf{(7.47)}$$

Substituting (7.46) into this expression, we get

$$\begin{aligned} A^{n+1} &= -c_{n-1}(-c_{n-1}A^{n-1} - \cdots - c_1 A - c_0 I_n) - \cdots - c_1 A^2 - c_0 A \\ &= (c_{n-1}^2 - c_{n-2})A^{n-1} + \cdots + (c_{n-1}c_1 - c_0)A + c_{n-1}c_0 I_n \end{aligned} \qquad \textbf{(7.48)}$$

Observe that A^{n+1} is expressed in terms of $A^{n-1}, A^{n-2}, \ldots, A$ and I_n. This process can be continued to prove that A to any power can be represented as the weighted sum of matrices involving A to powers not exceeding $n - 1$. As a result, functions of matrix A that can be expressed in power series form, say

$$f(A) = \alpha_0 I + \alpha_1 A + \alpha_2 A^2 + \cdots + \alpha_n A^n + \cdots = \sum_{k=0}^{\infty} \alpha_k A^k \qquad \textbf{(7.49)}$$

can be expressed as

$$f(A) = \beta_0 I + \beta_1 A + \beta_2 A^2 + \cdots + \beta_{n-1} A^{n-1} = \sum_{k=0}^{n-1} \beta_k A^k \qquad \textbf{(7.50)}$$

For a convenient method for finding the coefficients β_i we retrace the above steps starting with (7.44). Substituting λ for A in (7.46) through (7.50), we get

$$f(\lambda) = \beta_0 + \beta_1\lambda + \beta_2\lambda^2 + \cdots + \beta_{n-1}\lambda^{n-1} = \sum_{k=0}^{n-1} \beta_k \lambda^k \qquad \textbf{(7.51)}$$

Now (7.51) is valid for any λ that is a solution of the characteristic equation, that is, for any eigenvalues of the matrix A. When the eigenvalues are all distinct, (7.51) yields n equations in n unknowns:

$$\begin{aligned} f(\lambda_1) &= \beta_0 + \beta_1\lambda_1 + \cdots + \beta_{n-1}\lambda_1^{n-1} \\ f(\lambda_2) &= \beta_0 + \beta_1\lambda_2 + \cdots + \beta_{n-1}\lambda_2^{n-1} \\ &\vdots \\ f(\lambda_n) &= \beta_0 + \beta_1\lambda_n + \cdots + \beta_{n-1}\lambda_n^{n-1} \end{aligned} \qquad \textbf{(7.52)}$$

This system of equations (7.52) can be solved for the coefficients $\beta_0, \beta_1, \ldots, \beta_{n-1}$. (For the case of multiple eigenvalues, see Problem 7-31.)

EXAMPLE 7-22: Evaluate A^k for

$$A = \begin{bmatrix} 1 & 2 \\ 2 & 1 \end{bmatrix}$$

Solution: From Example 7-1, the characteristic equation of A is

$$c(\lambda) = \lambda^2 - 2\lambda - 3 = (\lambda + 1)(\lambda - 3) = 0$$

with eigenvalues $\lambda_1 = -1$, $\lambda_2 = 3$. Now from (7.50) we get

$$A^k = \beta_0 I + \beta_1 A$$

and from (7.52) we have

$$(-1)^k = \beta_0 - \beta_1$$
$$3^k = \beta_0 + 3\beta_1$$

Solving for β_0 and β_1, we get

$$\beta_0 = \tfrac{1}{4}3^k + \tfrac{3}{4}(-1)^k$$
$$\beta_1 = \tfrac{1}{4}3^k - \tfrac{1}{4}(-1)^k$$

Hence

$$A^k = \left[\frac{1}{4}3^k + \frac{3}{4}(-1)^k\right]\begin{bmatrix} 1 & 0 \\ 0 & 1 \end{bmatrix} + \left[\frac{1}{4}3^k - \frac{1}{4}(-1)^k\right]\begin{bmatrix} 1 & 2 \\ 2 & 1 \end{bmatrix}$$
$$= \begin{bmatrix} \frac{1}{2}3^k + \frac{1}{2}(-1)^k & \frac{1}{2}3^k - \frac{1}{2}(-1)^k \\ \frac{1}{2}3^k - \frac{1}{2}(-1)^k & \frac{1}{2}3^k + \frac{1}{2}(-1)^k \end{bmatrix} = \frac{1}{2}\begin{bmatrix} 3^k + (-1)^k & 3^k - (-1)^k \\ 3^k - (-1)^k & 3^k + (-1)^k \end{bmatrix}$$

EXAMPLE 7-23: Find A^{-1} for

$$A = \begin{bmatrix} 1 & 2 \\ 2 & 1 \end{bmatrix}$$

using **(a)** the Cayley-Hamilton theorem and **(b)** the result of Example 7-22.

Solution

(a) Since A satisfies its own characteristic equation, then

$$A^2 - 2A - 3I_2 = O$$

from which, by multiplying each term by A^{-1}, we get

$$A - 2I_2 - 3A^{-1} = O$$

Thus,

$$A^{-1} = \tfrac{1}{3}A - \tfrac{2}{3}I_2$$
$$= \frac{1}{3}\begin{bmatrix} 1 & 2 \\ 2 & 1 \end{bmatrix} - \frac{2}{3}\begin{bmatrix} 1 & 0 \\ 0 & 1 \end{bmatrix} = \begin{bmatrix} -\frac{1}{3} & \frac{2}{3} \\ \frac{2}{3} & -\frac{1}{3} \end{bmatrix}$$

(b) Setting $k = -1$ in the result of Example 7-22, we get

$$A^{-1} = \frac{1}{2}\begin{bmatrix} 3^{-1} + (-1)^{-1} & 3^{-1} - (-1)^{-1} \\ 3^{-1} - (-1)^{-1} & 3^{-1} + (-1)^{-1} \end{bmatrix}$$
$$= \frac{1}{2}\begin{bmatrix} \frac{1}{3} - 1 & \frac{1}{3} + 1 \\ \frac{1}{3} + 1 & \frac{1}{3} - 1 \end{bmatrix} = \frac{1}{2}\begin{bmatrix} -\frac{2}{3} & \frac{4}{3} \\ \frac{4}{3} & -\frac{2}{3} \end{bmatrix} = \begin{bmatrix} -\frac{1}{3} & \frac{2}{3} \\ \frac{2}{3} & -\frac{1}{3} \end{bmatrix}$$

SUMMARY

1. Let A be an $n \times n$ matrix. If $A\mathbf{x} = \lambda\mathbf{x}$, then λ is called an eigenvalue of A and $\mathbf{x}(\neq\mathbf{0})$ is called the corresponding eigenvector of A.
2. The eigenvalue λ of A can be found from the characteristic equation of A defined by

$$\det(\lambda I_n - A) = 0$$

3. Let $T: V \to V$ be a linear operator on a finite-dimensional vector space V. If $T\mathbf{v} = \lambda\mathbf{v}$, then λ is called an eigenvalue of T and $\mathbf{v}(\neq \mathbf{0})$ is called the corresponding eigenvector.
4. Let A be the matrix of T relative to a basis $\mathscr{B}$; then T and A have the same eigenvalues.
5. Similar matrices have the same eigenvalues.
6. Let A be an $n \times n$ matrix with eigenvalues $\lambda_1, \lambda_2, \ldots, \lambda_n$. Then

(1) $\det A = \lambda_1 \lambda_2 \ldots \lambda_n$.
(2) $\operatorname{tr} A = \lambda_1 + \lambda_2 + \ldots + \lambda_n$.
(3) A is singular if and only if 0 is an eigenvalue of A.
(4) A and A^T have the same eigenvalues.
(5) If A is nonsingular, then $\lambda_1^{-1}, \lambda_2^{-1}, \ldots, \lambda_n^{-1}$ are the eigenvalues of A^{-1}.

7. If $\lambda_1, \lambda_2, \ldots, \lambda_k$ are distinct eigenvalues of an $n \times n$ matrix A, then the corresponding eigenvectors $\mathbf{x}_1, \mathbf{x}_2, \ldots, \mathbf{x}_k$ are linearly independent.
8. An $n \times n$ matrix A is diagonalizable if A is similar to a diagonal matrix D.
9. An $n \times n$ matrix A is diagonalizable if and only if A has n linearly independent eigenvectors.
10. If an $n \times n$ matrix A has n distinct eigenvalues, then A is diagonalizable.
11. The eigenvalues of a symmetric matrix are all real, and eigenvectors corresponding to distinct eigenvalues are orthogonal to each other.
12. An $n \times n$ complex matrix A is Hermitian if

$$A^* = (\bar{A})^T = A$$

13. The eigenvalues of a Hermitian matrix are all real, and eigenvectors corresponding to distinct eigenvalues are orthogonal to each other.
14. The Cayley-Hamilton theorem says that every square matrix satisfies its own characteristic equation; that is, if $c(\lambda) = \det(\lambda I_n - A)$, then $c(A) = O$.

RAISE YOUR GRADES

Can you explain ...?

☑ how to find the eigenvalues and the corresponding eigenvectors of a matrix A
☑ under what condition a matrix A is diagonalizable
☑ a Hermitian matrix
☑ the properties of the eigenvalues and eigenvectors of a symmetric or Hermitian matrix
☑ the Cayley-Hamilton theorem and how to use it to find a function of a matrix

SOLVED PROBLEMS

Eigenvalues and Eigenvectors

PROBLEM 7-1 Find the eigenvalues and the corresponding eigenvectors of the matrix

$$A = \begin{bmatrix} 0 & 1 \\ -1 & 0 \end{bmatrix}$$

Solution Since

$$c(\lambda) = |\lambda I_2 - A| = \begin{vmatrix} \lambda & -1 \\ 1 & \lambda \end{vmatrix} = \lambda^2 + 1 = (\lambda - i)(\lambda + i)$$

if $\mathbf{x} \in R^2$, then there are no eigenvalues of A. If $\mathbf{x} \in C^2$, then $\lambda = i$ and $\lambda = -i$ are the eigenvalues of A.

When $\lambda = i$, the corresponding eigenvector $\mathbf{x}$ is found by $(\lambda I_2 - A)\mathbf{x} = \mathbf{0}$ with $\lambda = i$, or

$$\begin{bmatrix} i & -1 \\ 1 & i \end{bmatrix}\begin{bmatrix} x_1 \\ x_2 \end{bmatrix} = \begin{bmatrix} 0 \\ 0 \end{bmatrix}$$

which corresponds to

$$\begin{aligned} ix_1 - x_2 &= 0 \\ x_1 + ix_2 &= 0 \end{aligned}$$

and we get $x_2 = ix_1$. Thus the eigenvectors of A corresponding to $\lambda = i$ are all the nonzero multiples of

$$\mathbf{u} = \begin{bmatrix} 1 \\ i \end{bmatrix}$$

Similarly, when $\lambda = -i$, we have

$$\begin{bmatrix} -i & -1 \\ 1 & -i \end{bmatrix}\begin{bmatrix} x_1 \\ x_2 \end{bmatrix} = \begin{bmatrix} 0 \\ 0 \end{bmatrix}$$

or

$$\begin{aligned} -ix_1 - x_2 &= 0 \\ x_1 - ix_2 &= 0 \end{aligned}$$

from which we get $x_2 = -ix_1$. Thus the eigenvectors of A corresponding to $\lambda = -i$ are all the nonzero complex multiples of

$$\mathbf{v} = \begin{bmatrix} 1 \\ -i \end{bmatrix}$$

PROBLEM 7-2 Find the eigenvalues of

$$A = \begin{bmatrix} 1 & 2 & -1 \\ 0 & 3 & 4 \\ 0 & 0 & 5 \end{bmatrix}$$

Solution The characteristic polynomial of A is

$$c(\lambda) = |\lambda I_3 - A| = \begin{vmatrix} \lambda - 1 & -2 & 1 \\ 0 & \lambda - 3 & -4 \\ 0 & 0 & \lambda - 5 \end{vmatrix} = (\lambda - 1)(\lambda - 3)(\lambda - 5)$$

Note that the determinant is just the product of the diagonal entries. [see property (2.5)]. Thus the eigenvalues are $\lambda_1 = 1$, $\lambda_2 = 3$, and $\lambda_3 = 5$. Note that the eigenvalues are exactly the same as the main diagonal entries.

PROBLEM 7-3 Let $T: R^3 \to R^3$ be a linear operator defined by

$$T\left(\begin{bmatrix} x_1 \\ x_2 \\ x_3 \end{bmatrix}\right) = \begin{bmatrix} 2x_1 + x_2 \\ 2x_2 \\ 2x_1 + 3x_2 + x_3 \end{bmatrix}$$

Find the eigenvalues and eigenvectors of T.

Solution The standard basis of R^3 is $\{\mathbf{i},\mathbf{j},\mathbf{k}\}$. Then

$$T(\mathbf{i}) = T\left(\begin{bmatrix} 1 \\ 0 \\ 0 \end{bmatrix}\right) = \begin{bmatrix} 2 \\ 0 \\ 2 \end{bmatrix}, \qquad T(\mathbf{j}) = T\left(\begin{bmatrix} 0 \\ 1 \\ 0 \end{bmatrix}\right) = \begin{bmatrix} 1 \\ 2 \\ 3 \end{bmatrix}, \qquad T(\mathbf{k}) = T\left(\begin{bmatrix} 0 \\ 0 \\ 1 \end{bmatrix}\right) = \begin{bmatrix} 0 \\ 0 \\ 1 \end{bmatrix}$$

Thus the standard matrix A of T is

$$A = \begin{bmatrix} 2 & 1 & 0 \\ 0 & 2 & 0 \\ 2 & 3 & 1 \end{bmatrix}$$

The characteristic polynomial of A is

$$|\lambda I_3 - A| = \begin{bmatrix} \lambda - 2 & -1 & 0 \\ 0 & \lambda - 2 & 0 \\ -2 & -3 & \lambda - 1 \end{bmatrix} = (\lambda - 1)(\lambda - 2)^2$$

Thus the eigenvalues of T are $\lambda_1 = 1$, $\lambda_2 = 2$, and $\lambda_3 = 2$.

To determine the eigenvectors corresponding to $\lambda_1 = 1$, we consider

$$(\lambda I_3 - A)\mathbf{x} = \mathbf{0} \qquad \text{with } \lambda = \lambda_1 = 1$$

or

$$\begin{bmatrix} -1 & -1 & 0 \\ 0 & -1 & 0 \\ -2 & -3 & 0 \end{bmatrix} \begin{bmatrix} x_1 \\ x_2 \\ x_3 \end{bmatrix} = \begin{bmatrix} 0 \\ 0 \\ 0 \end{bmatrix}$$

or

$$\begin{aligned} -x_1 - x_2 &= 0 \\ - x_2 &= 0 \\ -2x_1 - 3x_2 &= 0 \end{aligned}$$

The solutions of this system are given by $x_1 = 0$, $x_2 = 0$, and $x_3 = t$, where t is an arbitrary parameter. Thus the eigenvectors corresponding to the eigenvalue 1 are all the nonzero real scalar multiples of

$$\mathbf{u} = \begin{bmatrix} 0 \\ 0 \\ 1 \end{bmatrix}$$

For the eigenvalues $\lambda_2 = \lambda_3 = 2$, we have

$$\begin{bmatrix} 0 & -1 & 0 \\ 0 & 0 & 0 \\ -2 & -3 & 1 \end{bmatrix} \begin{bmatrix} x_1 \\ x_2 \\ x_3 \end{bmatrix} = \begin{bmatrix} 0 \\ 0 \\ 0 \end{bmatrix}$$

or

$$\begin{aligned} - x_2 \qquad &= 0 \\ -2x_1 - 3x_2 + x_3 &= 0 \end{aligned}$$

The solutions of this system are given by $x_1 = s$, $x_2 = 0$, and $x_3 = 2x_1 = 2s$, where s is an arbitrary constant. Thus the eigenvectors corresponding to the eigenvalue 2 are all the nonzero real scalar multiples of

$$\mathbf{v} = \begin{bmatrix} 1 \\ 0 \\ 2 \end{bmatrix}$$

PROBLEM 7-4 Verify (7.8), that is,

$$\det A = \lambda_1 \lambda_2 \cdots \lambda_n$$

Solution From (7.4) and (7.5) we have

$$c(\lambda) = \det(\lambda I_n - A) = \lambda^n + c_1\lambda^{n-1} + \cdots + c_n$$
$$= (\lambda - \lambda_1)(\lambda - \lambda_2)\cdots(\lambda - \lambda_n)$$

for all λ. In particular, when $\lambda = 0$, we get

$$\det(-A) = (-\lambda_1)(-\lambda_2)\cdots(-\lambda_n) = (-1)^n\lambda_1\lambda_2\cdots\lambda_n$$

Since

$$\det(-A) = (-1)^n \det A$$

(see Problem 2-8), it follows that $\det A = \lambda_1\lambda_2\cdots\lambda_n$.

PROBLEM 7-5 Verify (7.9), that is,

$$\operatorname{tr}(A) = \lambda_1 + \lambda_2 + \cdots + \lambda_n$$

Solution Let

$$A = \begin{bmatrix} a_{11} & a_{12} & \cdots & a_{1n} \\ a_{21} & a_{22} & \cdots & a_{2n} \\ \vdots & \vdots & & \vdots \\ a_{n1} & a_{n2} & \cdots & a_{nn} \end{bmatrix}$$

Then

$$c(\lambda) = |\lambda I_n - A| = \begin{vmatrix} \lambda - a_{11} & -a_{12} & \cdots & -a_{1n} \\ -a_{21} & \lambda - a_{22} & \cdots & -a_{2n} \\ \vdots & \vdots & & \vdots \\ -a_{n1} & -a_{n2} & \cdots & \lambda - a_{nn} \end{vmatrix} \qquad \textbf{(a)}$$

$$= (\lambda - a_{11})(\lambda - a_{22})\cdots(\lambda - a_{nn})$$
$$+ \text{ terms with at most } (n-2) \text{ factors of the form } (\lambda - a_{ii})$$
$$= \lambda^n - (a_{11} + a_{22} + \cdots + a_{nn})\lambda^{n-1} + \text{terms of lower degree}$$

Now if $\lambda_1, \lambda_2, \ldots, \lambda_n$ are eigenvalues of A, then

$$c(\lambda) = (\lambda - \lambda_1)(\lambda - \lambda_2)\cdots(\lambda - \lambda_n) \qquad \textbf{(b)}$$
$$= \lambda^n - (\lambda_1 + \lambda_2 + \cdots + \lambda_n)\lambda^{n-1} + \text{terms of lower degree}$$

Comparing (a) and (b), we get

$$\operatorname{tr}(A) = a_{11} + a_{22} + \cdots + a_{nn} = \lambda_1 + \lambda_2 + \cdots + \lambda_n$$

PROBLEM 7-6 Show that 0 is an eigenvalue of A if and only if A is singular.

Solution From (7.8) we have

$$\det A = \lambda_1\lambda_2\cdots\lambda_n$$

where λ_i are the eigenvalues of A. Thus if 0 is an eigenvalue of A, then $\det A = 0$; hence A is singular. On the other hand, if A is singular, then $\det A = 0 = \lambda_1\lambda_2\cdots\lambda_n$, and at least one eigenvalue of A is zero.

PROBLEM 7-7 Show that A and A^T have the same eigenvalues.

Solution Since $\det(A) = \det(A^T)$ for any $n \times n$ matrix [see property (2.6)], we have

$$\det(\lambda I_n - A) = \det(\lambda I_n - A)^T$$
$$= \det(\lambda I_n^T - A^T) \qquad \text{[by (1.41)]}$$
$$= \det(\lambda I_n - A^T)$$

so that A and A^T have the same characteristic polynomial. Therefore A and A^T have the same eigenvalues.

PROBLEM 7-8 Let A be a nonsingular matrix. Show that if λ is an eigenvalue of A, then λ^{-1} is an eigenvalue of A^{-1}.

Solution If λ is an eigenvalue of A, then by definition, there is a nonzero vector $\mathbf{x}$ such that

$$A\mathbf{x} = \lambda\mathbf{x}$$

Multiplying both sides by A^{-1} from the left we get $\mathbf{x} = \lambda A^{-1}\mathbf{x}$. Hence

$$A^{-1}\mathbf{x} = \frac{1}{\lambda}\mathbf{x} = \lambda^{-1}\mathbf{x}$$

which indicates that λ^{-1} is an eigenvalue of A^{-1}.

PROBLEM 7-9 Show that similar matrices have the same characteristic polynomial and hence the same eigenvalues.

Solution If A and B are similar, then by definition (5.36), there is a nonsingular matrix P such that $B = P^{-1}AP$. Then

$$\begin{aligned}
\det(\lambda I_n - B) &= \det(\lambda I_n - P^{-1}AP) \\
&= \det(\lambda P^{-1}I_nP - P^{-1}AP) \\
&= \det[P^{-1}(\lambda I_n - A)P] \\
&= \det(P^{-1})\det(\lambda I_n - A)\det(P) \qquad \text{[by (2.7)]} \\
&= \det(\lambda I_n - A)
\end{aligned}$$

since $\det(P^{-1})\det(P) = 1$ by (2.9). Thus A and B have the same characteristic polynomial and hence, the same eigenvalues.

PROBLEM 7-10 Let A and B be similar matrices and $B = P^{-1}AP$. Show that $\mathbf{x}$ is an eigenvector of A corresponding to the eigenvalue λ if and only if $P^{-1}\mathbf{x}$ is an eigenvector of B corresponding to the eigenvalue λ.

Solution Let λ be an eigenvalue of A and $\mathbf{x}$ be the eigenvector corresponding to λ. Then $A\mathbf{x} = \lambda\mathbf{x}$. From $B = P^{-1}AP$, we get $A = PBP^{-1}$. Thus $PBP^{-1}\mathbf{x} = \lambda\mathbf{x}$. Multiplying both sides by P^{-1} from the left, we get $BP^{-1}\mathbf{x} = P^{-1}(\lambda\mathbf{x})$ or $B(P^{-1}\mathbf{x}) = \lambda(P^{-1}\mathbf{x})$, which indicates that $P^{-1}\mathbf{x}$ is the eigenvector of B corresponding to the eigenvalue λ.

On the other hand, if $P^{-1}\mathbf{x}$ is the eigenvector of B corresponding to an eigenvalue λ, then tracing backward, we can show that $A\mathbf{x} = \lambda\mathbf{x}$. That is, $\mathbf{x}$ is the eigenvector of A corresponding to the eigenvalue λ.

PROBLEM 7-11 Give an example to show that matrices having the same characteristic polynomial are not necessarily similar.

Solution Consider

$$I_2 = \begin{bmatrix} 1 & 0 \\ 0 & 1 \end{bmatrix} \quad \text{and} \quad B = \begin{bmatrix} 1 & 1 \\ 0 & 1 \end{bmatrix}$$

Since

$$|\lambda I_2 - I_2| = \begin{vmatrix} \lambda - 1 & 0 \\ 0 & \lambda - 1 \end{vmatrix} = (\lambda - 1)^2$$

$$|\lambda I_2 - B| = \begin{vmatrix} \lambda - 1 & -1 \\ 0 & \lambda - 1 \end{vmatrix} = (\lambda - 1)^2$$

I_2 and B have the same characteristic polynomial $(\lambda - 1)^2$; however, for all nonsingular 2×2 matrices P,

$$P^{-1}I_2P = P^{-1}P = I_2$$

Thus

$$P^{-1}I_2P \neq B$$

for any P. Hence I_2 and B are not similar.

Eigenspaces

PROBLEM 7-12 Given

$$A = \begin{bmatrix} -1 & 2 & 2 \\ 2 & -1 & 2 \\ 2 & 2 & -1 \end{bmatrix}$$

find the eigenvalues of A and the dimensions for the corresponding eigenspaces.

Solution

$$|\lambda I_3 - A| = \begin{vmatrix} \lambda + 1 & -2 & -2 \\ -2 & \lambda + 1 & -2 \\ -2 & -2 & \lambda + 1 \end{vmatrix} = \lambda^3 + 3\lambda^2 - 9\lambda - 27 = (\lambda - 3)(\lambda + 3)^2$$

Thus the eigenvalues of A are $\lambda_1 = 3$ and $\lambda_2 = -3$ with multiplicity 2.

Let

$$\mathbf{x} = \begin{bmatrix} x_1 \\ x_2 \\ x_3 \end{bmatrix}$$

be the eigenvector of A. Then we have $[\lambda I_3 - A]\mathbf{x} = \mathbf{0}$.

When $\lambda = 3$, we have

$$\begin{bmatrix} 4 & -2 & -2 \\ -2 & 4 & -2 \\ -2 & -2 & 4 \end{bmatrix} \begin{bmatrix} x_1 \\ x_2 \\ x_3 \end{bmatrix} = \begin{bmatrix} 0 \\ 0 \\ 0 \end{bmatrix}$$

or

$$\begin{bmatrix} 2 & -1 & -1 \\ -1 & 2 & -1 \\ -1 & -1 & 2 \end{bmatrix} \begin{bmatrix} x_1 \\ x_2 \\ x_3 \end{bmatrix} = \begin{bmatrix} 0 \\ 0 \\ 0 \end{bmatrix}$$

which corresponds to

$$\begin{aligned} 2x_1 - x_2 - x_3 &= 0 \\ -x_1 + 2x_2 - x_3 &= 0 \\ -x_1 - x_2 + 2x_3 &= 0 \end{aligned}$$

Solving for x_1, x_2, x_3, we get $x_1 = x_2 = x_3 = t$, where t is an arbitrary parameter. Hence $\mathbf{x}$ has the form

$$\mathbf{x} = \begin{bmatrix} t \\ t \\ t \end{bmatrix} = t \begin{bmatrix} 1 \\ 1 \\ 1 \end{bmatrix}$$

Thus the basis for the eigenspace of A corresponding to $\lambda = 3$ is

$$\mathbf{u} = \begin{bmatrix} 1 \\ 1 \\ 1 \end{bmatrix}$$

and the dimension of this eigenspace is 1.

When $\lambda = -3$, we have

$$\begin{bmatrix} -2 & -2 & -2 \\ -2 & -2 & -2 \\ -2 & -2 & -2 \end{bmatrix}\begin{bmatrix} x_1 \\ x_2 \\ x_3 \end{bmatrix} = \begin{bmatrix} 0 \\ 0 \\ 0 \end{bmatrix}$$

which corresponds to the single equation

$$x_1 + x_2 + x_3 = 0$$

Solving for x_1, x_2, x_3, we get $x_3 = s$, $x_2 = t$, $x_1 = -x_2 - x_3 = -t - s$, where t and s are arbitrary parameters. Thus $\mathbf{x}$ has the form

$$\mathbf{x} = \begin{bmatrix} -t-s \\ t \\ s \end{bmatrix} = t\begin{bmatrix} -1 \\ 1 \\ 0 \end{bmatrix} + s\begin{bmatrix} -1 \\ 0 \\ 1 \end{bmatrix}$$

Hence the bases for the eigenspace of A corresponding to $\lambda = -3$ are

$$\mathbf{v} = \begin{bmatrix} -1 \\ 1 \\ 0 \end{bmatrix} \quad \text{and} \quad \mathbf{w} = \begin{bmatrix} -1 \\ 0 \\ 1 \end{bmatrix}$$

and the dimension of this eigenspace is 2.

PROBLEM 7-13 Verify Theorem 7-2.1.

Solution The zero vector is in V_λ, so V_λ is nonempty. Let $\mathbf{v}_1, \mathbf{v}_2 \in V_\lambda$ and α any scalar. Then

$$T(\mathbf{v}_1 + \mathbf{v}_2) = T\mathbf{v}_1 + T\mathbf{v}_2 = \lambda\mathbf{v}_1 + \lambda\mathbf{v}_2 = \lambda(\mathbf{v}_1 + \mathbf{v}_2)$$

Hence $\mathbf{v}_1 + \mathbf{v}_2 \in V_\lambda$. Next,

$$T(\alpha\mathbf{v}_1) = \alpha T\mathbf{v}_1 = \alpha\lambda\mathbf{v}_1 = \lambda(\alpha\mathbf{v}_1)$$

Hence

$$\alpha\mathbf{v}_1 \in V_\lambda$$

Thus V_λ is closed under addition and scalar multiplication. Therefore by Theorem 3-4.1, V is a subspace of V.

PROBLEM 7-14 Verify Theorem 7-2.2.

Solution We shall prove this theorem by contradiction.

Assume that $\{\mathbf{x}_1, \mathbf{x}_2, \ldots, \mathbf{x}_k\}$ is linearly dependent. Now, since an eigenvector is, by definition, a nonzero vector, $\mathbf{x}_1 \neq \mathbf{0}$; then it is obvious that $\{\mathbf{x}_1\}$ is linearly independent. Next, let m be the largest integer such that $\{\mathbf{x}_1, \mathbf{x}_2, \ldots, \mathbf{x}_m\}$ is linearly independent. Evidently, $1 \leq m < k$ and the set $\{\mathbf{x}_1, \ldots, \mathbf{x}_m, \mathbf{x}_{m+1}\}$ is linearly dependent. Thus there are scalars $\alpha_1, \alpha_2, \ldots, \alpha_{m+1}$, not all zero, such that

$$\alpha_1\mathbf{x}_1 + \alpha_2\mathbf{x}_2 + \cdots + \alpha_m\mathbf{x}_m + \alpha_{m+1}\mathbf{x}_{m+1} = \mathbf{0} \tag{a}$$

Then

$$A(\alpha_1\mathbf{x}_1 + \alpha_2\mathbf{x}_2 + \cdots + \alpha_m\mathbf{x}_m + \alpha_{m+1}\mathbf{x}_{m+1}) = A\mathbf{0} = \mathbf{0}$$

or

$$\alpha_1 A\mathbf{x}_1 + \alpha_2 A\mathbf{x}_2 + \cdots + \alpha_m A\mathbf{x}_m + \alpha_{m+1} A\mathbf{x}_{m+1} = \mathbf{0}$$

or

$$\alpha_1\lambda_1\mathbf{x}_1 + \alpha_2\lambda_2\mathbf{x}_2 + \cdots + \alpha_m\lambda_m\mathbf{x}_m + \alpha_{m+1}\lambda_{m+1}\mathbf{x}_{m+1} = \mathbf{0} \tag{b}$$

Multiplying both sides of (a) by λ_{m+1} and subtracting the resulting equation from (b) yields

$$\alpha_1(\lambda_1 - \lambda_{m+1})\mathbf{x}_1 + \alpha_2(\lambda_2 - \lambda_{m+1})\mathbf{x}_2 + \cdots + \alpha_m(\lambda_m - \lambda_{m+1})\mathbf{x}_m = \mathbf{0} \tag{c}$$

Since $\{\mathbf{x}_1, \mathbf{x}_2, \ldots, \mathbf{x}_m\}$ is linearly independent, (c) implies that

$$\alpha_1(\lambda_1 - \lambda_{m+1}) = \alpha_2(\lambda_2 - \lambda_{m+1}) = \cdots = \alpha_m(\lambda_m - \lambda_{m+1}) = 0$$

Since $\lambda_i - \lambda_k \neq 0$ for $i \neq k$, it follows that

$$\alpha_1 = \alpha_2 = \cdots = \alpha_m = 0 \tag{d}$$

Substituting these values in (a) yields

$$\alpha_{m+1}\mathbf{x}_{m+1} = \mathbf{0}$$

Since $\mathbf{x}_{m+1} \neq \mathbf{0}$, we must have

$$\alpha_{m+1} = 0$$

Consequently, $\{\mathbf{x}_1, \ldots, \mathbf{x}_m, \mathbf{x}_{m+1}\}$ is linearly independent. This contradicts our assumption that $\{\mathbf{x}_1, \mathbf{x}_2, \ldots, \mathbf{x}_{m+1}\}$ is linearly dependent, which is based on the assumption that $\{\mathbf{x}_1, \mathbf{x}_2, \ldots, \mathbf{x}_k\}$ is linearly dependent. Therefore $\{\mathbf{x}_1, \mathbf{x}_2, \ldots, \mathbf{x}_k\}$ must be linearly independent.

Diagonalization

PROBLEM 7-15 Show that

$$A = \begin{bmatrix} 1 & 1 & 0 \\ 0 & 2 & 1 \\ 0 & 0 & 3 \end{bmatrix}$$

is diagonalizable; also, find P such that $P^{-1}AP = D$.

Solution

$$c(\lambda) = |\lambda I_3 - A| = \begin{bmatrix} \lambda - 1 & -1 & 0 \\ 0 & \lambda - 2 & -1 \\ 0 & 0 & \lambda - 3 \end{bmatrix} = (\lambda - 1)(\lambda - 2)(\lambda - 3) = 0$$

Hence the eigenvalues of A are $\lambda_1 = 1$, $\lambda_2 = 2$, $\lambda_3 = 3$ and they are all distinct. Thus by Theorem 7-3.2, A is diagonalizable.

Let

$$\mathbf{x} = \begin{bmatrix} x_1 \\ x_2 \\ x_3 \end{bmatrix}$$

be an eigenvector of A. Then for $\lambda = 1$ we have

$$\begin{bmatrix} 0 & -1 & 0 \\ 0 & -1 & -1 \\ 0 & 0 & -2 \end{bmatrix} \begin{bmatrix} x_1 \\ x_2 \\ x_3 \end{bmatrix} = \begin{bmatrix} 0 \\ 0 \\ 0 \end{bmatrix}$$

or

$$\begin{aligned} -x_2 \qquad &= 0 \\ -x_2 - x_3 &= 0 \\ -2x_3 &= 0 \end{aligned}$$

from which we get $x_1 = s$, $x_2 = x_3 = 0$, where s is arbitrary. Setting $s = 1$,

$$\mathbf{x} = \begin{bmatrix} 1 \\ 0 \\ 0 \end{bmatrix} = \mathbf{x}_1$$

For $\lambda = 2$,

$$\begin{bmatrix} 1 & -1 & 0 \\ 0 & 0 & -1 \\ 0 & 0 & -1 \end{bmatrix} \begin{bmatrix} x_1 \\ x_2 \\ x_3 \end{bmatrix} = \begin{bmatrix} 0 \\ 0 \\ 0 \end{bmatrix}$$

or

$$\begin{aligned} x_1 - x_2 &= 0 \\ -x_3 &= 0 \end{aligned}$$

from which we get $x_1 = x_2 = t$, $x_3 = 0$, where t is arbitrary. Setting $t = 1$,

$$\mathbf{x} = \begin{bmatrix} 1 \\ 1 \\ 0 \end{bmatrix} = \mathbf{x}_2$$

For $\lambda = 3$,

$$\begin{bmatrix} 2 & -1 & 0 \\ 0 & 1 & -1 \\ 0 & 0 & 0 \end{bmatrix} \begin{bmatrix} x_1 \\ x_2 \\ x_3 \end{bmatrix} = \begin{bmatrix} 0 \\ 0 \\ 0 \end{bmatrix}$$

or

$$\begin{aligned} 2x_1 - x_2 &= 0 \\ x_2 - x_3 &= 0 \end{aligned}$$

from which we get $x_1 = \frac{1}{2}r$, $x_2 = x_3 = r$, where r is arbitrary. Setting $r = 2$,

$$\mathbf{x} = \begin{bmatrix} 1 \\ 2 \\ 2 \end{bmatrix} = \mathbf{x}_3$$

Let

$$P = [\mathbf{x}_1, \mathbf{x}_2, \mathbf{x}_3] = \begin{bmatrix} 1 & 1 & 1 \\ 0 & 1 & 2 \\ 0 & 0 & 2 \end{bmatrix}$$

then

$$P^{-1} = \begin{bmatrix} 1 & -1 & \frac{1}{2} \\ 0 & 1 & -1 \\ 0 & 0 & \frac{1}{2} \end{bmatrix}$$

and

$$P^{-1}AP = \begin{bmatrix} 1 & -1 & \frac{1}{2} \\ 0 & 1 & -1 \\ 0 & 0 & \frac{1}{2} \end{bmatrix} \begin{bmatrix} 1 & 1 & 0 \\ 0 & 2 & 1 \\ 0 & 0 & 3 \end{bmatrix} \begin{bmatrix} 1 & 1 & 1 \\ 0 & 1 & 2 \\ 0 & 0 & 2 \end{bmatrix} = \begin{bmatrix} 1 & 0 & 0 \\ 0 & 2 & 0 \\ 0 & 0 & 3 \end{bmatrix}$$

PROBLEM 7-16 Determine whether the matrix of Problem 7-12,

$$A = \begin{bmatrix} -1 & 2 & 2 \\ 2 & -1 & 2 \\ 2 & 2 & -1 \end{bmatrix}$$

is diagonalizable.

Solution From Problem 7-12, the eigenvalues of A are $\lambda_1 = 3$ and $\lambda_2 = -3$ with multiplicity 2. There are, however, three linearly independent eigenvectors:

$$\begin{bmatrix} 1 \\ 1 \\ 1 \end{bmatrix}, \quad \begin{bmatrix} -1 \\ 1 \\ 0 \end{bmatrix}, \quad \begin{bmatrix} -1 \\ 0 \\ 1 \end{bmatrix}$$

Hence, by Theorem 7-3.1, A is diagonalizable.

As a matter of fact, let

$$P = \begin{bmatrix} 1 & -1 & -1 \\ 1 & 1 & 0 \\ 1 & 0 & 1 \end{bmatrix}$$

Then

$$P^{-1} = \frac{1}{3}\begin{bmatrix} 1 & 1 & 1 \\ -1 & 2 & -1 \\ 1 & -1 & 2 \end{bmatrix}$$

and

$$P^{-1}AP = \frac{1}{3}\begin{bmatrix} 1 & 1 & 1 \\ -1 & 2 & -1 \\ 1 & -1 & 2 \end{bmatrix}\begin{bmatrix} -1 & 2 & 2 \\ 2 & -1 & 2 \\ 2 & 2 & -1 \end{bmatrix}\begin{bmatrix} 1 & -1 & -1 \\ 1 & 1 & 0 \\ 1 & 0 & 1 \end{bmatrix} = \begin{bmatrix} 3 & 0 & 0 \\ 0 & -3 & 0 \\ 0 & 0 & -3 \end{bmatrix}$$

PROBLEM 7-17 Verify Theorem 7-3.1.

Solution Suppose that A has linearly independent eigenvectors $\mathbf{x}_1, \mathbf{x}_2, \ldots, \mathbf{x}_n$. Let λ_i be the eigenvalue of A corresponding to $\mathbf{x}_i$ for each i. (Note that some of the λ_i's may be equal; see Problem 7-12.) Let P be the matrix whose ith column vector is $\mathbf{x}_i$ (where $i = 1, 2, \ldots, n$), that is,

$$P = [\mathbf{x}_1, \mathbf{x}_2, \ldots, \mathbf{x}_n]$$

Then

$$\begin{aligned} AP &= A[\mathbf{x}_1, \mathbf{x}_2, \ldots, \mathbf{x}_n] \\ &= [A\mathbf{x}_1, A\mathbf{x}_2, \ldots, A\mathbf{x}_n] \\ &= [\lambda_1\mathbf{x}_1, \lambda_2\mathbf{x}_2, \ldots, \lambda_n\mathbf{x}_n] \\ &= [\mathbf{x}_1, \mathbf{x}_2, \ldots, \mathbf{x}_n]\begin{bmatrix} \lambda_1 & 0 & \cdots & 0 \\ 0 & \lambda_2 & \cdots & 0 \\ \vdots & \vdots & & \vdots \\ 0 & 0 & \cdots & \lambda_n \end{bmatrix} \\ &= PD \end{aligned}$$

where D is the diagonal matrix having the eigenvalues $\lambda_1, \lambda_2, \ldots, \lambda_n$ on the diagonal entries.

Since the column vectors of P are linearly independent, it follows that P is nonsingular (see Example 3-58), and hence $D = P^{-1}PD = P^{-1}AP$. Conversely, if A is diagonalizable, then there exists a nonsingular matrix P such that $P^{-1}AP = D$ or $AP = PD$. If $\mathbf{x}_1, \mathbf{x}_2, \ldots, \mathbf{x}_n$ are the column vectors of P, then

$$\begin{aligned} AP &= A[\mathbf{x}_1, \mathbf{x}_2, \ldots, \mathbf{x}_n] \\ &= [A\mathbf{x}_1, A\mathbf{x}_2, \ldots, A\mathbf{x}_n] \\ &= PD = [\mathbf{x}_1, \mathbf{x}_2, \ldots, \mathbf{x}_n]\begin{bmatrix} d_1 & 0 & \cdots & 0 \\ 0 & d_2 & \cdots & 0 \\ \vdots & \vdots & & \vdots \\ 0 & 0 & \cdots & d_n \end{bmatrix} \\ &= [d_1\mathbf{x}_1, d_2\mathbf{x}_2, \ldots, d_n\mathbf{x}_n] \end{aligned}$$

Thus

$$A\mathbf{x}_i = d_i\mathbf{x}_i = \lambda_i\mathbf{x}_i \qquad (d_i = \lambda_i, i = 1, 2, \ldots, n)$$

Hence for each i, $\lambda_i = d_i$ is an eigenvalue of A and $\mathbf{x}_i$ is an eigenvector corresponding to λ_i. Since P is nonsingular, the column vectors of P are linearly independent, and it follows that A has n linearly independent eigenvectors.

PROBLEM 7-18 Find a 2×2 matrix A whose eigenvalues are $\lambda_1 = 1$ and $\lambda_2 = 3$ with corresponding eigenvectors

$$\mathbf{x}_1 = \begin{bmatrix} 1 \\ 3 \end{bmatrix}, \qquad \mathbf{x}_2 = \begin{bmatrix} 2 \\ 5 \end{bmatrix}$$

Solution The matrix A must be diagonalizable by the matrix

$$P = [\mathbf{x}_1, \mathbf{x}_2] = \begin{bmatrix} 1 & 2 \\ 3 & 5 \end{bmatrix}$$

Since we have

$$P^{-1}AP = D = \begin{bmatrix} \lambda_1 & 0 \\ 0 & \lambda_2 \end{bmatrix} = \begin{bmatrix} 1 & 0 \\ 0 & 3 \end{bmatrix}$$

we can find A by $A = PDP^{-1}$. Now since

$$P^{-1} = \begin{bmatrix} 1 & 2 \\ 3 & 5 \end{bmatrix}^{-1} = \begin{bmatrix} -5 & 2 \\ 3 & -1 \end{bmatrix}$$

we get

$$A = PDP^{-1} = \begin{bmatrix} 1 & 2 \\ 3 & 5 \end{bmatrix}\begin{bmatrix} 1 & 0 \\ 0 & 3 \end{bmatrix}\begin{bmatrix} -5 & 2 \\ 3 & -1 \end{bmatrix} = \begin{bmatrix} 13 & -4 \\ 30 & -9 \end{bmatrix}$$

PROBLEM 7-19 If A is diagonalizable, that is, $P^{-1}AP = D$, then show that $A^k = PD^kP^{-1}$ for any integer k.

Solution Since $P^{-1}AP = D$, we get $A = PDP^{-1}$. Then

$$A^2 = (PDP^{-1})(PDP^{-1}) = PD^2P^{-1}$$
$$A^3 = A^2A = (PD^2P^{-1})(PDP^{-1}) = PD^3P^{-1}$$
$$\vdots$$

Now assume that $A^k = PD^kP^{-1}$ is true for $k = n$, that is, $A^n = PD^nP^{-1}$. Then

$$A^{n+1} = A^nA = (PD^nP^{-1})(PDP^{-1}) = PD^{n+1}P^{-1}$$

Hence $A^k = PD^kP^{-1}$ is true for any integer k.

PROBLEM 7-20 Evaluate A^k for

$$A = \begin{bmatrix} 1 & 2 \\ 2 & 1 \end{bmatrix}$$

by the diagonalization technique.

Solution From Example 7-6, we get $P^{-1}AP = D$, where

$$P = \begin{bmatrix} 1 & 1 \\ -1 & 1 \end{bmatrix}, \qquad P^{-1} = \frac{1}{2}\begin{bmatrix} 1 & -1 \\ 1 & 1 \end{bmatrix}, \qquad D = \begin{bmatrix} -1 & 0 \\ 0 & 3 \end{bmatrix}$$

Now by Problem 7-19 we have $A^k = PD^kP^{-1}$, and by (7.43)

$$D^k = \begin{bmatrix} (-1)^k & 0 \\ 0 & 3^k \end{bmatrix}$$

Hence

$$A^k = \frac{1}{2}\begin{bmatrix} 1 & 1 \\ -1 & 1 \end{bmatrix}\begin{bmatrix} (-1)^k & 0 \\ 0 & 3^k \end{bmatrix}\begin{bmatrix} 1 & -1 \\ 1 & 1 \end{bmatrix} = \frac{1}{2}\begin{bmatrix} (-1)^k + 3^k & -(-1)^k + 3^k \\ -(-1)^k + 3^k & (-1)^k + 3^k \end{bmatrix}$$

(cf. Example 7-22).

Symmetric Matrices and Hermitian Matrices

PROBLEM 7-21 Let A be a real $n \times n$ symmetric matrix. Show that A has a real nonzero eigenvector.

Solution We use the fact that if we view A as a linear operator of C^n into C^n, then A has a complex nonzero eigenvector $\mathbf{z} \in C^n$. Thus there exists $\lambda \in C$ such that

$$A\mathbf{z} = \lambda\mathbf{z}$$

and by Theorem 7-4.1, λ is real. Let $\mathbf{z} \in C^n$. Then we can express $\mathbf{z}$ in a unique way as

$$\mathbf{z} = \mathbf{x} + i\mathbf{y}$$

with real vectors $\mathbf{x}$, $\mathbf{y}$. For example, $\begin{bmatrix} 1+2i \\ 3-i \end{bmatrix} = \begin{bmatrix} 1 \\ 3 \end{bmatrix} + i\begin{bmatrix} 2 \\ -1 \end{bmatrix}$. Now

$$A\mathbf{z} = A\mathbf{x} + iA\mathbf{y}$$

and

$$A\mathbf{z} = \lambda\mathbf{z} = \lambda\mathbf{x} + i\lambda\mathbf{y}$$

Since λ is real, and since the real and imaginary parts of a complex vector are uniquely determined, it follows that

$$A\mathbf{x} = \lambda\mathbf{x} \qquad \text{and} \qquad A\mathbf{y} = \lambda\mathbf{y}.$$

Since $\mathbf{z} \neq \mathbf{0}$, at least one of $\mathbf{x}$ and $\mathbf{y}$ is not a zero vector; and so one of them is a nonzero eigenvector.

PROBLEM 7-22 Show that the matrix (of Problem 7-12)

$$A = \begin{bmatrix} -1 & 2 & 2 \\ 2 & -1 & 2 \\ 2 & 2 & -1 \end{bmatrix}$$

is orthogonally diagonalizable.

Solution It is seen that A is symmetric. From Problem 7-12, the eigenvalues of A are $\lambda_1 = 3$ and $\lambda_2 = -3$ with multiplicity 2, and the corresponding eigenvectors are

$$\mathbf{u} = \begin{bmatrix} 1 \\ 1 \\ 1 \end{bmatrix}, \qquad \mathbf{v} = \begin{bmatrix} -1 \\ 1 \\ 0 \end{bmatrix}, \qquad \mathbf{w} = \begin{bmatrix} -1 \\ 0 \\ 1 \end{bmatrix}$$

Although $\mathbf{u}$ is orthogonal to $\mathbf{v}$ and $\mathbf{w}$, we know that $\mathbf{v}$ and $\mathbf{w}$ are not orthogonal. Recalling that if

$$\mathbf{x} = \begin{bmatrix} x_1 \\ x_2 \\ x_3 \end{bmatrix}$$

is an eigenvector corresponding to $\lambda = -3$, we get

$$x_1 + x_2 + x_3 = 0$$

If we select

$$\mathbf{v} = \begin{bmatrix} -1 \\ 1 \\ 0 \end{bmatrix}$$

then we can select **w** such that **w** is orthogonal to **v**, that is, **w** satisfies

$$x_1 + x_2 + x_3 = 0 \qquad \text{and} \qquad -x_1 + x_2 = 0$$

It thus has the form $x_1 = s, x_2 = s, x_3 = -2s$. Setting $s = 1$, we get

$$\mathbf{w} = \begin{bmatrix} 1 \\ 1 \\ -2 \end{bmatrix}$$

Now let

$$\mathbf{u}_1 = \frac{1}{\|\mathbf{u}\|}\mathbf{u} = \frac{1}{\sqrt{3}}\begin{bmatrix} 1 \\ 1 \\ 1 \end{bmatrix}$$

$$\mathbf{u}_2 = \frac{1}{\|\mathbf{v}\|}\mathbf{v} = \frac{1}{\sqrt{2}}\begin{bmatrix} -1 \\ 1 \\ 0 \end{bmatrix}$$

$$\mathbf{u}_3 = \frac{1}{\|\mathbf{w}\|}\mathbf{w} = \frac{1}{\sqrt{6}}\begin{bmatrix} 1 \\ 1 \\ -2 \end{bmatrix}$$

and

$$P = [\mathbf{u}_1, \mathbf{u}_2, \mathbf{u}_3] = \begin{bmatrix} \frac{1}{\sqrt{3}} & -\frac{1}{\sqrt{2}} & \frac{1}{\sqrt{6}} \\ \frac{1}{\sqrt{3}} & \frac{1}{\sqrt{2}} & \frac{1}{\sqrt{6}} \\ \frac{1}{\sqrt{3}} & 0 & -\frac{2}{\sqrt{6}} \end{bmatrix}$$

Then

$$P^{-1} = \begin{bmatrix} \frac{1}{\sqrt{3}} & \frac{1}{\sqrt{3}} & \frac{1}{\sqrt{3}} \\ -\frac{1}{\sqrt{2}} & \frac{1}{\sqrt{2}} & 0 \\ \frac{1}{\sqrt{6}} & \frac{1}{\sqrt{6}} & -\frac{2}{\sqrt{6}} \end{bmatrix} = P^T$$

and P is an orthogonal matrix. Now

$$P^{-1}AP = \begin{bmatrix} \frac{1}{\sqrt{3}} & \frac{1}{\sqrt{3}} & \frac{1}{\sqrt{3}} \\ -\frac{1}{\sqrt{2}} & \frac{1}{\sqrt{2}} & 0 \\ \frac{1}{\sqrt{6}} & \frac{1}{\sqrt{6}} & -\frac{2}{\sqrt{6}} \end{bmatrix} \begin{bmatrix} -1 & 2 & 2 \\ 2 & -1 & 2 \\ 2 & 2 & -1 \end{bmatrix} \begin{bmatrix} \frac{1}{\sqrt{3}} & -\frac{1}{\sqrt{2}} & \frac{1}{\sqrt{6}} \\ \frac{1}{\sqrt{3}} & \frac{1}{\sqrt{2}} & \frac{1}{\sqrt{6}} \\ \frac{1}{\sqrt{3}} & 0 & -\frac{2}{\sqrt{6}} \end{bmatrix} = \begin{bmatrix} 3 & 0 & 0 \\ 0 & -3 & 0 \\ 0 & 0 & -3 \end{bmatrix}$$

Thus A is orthogonally diagonalizable.

PROBLEM 7-23 Let A be a Hermitian matrix; then **(a)** show that $\langle A\mathbf{x}, \mathbf{x} \rangle$ is real for all complex vectors **x**, and **(b)** verify this result with

$$A = \begin{bmatrix} 2 & 1-i \\ 1+i & 1 \end{bmatrix}$$

Solution

(a) Using (6.43), we can express

$$\langle A\mathbf{x},\mathbf{x}\rangle = \mathbf{x}^*A\mathbf{x}$$

Now

$$(\mathbf{x}^*A\mathbf{x})^* = \mathbf{x}^*A^*(\mathbf{x}^*)^* = \mathbf{x}^*A\mathbf{x} \qquad [\text{since } (\mathbf{x}^*)^* = \mathbf{x},\ A^* = A]$$

which indicates that the complex conjugate of the 1×1 matrix $\mathbf{x}^*A\mathbf{x}$ is the same as the original $\mathbf{x}^*A\mathbf{x}$. Hence $\langle A\mathbf{x},\mathbf{x}\rangle$ must be real.

(b) Let

$$\mathbf{x} = \begin{bmatrix} x_1 \\ x_2 \end{bmatrix} \in C^2$$

Then

$$\begin{aligned}
\langle A\mathbf{x},\mathbf{x}\rangle &= \mathbf{x}^*A\mathbf{x} \\
&= [\bar{x}_1, \bar{x}_2]\begin{bmatrix} 2 & 1-i \\ 1+i & 1 \end{bmatrix}\begin{bmatrix} x_1 \\ x_2 \end{bmatrix} \\
&= [\bar{x}_1, \bar{x}_2]\begin{bmatrix} 2x_1 + (1-i)x_2 \\ (1+i)x_1 + x_2 \end{bmatrix} \\
&= 2x_1\bar{x}_1 + (1-i)\bar{x}_1x_2 + (1+i)x_1\bar{x}_2 + x_2\bar{x}_2
\end{aligned}$$

Now

$$x_1\bar{x}_1 = |x_1|^2, \qquad x_2\bar{x}_2 = |x_2|^2 \qquad [\text{by (6.26)}]$$

and since $\overline{(1-i)\bar{x}_1x_2} = \overline{(1-i)}\bar{\bar{x}}_1\bar{x}_2 = (1+i)x_1\bar{x}_2$,

$$(1-i)\bar{x}_1x_2 + (1+i)x_1\bar{x}_2 = 2\mathcal{Re}[(1-i)\bar{x}_1x_2] \qquad [\text{by (6.23)}]$$

Hence

$$\langle A\mathbf{x},\mathbf{x}\rangle = 2|x_1|^2 + 2\mathcal{Re}[(1-i)\bar{x}_1x_2] + |x_2|^2$$

is real.

PROBLEM 7-24 Show that the determinant of any Hermitian matrix is real and illustrate this by using the matrix of Example 7-16 or Problem 7-23:

$$A = \begin{bmatrix} 2 & 1-i \\ 1+i & 1 \end{bmatrix}$$

Solution Let $\lambda_1, \lambda_2, \ldots, \lambda_n$ be the eigenvalues of a Hermitian matrix A. Then they are real by Theorem 7-4.3. Furthermore, by (7.8) we have $\det A = \lambda_1\lambda_2\ldots\lambda_n$. Hence $\det A$ is real. And if

$$A = \begin{bmatrix} 2 & 1-i \\ 1+i & 1 \end{bmatrix}$$

then A is Hermitian and its eigenvalues are $\lambda_1 = 0$, $\lambda_2 = 3$. Now

$$\begin{aligned}
\det A = \begin{vmatrix} 2 & 1-i \\ 1+i & 1 \end{vmatrix} &= (2)(1) - (1+i)(1-i) \\
&= 2 - (1 - i^2) = 2 - (1+1) = 0 = \lambda_1\lambda_2 = (0)(3) = 0
\end{aligned}$$

PROBLEM 7-25 Show that

$$A = \begin{bmatrix} 2 & 3+3i \\ 3-3i & 5 \end{bmatrix}$$

is unitarily diagonalizable.

Solution

$$A^* = (\bar{A})^T = \begin{bmatrix} 2 & 3-3i \\ 3+3i & 5 \end{bmatrix}^T = \begin{bmatrix} 2 & 3+3i \\ 3-3i & 5 \end{bmatrix} = A$$

Hence A is Hermitian.

$$c(\lambda) = |\lambda I_2 - A| = \begin{vmatrix} \lambda - 2 & -3-3i \\ -3+3i & \lambda - 5 \end{vmatrix}$$

$$= (\lambda - 2)(\lambda - 5) - (-3+3i)(-3-3i)$$

$$= \lambda^2 - 7\lambda - 8 = (\lambda + 1)(\lambda - 8)$$

Hence the eigenvalues of A are $\lambda_1 = -1$, $\lambda_2 = 8$.

Let $\mathbf{x} = \begin{bmatrix} x_1 \\ x_2 \end{bmatrix}$ be an eigenvector of A. When $\lambda = -1$, then

$$\begin{bmatrix} -3 & -3-3i \\ -3+3i & -6 \end{bmatrix} \begin{bmatrix} x_1 \\ x_2 \end{bmatrix} = \begin{bmatrix} 0 \\ 0 \end{bmatrix}$$

or

$$\begin{aligned} -3x_1 + (-3-3i)x_2 &= 0 \\ (-3+3i)x_1 - \quad 6x_2 &= 0 \end{aligned}$$

from which we get $x_1 = -(1+i)x_2$. Hence, setting $x_2 = -1$, we get

$$\mathbf{x} = \begin{bmatrix} 1+i \\ -1 \end{bmatrix} = \mathbf{x}_1$$

When $\lambda = 8$, then

$$\begin{bmatrix} 6 & -3-3i \\ -3+3i & 3 \end{bmatrix} \begin{bmatrix} x_1 \\ x_2 \end{bmatrix} = \begin{bmatrix} 0 \\ 0 \end{bmatrix}$$

or

$$\begin{aligned} 6x_1 + (-3-3i)x_2 &= 0 \\ (-3+3i)x_1 + \quad 3x_2 &= 0 \end{aligned}$$

from which we get $x_2 = (1-i)x_1$. Hence, setting $x = 1$, we get

$$\mathbf{x} = \begin{bmatrix} 1 \\ 1-i \end{bmatrix} = \mathbf{x}_2$$

Next, normalizing $\mathbf{x}_1$ and $\mathbf{x}_2$, we get

$$\mathbf{u}_1 = \frac{1}{\|\mathbf{x}_1\|}\mathbf{x}_1 = \frac{1}{\sqrt{3}} \begin{bmatrix} 1+i \\ -1 \end{bmatrix}$$

$$\mathbf{u}_2 = \frac{1}{\|\mathbf{x}_2\|}\mathbf{x}_2 = \frac{1}{\sqrt{3}} \begin{bmatrix} 1 \\ 1-i \end{bmatrix}$$

Let

$$P = [\mathbf{u}_1, \mathbf{u}_2] = \frac{1}{\sqrt{3}} \begin{bmatrix} 1+i & 1 \\ -1 & 1-i \end{bmatrix}$$

Then

$$P^{-1} = \frac{1}{\sqrt{3}} \begin{bmatrix} 1-i & -1 \\ 1 & 1+i \end{bmatrix} = P^*$$

and

$$P^{-1}AP = \frac{1}{3}\begin{bmatrix} 1-i & -1 \\ 1 & 1+i \end{bmatrix}\begin{bmatrix} 2 & 3+3i \\ 3-3i & 5 \end{bmatrix}\begin{bmatrix} 1+i & 1 \\ -1 & 1-i \end{bmatrix}$$

$$= \frac{1}{3}\begin{bmatrix} -3 & 0 \\ 0 & 24 \end{bmatrix} = \begin{bmatrix} -1 & 0 \\ 0 & 8 \end{bmatrix} = \begin{bmatrix} \lambda_1 & 0 \\ 0 & \lambda_2 \end{bmatrix}$$

Hence A is unitarily diagonalizable.

PROBLEM 7-26 Verify the spectrum theorem (7.37) with

$$A = \begin{bmatrix} 2 & 3+3i \\ 3-3i & 5 \end{bmatrix}$$

Solution From Problem 7-25, we have $\lambda_1 = -1$, $\lambda_2 = 8$ and

$$\mathbf{u}_1 = \frac{1}{\sqrt{3}}\begin{bmatrix} 1+i \\ -1 \end{bmatrix}, \qquad \mathbf{u}_2 = \frac{1}{\sqrt{3}}\begin{bmatrix} 1 \\ 1-i \end{bmatrix}$$

Now

$$\mathbf{u}_1\mathbf{u}_1^* = \frac{1}{3}\begin{bmatrix} 1+i \\ -1 \end{bmatrix}[1-i, -1] = \frac{1}{3}\begin{bmatrix} 2 & -1-i \\ -1+i & 1 \end{bmatrix}$$

$$\mathbf{u}_2\mathbf{u}_2^* = \frac{1}{3}\begin{bmatrix} 1 \\ 1-i \end{bmatrix}[1, 1+i] = \frac{1}{3}\begin{bmatrix} 1 & 1+i \\ 1-i & 2 \end{bmatrix}$$

$$\lambda_1\mathbf{u}_1\mathbf{u}_1^* + \lambda_2\mathbf{u}_2\mathbf{u}_2^* = (-1)\frac{1}{3}\begin{bmatrix} 2 & -1-i \\ -1+i & 1 \end{bmatrix} + (8)\frac{1}{3}\begin{bmatrix} 1 & 1+i \\ 1-i & 2 \end{bmatrix}$$

$$= \begin{bmatrix} 2 & 3+3i \\ 3-3i & 5 \end{bmatrix} = A$$

Hence $A = \lambda_1\mathbf{u}_1\mathbf{u}_1^* + \lambda_2\mathbf{u}_2\mathbf{u}_2^*$.

The Cayley-Hamilton Theorem

PROBLEM 7-27 Let

$$p(\lambda) = \lambda^2 + 2\lambda - 3 = (\lambda - 1)(\lambda + 3)$$

and

$$A = \begin{bmatrix} 1 & -1 \\ 2 & 0 \end{bmatrix}$$

Evaluate $p(A)$.

Solution

$$p(A) = A^2 + 2A - 3I_2$$

$$= \begin{bmatrix} 1 & -1 \\ 2 & 0 \end{bmatrix}^2 + 2\begin{bmatrix} 1 & -1 \\ 2 & 0 \end{bmatrix} - 3\begin{bmatrix} 1 & 0 \\ 0 & 1 \end{bmatrix}$$

$$= \begin{bmatrix} -1 & -1 \\ 2 & -2 \end{bmatrix} + \begin{bmatrix} 2 & -2 \\ 4 & 0 \end{bmatrix} - \begin{bmatrix} 3 & 0 \\ 0 & 3 \end{bmatrix} = \begin{bmatrix} -2 & -3 \\ 6 & -5 \end{bmatrix}$$

or

$$p(A) = (A - I_2)(A + 3I_2)$$
$$= \left(\begin{bmatrix} 1 & -1 \\ 2 & 0 \end{bmatrix} - \begin{bmatrix} 1 & 0 \\ 0 & 1 \end{bmatrix}\right)\left(\begin{bmatrix} 1 & -1 \\ 2 & 0 \end{bmatrix} + 3\begin{bmatrix} 1 & 0 \\ 0 & 1 \end{bmatrix}\right)$$
$$= \begin{bmatrix} 0 & -1 \\ 2 & -1 \end{bmatrix}\begin{bmatrix} 4 & -1 \\ 2 & 3 \end{bmatrix} = \begin{bmatrix} -2 & -3 \\ 6 & -5 \end{bmatrix}$$

PROBLEM 7-28 Verify the Cayley-Hamilton theorem (7.45).

Solution Let A be an $n \times n$ matrix. Then

$$C = \lambda I_n - A = \begin{bmatrix} \lambda - a_{11} & -a_{12} & \cdots & -a_{1n} \\ -a_{21} & \lambda - a_{22} & \cdots & -a_{2n} \\ \vdots & \vdots & & \vdots \\ -a_{n1} & -a_{n2} & \cdots & \lambda - a_{nn} \end{bmatrix} \quad \textbf{(a)}$$

$$\det C = |\lambda I_n - A| = c(\lambda) = \lambda^n + c_{n-1}\lambda^{n-1} + \cdots + c_1\lambda + c_0 \quad \textbf{(b)}$$

Since C is of order n, adj C will contain polynomials in λ of degree not higher than $n - 1$. Hence adj C can be expanded into a polynomial with matrix coefficients of degree at most $n - 1$;

$$\operatorname{adj} C = C_{n-1}\lambda^{n-1} + C_{n-2}\lambda^{n-2} + \cdots + C_1\lambda + C_0 \quad \textbf{(c)}$$

where each C_i is a matrix with scalar elements. Now by (2.18) we have

$$C^{-1} = \frac{1}{\det C}\operatorname{adj} C$$

or

$$I_n = C^{-1}C = \frac{1}{\det C}(\operatorname{adj} C)C$$

Thus

$$(\operatorname{adj} C)C = (\det C)I_n = c(\lambda)I_n \quad \textbf{(d)}$$

or

$$(\operatorname{adj} C)(\lambda I_n - A) = (\operatorname{adj} C)\lambda - (\operatorname{adj} C)A = c(\lambda)I_n \quad \textbf{(e)}$$

Hence, by (b) and (c),

$$C_{n-1}\lambda^n + C_{n-2}\lambda^{n-1} + \cdots + C_1\lambda^2 + C_0\lambda - C_{n-1}A\lambda^{n-1} - \cdots - C_1A\lambda - C_0A \quad \textbf{(f)}$$
$$= I_n\lambda^n + c_{n-1}I_n\lambda^{n-1} + \cdots + c_0I_n$$

Equating the coefficients of λ^i for both sides, we get

$$C_{n-1} = I_n$$
$$C_{n-2} - C_{n-1}A = c_{n-1}I_n$$
$$\vdots$$
$$C_0 - C_1A = c_1I_n$$
$$-C_0A = c_0I_n$$

Multiplying each of these equations by $A^n, A^{n-1}, \ldots, A, I_n$ from right, respectively, we get

$$C_{n-1}A^n = A^n$$
$$C_{n-2}A^{n-1} - C_{n-1}A^n = c_{n-1}A^{n-1}$$
$$\vdots$$
$$C_0A - C_1A^2 = c_1A$$
$$-C_0A = c_0I_n$$

Now if we add both sides, the terms on the left side cancel out, leaving the zero matrix, and the terms on the right side add up to

$$A^n + c_{n-1}A^{n-1} + \cdots + c_1 A + c_0 I_n = c(A) = O$$

PROBLEM 7-29 Using (7.50), evaluate A^k for

$$A = \begin{bmatrix} 2 & 0 \\ 1 & 1 \end{bmatrix}$$

Solution The characteristic equation of A is

$$c(\lambda) = |\lambda I_2 - A| = \begin{vmatrix} \lambda - 2 & 0 \\ -1 & \lambda - 1 \end{vmatrix} = (\lambda - 2)(\lambda - 1) = 0$$

with eigenvalues $\lambda_1 = 2$, $\lambda_2 = 1$. Now from (7.50), $A^k = \beta_0 I_2 + \beta_1 A$. Then from (7.51), we have

$$2^k = \beta_0 + 2\beta_1$$
$$1^k = \beta_0 + \beta_1$$

Solving for β_0 and β_1, we get

$$\beta_0 = 2 - 2^k, \qquad \beta_1 = 2^k - 1$$

Hence

$$A^k = (2 - 2^k)\begin{bmatrix} 1 & 0 \\ 0 & 1 \end{bmatrix} + (2^k - 1)\begin{bmatrix} 2 & 0 \\ 1 & 1 \end{bmatrix}$$
$$= \begin{bmatrix} 2 - 2^k & 0 \\ 0 & 2 - 2^k \end{bmatrix} + \begin{bmatrix} 2(2^k - 1) & 0 \\ 2^k - 1 & 2^k - 1 \end{bmatrix}$$
$$= \begin{bmatrix} 2^k & 0 \\ 2^k - 1 & 1 \end{bmatrix}$$

PROBLEM 7-30 Find A^{-1} for the matrix of Problem 7-29 (**a**) using the Cayley-Hamilton theorem and (**b**) using the result of Problem 7-29.

Solution From Problem 7-29, the characteristic equation of A is

$$c(\lambda) = \lambda^2 - 3\lambda + 2 = 0$$

(**a**) By the Cayley-Hamilton theorem, A satisfies its own characteristic equation; then

$$A^2 - 3A + 2I_2 = O$$

Multiplying both sides by A^{-1}, we get

$$A - 3I_2 + 2A^{-1} = O$$

Therefore

$$A^{-1} = \frac{3}{2}I_2 - \frac{1}{2}A = \frac{3}{2}\begin{bmatrix} 1 & 0 \\ 0 & 1 \end{bmatrix} - \frac{1}{2}\begin{bmatrix} 2 & 0 \\ 1 & 1 \end{bmatrix} = \begin{bmatrix} \frac{1}{2} & 0 \\ -\frac{1}{2} & 1 \end{bmatrix}$$

(**b**) Setting $k = -1$ in the result of Problem 7-29, we get

$$A^{-1} = \begin{bmatrix} 2^{-1} & 0 \\ 2^{-1} - 1 & 1 \end{bmatrix} = \begin{bmatrix} \frac{1}{2} & 0 \\ \frac{1}{2} - 1 & 1 \end{bmatrix} = \begin{bmatrix} \frac{1}{2} & 0 \\ -\frac{1}{2} & 1 \end{bmatrix}$$

***PROBLEM 7-31** [*For readers who have studied calculus.*] Find A^k for

$$A = \begin{bmatrix} 1 & 0 & 4 \\ 0 & 2 & 0 \\ 0 & 1 & 2 \end{bmatrix}$$

Solution The characteristic equation of A is

$$c(\lambda) = |\lambda I_3 - A| = \begin{vmatrix} \lambda - 1 & 0 & -4 \\ 0 & \lambda - 2 & 0 \\ 0 & -1 & \lambda - 2 \end{vmatrix} = (\lambda - 1)(\lambda - 2)^2 = 0$$

and the eigenvalues of A are $\lambda_1 = 1$, $\lambda_2 = 2$ with multiplicity 2. Now from (7.50), we have

$$f(A) = A^k = \beta_0 I_3 + \beta_1 A + \beta_2 A^2$$

Let $f(\lambda) = \lambda^k$ and $g(\lambda) = \beta_0 + \beta_1 \lambda + \beta_2 \lambda^2$. Then coefficients $\beta_0, \beta_1, \beta_2$ are found from the conditions

$$f(\lambda)|_{\lambda=1} = g(\lambda)|_{\lambda=1}$$
$$f(\lambda)|_{\lambda=2} = g(\lambda)|_{\lambda=2}$$
$$f'(\lambda)|_{\lambda=2} = g'(\lambda)|_{\lambda=2}$$

Hence

$$1^k = \beta_0 + \beta_1 + \beta_2$$
$$2^k = \beta_0 + 2\beta_1 + 4\beta_2$$
$$k2^{k-1} = \beta_1 + 4\beta_2$$

Solving this system for $\beta_0, \beta_1, \beta_2$ we get

$$\beta_0 = -3(2)^k + 2k(2)^{k-1} + 4$$
$$\beta_1 = 4(2)^k - 3k(2)^{k-1} - 4$$
$$\beta_2 = -(2)^k + k(2)^{k-1} + 1$$

Thus

$$A^k = \beta_0 I_3 + \beta_1 A + \beta_2 A^2$$
$$= [-3(2)^k + 2k(2)^{k-1} + 4] \begin{bmatrix} 1 & 0 & 0 \\ 0 & 1 & 0 \\ 0 & 0 & 1 \end{bmatrix} + [4(2)^k - 3k(2)^{k-1} - 4] \begin{bmatrix} 1 & 0 & 4 \\ 0 & 2 & 0 \\ 0 & 1 & 2 \end{bmatrix}$$
$$+ [-(2)^k + k(2)^{k-1} + 1] \begin{bmatrix} 1 & 4 & 12 \\ 0 & 4 & 0 \\ 0 & 4 & 4 \end{bmatrix}$$
$$= \begin{bmatrix} 1 & -4(2^k - k2^{k-1} - 1) & 4(2^k - 1) \\ 0 & 2^k & 0 \\ 0 & k2^{k-1} & 2^k \end{bmatrix}$$

Supplementary Exercises

PROBLEM 7-32 Find the eigenvalues and the eigenvectors of the matrix

$$A = \begin{bmatrix} 1 & 2 \\ -2 & 1 \end{bmatrix}$$

Answer $1 + 2i, \begin{bmatrix} 1 \\ i \end{bmatrix}$ $1 - 2i, \begin{bmatrix} 1 \\ -i \end{bmatrix}$

***PROBLEM 7-33** [*For readers who have studied calculus.*] Let $T: C[a, b] \to C[a, b]$ be the differential operator, such that $T[f(x)] = f'(x)$. Find **(a)** the eigenvalues and **(b)** the eigenvectors of T.

Answer **(a)** Any scalar s (real or complex) **(b)** e^{sx}

PROBLEM 7-34 Find the eigenvalues and eigenvectors of $D = [d_i \delta_{ij}]$

Answer $\lambda_i = d_i$, $\mathbf{x}_i = \mathbf{e}_i$ for $i = 1, 2, \ldots, n$, where $\{\mathbf{e}_i\}$ is the standard basis for R^n

PROBLEM 7-35 Show that if λ is an eigenvalue of A, then λ^2 is an eigenvalue of A^2.

PROBLEM 7-36 Show that the eigenspace of $T: V \to V$, V_λ is the kernel of $\lambda I - T$, where I is the identity operator and λ is an eigenvalue of T.

PROBLEM 7-37 **(a)** Determine whether the matrix

$$A = \begin{bmatrix} 0 & 1 & 0 \\ -1 & 0 & 0 \\ 0 & 0 & 1 \end{bmatrix}$$

is diagonalizable; **(b)** if yes, then find P that diagonalizes A.

Answer **(a)** yes **(b)** $P = \begin{bmatrix} 0 & 1 & 1 \\ 0 & i & -i \\ 1 & 0 & 0 \end{bmatrix}$

PROBLEM 7-38 Determine whether the matrix

$$A = \begin{bmatrix} -1 & 1 & 1 \\ -2 & 2 & 1 \\ -1 & 1 & 1 \end{bmatrix}$$

is diagonalizable.

Answer No

PROBLEM 7-39 Show that if A is Hermitian and P is unitary, then $P^{-1}AP$ is also Hermitian.

PROBLEM 7-40 Show that

$$(P^{-1}AP)^k = P^{-1}A^kP$$

for all positive integers k.

PROBLEM 7-41 Given

$$A = \begin{bmatrix} 2 & 0 \\ 1 & 1 \end{bmatrix}$$

find $\sqrt{A}$. [*Hint:* Use (7.50) and (7.52).]

Answer $\begin{bmatrix} \sqrt{2} & 0 \\ \sqrt{2} - 1 & 1 \end{bmatrix}$

8 PROJECTION AND APPROXIMATION

THIS CHAPTER IS ABOUT

- ☑ Projection
- ☑ Best Approximation
- ☑ Least-Squares Fitting of Data
- ☑ Least-Squares Solutions of Inconsistent Systems
- ☑ Fourier Series

8-1. Projection

Let U be a finite-dimensional subspace of an inner vector space V. Let $\{\mathbf{u}_1, \mathbf{u}_2, \ldots, \mathbf{u}_n\}$ be an orthonormal basis for U. Let $\mathbf{v} \in U$. Then by Theorem 4-5.1 we have

$$\mathbf{v} = \langle \mathbf{v},\mathbf{u}_1 \rangle \mathbf{u}_1 + \langle \mathbf{v},\mathbf{u}_2 \rangle \mathbf{u}_2 + \cdots + \langle \mathbf{v},\mathbf{u}_n \rangle \mathbf{u}_n \tag{8.1}$$

Now if $\mathbf{v} \in V$ and $\mathbf{v} \notin U$, the vector, $\text{proj}_U \mathbf{v}$, defined by

$$\text{proj}_U \mathbf{v} = \langle \mathbf{v},\mathbf{u}_1 \rangle \mathbf{u}_1 + \cdots + \langle \mathbf{v},\mathbf{u}_n \rangle \mathbf{u}_n \tag{8.2}$$

is called the *projection of* $\mathbf{v}$ *on* U (see Problem 5-8). Then the vector

$$\mathbf{v}' = \mathbf{v} - \text{proj}_U \mathbf{v}$$

is called the *component of* $\mathbf{v}$ *orthogonal to* U (Figure 8-1).

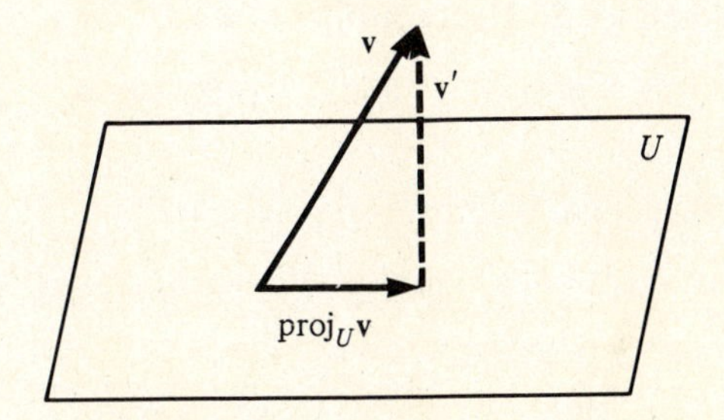

FIGURE 8-1. The projection of v on U.

EXAMPLE 8-1: Show that

$$\begin{aligned} \mathbf{v}' &= \mathbf{v} - \text{proj}_U \mathbf{v} \\ &= \mathbf{v} - (\langle \mathbf{v},\mathbf{u}_1 \rangle \mathbf{u}_1 + \langle \mathbf{v},\mathbf{u}_2 \rangle \mathbf{u}_2 + \cdots + \langle \mathbf{v},\mathbf{u}_n \rangle \mathbf{u}_n) \end{aligned} \tag{8.3}$$

is orthogonal to every vector in U.

Solution: Let $\mathbf{u} \in U$. Then by (8.1)

$$\mathbf{u} = \langle \mathbf{u},\mathbf{u}_1 \rangle \mathbf{u}_1 + \langle \mathbf{u},\mathbf{u}_2 \rangle \mathbf{u}_2 + \cdots + \langle \mathbf{u},\mathbf{u}_n \rangle \mathbf{u}_n$$

Now by (8.3), we have

$$\begin{aligned}
\langle \mathbf{v}',\mathbf{u}\rangle &= \langle \mathbf{v},\mathbf{u}\rangle - \langle \mathbf{v},\mathbf{u}_1\rangle\langle \mathbf{u}_1,\mathbf{u}\rangle - \langle \mathbf{v},\mathbf{u}_2\rangle\langle \mathbf{u}_2,\mathbf{u}\rangle \\
&\quad - \cdots - \langle \mathbf{v},\mathbf{u}_n\rangle\langle \mathbf{u}_n,\mathbf{u}\rangle \\
&= \langle \mathbf{v},(\langle \mathbf{u},\mathbf{u}_1\rangle\mathbf{u}_1 + \langle \mathbf{u},\mathbf{u}_2\rangle\mathbf{u}_2 + \cdots + \langle \mathbf{u},\mathbf{u}_n\rangle\mathbf{u}_n)\rangle \\
&\quad - \langle \mathbf{v},\mathbf{u}_1\rangle\langle \mathbf{u}_1,\mathbf{u}\rangle - \langle \mathbf{v},\mathbf{u}_2\rangle\langle \mathbf{u}_2,\mathbf{u}\rangle - \cdots - \langle \mathbf{v},\mathbf{u}_n\rangle\langle \mathbf{u}_n,\mathbf{u}\rangle \\
&= \overline{\langle \mathbf{u},\mathbf{u}_1\rangle}\langle \mathbf{v},\mathbf{u}_1\rangle + \overline{\langle \mathbf{u},\mathbf{u}_2\rangle}\langle \mathbf{v},\mathbf{u}_2\rangle + \cdots + \overline{\langle \mathbf{u},\mathbf{u}_n\rangle}\langle \mathbf{v},\mathbf{u}_n\rangle \\
&\quad - \langle \mathbf{v},\mathbf{u}_1\rangle\langle \mathbf{u}_1,\mathbf{u}\rangle - \langle \mathbf{v},\mathbf{u}_2\rangle\langle \mathbf{u}_2,\mathbf{u}\rangle - \cdots - \langle \mathbf{v},\mathbf{u}_n\rangle\langle \mathbf{u}_n,\mathbf{u}\rangle \\
&= \langle \mathbf{u}_1,\mathbf{u}\rangle\langle \mathbf{v},\mathbf{u}_1\rangle + \langle \mathbf{u}_2,\mathbf{u}\rangle\langle \mathbf{v},\mathbf{u}_2\rangle + \cdots + \langle \mathbf{u}_n,\mathbf{u}\rangle\langle \mathbf{v},\mathbf{u}_n\rangle \\
&\quad - \langle \mathbf{v},\mathbf{u}_1\rangle\langle \mathbf{u}_1,\mathbf{u}\rangle - \langle \mathbf{v},\mathbf{u}_2\rangle\langle \mathbf{u}_2,\mathbf{u}\rangle - \cdots - \langle \mathbf{v},\mathbf{u}_n\rangle\langle \mathbf{u}_n,\mathbf{u}\rangle \\
&= 0
\end{aligned}$$

since $\langle \mathbf{v}, \langle \mathbf{u},\mathbf{u}_i\rangle \mathbf{u}_i\rangle = \overline{\langle \mathbf{u},\mathbf{u}_i\rangle}\langle \mathbf{v},\mathbf{u}_i\rangle$ [by (6.41)] and $\overline{\langle \mathbf{u},\mathbf{u}_i\rangle} = \langle \mathbf{u}_i,\mathbf{u}\rangle$ [by (6.35)].

Thus, we conclude that $\mathbf{v}'$ is orthogonal to every vector $\mathbf{u}$ in U (cf. Problem 4-31).

8-2. Best Approximation

Let U be a finite-dimensional subspace of an inner product space V. If $\mathbf{v}$ is any vector in V, then $\mathbf{u}^\circ \in U$ is said to be the best approximation to $\mathbf{v}$ from U in the sense that

$$\|\mathbf{v} - \mathbf{u}^\circ\| < \|\mathbf{v} - \mathbf{u}\| \tag{8.4}$$

for every vector $\mathbf{u} \in U$ different from $\mathbf{u}^\circ$.

EXAMPLE 8-2: Show that the best approximation $\mathbf{u}^\circ$ of U to $\mathbf{v} \in V$ in the sense of (8.4) is given by

$$\mathbf{u}^\circ = \text{proj}_U\, \mathbf{v} \tag{8.5}$$

Solution: Let $\mathbf{u}$ be any vector in U; then $\mathbf{v} - \mathbf{u}$ can be expressed as

$$\mathbf{v} - \mathbf{u} = (\mathbf{v} - \text{proj}_U\, \mathbf{v}) + (\text{proj}_U\, \mathbf{v} - \mathbf{u}) \tag{8.6}$$

By definition (8.2), $\text{proj}_U\, \mathbf{v} \in U$ and $\mathbf{u} \in U$. Hence $\text{proj}_U\, \mathbf{v} - \mathbf{u} \in U$. Then by the result of Example 8-1

$$\mathbf{v} - \text{proj}_U\, \mathbf{v} \qquad \text{and} \qquad \text{proj}_U\, \mathbf{v} - \mathbf{u}$$

are orthogonal.

Using the generalized Pythagorean theorem (4.65), we have

$$\|\mathbf{v} - \mathbf{u}\|^2 = \|\mathbf{v} - \text{proj}_U\, \mathbf{v}\|^2 + \|\text{proj}_U\, \mathbf{v} - \mathbf{u}\|^2 \tag{8.7}$$

If $\text{proj}_U\, \mathbf{v} \neq \mathbf{u}$, then $\|\text{proj}_U\, \mathbf{v} - \mathbf{u}\|^2 > 0$, so that

$$\|\mathbf{v} - \text{proj}_U\, \mathbf{v}\|^2 < \|\mathbf{v} - \mathbf{u}\|^2 \tag{8.8}$$

or

$$\|\mathbf{v} - \text{proj}_U\, \mathbf{v}\| < \|\mathbf{v} - \mathbf{u}\| \tag{8.9}$$

Hence we conclude that

$$\mathbf{u}^\circ = \text{proj}_U\, \mathbf{v}.$$

8-3. Least-Squares Fitting of Data

Given the data points $(x_1, y_1), (x_2, y_2), \ldots, (x_n, y_n)$, we would like to find the equation of a line

$$y = mx + b \tag{8.10}$$

that comes close to fitting the given data points. If the line actually contains the data points, we would have the following system of equations for m and b:

$$\begin{aligned} mx_1 + b &= y_1 \\ mx_2 + b &= y_2 \\ \vdots \quad \vdots \; &\;\; \vdots \\ mx_n + b &= y_n \end{aligned} \tag{8.11}$$

Then, if we let

$$\mathbf{x} = \begin{bmatrix} x_1 \\ x_2 \\ \vdots \\ x_n \end{bmatrix}, \qquad \mathbf{y} = \begin{bmatrix} y_1 \\ y_2 \\ \vdots \\ y_n \end{bmatrix}, \qquad \mathbf{a} = \begin{bmatrix} 1 \\ 1 \\ \vdots \\ 1 \end{bmatrix} \tag{8.12}$$

the system of equations (8.11) for m and b can be written as

$$m\mathbf{x} + b\mathbf{a} = \mathbf{y} \tag{8.13}$$

If the data points do not lie on a line, then the system (8.11) is inconsistent and does not have a solution, which means that $\mathbf{y} \notin S\{\mathbf{x},\mathbf{a}\}$; that is, $\mathbf{y}$ does not belong to the span of $\mathbf{x}$ and $\mathbf{a}$ (see Theorem 3-9.3).

EXAMPLE 8-3: Show that the solution (m,b) of (8.11), which minimizes the following mean-square error,

$$\mathscr{E} = (y_1 - (mx_1 + b))^2 + \cdots + (y_n - (mx_n + b))^2 \tag{8.14}$$

is given by

$$m\mathbf{x} + b\mathbf{a} = \text{proj}_S\,\mathbf{y} \tag{8.15}$$

where S is the span of $\{\mathbf{x},\mathbf{a}\}$.

Solution: It is seen that the mean-square error $\mathscr{E}$ can be expressed as

$$\mathscr{E} = \|\mathbf{y} - (m\mathbf{x} + b\mathbf{a})\|^2 \tag{8.16}$$

Minimization of $\mathscr{E}$ is equivalent to the minimization of

$$\|\mathbf{y} - (m\mathbf{x} + b\mathbf{a})\| \tag{8.17}$$

which is the distance from $\mathbf{y}$ to S. That is, if $m'\mathbf{x} + b'\mathbf{a}$ is an arbitrary vector in S, then

$$\|\mathbf{y} - (m\mathbf{x} + b\mathbf{a})\| \le \|\mathbf{y} - (m'\mathbf{x} + b'\mathbf{a})\| \tag{8.18}$$

(see Figure 8-2). Therefore, by (8.5), (m,b) is given by $m\mathbf{x} + b\mathbf{a} = \text{proj}_S\,\mathbf{y}$.

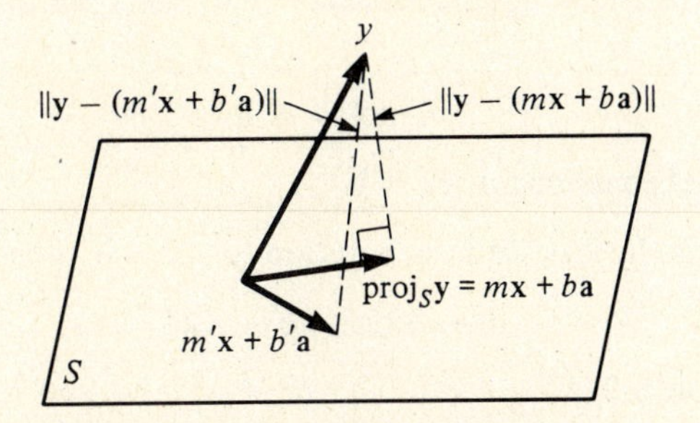

FIGURE 8-2. The minimum distance from y to S.

With (m,b) obtained by (8.15), the resulting line of approximation $y = mx + b$, shown in Figure 8-3, is called the **least-squares line** of approximation. The distance from a data point (x_i,y_i) to the line $y = mx + b$ is given by $|y_i - (mx_i + b)|$, and thus the number $\mathscr{E}$ of (8.14) is the sum of the squares of the distances. Clearly, if $\mathscr{E}$ is small, the distance from any data point to the line is small, and the line would be considered as a best approximation to the data point.

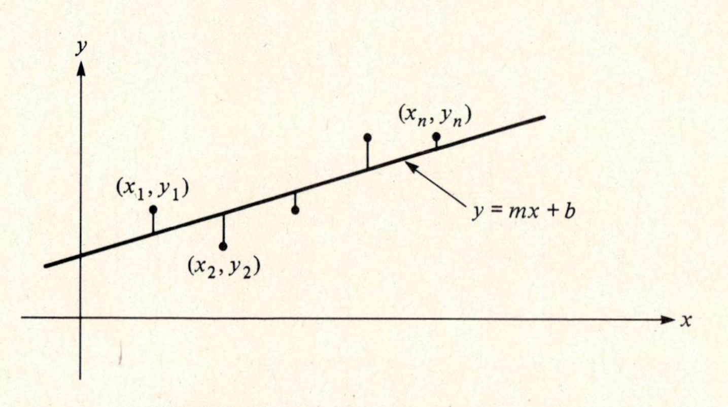

FIGURE 8-3. The least-squares fitting of data.

In the field of statistics, the fitting of a line to given data is called **linear regression** and the line is called the **linear regression line** for the data.

8-4. Least-Squares Solutions of Inconsistent Systems

A. Normal equation for $A\mathbf{x} = \mathbf{b}$

Let A be an $m \times n$ matrix $(m > n)$ and $\mathbf{b} \in R^m$. Consider $U = \{A\mathbf{x} | \mathbf{x} \in R^n\}$, which is the column space of A. It is clear that U is the subspace of R^n. If $\mathbf{b} \in U$, then the system of equations $A\mathbf{x} = \mathbf{b}$ is consistent (see Theorem 3-9.3).

If $\mathbf{b} \notin U$, then the system of equations

$$A\mathbf{x} = \mathbf{b} \tag{8.19}$$

is not consistent. In this case

$$\mathscr{E}(\mathbf{x}) = \mathbf{b} - A\mathbf{x} \tag{8.20}$$

is called an **error vector** (or *residual*). The distance between $\mathbf{b}$ and $A\mathbf{x}$ is given by

$$\|\mathbf{b} - A\mathbf{x}\| = \|\mathscr{E}(\mathbf{x})\| \tag{8.21}$$

A vector $\hat{\mathbf{x}} \in R^n$ that minimizes $\|\mathscr{E}(\mathbf{x})\|$—or equivalently, $\|\mathscr{E}(\mathbf{x})\|^2$—is called the **least-squares solution** to the inconsistent system $A\mathbf{x} = \mathbf{b}$.

EXAMPLE 8-4: Show that the least-squares solution $\hat{\mathbf{x}}$ to the system $A\mathbf{x} = \mathbf{b}$ satisfies the following equation:

$$A^T A\hat{\mathbf{x}} = A^T\mathbf{b} \tag{8.22}$$

which is known as the **normal equation** for $A\mathbf{x} = \mathbf{b}$.

Solution: By definition, if $\hat{\mathbf{x}}$ is the least-squares solution to $A\mathbf{x} = \mathbf{b}$, then

$$\|\mathbf{b} - A\hat{\mathbf{x}}\| \leq \|\mathbf{b} - A\mathbf{x}\|$$

Then by the result of Example 8-2 (8.5),

$$A\hat{\mathbf{x}} = \text{proj}_U \mathbf{b} \tag{8.23}$$

where U is the column space of A.

Since $\mathbf{b} - \text{proj}_U \mathbf{b} = \mathbf{b} - A\hat{\mathbf{x}}$ is orthogonal to every vector of $U = \{A\mathbf{x} | \mathbf{x} \in R^n\}$, we have $\langle A\mathbf{x}, \mathbf{b} - A\hat{\mathbf{x}} \rangle = 0$. Noting that $\langle A\mathbf{u}, \mathbf{v} \rangle = \langle \mathbf{u}, A^T\mathbf{v} \rangle$ [see Example 7-9 (7.21)], we get

$$\begin{aligned} \langle A\mathbf{x}, \mathbf{b} - A\hat{\mathbf{x}} \rangle &= \langle \mathbf{x}, A^T(\mathbf{b} - A\hat{\mathbf{x}}) \rangle \\ &= \langle \mathbf{x}, A^T\mathbf{b} - A^TA\hat{\mathbf{x}} \rangle = 0 \end{aligned} \tag{8.24}$$

for every $\mathbf{x} \in R^n$. This is true if and only if [see (4.48)]

$$A^T\mathbf{b} - A^TA\hat{\mathbf{x}} = \mathbf{0} \tag{8.25}$$

or

$$A^TA\hat{\mathbf{x}} = A^T\mathbf{b}$$

which is the desired result.

B. Generalized inverse

If A^TA is nonsingular, then from (8.22) the least-squares solution $\hat{\mathbf{x}}$ to $A\mathbf{x} = \mathbf{b}$ can be written as

$$\hat{\mathbf{x}} = (A^TA)^{-1}A^T\mathbf{b} = A^-\mathbf{b} \tag{8.26}$$

where

$$A^- = (A^TA)^{-1}A^T \tag{8.27}$$

is called the **generalized inverse** of A.

Note that if A is an $m \times n$ matrix, then A^T is an $n \times m$ matrix so that A^TA and $(A^TA)^{-1}$ are $n \times n$ matrices. Then $(A^TA)^{-1}A^T$ is an $n \times m$ matrix. Thus the generalized inverse A^- of an $m \times n$ matrix A is an $n \times m$ matrix.

EXAMPLE 8-5: Show that if A is an $m \times n$ matrix, then

$$A^-A = I_n \tag{8.28}$$

Solution: By definitions (8.27) and (1.37),

$$A^-A = (A^TA)^{-1}A^TA = (A^TA)^{-1}(A^TA) = I_n$$

Thus we identify A^- as a left inverse of A (see Problem 3-62).

C. Projection matrix

From (8.23) and (8.26) we have

$$\text{proj}_U \mathbf{b} = A\hat{\mathbf{x}} = A(A^TA)^{-1}A^T\mathbf{b} = AA^-\mathbf{b} = P\mathbf{b} \tag{8.29}$$

where

$$P = A(A^TA)^{-1}A^T = AA^- \tag{8.30}$$

is called the **projection matrix**. Hence the projection of $\mathbf{b}$ onto the column space of A is obtained by multiplying the vector $\mathbf{b}$ by the matrix $P = AA^-$. This is a lot simpler than the use of (8.2).

*8-5. Fourier Series [*For readers who have studied calculus.*]

A. Least-squares approximation

Let $\mathbf{f} \in C[a,b]$ and $\mathbf{g} \in U$, where U is a subspace of $C[a,b]$. If we approximate $\mathbf{f} = f(x)$ by $\mathbf{g} = g(x)$, then the mean-square error is defined by

$$\text{mean-square error} = \int_a^b [f(x) - g(x)]^2\,dx \tag{8.31}$$

If $\mathbf{g} = g(x)$ in U minimizes this mean-square error, then the function $\mathbf{g} = g(x)$ is called the **least-squares approximation** to $\mathbf{f} = f(x)$ from U.

EXAMPLE 8-6: If the inner product on $C[a,b]$ is defined by

$$\langle \mathbf{f},\mathbf{g} \rangle = \int_a^b f(x)g(x)\,dx \tag{8.32}$$

then show that the mean-square error of (8.31) can be expressed as

$$\text{mean-square error} = \|\mathbf{f} - \mathbf{g}\|^2 \tag{8.33}$$

Solution: By (4.51),

$$\|\mathbf{f} - \mathbf{g}\|^2 = \langle \mathbf{f} - \mathbf{g}, \mathbf{f} - \mathbf{g} \rangle = \int_a^b [f(x) - g(x)]^2\,dx = \text{mean-square error}$$

B. Fourier series

EXAMPLE 8-7: Let

$$\mathbf{g}_0 = g_0(x) = \frac{1}{\sqrt{2\pi}}$$

$$\mathbf{g}_1 = g_1(x) = \frac{1}{\sqrt{\pi}}\cos x \qquad \mathbf{g}_2 = g_2(x) = \frac{1}{\sqrt{\pi}}\sin x$$

$$\mathbf{g}_3 = g_3(x) = \frac{1}{\sqrt{\pi}}\cos 2x \qquad \mathbf{g}_4 = g_4(x) = \frac{1}{\sqrt{\pi}}\sin 2x$$

$$\vdots \qquad\qquad \vdots$$

$$\mathbf{g}_{2k-1} = g_{2k-1}(x) = \frac{1}{\sqrt{\pi}}\cos kx \qquad \mathbf{g}_{2k} = g_{2k}(x) = \frac{1}{\sqrt{\pi}}\sin kx$$

where k is a positive integer.

Show that $\mathscr{B} = \{\mathbf{g}_0, \mathbf{g}_1, \ldots, \mathbf{g}_{2k}\}$ forms an orthonormal set in $C[-\pi, \pi]$ having the inner product

$$\langle \mathbf{g}_i, \mathbf{g}_j \rangle = \int_{-\pi}^{\pi} g_i(x)g_j(x)\,dx$$

Solution: Using the trigonometric identities

$$\cos A \cos B = \tfrac{1}{2}[\cos(A - B) + \cos(A + B)]$$

$$\sin A \sin B = \tfrac{1}{2}[\cos(A - B) - \cos(A + B)]$$

$$\sin A \cos B = \tfrac{1}{2}[\sin(A - B) + \sin(A + B)]$$

we can show that if k, m, and n are positive integers, then

$$\int_{-\pi}^{\pi} \cos kx\,dx = 0$$

$$\int_{-\pi}^{\pi} \sin kx\,dx = 0$$

$$\int_{-\pi}^{\pi} \cos mx \cos nx\,dx = \begin{cases} 0 & \text{if } m \neq n \\ \pi & \text{if } m = n \end{cases}$$

$$\int_{-\pi}^{\pi} \sin mx \sin nx\,dx = \begin{cases} 0 & \text{if } m \neq n \\ \pi & \text{if } m = n \end{cases}$$

$$\int_{-\pi}^{\pi} \cos mx \sin nx\,dx = 0$$

Thus $\langle \mathbf{g}_i,\mathbf{g}_j\rangle = 0$ if $i \neq j$ and

$$\langle \mathbf{g}_0,\mathbf{g}_0\rangle = \int_{-\pi}^{\pi}\left(\frac{1}{\sqrt{2\pi}}\right)\left(\frac{1}{\sqrt{2\pi}}\right)dx = \frac{1}{2\pi}\int_{-\pi}^{\pi}dx = 1$$

$$\langle \mathbf{g}_{2k-1},\mathbf{g}_{2k-1}\rangle = \int_{-\pi}^{\pi}\left(\frac{1}{\sqrt{\pi}}\cos kx\right)\left(\frac{1}{\sqrt{\pi}}\cos kx\right)dx$$

$$= \frac{1}{\pi}\int_{-\pi}^{\pi}\cos^2 kx\,dx = \frac{1}{\pi}(\pi) = 1$$

$$\langle \mathbf{g}_{2k},\mathbf{g}_{2k}\rangle = \int_{-\pi}^{\pi}\left(\frac{1}{\sqrt{\pi}}\sin kx\right)\left(\frac{1}{\sqrt{\pi}}\sin kx\right)dx$$

$$= \frac{1}{\pi}\int_{-\pi}^{\pi}\sin^2 kx\,dx = \frac{1}{\pi}(\pi) = 1$$

Hence

$$\langle \mathbf{g}_i,\mathbf{g}_j\rangle = \delta_{ij} = \begin{cases} 1 & i = j \\ 0 & i \neq j \end{cases}$$

and $\mathscr{B}$ is an orthonormal set.

EXAMPLE 8-8: Let U be a finite-dimensional subspace of $C[-\pi,\pi]$ spanned by $\mathscr{B} = \{\mathbf{g}_0, \mathbf{g}_1, \ldots, \mathbf{g}_{2k}\}$ of Example 8-7. Let $\mathbf{f} = f(x) \in C[-\pi,\pi]$. Find the least approximation to $\mathbf{f}$ from U.

Solution: From (8.5) and (8.2) the least approximation of $\mathbf{f}$ is given by

$$\text{proj}_U \mathbf{f} = \langle \mathbf{f},\mathbf{g}_0\rangle \mathbf{g}_0 + \langle \mathbf{f},\mathbf{g}_1\rangle \mathbf{g}_1 + \cdots + \langle \mathbf{f},\mathbf{g}_{2k}\rangle \mathbf{g}_{2k} \tag{8.34}$$

$$= \frac{a_0}{2} + a_1\cos x + b_1 \sin x + \cdots + a_k \cos kx + b_k \sin kx$$

$$= \frac{a_0}{2} + \sum_{n=1}^{k}(a_n \cos nx + b_n \sin nx)$$

where

$$a_0 = \frac{1}{\pi}\int_{-\pi}^{\pi} f(x)\,dx \tag{8.35}$$

$$a_n = \frac{1}{\pi}\int_{-\pi}^{\pi} f(x)\cos nx\,dx, \quad n = 1, 2, \ldots, k \tag{8.36}$$

$$b_n = \frac{1}{\pi}\int_{-\pi}^{\pi} f(x)\sin nx\,dx, \quad n = 1, 2, \ldots, k \tag{8.37}$$

The right side of Equation (8.34) is called a **finite Fourier approximation** for $\mathbf{f} = f(x)$ and $a_0/2, a_1, \ldots, a_n, b_1, \ldots, b_n$ are called the **Fourier coefficients** of $\mathbf{f}$ (see Example 4-34).

It is natural to expect that the approximation to $\mathbf{f}$ given in (8.34) will improve as k increases. It can be proved that the mean-square error approaches zero as $k \to \infty$; that is,

$$\lim_{k\to\infty}\left\| f(x) - \left[\frac{a_0}{2} + \sum_{n=1}^{k}(a_n\cos nx + b_n \sin nx)\right]\right\| = 0 \tag{8.38}$$

Thus, we can write

$$f(x) = \frac{a_0}{2} + \sum_{n=1}^{\infty}(a_n \cos nx + b_n \sin nx) \tag{8.39}$$

The right side of this equation is called the **Fourier series** for $f(x)$ on $[-\pi,\pi]$.

SUMMARY

1. If U is a subspace of V and if the vector $\mathbf{v}$ is not in U, then the vector

$$\mathbf{v}' = \mathbf{v} - \text{proj}_U \mathbf{v}$$

is orthogonal to every vector in U.

2. The best approximation $\mathbf{u}^\circ$ of U to $\mathbf{v}$ in the sense that

$$\|\mathbf{v} - \mathbf{u}^\circ\| < \|\mathbf{v} - \mathbf{u}\|$$

for every vector $\mathbf{u}$ in U is given by

$$\mathbf{u}^\circ = \text{proj}_U \mathbf{v}$$

3. The slope m and intercept b of the least-squares line of approximation, $y = mx + b$, for the data points $(x_1,y_1), \ldots, (x_n,y_n)$ are given by

$$m\mathbf{x} + b\mathbf{a} = \text{proj}_S \mathbf{y}$$

where

$$\mathbf{x} = \begin{bmatrix} x_1 \\ x_2 \\ \vdots \\ x_n \end{bmatrix}, \qquad \mathbf{y} = \begin{bmatrix} y_1 \\ y_2 \\ \vdots \\ y_n \end{bmatrix}, \qquad \mathbf{a} = \begin{bmatrix} 1 \\ 1 \\ \vdots \\ 1 \end{bmatrix}$$

and S is the span of $\{\mathbf{x},\mathbf{a}\}$.

4. The least-squares solution $\hat{\mathbf{x}}$ to an inconsistent system $A\mathbf{x} = \mathbf{b}$ satisfies the normal equation

$$A^T A\hat{\mathbf{x}} = A^T\mathbf{b}$$

5. A generalized inverse A^- of an $m \times n$ matrix A is an $n \times m$ matrix defined by

$$A^- = (A^T A)^{-1} A^T$$

6. A projection matrix P is defined by

$$P = AA^- = A(A^T A)^{-1} A^T$$

and

$$\text{proj}_U \mathbf{b} = P\mathbf{b}$$

where U is the column space of A.

7. The least-squares approximation of $f(x) \in C[-\pi,\pi]$ is given by a finite Fourier series given by

$$\frac{a_0}{2} + \sum_{n=1}^{k} (a_n \cos nx + b_n \sin nx)$$

where

$$a_0 = \frac{1}{\pi}\int_{-\pi}^{\pi} f(x)\,dx$$

$$a_n = \frac{1}{\pi}\int_{-\pi}^{\pi} f(x)\cos nx\,dx$$

$$b_n = \frac{1}{\pi}\int_{-\pi}^{\pi} f(x)\sin nx\,dx$$

RAISE YOUR GRADES

Can you explain ...?

☑ how to find the best approximation from the concept of minimum distance
☑ how to find the slope and intercept of a least-squares line of fitting data points

☑ how to find the least-squares solution to an inconsistent system
☑ how to find a Fourier series to approximate a function

SOLVED PROBLEMS

Best Approximation

PROBLEM 8-1 Let U be the subspace of R^3 spanned by $\{\mathbf{i},\mathbf{j}\}$. Find the vector of U that best approximates the vector $\mathbf{v} = \begin{bmatrix} 1 \\ 1 \\ 1 \end{bmatrix}$.

Solution The projection of $\mathbf{v}$ on U is given by

$$\text{proj}_U \mathbf{v} = \langle \mathbf{v},\mathbf{i} \rangle \mathbf{i} + \langle \mathbf{v},\mathbf{j} \rangle \mathbf{j} = 1 \begin{bmatrix} 1 \\ 0 \\ 0 \end{bmatrix} + 1 \begin{bmatrix} 0 \\ 1 \\ 0 \end{bmatrix} = \begin{bmatrix} 1 \\ 1 \\ 0 \end{bmatrix}$$

Thus $\begin{bmatrix} 1 \\ 1 \\ 0 \end{bmatrix}$ is the best approximation to $\begin{bmatrix} 1 \\ 1 \\ 1 \end{bmatrix}$.

Least-Squares Fitting of Data

PROBLEM 8-2 Show that the slope m and intercept b of the least-squares line of approximation, $y = mx + b$, for the data points $(x_i,y_i), \ldots, (x_n,y_n)$ are given by

$$m = \frac{n\langle \mathbf{x},\mathbf{y} \rangle - \langle \mathbf{a},\mathbf{x} \rangle \langle \mathbf{a},\mathbf{y} \rangle}{n\langle \mathbf{x},\mathbf{x} \rangle - \langle \mathbf{a},\mathbf{x} \rangle^2} \qquad \textbf{(a)}$$

$$b = \frac{\langle \mathbf{x},\mathbf{x} \rangle \langle \mathbf{a},\mathbf{y} \rangle - \langle \mathbf{x},\mathbf{y} \rangle \langle \mathbf{a},\mathbf{x} \rangle}{n\langle \mathbf{x},\mathbf{x} \rangle - \langle \mathbf{a},\mathbf{x} \rangle^2} \qquad \textbf{(b)}$$

where

$$\mathbf{a} = \begin{bmatrix} 1 \\ 1 \\ \vdots \\ 1 \end{bmatrix}, \qquad \mathbf{x} = \begin{bmatrix} x_1 \\ x_2 \\ \vdots \\ x_n \end{bmatrix}, \qquad \mathbf{y} = \begin{bmatrix} y_1 \\ y_2 \\ \vdots \\ y_n \end{bmatrix}$$

Solution Since (m,b) satisfies (8.15), that is,

$$m\mathbf{x} + b\mathbf{a} = \text{proj}_S \mathbf{y}$$

the vector

$$\mathbf{y} - (\text{proj}\, \mathbf{y}) = \mathbf{y} - (m\mathbf{x} + b\mathbf{a})$$

is orthogonal to $\mathbf{x}$ and $\mathbf{a}$. Thus, we get

$$\langle \mathbf{x}, \mathbf{y} - (m\mathbf{x} + b\mathbf{a}) \rangle = 0$$

$$\langle \mathbf{a}, \mathbf{y} - (m\mathbf{x} + b\mathbf{a}) \rangle = 0$$

which can be written as

$$\langle \mathbf{x}, m\mathbf{x} + b\mathbf{a} \rangle = \langle \mathbf{x}, \mathbf{y} \rangle$$
$$\langle \mathbf{a}, m\mathbf{x} + b\mathbf{a} \rangle = \langle \mathbf{a}, \mathbf{y} \rangle$$

Using (4.40), we get

$$\langle \mathbf{x}, \mathbf{x} \rangle m + \langle \mathbf{x}, \mathbf{a} \rangle b = \langle \mathbf{x}, \mathbf{y} \rangle \qquad \text{(c)}$$
$$\langle \mathbf{a}, \mathbf{x} \rangle m + \langle \mathbf{a}, \mathbf{a} \rangle b = \langle \mathbf{a}, \mathbf{y} \rangle$$

Solving (c) for m and b by Cramer's rule (2.26), we obtain

$$m = \frac{\begin{vmatrix} \langle \mathbf{x},\mathbf{y} \rangle & \langle \mathbf{x},\mathbf{a} \rangle \\ \langle \mathbf{a},\mathbf{y} \rangle & \langle \mathbf{a},\mathbf{a} \rangle \end{vmatrix}}{\begin{vmatrix} \langle \mathbf{x},\mathbf{x} \rangle & \langle \mathbf{x},\mathbf{a} \rangle \\ \langle \mathbf{a},\mathbf{x} \rangle & \langle \mathbf{a},\mathbf{a} \rangle \end{vmatrix}} = \frac{n\langle \mathbf{x},\mathbf{y} \rangle - \langle \mathbf{a},\mathbf{x} \rangle \langle \mathbf{a},\mathbf{y} \rangle}{n\langle \mathbf{x},\mathbf{x} \rangle - \langle \mathbf{a},\mathbf{x} \rangle^2}$$

$$b = \frac{\begin{vmatrix} \langle \mathbf{x},\mathbf{x} \rangle & \langle \mathbf{x},\mathbf{y} \rangle \\ \langle \mathbf{a},\mathbf{x} \rangle & \langle \mathbf{a},\mathbf{y} \rangle \end{vmatrix}}{\begin{vmatrix} \langle \mathbf{x},\mathbf{x} \rangle & \langle \mathbf{x},\mathbf{a} \rangle \\ \langle \mathbf{a},\mathbf{x} \rangle & \langle \mathbf{a},\mathbf{a} \rangle \end{vmatrix}} = \frac{\langle \mathbf{x},\mathbf{x} \rangle \langle \mathbf{a},\mathbf{y} \rangle - \langle \mathbf{x},\mathbf{y} \rangle \langle \mathbf{a},\mathbf{x} \rangle}{n\langle \mathbf{x},\mathbf{x} \rangle - \langle \mathbf{a},\mathbf{x} \rangle^2}$$

since $\langle \mathbf{x},\mathbf{a} \rangle = \langle \mathbf{a},\mathbf{x} \rangle$ and $\langle \mathbf{a},\mathbf{a} \rangle = n$.

PROBLEM 8-3 Find the least-squares line of approximation for the data points (0, 1), (3, 4), (6, 5). Then find the value of y when $x = 8$.

Solution We have $n = 3$, and

$$\mathbf{x} = \begin{bmatrix} 0 \\ 3 \\ 6 \end{bmatrix}, \qquad \mathbf{y} = \begin{bmatrix} 1 \\ 4 \\ 5 \end{bmatrix}, \qquad \mathbf{a} = \begin{bmatrix} 1 \\ 1 \\ 1 \end{bmatrix}$$

Thus

$$\langle \mathbf{x},\mathbf{x} \rangle = 0^2 + 3^2 + 6^2 = 45$$
$$\langle \mathbf{a},\mathbf{x} \rangle = 0 + 3 + 6 = 9$$
$$\langle \mathbf{a},\mathbf{y} \rangle = 1 + 4 + 5 = 10$$
$$\langle \mathbf{x},\mathbf{y} \rangle = (0)(1) + (3)(4) + (6)(5) = 42$$

Substituting these values in (a) and (b) of Problem 8-2, we get

$$m = \frac{3(42) - (9)(10)}{3(45) - (9)^2} = \frac{36}{54} = \frac{2}{3}$$

$$b = \frac{(45)(10) - (42)(9)}{3(45) - (9)^2} = \frac{72}{54} = \frac{4}{3}$$

Thus the least-squares line is given by $y = \frac{2}{3}x + \frac{4}{3}$. For $x = 8$, we get $y = \frac{2}{3}(8) + \frac{4}{3} = \frac{20}{3} \simeq 6.67$.

Least-Squares Solutions of Inconsistent Systems

PROBLEM 8-4 Given

$$A = \begin{bmatrix} 1 & 2 \\ 1 & 1 \\ 0 & 1 \end{bmatrix} \quad \text{and} \quad \mathbf{b} = \begin{bmatrix} 1 \\ 2 \\ 2 \end{bmatrix}$$

show that $A\mathbf{x} = \mathbf{b}$ is an inconsistent system. Then find the least-squares solution of this inconsistent system. Also find the projection of $\mathbf{b}$ onto the column space of A.

Solution Let $\mathbf{x} = \begin{bmatrix} x_1 \\ x_2 \end{bmatrix}$. Then for given A and $\mathbf{b}$, $A\mathbf{x} = \mathbf{b}$ becomes

$$\begin{bmatrix} 1 & 2 \\ 1 & 1 \\ 0 & 1 \end{bmatrix} \begin{bmatrix} x_1 \\ x_2 \end{bmatrix} = \begin{bmatrix} 1 \\ 2 \\ 2 \end{bmatrix} \tag{a}$$

or

$$\begin{aligned} x_1 + 2x_2 &= 1 \\ x_1 + x_2 &= 2 \\ x_2 &= 2 \end{aligned} \tag{b}$$

By backward substitution, we get $x_2 = 2$, $x_1 = 0$, and $2(x_2) = 2(2) = 4 \neq 1$. Thus the system (b), or system (a), is inconsistent.

Next

$$A^T A = \begin{bmatrix} 1 & 1 & 0 \\ 2 & 1 & 1 \end{bmatrix} \begin{bmatrix} 1 & 2 \\ 1 & 1 \\ 0 & 1 \end{bmatrix} = \begin{bmatrix} 2 & 3 \\ 3 & 6 \end{bmatrix}$$

$$A^T \mathbf{b} = \begin{bmatrix} 1 & 1 & 0 \\ 2 & 1 & 1 \end{bmatrix} \begin{bmatrix} 1 \\ 2 \\ 2 \end{bmatrix} = \begin{bmatrix} 3 \\ 6 \end{bmatrix}$$

Then the normal equation for $A\mathbf{x} = \mathbf{b}$ (8.22)

$$A^T A \hat{\mathbf{x}} = A^T \mathbf{b}$$

becomes

$$\begin{bmatrix} 2 & 3 \\ 3 & 6 \end{bmatrix} \begin{bmatrix} \hat{x}_1 \\ \hat{x}_2 \end{bmatrix} = \begin{bmatrix} 3 \\ 6 \end{bmatrix}$$

Now

$$\begin{bmatrix} \hat{x}_1 \\ \hat{x}_2 \end{bmatrix} = \begin{bmatrix} 2 & 3 \\ 3 & 6 \end{bmatrix}^{-1} \begin{bmatrix} 3 \\ 6 \end{bmatrix} = \frac{1}{3} \begin{bmatrix} 6 & -3 \\ -3 & 2 \end{bmatrix} \begin{bmatrix} 3 \\ 6 \end{bmatrix} = \begin{bmatrix} 0 \\ 1 \end{bmatrix}$$

Hence the least-squares solution of $A\mathbf{x} = \mathbf{b}$ is

$$\hat{\mathbf{x}} = \begin{bmatrix} 0 \\ 1 \end{bmatrix}$$

The projection of $\mathbf{b}$ onto the column space of A is

$$\text{proj}_U \mathbf{b} = A\hat{\mathbf{x}} = \begin{bmatrix} 1 & 2 \\ 1 & 1 \\ 0 & 1 \end{bmatrix} \begin{bmatrix} 0 \\ 1 \end{bmatrix} = \begin{bmatrix} 2 \\ 1 \\ 1 \end{bmatrix}$$

PROBLEM 8-5 Given

$$A = \begin{bmatrix} 1 & -1 \\ 0 & 1 \\ -1 & 2 \end{bmatrix}$$

(a) find the generalized inverse A^-, and **(b)** verify by direct computation that $A^- A = I_2$.

Solution

(a) Since

$$A^TA = \begin{bmatrix} 1 & 0 & -1 \\ -1 & 1 & 2 \end{bmatrix}\begin{bmatrix} 1 & -1 \\ 0 & 1 \\ -1 & 2 \end{bmatrix} = \begin{bmatrix} 2 & -3 \\ -3 & 6 \end{bmatrix}$$

then

$$(A^TA)^{-1} = \begin{bmatrix} 2 & -3 \\ -3 & 6 \end{bmatrix}^{-1} = \frac{1}{3}\begin{bmatrix} 6 & 3 \\ 3 & 2 \end{bmatrix}$$

by (8.27),

$$A^- = (A^TA)^{-1}A^T = \frac{1}{3}\begin{bmatrix} 6 & 3 \\ 3 & 2 \end{bmatrix}\begin{bmatrix} 1 & 0 & -1 \\ -1 & 1 & 2 \end{bmatrix} = \frac{1}{3}\begin{bmatrix} 3 & 3 & 0 \\ 1 & 2 & 1 \end{bmatrix}$$

(b)

$$A^-A = \frac{1}{3}\begin{bmatrix} 3 & 3 & 0 \\ 1 & 2 & 1 \end{bmatrix}\begin{bmatrix} 1 & -1 \\ 0 & 1 \\ -1 & 2 \end{bmatrix} = \frac{1}{3}\begin{bmatrix} 3 & 0 \\ 0 & 3 \end{bmatrix} = \begin{bmatrix} 1 & 0 \\ 0 & 1 \end{bmatrix} = I_2$$

PROBLEM 8-6 Show that if A is an $n \times n$ matrix and nonsingular, then

$$A^- = A^{-1}$$

Solution Since A is nonsingular, A^{-1} exists. Then by (1.38)

$$A^- = (A^TA)^{-1}A^T = A^{-1}(A^T)^{-1}A^T = A^{-1}I_n = A^{-1}$$

PROBLEM 8-7 Show that the slope m and intercept b of the least-squares line of approximation $y = mx + b$ for the noncollinear data points $(x_1,y_1), (x_2,y_2), \ldots, (x_n,y_n)$ are given by

$$\begin{bmatrix} m \\ b \end{bmatrix} = A^-\mathbf{y} = (A^TA)^{-1}A^T\mathbf{y}$$

where

$$A = \begin{bmatrix} x_1 & 1 \\ x_2 & 1 \\ \vdots & \vdots \\ x_n & 1 \end{bmatrix} \quad \text{and} \quad \mathbf{y} = \begin{bmatrix} y_1 \\ y_2 \\ \vdots \\ y_n \end{bmatrix}$$

Solution Note that the system of equations (8.11)

$$\begin{aligned} mx_1 + b &= y_1 \\ mx_2 + b &= y_2 \\ \vdots \quad & \quad \vdots \\ mx_n + b &= y_n \end{aligned}$$

can be expressed as

$$\begin{bmatrix} x_1 & 1 \\ x_2 & 1 \\ \vdots & \vdots \\ x_n & 1 \end{bmatrix}\begin{bmatrix} m \\ b \end{bmatrix} = \begin{bmatrix} y_1 \\ y_2 \\ \vdots \\ y_n \end{bmatrix} \tag{a}$$

or

$$A\mathbf{u} = \mathbf{y} \tag{b}$$

where

$$A = \begin{bmatrix} x_1 & 1 \\ x_2 & 1 \\ \vdots & \vdots \\ x_n & 1 \end{bmatrix}, \qquad \mathbf{u} = \begin{bmatrix} m \\ b \end{bmatrix}, \qquad \mathbf{y} = \begin{bmatrix} y_1 \\ y_2 \\ \vdots \\ y_n \end{bmatrix}$$

If the data points are noncollinear, then the system (b) is inconsistent and the least-squares line of approximation is determined by the least-squares solution of (b). Thus, $\mathbf{u}$ satisfies the normal equation $A^T A\mathbf{u} = A^T\mathbf{y}$ and by (8.26) we get

$$\mathbf{u} = \begin{bmatrix} m \\ b \end{bmatrix} = A^{-}\mathbf{y} = (A^T A)^{-1} A^T \mathbf{y}.$$

PROBLEM 8-8 Using the results of Problem 8-7, rework Problem 8-3.

Solution The data points are (0, 1), (3, 4), (6, 5). Thus

$$A = \begin{bmatrix} x_1 & 1 \\ x_2 & 1 \\ x_3 & 1 \end{bmatrix} = \begin{bmatrix} 0 & 1 \\ 3 & 1 \\ 6 & 1 \end{bmatrix}, \qquad \mathbf{y} = \begin{bmatrix} 1 \\ 4 \\ 5 \end{bmatrix}, \qquad \mathbf{u} = \begin{bmatrix} m \\ b \end{bmatrix}$$

Now

$$A^T A = \begin{bmatrix} 0 & 3 & 6 \\ 1 & 1 & 1 \end{bmatrix} \begin{bmatrix} 0 & 1 \\ 3 & 1 \\ 6 & 1 \end{bmatrix} = \begin{bmatrix} 45 & 9 \\ 9 & 3 \end{bmatrix}$$

So the normal equation $A^T A\mathbf{u} = A^T\mathbf{y}$ becomes

$$\begin{bmatrix} 45 & 9 \\ 9 & 3 \end{bmatrix} \begin{bmatrix} m \\ b \end{bmatrix} = \begin{bmatrix} 0 & 3 & 6 \\ 1 & 1 & 1 \end{bmatrix} \begin{bmatrix} 1 \\ 4 \\ 5 \end{bmatrix} = \begin{bmatrix} 42 \\ 10 \end{bmatrix}$$

Thus

$$\begin{bmatrix} m \\ b \end{bmatrix} = \begin{bmatrix} 45 & 9 \\ 9 & 3 \end{bmatrix}^{-1} \begin{bmatrix} 42 \\ 10 \end{bmatrix} = \frac{1}{54} \begin{bmatrix} 3 & -9 \\ -9 & 45 \end{bmatrix} \begin{bmatrix} 42 \\ 10 \end{bmatrix} = \begin{bmatrix} \frac{2}{3} \\ \frac{4}{3} \end{bmatrix}$$

and the least-squares line is given by $y = \frac{2}{3}x + \frac{4}{3}$.

PROBLEM 8-9 Suppose that the data points (x_1, y_1), (x_2, y_2), …, (x_n, y_n) have a general parabolic appearance. Find the coefficients a, b, and c, such that the quadratic equation

$$y = ax^2 + bx + c \tag{a}$$

provides a least-squares approximation to the data points.

Solution If we substitute each data point into equation (a) we get the system

$$\begin{aligned} ax_1^2 + bx_1 + c &= y_1 \\ ax_2^2 + bx_2 + c &= y_2 \\ \vdots \quad \vdots \quad \vdots \quad &\vdots \\ ax_n^2 + bx_n + c &= y_n \end{aligned} \tag{b}$$

If we let

$$A = \begin{bmatrix} x_1^2 & x_1 & 1 \\ x_2^2 & x_2 & 1 \\ \vdots & \vdots & \vdots \\ x_n^2 & x_n & 1 \end{bmatrix}, \qquad \mathbf{u} = \begin{bmatrix} a \\ b \\ c \end{bmatrix}, \qquad \mathbf{y} = \begin{bmatrix} y_1 \\ y_2 \\ \vdots \\ y_n \end{bmatrix} \tag{c}$$

then (b) can be written as

$$A\mathbf{u} = \mathbf{y} \tag{d}$$

If the given data points do not lie on a single quadratic curve, then (c) will be inconsistent and the method of least-squares approximation can be applied. Then by (8.26), we get

$$\begin{bmatrix} a \\ b \\ c \end{bmatrix} = A^{-}\mathbf{y} = (A^TA)^{-1}A^T\mathbf{y}$$

PROBLEM 8-10 Find the least-squares quadratic approximation to the data points $(-1, 3)$, $(0, -1)$, $(1, 0)$, $(2, 3)$.

Solution Let $y = ax^2 + bx + c$ be the least-squares quadratic approximation to the data points. For these data points, $x_1 = -1$, $x_2 = 0$, $x_3 = 1$, $x_4 = 2$ and $y_1 = 3$, $y_2 = -1$, $y_3 = 0$, $y_4 = 3$. Thus, using the results of Problem 8-9, we have

$$A = \begin{bmatrix} x_1^2 & x_1 & 1 \\ x_2^2 & x_2 & 1 \\ x_3^2 & x_3 & 1 \\ x_4^2 & x_4 & 1 \end{bmatrix} = \begin{bmatrix} 1 & -1 & 1 \\ 0 & 0 & 1 \\ 1 & 1 & 1 \\ 4 & 2 & 1 \end{bmatrix}, \qquad \mathbf{u} = \begin{bmatrix} a \\ b \\ c \end{bmatrix}, \qquad \mathbf{y} = \begin{bmatrix} 3 \\ -1 \\ 0 \\ 3 \end{bmatrix}$$

Now

$$A^TA = \begin{bmatrix} 1 & 0 & 1 & 4 \\ -1 & 0 & 1 & 2 \\ 1 & 1 & 1 & 1 \end{bmatrix} \begin{bmatrix} 1 & -1 & 1 \\ 0 & 0 & 1 \\ 1 & 1 & 1 \\ 4 & 2 & 1 \end{bmatrix} = \begin{bmatrix} 18 & 8 & 6 \\ 8 & 6 & 2 \\ 6 & 2 & 4 \end{bmatrix}$$

$$(A^TA)^{-1} = \frac{1}{20}\begin{bmatrix} 5 & -5 & -5 \\ -5 & 9 & 3 \\ -5 & 3 & 11 \end{bmatrix}, \qquad A^T\mathbf{y} = \begin{bmatrix} 1 & 0 & 1 & 4 \\ -1 & 0 & 1 & 2 \\ 1 & 1 & 1 & 1 \end{bmatrix} \begin{bmatrix} 3 \\ -1 \\ 0 \\ 3 \end{bmatrix} = \begin{bmatrix} 15 \\ 3 \\ 5 \end{bmatrix}$$

Then by (8.26) we get

$$\begin{bmatrix} a \\ b \\ c \end{bmatrix} = A^{-}\mathbf{y} = (A^TA)^{-1}A^T\mathbf{y} = \frac{1}{20}\begin{bmatrix} 5 & -5 & -5 \\ -5 & 9 & 3 \\ -5 & 3 & 11 \end{bmatrix} \begin{bmatrix} 15 \\ 3 \\ 5 \end{bmatrix} = \frac{1}{20}\begin{bmatrix} 35 \\ -33 \\ -11 \end{bmatrix} = \begin{bmatrix} 1.75 \\ -1.65 \\ -0.55 \end{bmatrix}$$

Thus the least-squares quadratic approximation is

$$y = 1.75x^2 - 1.65x - 0.55$$

PROBLEM 8-11 Show that the projection matrix $P = A(A^TA)^{-1}A^T$ of (8.30) has the following two basic properties:

(a) $P^2 = P$ (idempotency)
(b) $P^T = P$ (symmetry)

Solution

(a)

$$P^2 = PP = A(A^TA)^{-1}A^TA(A^TA)^{-1}A^T = A(A^TA)^{-1}A^T = P$$

It is easy to prove $P^2 = P$ geometrically. If we start with any vector **b**, the vector $P\mathbf{b}$ lies in the subspace onto which we are projecting. Then when we project again, producing $P(P\mathbf{b}) = P^2\mathbf{b}$, nothing is changed, since the vector $P\mathbf{b}$ is already in the subspace and $P\mathbf{b} = P^2\mathbf{b}$ for every **b**. Thus $P^2 = P$.

(b)

$$\begin{aligned}
P^T &= (A(A^TA)^{-1}A^T)^T \\
&= (A^T)^T[(A^TA)^{-1}]^TA^T && \text{[by Problem 1-28]} \\
&= A[(A^TA)^T]^{-1}]A^T && \text{[by (1.45)]} \\
&= A(A^TA)^{-1}A^T && \text{[by (1.43)]} \\
&= P
\end{aligned}$$

*Fourier Series [*For readers who have studied calculus.*]

PROBLEM 8-12 Find the least-squares approximation of $f(x) = x$ on $[-\pi,\pi]$ by a function of the form

$$\frac{a_0}{2} + a_1 \cos x + a_2 \cos 2x + b_1 \sin x + b_2 \sin 2x$$

Solution By (8.35)–(8.37),

$$a_0 = \frac{1}{\pi}\int_{-\pi}^{\pi} f(x)\,dx = \frac{1}{\pi}\int_{-\pi}^{\pi} x\,dx = \frac{1}{\pi}\frac{x^2}{2}\bigg|_{-\pi}^{\pi} = 0$$

For $n = 1, 2$, integration by parts yields

$$\begin{aligned}
a_n &= \frac{1}{\pi}\int_{-\pi}^{\pi} f(x)\cos nx\,dx \\
&= \frac{1}{\pi}\int_{-\pi}^{\pi} x\cos nx\,dx \\
&= \frac{1}{\pi}\left[\frac{x}{n}\sin nx\bigg|_{-\pi}^{\pi} - \frac{1}{n}\int_{-\pi}^{\pi}\sin nx\,dx\right] = \frac{1}{n\pi}\frac{1}{n}\cos nx\bigg|_{-\pi}^{\pi} = 0 \\
b_n &= \frac{1}{\pi}\int_{-\pi}^{\pi} f(x)\sin nx\,dx \\
&= \frac{1}{\pi}\int_{-\pi}^{\pi} x\sin nx\,dx \\
&= \frac{1}{\pi}\left[-\frac{x}{n}\cos nx\bigg|_{-\pi}^{\pi} + \frac{1}{n}\int_{-\pi}^{\pi}\cos nx\,dx\right] \\
&= \frac{1}{n\pi}\left[-2\pi + \frac{1}{n}\sin nx\bigg|_{-\pi}^{\pi}\right] = -\frac{2}{n}
\end{aligned}$$

Thus, the least-squares approximation to x on $[-\pi,\pi]$ is

$$x \simeq -2\sin x - \sin 2x$$

Supplementary Exercises

PROBLEM 8-13 Let U be the subspace of R^3 spanned by $\{\mathbf{v}_1,\mathbf{v}_2\}$ where

$$\mathbf{v}_1 = \begin{bmatrix} 1 \\ 1 \\ 0 \end{bmatrix}, \qquad \mathbf{v}_2 = \begin{bmatrix} 0 \\ 1 \\ 1 \end{bmatrix}$$

Find the vector of U that best approximates $\mathbf{v} = \begin{bmatrix} 2 \\ 5 \\ 4 \end{bmatrix}$.

Answer $\begin{bmatrix} \frac{5}{3} \\ \frac{16}{3} \\ \frac{11}{3} \end{bmatrix}$

PROBLEM 8-14 Find the least-squares line of approximation for the data points (0, 2), (3, 5), (5, 6).

Answer $y = \dfrac{31}{38}x + \dfrac{82}{38}$

PROBLEM 8-15 Rework Problem 8-14 using the method of the least-squares solution of inconsistent systems.

PROBLEM 8-16 Show that if P is a projection matrix, then $I - P$ is also a projection matrix where I is an identity matrix. [*Hint:* Use the relationships (a) and (b) of Problem 8-11.]

***PROBLEM 8-17** Compute the Fourier series for $f(x) = x$ for over $[0,2\pi]$.

Answer $\pi - \sum_{n=1}^{\infty} \dfrac{2}{n} \sin nx$

9 APPLICATIONS

THIS CHAPTER IS ABOUT

☑ **Markov Analysis**
☑ **Difference Equations**
☑ **Differential Equations**

9-1. Markov Analysis

A. Markov chain

DEFINITION A **Markov chain** is a sequence of n experiments in which each experiment has m possible outcomes $a_1, a_2, \ldots, a_m$ called *states* and the likelihood, or probability, that a particular state occurs depends only on the outcome of the preceding experiment.

Let p_{ij} represent the probability that the state (outcome) a_j occurs on any given experiment, providing that state a_i occurred on the preceding experiment. These probabilities are called the **transition probabilities**. We assume that these probabilities are nonnegative numbers between 0 and 1; that is,

$$0 \leq p_{ij} \leq 1 \tag{9.1}$$

and

$$p_{i1} + p_{i2} + \cdots + p_{im} = 1 \tag{9.2}$$

A row vector with nonnegative entries such that the sum of the entries is 1 is called a **probability vector**.

A matrix $S = [p_{ij}]$ is called a **stochastic matrix**, or **Markov matrix**, if it is square and its row vectors are probability vectors; that is,

$$S = \begin{bmatrix} p_{11} & p_{12} & \cdots & p_{1m} \\ p_{21} & p_{22} & \cdots & p_{2m} \\ \vdots & \vdots & & \vdots \\ p_{m1} & p_{m2} & \cdots & p_{mm} \end{bmatrix} \tag{9.3}$$

EXAMPLE 9-1: Let S be the stochastic matrix for a Markov chain. Show that the probability vector

$$\mathbf{p}^{(k)} = [p_1^{(k)}, p_2^{(k)}, \ldots, p_m^{(k)}]$$

after k steps is given by

$$\mathbf{p}^{(k)} = \mathbf{p}^{(0)} S^k \tag{9.4}$$

where $\mathbf{p}^{(0)} = [p_1^{(0)}, p_2^{(0)}, \ldots, p_m^{(0)}]$ is the initial probability vector.

Solution: Let $p_j^{(k)}$ be the probability that after k steps the state is at the jth state. Then the probability vector

$$\mathbf{p}^{(k)} = [p_1^{(k)}, p_2^{(k)}, \ldots, p_m^{(k)}] \tag{9.5}$$

represents the probability distribution for this situation.

Now, the probability of being in state j after k steps is the sum of m terms composed of the probability of being in any of the m states after $k-1$ steps multiplied by the probability of transition from that state to state j at the kth step. Thus we have

$$p_j^{(k)} = p_1^{(k-1)}p_{1j} + p_2^{(k-1)}p_{2j} + \cdots + p_m^{(k-1)}p_{mj} \tag{9.6}$$

Noting that if we let

$$\mathbf{p}^{(k-1)} = [p_1^{(k-1)}, p_2^{(k-1)}, \ldots, p_m^{(k-1)}] \tag{9.7}$$

then, in view of (9.6), we can write

$$\mathbf{p}^{(k)} = \mathbf{p}^{(k-1)} \begin{bmatrix} p_{11} & \cdots & p_{1j} & \cdots & p_{1m} \\ p_{21} & \cdots & p_{2j} & \cdots & p_{2m} \\ \vdots & & \vdots & & \vdots \\ p_{m1} & \cdots & p_{mj} & \cdots & p_{mm} \end{bmatrix} \tag{9.8}$$

or

$$\mathbf{p}^{(k)} = \mathbf{p}^{(k-1)}S \tag{9.9}$$

Setting $k = 1, 2, \ldots$, we get

$$\begin{aligned} \mathbf{p}^{(1)} &= \mathbf{p}^{(0)}S \\ \mathbf{p}^{(2)} &= \mathbf{p}^{(1)}S = (\mathbf{p}^{(0)}S)S = \mathbf{p}^{(0)}S^2 \\ &\vdots \\ \mathbf{p}^{(k)} &= \mathbf{p}^{(k-1)}S = \mathbf{p}^{(0)}S^k \end{aligned}$$

THEOREM 9-1.1 If S is a stochastic matrix, then S^k is again stochastic for any positive integer k. If $\mathbf{p}^{(0)}$ is a probability vector, then $\mathbf{p}^{(k)} = \mathbf{p}^{(0)}S^k$ is also a probability vector.

For the proof of Theorem 9-1.1, see Problems 9-3 and 9-4.

EXAMPLE 9-2: Given

$$\mathbf{p}^{(0)} = [0.9, 0.1], \qquad S = \begin{bmatrix} 0.8 & 0.2 \\ 0.3 & 0.7 \end{bmatrix}$$

show that S^4 is a stochastic matrix and $\mathbf{p}^{(4)} = \mathbf{p}^{(0)}S^4$ is a probability vector.

Solution

$$S^2 = SS = \begin{bmatrix} 0.8 & 0.2 \\ 0.3 & 0.7 \end{bmatrix}\begin{bmatrix} 0.8 & 0.2 \\ 0.3 & 0.7 \end{bmatrix} = \begin{bmatrix} 0.7 & 0.3 \\ 0.45 & 0.45 \end{bmatrix}$$

$$S^4 = S^2S^2 = \begin{bmatrix} 0.7 & 0.3 \\ 0.45 & 0.45 \end{bmatrix}\begin{bmatrix} 0.7 & 0.3 \\ 0.45 & 0.45 \end{bmatrix} = \begin{bmatrix} 0.625 & 0.375 \\ 0.5625 & 0.4375 \end{bmatrix}$$

$$\mathbf{p}^{(0)}S^4 = [0.9, 0.1]\begin{bmatrix} 0.625 & 0.375 \\ 0.5625 & 0.4375 \end{bmatrix} = [0.61875, 0.38125]$$

We see that S^4 has row sums equal to 1, so it is a stochastic matrix; and the row sum of $\mathbf{p}^{(0)}S^4$ is also equal to 1, so it is a probability vector.

B. Distribution vector and transition matrix

Taking the transpose of (9.9), we get

$$[\mathbf{p}^{(k)}]^T = [\mathbf{p}^{(k-1)}S]^T = S^T[\mathbf{p}^{(k-1)}]^T \tag{9.10}$$

Now if we let $\mathbf{x}_k = [\mathbf{p}^{(k)}]^T$ and $A = S^T$, then (9.10) can be rewritten as

$$\mathbf{x}_k = A\mathbf{x}_{k-1} \tag{9.11}$$

Setting $k = 1, 2, \ldots$, we get

$$\begin{aligned} \mathbf{x}_1 &= A\mathbf{x}_0 \\ \mathbf{x}_2 &= A\mathbf{x}_1 = A(A\mathbf{x}_0) = A^2\mathbf{x}_0 \\ &\vdots \\ \mathbf{x}_k &= A\mathbf{x}_{k-1} = A^k\mathbf{x}_0 \end{aligned}$$

Then it is appropriate to call $\mathbf{x}_k$ the k*th stage distribution vector*, $\mathbf{x}_0$ the *initial distribution vector*, and A a *transition matrix*.

note: In the literature S is called a transition matrix. Thus the sum of the entries of the distribution vector is equal to 1 and each column of a transition matrix A adds up to 1 and $a_{ij} \geq 0$.

EXAMPLE 9-3: A certain product is made by two competing companies, B and C, that control the entire market. Currently, B has $\frac{3}{5}$ of the total market and C has $\frac{2}{5}$ of the total market. Each year, B loses $\frac{2}{3}$ of its market share to C while C loses $\frac{1}{2}$ of its share to B. Find the transition matrix—the relative proportion of the market that each holds after 1 and 2 years.

Solution: Let $\mathbf{x}_0 = \begin{bmatrix} b_0 \\ c_0 \end{bmatrix}$ represent the initial distribution of the market. Since B controls $\frac{3}{5}$ and C controls $\frac{2}{5}$ of the market, we get

$$\mathbf{x}_0 = \begin{bmatrix} b_0 \\ c_0 \end{bmatrix} = \begin{bmatrix} \frac{3}{5} \\ \frac{2}{5} \end{bmatrix} = \begin{bmatrix} 0.6 \\ 0.4 \end{bmatrix}$$

Note that the column sum of $\mathbf{x}_0$ is equal to 1. Thus $\mathbf{x}_0$ is a distribution vector.

Let $\mathbf{x}_1 = \begin{bmatrix} b_1 \\ c_1 \end{bmatrix}$ represent the distribution of the market after 1 year. Since B retains $\frac{1}{3}$ of its share and gains $\frac{1}{2}$ of C's share, we have

$$b_1 = \tfrac{1}{3}b_0 + \tfrac{1}{2}c_0$$

Similarly, C gets $\frac{2}{3}$ of B's share and retains $\frac{1}{2}$ of its share, so we get

$$c_1 = \tfrac{2}{3}b_0 + \tfrac{1}{2}c_0$$

Thus we have

$$\mathbf{x}_1 = \begin{bmatrix} \frac{1}{3} & \frac{1}{2} \\ \frac{2}{3} & \frac{1}{2} \end{bmatrix} \mathbf{x}_0 = A\mathbf{x}_0$$

where

$$A = \begin{bmatrix} \frac{1}{3} & \frac{1}{2} \\ \frac{2}{3} & \frac{1}{2} \end{bmatrix}$$

Note that A has a column sum equal to 1; thus A is the transition matrix.

Now, using

$$\mathbf{x}_0 = \begin{bmatrix} \frac{3}{5} \\ \frac{2}{5} \end{bmatrix}$$

we have

$$\mathbf{x}_1 = A\mathbf{x}_0 = \begin{bmatrix} \frac{1}{3} & \frac{1}{2} \\ \frac{2}{3} & \frac{1}{2} \end{bmatrix} \begin{bmatrix} \frac{3}{5} \\ \frac{2}{5} \end{bmatrix} = \begin{bmatrix} \frac{2}{5} \\ \frac{3}{5} \end{bmatrix} = \begin{bmatrix} 0.4 \\ 0.6 \end{bmatrix} = \begin{bmatrix} b_1 \\ c_1 \end{bmatrix}$$

Thus after 1 year, B controls 40% of the market and C controls 60%. We see that after 1 year B has lost and C has gained significantly. Note that $b_1 + c_1 = 1$.

Next,

$$\mathbf{x}_2 = A\mathbf{x}_1 = \begin{bmatrix} \frac{1}{3} & \frac{1}{2} \\ \frac{2}{3} & \frac{1}{2} \end{bmatrix} \begin{bmatrix} \frac{2}{5} \\ \frac{3}{5} \end{bmatrix} = \begin{bmatrix} \frac{13}{30} \\ \frac{17}{30} \end{bmatrix} = \begin{bmatrix} 0.433 \\ 0.567 \end{bmatrix} = \begin{bmatrix} b_2 \\ c_2 \end{bmatrix}$$

Thus after 2 years, B controls 43.3% of the market and C controls 56.7%. We see that B regains a little in the second year.

C. Steady-state vector

Suppose we have a Markov chain with transition matrix A. A distribution vector $\mathbf{s}$ is called a **steady-state vector** (or a **stable vector**) for A if it satisfies

$$A\mathbf{s} = \mathbf{s} \tag{9.12}$$

Thus if A is applied to a steady-state vector, there is no change in the state distribution. Equation (9.12) indicates that the steady-state vector $\mathbf{s}$ is the eigenvector of A corresponding to the eigenvalue 1.

EXAMPLE 9-4: Find the condition for the existence of a steady-state vector for a given transition matrix A.

Solution: Definition (9.12) can be rewritten as

$$(A - I)\mathbf{s} = \mathbf{0} \tag{9.13}$$

where I is the identity matrix of the same order as A. Then from Corollary 1-4.5, (9.13) will have a nonzero $\mathbf{s}$ if and only if $(A - I)$ is singular; that is,

$$\det(A - I) = 0 \tag{9.14}$$

which is the required condition.

EXAMPLE 9-5: Find the steady-state vector for the transition matrix A of Example 9-3.

Solution: From Example 9-3,

$$A = \begin{bmatrix} \frac{1}{3} & \frac{1}{2} \\ \frac{2}{3} & \frac{1}{2} \end{bmatrix}$$

Then

$$A - I = \begin{bmatrix} \frac{1}{3} & \frac{1}{2} \\ \frac{2}{3} & \frac{1}{2} \end{bmatrix} - \begin{bmatrix} 1 & 0 \\ 0 & 1 \end{bmatrix} = \begin{bmatrix} -\frac{2}{3} & \frac{1}{2} \\ \frac{2}{3} & -\frac{1}{2} \end{bmatrix}$$

and

$$\det(A - I) = \begin{vmatrix} -\frac{2}{3} & \frac{1}{2} \\ \frac{2}{3} & -\frac{1}{2} \end{vmatrix} = 0$$

Hence a steady-state vector exists. Let $\mathbf{s} = \begin{bmatrix} s_1 \\ s_2 \end{bmatrix}$ be the steady-state vector. Then $(A - I)\mathbf{s} = \mathbf{0}$ becomes

$$\begin{bmatrix} -\frac{2}{3} & \frac{1}{2} \\ \frac{2}{3} & -\frac{1}{2} \end{bmatrix} \begin{bmatrix} s_1 \\ s_2 \end{bmatrix} = \begin{bmatrix} 0 \\ 0 \end{bmatrix}$$

which reduces to only one equation,

$$\tfrac{2}{3}s_1 - \tfrac{1}{2}s_2 = 0$$

However, **s** is a distribution vector, so we must have $s_1, s_2 \geq 0$ and

$$s_1 + s_2 = 1$$

Thus, solving the above two equations for s_1 and s_2, we get

$$s_1 = \tfrac{3}{7}, \qquad s_2 = \tfrac{4}{7}$$

Hence the steady-state vector is $\begin{bmatrix} \frac{3}{7} \\ \frac{4}{7} \end{bmatrix}$.

D. Regular Markov chain

A Markov chain is called *regular* if some power A^k of the transition matrix A has no zero entries. For regular Markov chains, regardless of the initial distribution, the distribution will eventually approach a steady state. Furthermore, each column of A will approach the steady-state vector as k increases. This is illustrated in Problem 9-5.

9-2. Difference Equations

A. Linear difference equations of *n*th order

DEFINITION An equation that relates differences of an unknown function is known as a **difference equation.**

A *linear difference equation of* n*th order* with constant coefficients is an equation of the form

$$y(k + n) + a_1 y(k + n - 1) + \cdots + a_n y(k) = f(k) \tag{9.15}$$

where k is an integer variable, $f(k)$ is a known sequence, and $a_1, \ldots, a_n$ are given constants. A solution of this equation is any sequence $y(k)$ that satisfies equation (9.15). Equation (9.15) has many solutions; however, it has a unique solution if the values of $y(k)$ for k from 0 to $n - 1$,

$$y(0), y(1), \ldots, y(n - 1)$$

are given. These numbers are called *initial conditions.*

If $f(k) = 0$ in equation (9.15), that is,

$$y(k + n) + a_1 y(k + n - 1) + \cdots + a_n y(k) = 0 \tag{9.16}$$

then equation (9.16) is called the *homogeneous* n*th-order difference equation.*

B. Matrix representation of a linear difference equation

EXAMPLE 9-6: Show that the nth-order difference equation of (9.15) can be rewritten as a system of n first-order difference equations.

Solution: Let

$$\begin{aligned} x_1(k) &= y(k) \\ x_2(k) &= y(k + 1) \\ &\vdots \\ x_n(k) &= y(k + n - 1) \end{aligned} \tag{9.17}$$

Then we get

$$\begin{aligned} x_1(k + 1) &= x_2(k) \\ x_2(k + 1) &= x_3(k) \\ &\vdots \\ x_n(k + 1) &= -a_n x_1(k) - a_{n-1} x_2(k) - \cdots - a_1 x_n(k) + f(k) \end{aligned} \tag{9.18}$$

The first $n - 1$ equations come from (9.17) and the last equation comes from (9.15) and (9.17). Thus we obtained a system of n first-order difference equations.

The system (9.18) can be written in matrix form as

$$\begin{bmatrix} x_1(k+1) \\ x_2(k+1) \\ \vdots \\ x_n(k+1) \end{bmatrix} = \begin{bmatrix} 0 & 1 & 0 & \cdots & 0 \\ 0 & 0 & 1 & \cdots & 0 \\ \vdots & \vdots & \vdots & & \vdots \\ -a_n & -a_{n-1} & -a_{n-2} & \cdots & -a_1 \end{bmatrix} \begin{bmatrix} x_1(k) \\ x_2(k) \\ \vdots \\ x_n(k) \end{bmatrix} + \begin{bmatrix} 0 \\ 0 \\ \vdots \\ 1 \end{bmatrix} f(k) \tag{9.19}$$

or

$$\mathbf{x}(k+1) = A\mathbf{x}(k) + \mathbf{b}f(k) \tag{9.20}$$

where

$$\mathbf{x}(k) = \begin{bmatrix} x_1(k) \\ x_2(k) \\ \vdots \\ x_n(k) \end{bmatrix}, \qquad A = \begin{bmatrix} 0 & 1 & 0 & \cdots & 0 \\ 0 & 0 & 1 & \cdots & 0 \\ \vdots & \vdots & \vdots & & \vdots \\ -a_n & -a_{n-1} & -a_{n-2} & \cdots & -a_1 \end{bmatrix}, \qquad \mathbf{b} = \begin{bmatrix} 0 \\ 0 \\ \vdots \\ 1 \end{bmatrix} \tag{9.21}$$

note: Equation (9.20) is sometimes referred to as the *state equation* representation of the nth-order difference equation (9.15), and the initial conditions vector $\mathbf{x}(0)$ is called the *initial state*.

C. Solution of a matrix difference equation

EXAMPLE 9-7: Consider a homogeneous matrix difference equation

$$\mathbf{x}(k+1) = A\mathbf{x}(k) \tag{9.22}$$

Show that, given $\mathbf{x}(0)$, the solution of equation (9.22) is given by

$$\mathbf{x}(k) = A^k\mathbf{x}(0) \tag{9.23}$$

Solution: Solving equation (9.22) iteratively, we get

$$\mathbf{x}(1) = A\mathbf{x}(0)$$

$$\mathbf{x}(2) = A\mathbf{x}(1) = A[A\mathbf{x}(0)] = A^2\mathbf{x}(0)$$

$$\vdots$$

$$\mathbf{x}(k) = A^k x(0)$$

Assume that $\mathbf{x}(k) = A^k\mathbf{x}(0)$ is true for $k = n$; that is,

$$\mathbf{x}(n) = A^n\mathbf{x}(0)$$

Then

$$\mathbf{x}(n+1) = A\mathbf{x}(n) = A[A^n\mathbf{x}(0)] = A^{n+1}\mathbf{x}(0)$$

Thus $\mathbf{x}(k) = A^k\mathbf{x}(0)$ is true for any integer k.

Note that the Markov chain described by equation (9.11) is the first-order matrix difference equation.

EXAMPLE 9-8: Consider a nonhomogeneous matrix difference equation

$$\mathbf{x}(k+1) = A\mathbf{x}(k) + \mathbf{b}f(k) \tag{9.24}$$

Show that, given $\mathbf{x}(0)$, the solution of equation (9.24) is given by

$$\mathbf{x}(k) = A^k\mathbf{x}(0) + \sum_{m=0}^{k-1} A^{k-1-m}\mathbf{b}f(m) \tag{9.25}$$

Solution: We solve equation (9.24) iteratively; that is,

$$\begin{aligned}
\mathbf{x}(1) &= A\mathbf{x}(0) + \mathbf{b}f(0) \\
\mathbf{x}(2) &= A\mathbf{x}(1) + \mathbf{b}f(1) \\
&= A[A\mathbf{x}(0) + \mathbf{b}f(0)] + \mathbf{b}f(1) \\
&= A^2\mathbf{x}(0) + A\mathbf{b}f(0) + \mathbf{b}f(1) \\
\mathbf{x}(3) &= A\mathbf{x}(2) + \mathbf{b}f(2) \\
&= A[A^2\mathbf{x}(0) + A\mathbf{b}f(0) + \mathbf{b}f(1)] + \mathbf{b}f(2) \\
&= A^3\mathbf{x}(0) + A^2\mathbf{b}f(0) + A\mathbf{b}f(1) + \mathbf{b}f(2) \\
&\vdots \\
\mathbf{x}(k) &= A\mathbf{x}(k-1) + \mathbf{b}f(k-1) \\
&= A^k\mathbf{x}(0) + A^{k-1}\mathbf{b}f(0) + A^{k-2}\mathbf{b}f(1) + \cdots + \mathbf{b}f(k-1) \\
&= A^k\mathbf{x}(0) + \sum_{m=0}^{k-1} A^{k-1-m}\mathbf{b}f(m)
\end{aligned}$$

where $A^0 \equiv I$ (identity matrix).

EXAMPLE 9-9: Solve

$$y(k+2) - 5y(k+1) + 6y(k) = 0$$

with the initial conditions $y(0) = 4$, $y(1) = 6$.

Solution: Let

$$\begin{aligned}
x_1(k) &= y(k) \\
x_2(k) &= y(k+1)
\end{aligned}$$

Then

$$\begin{aligned}
x_1(k+1) &= x_2(k) \\
x_2(k+1) &= -6y(k) + 5y(k+1) \\
&= -6x_1(k) + 5x_2(k)
\end{aligned}$$

Writing in matrix form, we get

$$\begin{bmatrix} x_1(k+1) \\ x_2(k+1) \end{bmatrix} = \begin{bmatrix} 0 & 1 \\ -6 & 5 \end{bmatrix} \begin{bmatrix} x_1(k) \\ x_2(k) \end{bmatrix}$$

or

$$\mathbf{x}(k+1) = A\mathbf{x}(k)$$

with

$$\mathbf{x}(0) = \begin{bmatrix} x_1(0) \\ x_2(0) \end{bmatrix} = \begin{bmatrix} y(0) \\ y(1) \end{bmatrix} = \begin{bmatrix} 4 \\ 6 \end{bmatrix}$$

Hence, by (9.23),

$$\mathbf{x}(k) = A^k\mathbf{x}(0) = \begin{bmatrix} 0 & 1 \\ -6 & 5 \end{bmatrix}^k \begin{bmatrix} 4 \\ 6 \end{bmatrix}$$

We need to find A^k. So we write the characteristic equation of A:

$$|\lambda I_2 - A| = \begin{vmatrix} \lambda & -1 \\ 6 & \lambda - 5 \end{vmatrix} = \lambda(\lambda - 5) + 6 = \lambda^2 - 5\lambda + 6 = (\lambda - 2)(\lambda - 3) = 0$$

This gives us the eigenvalues of A, which are $\lambda_1 = 2$, $\lambda_2 = 3$. Then by the technique of (7.50) and (7.52), we have

$$A^k = \alpha_1 A + \alpha_0 I$$

and

$$\lambda_i^k = \alpha_1 \lambda_i + \alpha_0 \qquad (i = 1, 2)$$

Thus,

$$2^k = 2\alpha_1 + \alpha_0$$

$$3^k = 3\alpha_1 + \alpha_0$$

Solving for α_1 and α_0, we get

$$\alpha_1 = (3)^k - (2)^k, \qquad \alpha_0 = -2(3)^k + 3(2)^k$$

Hence

$$\begin{bmatrix} 0 & 1 \\ -6 & 5 \end{bmatrix}^k = [(3)^k - (2)^k]\begin{bmatrix} 0 & 1 \\ -6 & 5 \end{bmatrix} + [-2(3)^k + 3(2)^k]\begin{bmatrix} 1 & 0 \\ 0 & 1 \end{bmatrix}$$

$$= \begin{bmatrix} -2(3)^k + 3(2)^k & (3)^k - (2)^k \\ -6(3)^k + 6(2)^k & 3(3)^k - 2(2)^k \end{bmatrix}$$

and

$$x(k) = \begin{bmatrix} x_1(k) \\ x_2(k) \end{bmatrix} = \begin{bmatrix} -2(3)^k + 3(2)^k & (3)^k - (2)^k \\ -6(3)^k + 6(2)^k & 3(3)^k - 2(2)^k \end{bmatrix}\begin{bmatrix} 4 \\ 6 \end{bmatrix}$$

$$= \begin{bmatrix} -2(3)^k + 6(2)^k \\ -6(3)^k + 12(2)^k \end{bmatrix}$$

Since $y(k) = x_1(k)$, we get $y(k) = -2(3)^k + 6(2)^k$.

---

*9-3. Differential Equations [*For readers who have studied calculus.*]

A. A linear differential equation of *n*th order

DEFINITION **A linear differential equation of *n*th order** with constant coefficients is an equation of the form

$$y^{(n)}(t) + a_1 y^{(n-1)}(t) + \cdots + a_n y(t) = f(t) \tag{9.26}$$

where $f(t)$ is a known function and $a_1, \ldots, a_n$ are given constants. A solution of this equation is any function $y(t)$ that satisfies equation (9.26). Equation (9.26) has many solutions; however, it has a unique solution if the initial values of $y(t)$ and its first $n - 1$ derivatives

$$y(0), y'(0), \ldots, y^{(n-1)}(0) \tag{9.27}$$

are given. These numbers are called *initial conditions.*

If $f(t) = 0$ in equation (9.26), that is,

$$y^{(n)}(t) + a_1 y^{(n-1)}(t) + \cdots + a_n y(t) = 0 \tag{9.28}$$

then equation (9.28) is called the *homogeneous nth-order differential equation.*

B. Matrix representation of a linear differential equation

EXAMPLE 9-10: Show that the nth-order differential equation of (9.26) can be rewritten as a system of n first-order differential equations.

Solution: Let

$$\begin{aligned} x_1(t) &= y(t) \\ x_2(t) &= y'(t) \\ &\vdots \\ x_n(t) &= y^{(n-1)}(t) \end{aligned} \tag{9.29}$$

Then we get

$$\begin{aligned} x_1'(t) &= y'(t) = x_2(t) \\ x_2'(t) &= y''(t) = x_3(t) \\ &\vdots \\ x_n'(t) &= y^{(n)}(t) = -a_n x_1(t) - a_{n-1} x_2(t) - \cdots - a_1 x_n(t) + f(t) \end{aligned} \tag{9.30}$$

[The last equation comes from (9.26).] Thus we have obtained a system of n first-order differential equations.

note: The system (9.30) can be written in matrix form as

$$\begin{bmatrix} x_1'(t) \\ x_2'(t) \\ \vdots \\ x_n'(t) \end{bmatrix} = \begin{bmatrix} 0 & 1 & 0 & \cdots & 0 \\ 0 & 0 & 1 & \cdots & 0 \\ \vdots & \vdots & \vdots & & \vdots \\ -a_n & -a_{n-1} & -a_{n-2} & \cdots & -a_1 \end{bmatrix} \begin{bmatrix} x_1(t) \\ x_2(t) \\ \vdots \\ x_n(t) \end{bmatrix} + \begin{bmatrix} 0 \\ 0 \\ \vdots \\ 1 \end{bmatrix} f(t) \tag{9.31}$$

or

$$\mathbf{x}'(t) = A\mathbf{x}(t) + \mathbf{b}f(t) \tag{9.32}$$

where

$$\mathbf{x}(t) = \begin{bmatrix} x_1(t) \\ x_2(t) \\ \vdots \\ x_n(t) \end{bmatrix}, \quad \mathbf{x}'(t) = \begin{bmatrix} x_1'(t) \\ x_2'(t) \\ \vdots \\ x_n'(t) \end{bmatrix}, \quad A = \begin{bmatrix} 0 & 1 & 0 & \cdots & 0 \\ 0 & 0 & 1 & \cdots & 0 \\ \vdots & \vdots & \vdots & & \vdots \\ -a_n & -a_{n-1} & -a_{n-2} & \cdots & -a_1 \end{bmatrix}, \quad \mathbf{b} = \begin{bmatrix} 0 \\ 0 \\ \vdots \\ 1 \end{bmatrix} \tag{9.33}$$

Equation (9.32) is sometimes referred to as the *state equation* representation of the nth-order differential equation (9.26), and the initial conditions vector $\mathbf{x}(0)$ is called the *initial state.*

C. Solution of a matrix differential equation

DEFINITION We recall that the exponential function can be defined as the power series

$$e^{\lambda} = 1 + \lambda + \frac{\lambda^2}{2!} + \cdots + \frac{\lambda^k}{k!} + \cdots \tag{9.34}$$

Now we use this expansion to define the matrix function (see Section 7-5)

$$\begin{aligned} e^{At} &= I_n + At + \frac{(At)^2}{2!} + \cdots + \frac{(At)^k}{k!} + \cdots \\ &= I_n + At + \frac{A^2t^2}{2!} + \cdots + \frac{A^kt^k}{k!} + \cdots \end{aligned} \tag{9.35}$$

Note that, since powers of the matrix A are $n \times n$ matrices, the right side of (9.35) will be an $n \times n$ matrix if the series converges. Also note that $e^{O} = I_n$.

EXAMPLE 9-11: Show that if e^{At} converges for all t, then

$$\frac{d}{dt}(e^{At}) = Ae^{At} \tag{9.36}$$

Solution: Since e^{At} converges for all t, (9.35) can be differentiated term by term. So we get

$$\begin{aligned}\frac{d}{dt}(e^{At}) &= \frac{d}{dt}\left[I_n + At + \frac{A^2t^2}{2!} + \cdots + \frac{A^kt^k}{k!} + \cdots\right] \\ &= 0 + A + A^2t + \cdots + \frac{A^kt^{k-1}}{(k-1)!} + \cdots \\ &= A\left[I_n + At + \cdots + \frac{A^{k-1}t^{k-1}}{(k-1)!} + \cdots\right] \\ &= Ae^{At}\end{aligned}$$

EXAMPLE 9-12: Consider a homogeneous matrix differential equation

$$\mathbf{x}'(t) = A\mathbf{x}(t) \qquad \textbf{(9.37)}$$

Show that given initial conditions $\mathbf{x}(0)$, the solution of equation (9.37) is given by

$$\mathbf{x}(t) = e^{At}\mathbf{x}(0) \qquad \textbf{(9.38)}$$

Solution: From (9.36), we have

$$\begin{aligned}\mathbf{x}'(t) &= \frac{d}{dt}[\mathbf{x}(t)] \\ &= \frac{d}{dt}[e^{At}\mathbf{x}(0)] \\ &= \frac{d}{dt}[e^{At}]\mathbf{x}(0) \\ &= [Ae^{At}]\mathbf{x}(0) \\ &= A[e^{At}\mathbf{x}(\mathbf{0})] = A\mathbf{x}(t)\end{aligned}$$

and $\mathbf{x}(0) = e^{O}\mathbf{x}(0) = I\mathbf{x}(0) = \mathbf{x}(0)$. Thus $\mathbf{x}(t) = e^{At}\mathbf{x}(0)$ is the solution of equation (9.37) with the given initial conditions.

EXAMPLE 9-13: Solve the following second-order differential equation

$$y''(t) + 3y'(t) + 2y(t) = 0$$

with the initial conditions $y(0) = 0$, $y'(0) = 1$.

Solution: Let

$$\begin{aligned}x_1(t) &= y(t) \\ x_2(t) &= y'(t)\end{aligned}$$

Then

$$\begin{aligned}x_1'(t) &= y'(t) = x_2(t) \\ x_2'(t) &= y''(t) = -2y(t) - 3y'(t) \\ &= -2x_1(t) - 3x_2(t)\end{aligned}$$

Thus, we get

$$\begin{bmatrix} x_1'(t) \\ x_2'(t) \end{bmatrix} = \begin{bmatrix} 0 & 1 \\ -2 & -3 \end{bmatrix}\begin{bmatrix} x_1(t) \\ x_2(t) \end{bmatrix}$$

or

$$\mathbf{x}'(t) = A\mathbf{x}(t)$$

with

$$\mathbf{x}(0) = \begin{bmatrix} x_1(0) \\ x_2(0) \end{bmatrix} = \begin{bmatrix} y(0) \\ y'(0) \end{bmatrix} = \begin{bmatrix} 0 \\ 1 \end{bmatrix}$$

Hence, by (9.38),

$$\mathbf{x}(t) = e^{At}\mathbf{x}(0)$$

We need to find e^{At}. So we write the characteristic equation of A:

$$|\lambda I_2 - A| = \begin{vmatrix} \lambda & -1 \\ 2 & \lambda + 3 \end{vmatrix} = \lambda(\lambda + 3) + 2 = \lambda^2 + 3\lambda + 2 = (\lambda + 1)(\lambda + 2) = 0$$

Thus the eigenvalues of A are $\lambda_1 = -1$ and $\lambda_2 = -2$. Now, by the technique of (7.50) and (7.52), we have

$$e^{At} = \alpha_1 A + \alpha_0 I_2$$

and

$$e^{\lambda_i t} = \alpha_1 \lambda_i + \alpha_0 \qquad (i = 1, 2)$$

Thus

$$e^{-t} = -\alpha_1 + \alpha_0$$
$$e^{-2t} = -2\alpha_1 + \alpha_0$$

Solving for α_1 and α_0, we get

$$\alpha_1 = e^{-t} - e^{-2t}, \qquad \alpha_0 = 2e^{-t} - e^{-2t}$$

Hence

$$e^{At} = (e^{-t} - e^{-2t})\begin{bmatrix} 0 & 1 \\ -2 & -3 \end{bmatrix} + (2e^{-t} - e^{-2t})\begin{bmatrix} 1 & 0 \\ 0 & 1 \end{bmatrix}$$
$$= \begin{bmatrix} 2e^{-t} - e^{-2t} & e^{-t} - e^{-2t} \\ -2e^{-t} + 2e^{-2t} & -e^{-t} + 2e^{-2t} \end{bmatrix}$$

and

$$\mathbf{x}(t) = \begin{bmatrix} x_1(t) \\ x_2(t) \end{bmatrix} = e^{At}\mathbf{x}(0) = \begin{bmatrix} 2e^{-t} - e^{-2t} & e^{-t} - e^{-2t} \\ -2e^{-t} + 2e^{-2t} & -e^{-t} + 2e^{-2t} \end{bmatrix}\begin{bmatrix} 0 \\ 1 \end{bmatrix}$$
$$= \begin{bmatrix} e^{-t} - e^{-2t} \\ -e^{-t} + 2e^{-2t} \end{bmatrix}$$

Since $y(t) = x_1(t)$, we get $y(t) = e^{-t} - e^{-2t}$.

SUMMARY

1. A row vector $\mathbf{p}$ with nonnegative entries such that the sum of the entries is 1 is called a probability vector.
2. A matrix S is called a stochastic matrix if it is square and its row vectors are probability vectors.
3. If S is a stochastic matrix, then S^k is again stochastic for any positive integer k. If $\mathbf{p}^{(0)}$ is a probability vector, then $\mathbf{p}^{(k)} = \mathbf{p}^{(0)}S^k$ is also a probability vector.
4. A distribution vector $\mathbf{s}$ is the transpose of a probability vector $\mathbf{p}$, and a transition matrix A is the transpose of a stochastic matrix S; that is,

$$\mathbf{s} = \mathbf{p}^T, \qquad A = S^T$$

5. A distribution vector $\mathbf{s}$ is called a steady-state vector (or a stable vector) if it satisfies $A\mathbf{s} = \mathbf{s}$.
6. A Markov chain is called regular if some power A^k of the transition matrix A has no zero entries.
7. A linear difference equation of nth order with constant coefficients

$$y(k + n) + a_1 y(k + n - 1) + \cdots + a_n y(k) = f(k)$$

can be expressed as a system of n first-order difference equations

$$\mathbf{x}(k+1) = A\mathbf{x}(k) + \mathbf{b}f(k)$$

where

$$\mathbf{x}(k) = \begin{bmatrix} x_1(k) \\ x_2(k) \\ \vdots \\ x_n(k) \end{bmatrix} = \begin{bmatrix} y(k) \\ y(k+1) \\ \vdots \\ y(k+n-1) \end{bmatrix}, \quad A = \begin{bmatrix} 0 & 1 & 0 & \cdots & 0 \\ 0 & 0 & 1 & \cdots & 0 \\ \vdots & \vdots & \vdots & & \vdots \\ -a_n & -a_{n-1} & -a_{n-2} & \cdots & -a_1 \end{bmatrix}, \quad \mathbf{b} = \begin{bmatrix} 0 \\ 0 \\ \vdots \\ 1 \end{bmatrix}$$

8. The solution to a homogeneous matrix difference equation

$$\mathbf{x}(k+1) = A\mathbf{x}(k)$$

with the initial condition $\mathbf{x}(0)$ is given by

$$\mathbf{x}(k) = A^k\mathbf{x}(0)$$

9. The solution to a nonhomogeneous matrix difference equation

$$\mathbf{x}(k+1) = A\mathbf{x}(k) + \mathbf{b}f(k)$$

with the initial conditions $\mathbf{x}(0)$ is given by

$$\mathbf{x}(k) = A^k\mathbf{x}(0) + \sum_{m=0}^{k-1} A^{k-1-m}\mathbf{b}f(m)$$

10. A linear differential equation of nth order with constant coefficients

$$y^{(n)}(t) + a_1 y^{(n-1)}(t) + \cdots + a_n y(t) = f(t)$$

can be expressed as a system of n first-order differential equations

$$\mathbf{x}'(t) = A\mathbf{x}(t) + \mathbf{b}f(t)$$

where

$$\mathbf{x}(t) = \begin{bmatrix} x_1(t) \\ x_2(t) \\ \vdots \\ x_n(t) \end{bmatrix} = \begin{bmatrix} y(t) \\ y'(t) \\ \vdots \\ y^{(n-1)}(t) \end{bmatrix}, \qquad \mathbf{x}'(t) = \begin{bmatrix} x_1'(t) \\ x_2'(t) \\ \vdots \\ x_n'(t) \end{bmatrix},$$

$$A = \begin{bmatrix} 0 & 1 & 0 & \cdots & 0 \\ 0 & 0 & 1 & \cdots & 0 \\ \vdots & \vdots & \vdots & & \vdots \\ -a_n & -a_{n-1} & -a_{n-2} & \cdots & -a_1 \end{bmatrix}, \qquad \mathbf{b} = \begin{bmatrix} 0 \\ 0 \\ \vdots \\ 1 \end{bmatrix}$$

11. The solution to a homogeneous matrix differential equation

$$\mathbf{x}'(t) = A\mathbf{x}(t)$$

with the initial conditions $\mathbf{x}(0)$ is given by

$$\mathbf{x}(t) = e^{At}\mathbf{x}(0)$$

where

$$e^{At} \equiv \sum_{k=0}^{\infty} \frac{(At)^k}{k!}, \qquad A^0 = I$$

RAISE YOUR GRADES

Can you explain . . . ?

☑ how to decide whether a given matrix is stochastic
☑ how to find a distribution vector after k steps given the initial distribution
☑ how to find the steady-state distribution vector for a given transition matrix
☑ the regular Markov chain
☑ how to find the power of the transition matrix
☑ how to transform a linear nth-order difference equation into a system of n first-order difference equations
☑ how to find the solution of a first-order matrix difference equation
☑ how to transform a linear nth-order differential equation into a system of n first-order differential equations
☑ how to find the solution of a first-order matrix differential equation

SOLVED PROBLEMS

Markov Analysis

PROBLEM 9-1 Determine which of the following row vectors are probability vectors:

$$\mathbf{r}_1 = [1, 0], \qquad \mathbf{r}_2 = [\tfrac{1}{3}, \tfrac{3}{7}], \qquad \mathbf{r}_3 = [\tfrac{2}{5}, \tfrac{3}{5}], \qquad \mathbf{r}_4 = [0.2, 0.4, 0.4]$$

Solution Each row sum of $\mathbf{r}_1$, $\mathbf{r}_3$, $\mathbf{r}_4$ is 1 while the row sum of $\mathbf{r}_2$ is not 1; so $\mathbf{r}_1$, $\mathbf{r}_3$, $\mathbf{r}_4$ are probability vectors and $\mathbf{r}_2$ is not.

PROBLEM 9-2 Determine which of the following matrices are stochastic matrices:

$$A = \begin{bmatrix} \frac{1}{2} & \frac{1}{2} \\ \frac{1}{3} & \frac{2}{3} \end{bmatrix}, \qquad B = \begin{bmatrix} 0 & 1 \\ 1 & 0 \end{bmatrix}, \qquad C = \begin{bmatrix} 1 & 0.4 \\ 0 & 0.6 \end{bmatrix}, \qquad D = \begin{bmatrix} 0.8 & 0.1 & 0.1 \\ 0.2 & 0.7 & 0.1 \\ 0.1 & 0.3 & 0.6 \end{bmatrix}$$

Solution The sum of the entries for each row in A, B, and D is 1, while in C it is not; A, B, and D are stochastic matrices but C is not. (Note that C^T is a stochastic matrix.)

PROBLEM 9-3 Show that if S is a stochastic matrix, then S^k is again stochastic for any positive integer k.

Solution Let

$$S = \begin{bmatrix} p_{11} & p_{12} & \cdots & p_{1m} \\ p_{21} & p_{22} & \cdots & p_{2m} \\ \vdots & \vdots & & \vdots \\ p_{m1} & p_{m2} & \cdots & p_{mm} \end{bmatrix} \tag{a}$$

The fact that S is stochastic can be expressed as

$$\begin{bmatrix} p_{11} & p_{12} & \cdots & p_{1m} \\ p_{21} & p_{22} & \cdots & p_{2m} \\ \vdots & \vdots & & \vdots \\ p_{m1} & p_{m2} & \cdots & p_{mm} \end{bmatrix} \begin{bmatrix} 1 \\ 1 \\ \vdots \\ 1 \end{bmatrix} = \begin{bmatrix} p_{11} + p_{12} + \cdots + p_{1m} \\ p_{21} + p_{22} + \cdots + p_{2m} \\ \vdots \quad \vdots \quad \vdots \\ p_{m1} + p_{m2} + \cdots + p_{mm} \end{bmatrix} = \begin{bmatrix} 1 \\ 1 \\ \vdots \\ 1 \end{bmatrix} \tag{b}$$

in view of (9.2). Thus we have

$$S\mathbf{a} = \mathbf{a} \tag{c}$$

where

$$\mathbf{a} = \begin{bmatrix} 1 \\ 1 \\ \vdots \\ 1 \end{bmatrix}$$

Then if we premultiply both sides of (c) by S, we get

$$S^2\mathbf{a} = S\mathbf{a} = \mathbf{a} \tag{d}$$

which shows that S^2 is stochastic. And repeated premultiplication by S shows that

$$S^k\mathbf{a} = \mathbf{a} \tag{e}$$

Thus S^k is stochastic for any positive integer k.

PROBLEM 9-4 Show that if S is a stochastic matrix and $\mathbf{p}^{(0)}$ is a probability vector, then $\mathbf{p}^{(k)} = \mathbf{p}^{(0)}S$ is also a probability vector.

Solution If $\mathbf{p}^{(0)}$ is a probability vector, then $\mathbf{p}^{(k)} = \mathbf{p}^{(0)}S^k$ has nonnegative entries, and by the result of Problem 9-3 [see (e)] we have

$$\mathbf{p}^{(k)}\mathbf{a} = \mathbf{p}^{(0)}S^k\mathbf{a} = \mathbf{p}^{(0)}\mathbf{a} = [p_1^{(0)}, p_2^{(0)}, \ldots, p_m^{(0)}] \begin{bmatrix} 1 \\ 1 \\ \vdots \\ 1 \end{bmatrix}$$

$$= p_1^{(0)} + p_2^{(0)} + \cdots + p_m^{(0)} = 1$$

Hence

$$\mathbf{p}^{(k)}\mathbf{a} = [p_1^{(k)}, p_2^{(k)}, \ldots, p_m^{(k)}] \begin{bmatrix} 1 \\ 1 \\ \vdots \\ 1 \end{bmatrix} = p_1^{(k)} + p_2^{(k)} + \cdots + p_m^{(k)} = 1$$

which shows that $\mathbf{p}^{(k)}$ is a probability vector.

PROBLEM 9-5 For the transition matrix of Example 9-3, calculate A^k and show that $A^k\mathbf{x}_0$ and the column vectors of A^k approach the steady-state vector obtained in Example 9-5, as $k \to \infty$.

Solution

$$A = \begin{bmatrix} \frac{1}{3} & \frac{1}{2} \\ \frac{2}{3} & \frac{1}{2} \end{bmatrix}$$

First, let us find the eigenvalues and the eigenvectors of A. The characteristic equation of A is

$$|\lambda I_2 - A| = \begin{vmatrix} \lambda - \frac{1}{3} & -\frac{1}{2} \\ -\frac{2}{3} & \lambda - \frac{1}{2} \end{vmatrix} = (\lambda - \tfrac{1}{3})(\lambda - \tfrac{1}{2}) - \tfrac{1}{3}$$

$$= \lambda^2 - \tfrac{5}{6}\lambda - \tfrac{1}{6} = (\lambda - 1)(\lambda + \tfrac{1}{6}) = 0$$

Thus the eigenvalues of A are $\lambda_1 = 1$, and $\lambda_2 = -\frac{1}{6}$.

Let $\mathbf{x} = \begin{bmatrix} x_1 \\ x_2 \end{bmatrix}$ be an eigenvector of A. Then $(\lambda I_2 - A)\mathbf{x} = \mathbf{0}$. The eigenvector corresponding to $\lambda = 1$ is found in Example 9-5 as $\begin{bmatrix} \frac{3}{7} \\ \frac{4}{7} \end{bmatrix}$. And for $\lambda = -\frac{1}{6}$, we have

$$\begin{bmatrix} -\frac{1}{2} & -\frac{1}{2} \\ -\frac{2}{3} & -\frac{2}{3} \end{bmatrix} \begin{bmatrix} x_1 \\ x_2 \end{bmatrix} = \begin{bmatrix} 0 \\ 0 \end{bmatrix}$$

from which we have $x_1 + x_2 = 0$, or $x_2 = -x_1$. Thus setting $x_1 = 1$, we get the eigenvector of A corresponding to $\lambda = -\frac{1}{6}$ as $\begin{bmatrix} 1 \\ -1 \end{bmatrix}$.

Let

$$P = \begin{bmatrix} \frac{3}{7} & 1 \\ \frac{4}{7} & -1 \end{bmatrix}$$

then

$$P^{-1} = \begin{bmatrix} 1 & 1 \\ \frac{4}{7} & -\frac{3}{7} \end{bmatrix}$$

and

$$P^{-1}AP = \begin{bmatrix} 1 & 1 \\ \frac{4}{7} & -\frac{3}{7} \end{bmatrix} \begin{bmatrix} \frac{1}{3} & \frac{1}{2} \\ \frac{2}{3} & \frac{1}{2} \end{bmatrix} \begin{bmatrix} \frac{3}{7} & 1 \\ \frac{4}{7} & -1 \end{bmatrix} = \begin{bmatrix} 1 & 0 \\ 0 & -\frac{1}{6} \end{bmatrix} = 0$$

Hence

$$A = PDP^{-1}$$

and

$$A^k = PD^kP^{-1} = \begin{bmatrix} \frac{3}{7} & 1 \\ \frac{4}{7} & -1 \end{bmatrix} \begin{bmatrix} 1 & 0 \\ 0 & -\frac{1}{6} \end{bmatrix}^k \begin{bmatrix} 1 & 1 \\ \frac{4}{7} & -\frac{3}{7} \end{bmatrix}$$

$$= \begin{bmatrix} \frac{3}{7} & 1 \\ \frac{4}{7} & -1 \end{bmatrix} \begin{bmatrix} 1^k & 0 \\ 0 & (-\frac{1}{6})^k \end{bmatrix} \begin{bmatrix} 1 & 1 \\ \frac{4}{7} & -\frac{3}{7} \end{bmatrix}$$

$$= \begin{bmatrix} \frac{3}{7} & \frac{3}{7} \\ \frac{4}{7} & \frac{4}{7} \end{bmatrix} + (-\tfrac{1}{6})^k \begin{bmatrix} \frac{4}{7} & -\frac{3}{7} \\ -\frac{4}{7} & \frac{3}{7} \end{bmatrix}$$

Thus when $k \to \infty$, $(-\frac{1}{6})^k \to 0$, and we get

$$\lim_{k \to \infty} A^k = \begin{bmatrix} \frac{3}{7} & \frac{3}{7} \\ \frac{4}{7} & \frac{4}{7} \end{bmatrix}$$

and

$$\lim_{k \to \infty} A^k\mathbf{x}_0 = \left(\lim_{k \to \infty} A^k\right)\mathbf{x}_0 = \begin{bmatrix} \frac{3}{7} & \frac{3}{7} \\ \frac{4}{7} & \frac{4}{7} \end{bmatrix} \begin{bmatrix} \frac{3}{5} \\ \frac{2}{5} \end{bmatrix} = \begin{bmatrix} \frac{3}{7} \\ \frac{4}{7} \end{bmatrix}$$

It is seen that $A^k\mathbf{x}_0$ and each column of A approach the steady-state vector as $k \to \infty$.

PROBLEM 9-6 Consider

$$A = \begin{bmatrix} 0 & 1 \\ 1 & 0 \end{bmatrix} \quad \text{and} \quad \mathbf{x}_0 = \begin{bmatrix} \frac{1}{3} \\ \frac{2}{3} \end{bmatrix}$$

Find the steady-state vector for A; then find $A\mathbf{x}_0, A^2\mathbf{x}_0, A^3\mathbf{x}_0, \ldots$. Compare the results.

Solution Let $\mathbf{s} = \begin{bmatrix} s_1 \\ s_2 \end{bmatrix}$ be the steady-state vector. Then $\mathbf{s}$ must satisfy (9.13); that is,

$$(A - I_2)\mathbf{s} = \mathbf{0}$$

Thus

$$\begin{bmatrix} -1 & 1 \\ 1 & -1 \end{bmatrix}\begin{bmatrix} s_1 \\ s_2 \end{bmatrix} = \begin{bmatrix} 0 \\ 0 \end{bmatrix}$$

or

$$-s_1 + s_2 = 0$$

So $s_1 = s_2 = \frac{1}{2}$, since $s_1 + s_2 = 1$. Hence the steady-state vector is $\mathbf{s} = \begin{bmatrix} \frac{1}{2} \\ \frac{1}{2} \end{bmatrix}$.

Now

$$A = \begin{bmatrix} 0 & 1 \\ 1 & 0 \end{bmatrix}, \qquad A^2 = \begin{bmatrix} 0 & 1 \\ 1 & 0 \end{bmatrix}\begin{bmatrix} 0 & 1 \\ 1 & 0 \end{bmatrix} = \begin{bmatrix} 1 & 0 \\ 0 & 1 \end{bmatrix} = I_2$$

Then

$$A^3 = AA^2 = AI_2 = A, \qquad A^4 = AA^3 = AA = A^2 = I_2, \ldots$$

That is, the powers of A alternate between A and I_2. Thus we get

$$A\mathbf{x}_0 = \begin{bmatrix} 0 & 1 \\ 1 & 0 \end{bmatrix}\begin{bmatrix} \frac{1}{3} \\ \frac{2}{3} \end{bmatrix} = \begin{bmatrix} \frac{2}{3} \\ \frac{1}{3} \end{bmatrix}$$

$$A^2\mathbf{x}_0 = I_2\mathbf{x}_0 = \begin{bmatrix} \frac{1}{3} \\ \frac{2}{3} \end{bmatrix}, \qquad A^3\mathbf{x}_0 = A\mathbf{x}_0 = \begin{bmatrix} \frac{2}{3} \\ \frac{1}{3} \end{bmatrix}, \ldots$$

Hence the steady-state vector $\mathbf{s}$ can never be reached from the initial distribution of $\begin{bmatrix} \frac{1}{3} \\ \frac{2}{3} \end{bmatrix}$. Note that the transition matrix A is not regular.

Difference Equations

PROBLEM 9-7 Show that the linear difference equation of nth order with constant coefficients (9.15) can be rewritten as

$$y(k) + a_1 y(k-1) + \cdots + a_n y(k-n) = g(k)$$

Solution Writing equation (9.15)

$$y(k+n) + a_1 y(k+n-1) + \cdots + a_n y(k) = f(k)$$

and setting $k + n = m$, we have

$$y(m) + a_1 y(m-1) + \cdots + a_n y(m-n) = f(m-n)$$

Now changing m back to k, we get

$$y(k) + a_1 y(k-1) + \cdots + a_n y(k-n) = f(k-n) = g(k)$$

PROBLEM 9-8 Given

$$y(k+1) - 2y(k) = 0$$

with $y(0) = 1$, find $y(k)$ recursively.

Solution Rewriting the given equation

$$y(k+1) = 2y(k)$$

and setting $k = 0, 1, 2, \ldots$, we get

$$\begin{aligned} y(1) &= 2y(0) = 2(1) = 2 \\ y(2) &= 2y(1) = 2(2) = 2^2 \\ y(3) &= 2y(2) = 2(2^2) = 2^3 \\ &\vdots \end{aligned}$$

Continuing this process recursively, it is easy to see for all $k \geq 0$ that $y(k) = 2^k$.

***PROBLEM 9-9** [*For readers who have studied calculus.*] Solve

$$y(k) - 3y(k-1) + 3y(k-2) - y(k-3) = 0$$

with the initial conditions $y(-1) = 1$, $y(-2) = -1$, $y(-3) = 0$.

Solution Let

$$\begin{aligned} x_1(k) &= y(k-3) \\ x_2(k) &= y(k-2) \\ x_3(k) &= y(k-1) \end{aligned}$$

Then

$$\begin{aligned} x_1(k+1) &= x_2(k) \\ x_2(k+1) &= x_3(k) \\ x_3(k+1) &= y(k) = y(k-3) - 3y(k-2) + 3y(k-1) \\ &= x_1(k) - 3x_2(k) + 3x_3(k) \end{aligned}$$

Writing in matrix form, we get

$$\begin{bmatrix} x_1(k+1) \\ x_2(k+1) \\ x_3(k+1) \end{bmatrix} = \begin{bmatrix} 0 & 1 & 0 \\ 0 & 0 & 1 \\ 1 & -3 & 3 \end{bmatrix} \begin{bmatrix} x_1(k) \\ x_2(k) \\ x_3(k) \end{bmatrix}$$

or

$$\mathbf{x}(k+1) = A\mathbf{x}(k)$$

with

$$\mathbf{x}(0) = \begin{bmatrix} x_1(0) \\ x_2(0) \\ x_3(0) \end{bmatrix} = \begin{bmatrix} y(-3) \\ y(-2) \\ y(-1) \end{bmatrix} = \begin{bmatrix} 0 \\ -1 \\ 1 \end{bmatrix}$$

Thus by (9.23),

$$\mathbf{x}(k) = A^k\mathbf{x}(0) = \begin{bmatrix} 0 & 1 & 0 \\ 0 & 0 & 1 \\ 1 & -3 & 3 \end{bmatrix}^k \begin{bmatrix} 0 \\ -1 \\ 1 \end{bmatrix}$$

Now the characteristic equation of A is

$$|\lambda I_3 - A| = \begin{vmatrix} \lambda & -1 & 0 \\ 0 & \lambda & -1 \\ -1 & 3 & \lambda - 3 \end{vmatrix} = \lambda^2(\lambda - 3) + 3\lambda - 1 = \lambda^3 - 3\lambda^2 + 3\lambda - 1 = (\lambda - 1)^3$$

Thus the eigenvalue of A is $\lambda = 1$ with multiplicity 3.

Then by the technique of (7.50) and Problem 7-31 we have

$$A^k = \beta_2 A^2 + \beta_1 A + \beta_0 I_3$$

Let

$$f(\lambda) = \lambda^k$$

and

$$g(\lambda) = \beta_2 \lambda^2 + \beta_1 \lambda + \beta_0$$

Then coefficients $\beta_2, \beta_1, \beta_0$ are found from the conditions

$$f(\lambda)|_{\lambda=1} = g(\lambda)|_{\lambda=1}$$
$$f'(\lambda)|_{\lambda=1} = g'(\lambda)|_{\lambda=1}$$
$$f''(\lambda)|_{\lambda=1} = g''(\lambda)|_{\lambda=1}$$

Hence

$$1^k = \beta_2 + \beta_1 + \beta_0$$
$$k(1)^{k-1} = 2\beta_2 + \beta_1$$
$$k(k-1)(1)^{k-2} = 2\beta_2$$

Solving for $\beta_2, \beta_1, \beta_0$, we get

$$\beta_2 = \tfrac{1}{2}k^2 - \tfrac{1}{2}k, \qquad \beta_1 = -k^2 + 2k, \qquad \beta_0 = \tfrac{1}{2}k^2 - \tfrac{3}{2}k + 1$$

Thus

$$A^k = (\tfrac{1}{2}k^2 - \tfrac{1}{2}k)\begin{bmatrix} 0 & 0 & 1 \\ 1 & -3 & 3 \\ 3 & -8 & 6 \end{bmatrix} + (-k^2 + 2k)\begin{bmatrix} 0 & 1 & 0 \\ 0 & 0 & 1 \\ 1 & -3 & 3 \end{bmatrix} + (\tfrac{1}{2}k^2 - \tfrac{3}{2}k + 1)\begin{bmatrix} 1 & 0 & 0 \\ 0 & 1 & 0 \\ 0 & 0 & 1 \end{bmatrix}$$
$$= \begin{bmatrix} \tfrac{1}{2}k^2 - \tfrac{3}{2}k + 1 & -k^2 + 2k & \tfrac{1}{2}k^2 - \tfrac{1}{2}k \\ \tfrac{1}{2}k^2 - \tfrac{1}{2}k & \tfrac{1}{2}k^2 - \tfrac{3}{2}k + 1 & \tfrac{1}{2}k^2 + \tfrac{1}{2}k \\ \tfrac{1}{2}k^2 + \tfrac{1}{2}k & -k^2 - 2k & \tfrac{1}{2}k^2 + \tfrac{3}{2}k + 1 \end{bmatrix}$$

and

$$\mathbf{x}(k) = \begin{bmatrix} x_1(k) \\ x_2(k) \\ x_3(k) \end{bmatrix} = \begin{bmatrix} \tfrac{1}{2}k^2 - \tfrac{3}{2}k + 1 & -k^2 + 2k & \tfrac{1}{2}k^2 - \tfrac{1}{2}k \\ \tfrac{1}{2}k^2 - \tfrac{1}{2}k & \tfrac{1}{2}k^2 - \tfrac{3}{2}k + 1 & \tfrac{1}{2}k^2 + \tfrac{1}{2}k \\ \tfrac{1}{2}k^2 + \tfrac{1}{2}k & -k^2 - 2k & \tfrac{1}{2}k^2 + \tfrac{3}{2}k + 1 \end{bmatrix}\begin{bmatrix} 0 \\ -1 \\ 1 \end{bmatrix} = \begin{bmatrix} \tfrac{3}{2}k^2 - \tfrac{5}{2}k \\ 2k - 1 \\ \tfrac{3}{2}k^2 + \tfrac{7}{2}k + 1 \end{bmatrix}$$

Since $x_3(k+1) = y(k)$, we have

$$y(k) = \tfrac{3}{2}(k+1)^2 + \tfrac{7}{2}(k+1) + 1 = \tfrac{3}{2}k^2 + \tfrac{13}{2}k + 6$$

PROBLEM 9-10 Solve

$$y(k) - \tfrac{3}{4}y(k-1) + \tfrac{1}{8}y(k-2) = (\tfrac{1}{2})^k$$

with $y(-1) = 0$, $y(-2) = 0$.

Solution Let

$$x_1(k) = y(k-2)$$
$$x_2(k) = y(k-1)$$

Then

$$x_1(k+1) = x_2(k)$$
$$x_2(k+1) = y(k) = -\tfrac{1}{8}y(k-2) + \tfrac{3}{4}y(k-1) + (\tfrac{1}{2})^k$$
$$= -\tfrac{1}{8}x_1(k) + \tfrac{3}{4}x_2(k) + (\tfrac{1}{2})^k$$

Writing in matrix form

$$\begin{bmatrix} x_1(k+1) \\ x_2(k+1) \end{bmatrix} = \begin{bmatrix} 0 & 1 \\ -\frac{1}{8} & \frac{3}{4} \end{bmatrix} \begin{bmatrix} x_1(k) \\ x_2(k) \end{bmatrix} + \begin{bmatrix} 0 \\ 1 \end{bmatrix} \left(\frac{1}{2}\right)^k$$

or

$$\mathbf{x}(k+1) = A\mathbf{x}(k) + \mathbf{b}f(k)$$

where

$$A = \begin{bmatrix} 0 & 1 \\ -\frac{1}{8} & \frac{3}{4} \end{bmatrix}, \qquad \mathbf{b} = \begin{bmatrix} 0 \\ 1 \end{bmatrix}, \qquad f(k) = (\tfrac{1}{2})^k$$

We need to find A^k. The characteristic equation of A is

$$|\lambda I_2 - A| = \begin{vmatrix} \lambda & -1 \\ \frac{1}{8} & \lambda - \frac{3}{4} \end{vmatrix} = \lambda\left(\lambda - \frac{3}{4}\right) + \frac{1}{8} = \lambda^2 - \frac{3}{4}\lambda + \frac{1}{8} = \left(\lambda - \frac{1}{4}\right)\left(\lambda - \frac{1}{2}\right) = 0$$

thus the eigenvalues of A are $\lambda_1 = \frac{1}{4}$, and $\lambda_2 = \frac{1}{2}$. Then, by the technique of (7.50) and (7.52), we have

$$A^k = \alpha_1 A + \alpha_0 I_2$$
$$\lambda_i^k = \alpha_1 \lambda_i + \alpha_0 \qquad (i = 1, 2)$$

Thus

$$(\tfrac{1}{4})^k = \tfrac{1}{4}\alpha_1 + \alpha_0$$
$$(\tfrac{1}{2})^k = \tfrac{1}{2}\alpha_1 + \alpha_0$$

Solving for α_1, α_0, we get

$$\alpha_1 = 4[(\tfrac{1}{2})^k - (\tfrac{1}{4})^k], \qquad \alpha_0 = 2(\tfrac{1}{4})^k - (\tfrac{1}{2})^k$$

So

$$A^k = 4[(\tfrac{1}{2})^k - (\tfrac{1}{4})^k]\begin{bmatrix} 0 & 1 \\ -\frac{1}{8} & \frac{3}{4} \end{bmatrix} + [2(\tfrac{1}{4})^k - (\tfrac{1}{2})^k]\begin{bmatrix} 1 & 0 \\ 0 & 1 \end{bmatrix} = \begin{bmatrix} -(\frac{1}{2})^k + 2(\frac{1}{4})^k & 4(\frac{1}{2})^k - 4(\frac{1}{4})^k \\ -\frac{1}{2}(\frac{1}{2})^k + \frac{1}{2}(\frac{1}{4})^k & 2(\frac{1}{2})^k - (\frac{1}{4})^k \end{bmatrix}$$

or

$$A^k = \left(\frac{1}{2}\right)^k \begin{bmatrix} -1 & 4 \\ -\frac{1}{2} & 2 \end{bmatrix} + \left(\frac{1}{4}\right)^k \begin{bmatrix} 2 & -4 \\ \frac{1}{2} & -1 \end{bmatrix}$$

Since

$$\mathbf{x}(0) = \begin{bmatrix} x_1(0) \\ x_2(0) \end{bmatrix} = \begin{bmatrix} y(-2) \\ y(-1) \end{bmatrix} = \begin{bmatrix} 0 \\ 0 \end{bmatrix}$$

by (9.25), we have

$$\mathbf{x}(k) = \sum_{m=0}^{k-1} A^{k-1-m}\mathbf{b}f(m)$$

Now

$$y(k) = -\tfrac{1}{8}x_1(k) + \tfrac{3}{4}x_2(k) + (\tfrac{1}{2})^k$$

can be written in matrix form as

$$y(k) = \left[-\frac{1}{8}, \frac{3}{4}\right]\begin{bmatrix} x_1(k) \\ x_2(k) \end{bmatrix} + \left(\frac{1}{2}\right)^k = C\mathbf{x}(k) + \left(\frac{1}{2}\right)^k$$

where $C = [-\frac{1}{8}, \frac{3}{4}]$. Thus

$$y(k) = C\left[\sum_{m=0}^{k-1} A^{k-1-m}\mathbf{b}f(m)\right] + \left(\frac{1}{2}\right)^k$$

$$= \sum_{m=0}^{k-1} CA^{k-1-m}\mathbf{b}\left(\frac{1}{2}\right)^m + \left(\frac{1}{2}\right)^k$$

Now

$$CA^{k-1-m}\mathbf{b} = \left[-\frac{1}{8} \;\; \frac{3}{4}\right]\left\{\left(\frac{1}{2}\right)^{k-1-m}\begin{bmatrix} -1 & 4 \\ -\frac{1}{2} & 2 \end{bmatrix} + \left(\frac{1}{4}\right)^{k-1-m}\begin{bmatrix} 2 & -4 \\ \frac{1}{2} & -1 \end{bmatrix}\right\}\begin{bmatrix} 0 \\ 1 \end{bmatrix}$$

$$= 2(\tfrac{1}{2})^{k-m} - (\tfrac{1}{4})^{k-m}$$

Hence

$$y(k) = \sum_{m=0}^{k-1}\left[2\left(\frac{1}{2}\right)^{k-m} - \left(\frac{1}{4}\right)^{k-m}\right]\left(\frac{1}{2}\right)^m + \left(\frac{1}{2}\right)^k$$

$$= \sum_{m=0}^{k-1}\left[2\left(\frac{1}{2}\right)^k - \left(\frac{1}{4}\right)^k(2)^m\right] + \left(\frac{1}{2}\right)^k$$

$$= 2\left(\frac{1}{2}\right)^k\sum_{m=0}^{k-1} 1 - \left(\frac{1}{4}\right)^k\sum_{m=0}^{k-1} 2^m + \left(\frac{1}{2}\right)^k$$

$$= 2\left(\frac{1}{2}\right)^k k - \left(\frac{1}{4}\right)^k\left(\frac{1-2^k}{1-2}\right) + \left(\frac{1}{2}\right)^k$$

$$= 2k(\tfrac{1}{2})^k + (\tfrac{1}{4})^k$$

where we used the geometric series summation formula $\sum_{m=0}^{n} r^m = 1 - r^{n+1}/1 - r$.

*Differential Equations [*For readers who have studied calculus.*]

PROBLEM 9-11 Solve the simplest first-order differential equation

$$x'(t) = ax(t) \tag{a}$$

with the initial condition $x(0) = c$.

Solution Let $x(t) = \alpha e^{at}$, where α is an arbitrary constant. Now

$$x'(t) = \frac{d}{dt}(\alpha e^{at}) = \alpha\frac{d}{dt}(e^{at})$$

$$= \alpha(ae^{at}) = a(\alpha e^{at}) = ax(t)$$

Thus $x(t) = \alpha e^{at}$ satisfies equation (a).

Setting $t = 0$, we get

$$x(0) = c = \alpha e^0 = \alpha$$

Hence

$$x(t) = ce^{at} \tag{b}$$

is the solution of equation (a) satisfying the given initial condition. Note that equation (b) can be rewritten as

$$x(t) = e^{at}x(0) \tag{c}$$

[cf. (9.38)].

PROBLEM 9-12 Solve the second-order differential equation

$$y''(t) + 2y'(t) + 2y(t) = 0$$

with $y(0) = 0$, $y'(0) = 1$

Solution Let

$$x_1(t) = y(t)$$
$$x_2(t) = y'(t)$$

Then

$$x_1'(t) = y'(t) = x_2(t)$$
$$x_2'(t) = y''(t) = -2y(t) - 2y'(t)$$
$$= -2x_1(t) - 2x_2(t)$$

Writing in matrix form, we get

$$\begin{bmatrix} x_1'(t) \\ x_2'(t) \end{bmatrix} = \begin{bmatrix} 0 & 1 \\ -2 & -2 \end{bmatrix} \begin{bmatrix} x_1(t) \\ x_2(t) \end{bmatrix}$$

or

$$\mathbf{x}'(t) = A\mathbf{x}(t)$$

with

$$\mathbf{x}(0) = \begin{bmatrix} x_1(0) \\ x_2(0) \end{bmatrix} = \begin{bmatrix} y(0) \\ y'(0) \end{bmatrix} = \begin{bmatrix} 0 \\ 1 \end{bmatrix}$$

Thus by (9.38), $\mathbf{x}(t) = e^{At}\mathbf{x}(0)$, and we need to find e^{At}.

The characteristic equation of A is

$$|\lambda I_2 - A| = \begin{vmatrix} \lambda & -1 \\ 2 & \lambda + 2 \end{vmatrix} = \lambda(\lambda + 2) + 2 = \lambda^2 + 2\lambda + 2 = 0$$

Thus the eigenvalues of A are $\lambda_1 = -1 + i$ and $\lambda_2 = -1 - i$. Now

$$e^{At} = \alpha_1 A + \alpha_0 I_2$$

and

$$e^{\lambda_k t} = \alpha_1 \lambda_k + \alpha_0 \qquad (k = 1, 2)$$

Thus

$$e^{(-1+i)t} = (-1 + i)\alpha_1 + \alpha_0$$
$$e^{(-1-i)t} = (-1 - i)\alpha_1 + \alpha_0$$

Solving for α_1 and α_0, we get

$$\alpha_1 = e^{-t}\left(\frac{e^{it} - e^{-it}}{2i}\right) = e^{-t}\sin t$$

$$\alpha_0 = e^{-t}\left(\frac{e^{it} + e^{-it}}{2}\right) + \alpha_1 = e^{-t}(\cos t + \sin t)$$

Thus

$$e^{At} = e^{-t}\sin t \begin{bmatrix} 0 & 1 \\ -2 & -2 \end{bmatrix} + e^{-t}(\cos t + \sin t) \begin{bmatrix} 1 & 0 \\ 0 & 1 \end{bmatrix}$$

$$= \begin{bmatrix} e^{-t}(\cos t + \sin t) & e^{-t}\sin t \\ -2e^{-t}\sin t & e^{-t}(\cos t - \sin t) \end{bmatrix}$$

or

$$e^{At} = e^{-t}\begin{bmatrix} \cos t + \sin t & \sin t \\ -2\sin t & \cos t - \sin t \end{bmatrix}$$

So

$$\mathbf{x}(t) = e^{At}\mathbf{x}(0) = e^{-t}\begin{bmatrix} \cos t + \sin t & \sin t \\ -2\sin t & \cos t - \sin t \end{bmatrix}\begin{bmatrix} 0 \\ 1 \end{bmatrix}$$

$$= e^{-t}\begin{bmatrix} \sin t \\ \cos t - \sin t \end{bmatrix} = \begin{bmatrix} e^{-t}\sin t \\ e^{-t}(\cos t - \sin t) \end{bmatrix}$$

Since $y(t) = x_1(t)$, we get $y(t) = e^{-t}\sin t$.

PROBLEM 9-13

(a) Under what condition on two square matrices A and B is it true that $e^{A+B} = e^A e^B$?
(b) Using the result of part **(a)**, show that $(e^A)^{-1} = e^{-A}$.

Solution

(a) By (9.34) we have

$$e^A = \sum_{k=0}^{\infty} \frac{1}{k!}A^k, \qquad e^B = \sum_{k=0}^{\infty} \frac{1}{k!}B^k$$

Then

$$\begin{aligned} e^A e^B &= (I + A + \tfrac{1}{2}A^2 + \tfrac{1}{6}A^3 + \cdots)(I + B + \tfrac{1}{2}B^2 + \tfrac{1}{6}B^3 + \cdots) \qquad \text{(a)} \\ &= I + (A + B) + (\tfrac{1}{2}A^2 + AB + \tfrac{1}{2}B^2) + (\tfrac{1}{6}A^3 + \tfrac{1}{2}A^2B + \tfrac{1}{2}AB^2 + \tfrac{1}{6}B^3) + \cdots \end{aligned}$$

and

$$\begin{aligned} e^{A+B} &= I + (A + B) + \tfrac{1}{2}(A + B)^2 + \tfrac{1}{6}(A + B)^3 + \cdots \qquad \text{(b)} \\ &= I + (A + B) + \tfrac{1}{2}(A^2 + AB + BA + B^2) \\ &\quad + \tfrac{1}{6}(A^3 + A^2B + ABA + AB^2 + BA^2 + BAB + B^2A + B^3) + \cdots \end{aligned}$$

Equations (a) and (b) will be identical if and only if $AB = BA$. Thus $e^{A+B} = e^A e^B$ if and only if A and B commute.

(b) By (9.34) we have

$$e^{-A} = \sum_{k=0}^{\infty} \frac{1}{k!}(-A)^k = \sum_{k=0}^{\infty} \frac{(-1)^k}{k!}A^k$$

and

$$A^m e^{-A} = A^m\left(\sum_{k=0}^{\infty} \frac{(-1)^k}{k!}A^k\right) = \sum_{k=0}^{\infty} \frac{(-1)^k}{k!}A^{k+m} = \left(\sum_{k=0}^{\infty} \frac{(-1)^k}{k!}A^k\right)A^m = e^{-A}A^m$$

for any m. Thus A^m and e^{-A} commute and so e^A and e^{-A} also commute.

Then, using the result of part **(a)**, we have

$$e^{-A}e^A = e^A e^{-A} = e^{A-A} = e^O = I$$

Hence by definition (1.37),

$$e^{-A} = (e^A)^{-1}$$

PROBLEM 9-14 Consider a nonhomogeneous matrix differential equation

$$\mathbf{x}'(t) = A\mathbf{x}(t) + \mathbf{b}f(t) \qquad \text{(a)}$$

Show that, given $\mathbf{x}(0)$, the solution of equation (a) is given by

$$\mathbf{x}(t) = e^{At}\mathbf{x}(0) + \int_0^t e^{A(t-\tau)}\mathbf{b}f(\tau)\,dt \qquad \text{(b)}$$

where

$$\int \mathbf{v}(\tau)\,d\tau = \begin{bmatrix} \int v_1(\tau)\,d\tau \\ \int v_2(\tau)\,d\tau \\ \vdots \\ \int v_n(\tau)\,d\tau \end{bmatrix} \qquad \text{if } \mathbf{v}(\tau) = \begin{bmatrix} v_1(\tau) \\ v_2(\tau) \\ \vdots \\ v_n(\tau) \end{bmatrix}$$

Solution The complete solution of equation (a) consists of a homogeneous solution $e^{At}\mathbf{x}(0)$ and a particular solution $\mathbf{x}_p(t)$.

Assume that $\mathbf{x}_p(t)$ is of the form

$$\mathbf{x}_p(t) = e^{At}\mathbf{q}(t) \tag{c}$$

where $\mathbf{q}(t)$ is an unknown function to be determined. Substituting equation (c) into equation (a), we get

$$\mathbf{x}_p'(t) = A\mathbf{x}_p(t) + \mathbf{b}f(t)$$

Now

$$\begin{aligned} \mathbf{x}_p'(t) = \frac{d}{dt}\mathbf{x}_p(t) &= \frac{d}{dt}[e^{At}\mathbf{q}(t)] \\ &= \frac{d}{dt}(e^{At})\mathbf{q}(t) + e^{At}\frac{d}{dt}\mathbf{q}(t) \\ &= Ae^{At}\mathbf{q}(t) + e^{At}\mathbf{q}'(t) \end{aligned}$$

So

$$Ae^{At}\mathbf{q}(t) + e^{At}\mathbf{q}'(t) = Ae^{At}\mathbf{q}(t) + \mathbf{b}f(t)$$

Thus

$$e^{At}\mathbf{q}'(t) = \mathbf{b}f(t)$$

Multiplying both sides by e^{-At} from the left (see Problem 9-13b), we get

$$\mathbf{q}'(t) = e^{-At}\mathbf{b}f(t)$$

Integrating, we obtain

$$\mathbf{q}(t) = \mathbf{q}(0) + \int_0^t e^{-A\tau}\mathbf{b}f(\tau)\,d\tau$$

Thus, the particular solution $\mathbf{x}_p(t)$ is

$$\mathbf{x}_p(t) = e^{At}\mathbf{q}(t) = e^{At}\mathbf{q}(0) + \int_0^t e^{A(t-\tau)}\mathbf{b}f(\tau)\,d\tau$$

Hence the complete solution $\mathbf{x}(t)$ of equation (a) is

$$\begin{aligned} \mathbf{x}(t) &= e^{At}\mathbf{x}(0) + \mathbf{x}_p(t) \\ &= e^{At}\mathbf{x}(0) + e^{At}\mathbf{q}(0) + \int_0^t e^{A(t-\tau)}\mathbf{b}f(\tau)\,d\tau \end{aligned}$$

To evaluate $\mathbf{q}(0)$ we set $t = 0$ in the above expression,

$$\begin{aligned} \mathbf{x}(0) &= e^{O}\mathbf{x}(0) + e^{O}\mathbf{q}(0) + \int_0^0 e^{A(0-\tau)}\mathbf{b}f(\tau)\,d\tau \\ &= \mathbf{x}(0) + \mathbf{q}(0) \end{aligned}$$

since $e^{O} = I$. Hence $\mathbf{q}(0) = \mathbf{0}$. Thus, finally, the complete solution of equation (a) is

$$\mathbf{x}(t) = e^{At}\mathbf{x}(0) + \int_0^t e^{A(t-\tau)}\mathbf{b}f(\tau)\,d\tau$$

PROBLEM 9-15 Solve the nonhomogeneous second-order differential equation

$$y''(t) + 3y'(t) + 2y(t) = u(t)$$

where $u(t)$ is a unit step function defined by

$$u(t) = \begin{cases} 1 & t > 0 \\ 0 & t < 0 \end{cases}$$

with the initial conditions $y(0) = y'(0) = 0$.

Solution Let $x_1(t) = y(t)$, $x_2(t) = y'(t)$, then, following the steps of Example 9-13, we have

$$\begin{bmatrix} x_1'(t) \\ x_2'(t) \end{bmatrix} = \begin{bmatrix} 0 & 1 \\ -2 & -3 \end{bmatrix} \begin{bmatrix} x_1(t) \\ x_2(t) \end{bmatrix} + \begin{bmatrix} 0 \\ 1 \end{bmatrix} u(t)$$

or

$$\mathbf{x}'(t) = A\mathbf{x}(t) + b\mathbf{u}(t)$$

Thus, by the result of Problem 9-14,

$$\mathbf{x}(t) = e^{At}\mathbf{x}(0) + \int_0^t e^{A(t-\tau)}\mathbf{b}u(\tau)\,d\tau$$

$$= \int_0^t e^{A(t-\tau)}\mathbf{b}u(\tau)\,d\tau$$

since $\mathbf{x}(0) = \mathbf{0}$. Now, from Example 9-13,

$$e^{At} = \begin{bmatrix} 2e^{-t} - e^{-2t} & e^{-t} - e^{-2t} \\ -2e^{-t} + 2e^{-2t} & -e^{-t} + 2e^{-2t} \end{bmatrix}$$

Thus

$$e^{A(t-\tau)}\mathbf{b} = \begin{bmatrix} 2e^{-(t-\tau)} - e^{-2(t-\tau)} & e^{-(t-\tau)} - e^{-2(t-\tau)} \\ -2e^{-(t-\tau)} + 2e^{-2(t-\tau)} & -e^{-(t-\tau)} + 2e^{-2(t-\tau)} \end{bmatrix} \begin{bmatrix} 0 \\ 1 \end{bmatrix}$$

$$= \begin{bmatrix} e^{-(t-\tau)} - e^{-2(t-\tau)} \\ -e^{-(t-\tau)} + 2e^{-2(t-\tau)} \end{bmatrix}$$

So

$$\mathbf{x}(t) = \int_0^t e^{A(t-\tau)}\mathbf{b}u(\tau)\,d\tau$$

$$= \begin{bmatrix} \displaystyle\int_0^t [e^{-(t-\tau)} - e^{-2(t-\tau)}]\,d\tau \\ \displaystyle\int_0^t [-e^{-(t-\tau)} + 2e^{-2(t-\tau)}]\,d\tau \end{bmatrix} = \begin{bmatrix} \frac{1}{2} - e^{-t} + \frac{1}{2}e^{-2t} \\ e^{-t} - 2^{-2t} \end{bmatrix}$$

Since $y(t) = x_1(t)$, we get $y(t) = \frac{1}{2} - e^{-t} + \frac{1}{2}e^{-2t}$.

Supplementary Exercises

PROBLEM 9-16 A transportation survey shows that each year 25% of train commuters switch to private autos while 40% of those driving autos switch to the train. The train line has 70% of all commuters at the present time. Find the transition matrix and the distribution after 1 and 2 years.

Answer $A = \begin{bmatrix} 0.75 & 0.4 \\ 0.25 & 0.6 \end{bmatrix}$, $\begin{bmatrix} 0.645 \\ 0.355 \end{bmatrix}$, $\begin{bmatrix} 0.626 \\ 0.374 \end{bmatrix}$

PROBLEM 9-17 Find the steady-state distribution vector for the transition matrix of Problem 9-16.

Answer $\begin{bmatrix} 0.615 \\ 0.385 \end{bmatrix}$

PROBLEM 9-18 Let

$$A = \begin{bmatrix} 0.75 & 0.4 \\ 0.25 & 0.6 \end{bmatrix}$$

Find A^k and show that each column of A^k becomes the steady-state vector of Problem 9-17 as $k \to \infty$.

Answer $A^k = \begin{bmatrix} \frac{8}{13} + \frac{5}{13}\left(\frac{7}{20}\right)^k & \frac{8}{13} - \frac{8}{13}\left(\frac{7}{20}\right)^k \\ \frac{5}{13} - \frac{5}{13}\left(\frac{7}{20}\right)^k & \frac{5}{13} + \frac{8}{13}\left(\frac{7}{20}\right)^k \end{bmatrix}$

PROBLEM 9-19 Let p and q be real numbers with $0 < p < 1$ and $0 < q < 1$. Show that any 2×2 transition matrix will have the form

$$A = \begin{bmatrix} 1 - p & q \\ p & 1 - q \end{bmatrix}$$

Also, find the steady-state vector for A. Discuss the case of $p = q$.

Answer

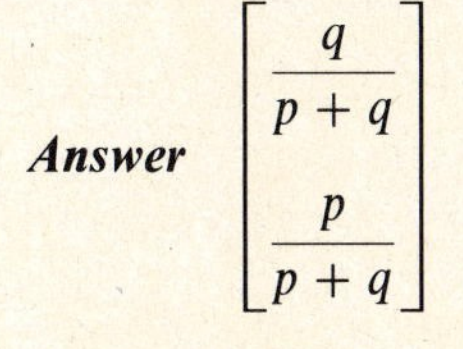

$$\begin{bmatrix} \dfrac{q}{p+q} \\ \dfrac{p}{p+q} \end{bmatrix}$$

PROBLEM 9-20 Solve

$$y(k) - 2y(k - 1) = k$$

with $y(-1) = 0$.

Answer $y(k) = 2(2)^k - 2 - k$

PROBLEM 9-21 Solve

$$y(k) - 7y(k - 1) + 10y(k - 2) = 0$$

with $y(-1) = 16$, $y(-2) = 5$.

Answer $y(k) = 12(2)^k + 50(5)^k$

PROBLEM 9-22 Solve

$$y(k + 2) - 5y(k + 1) - 6y(k) = 0$$

with $y(0) = 3$, $y(1) = 11$.

Answer $y(k) = 2(6)^k - (-1)^k$

PROBLEM 9-23 Solve

$$y''(t) + 3y'(t) + 2y(t) = 0$$

with $y(0) = y'(0) = 1$.

Answer $y(t) = 3e^{-t} - 2e^{-2t}$

PROBLEM 9-24 Solve

$$y''(t) + 3y'(t) + 2y(t) = e^{-t}$$

with $y(0) = y'(0) = 0$.

Answer $y(t) = e^{-2t} + (-1 + t)e^{-t}$

FINAL EXAM

1. For which values of λ does the following system of equations have nontrivial solutions?

$$(\lambda - 3)x + \quad y = 0$$
$$x + (\lambda - 3)y = 0$$

2. Show that the vectors x^2, $x^2 - 1$, $x + 2$ span the vector space P_2.

3. Let R^4 have the Euclidean inner product and

$$\mathbf{u} = \begin{bmatrix} 1 \\ k \\ 1 \\ k \end{bmatrix}, \qquad \mathbf{v} = \begin{bmatrix} 6 \\ 5 \\ 0 \\ k \end{bmatrix}$$

For which values of k are $\mathbf{u}$ and $\mathbf{v}$ orthogonal?

4. Let $T: P_1 \to P_2$ be the linear transformation defined by $T(p(x)) = xp(x)$. (**a**) Find the matrix of T relative to the bases $\mathscr{B} = \{\mathbf{u}_1, \mathbf{u}_2\}$ and $\mathscr{B}' = \{\mathbf{v}_1, \mathbf{v}_2, \mathbf{v}_3\}$, where $\mathbf{u}_1 = 1, \mathbf{u}_2 = x, \mathbf{v}_1 = 1, \mathbf{v}_2 = x, \mathbf{v}_3 = x^2$. (**b**) Use the matrix obtained in (**a**) to compute $T\mathbf{p}$ where $\mathbf{p} = 2 - x$.

5. Let $T: R^2 \to R^2$ be defined by

$$T\left(\begin{bmatrix} x_1 \\ x_2 \end{bmatrix}\right) = \begin{bmatrix} -2x_1 + 4x_2 \\ x_1 + \ x_2 \end{bmatrix}$$

Find (**a**) the standard matrix for T and (**b**) a basis $\mathscr{B}' = \{\mathbf{u}_1, \mathbf{u}_2\}$ for R^2 relative to which the matrix of T is diagonal.

Solutions to Final Exam

1. The given homogeneous system of equations can be expressed in a matrix form as

$$\begin{bmatrix} \lambda - 3 & 1 \\ 1 & \lambda - 3 \end{bmatrix} \begin{bmatrix} x \\ y \end{bmatrix} = \begin{bmatrix} 0 \\ 0 \end{bmatrix}$$

For the above system to have nontrivial solutions, the determinant of the coefficient matrix must be zero, i.e.,

$$\begin{vmatrix} \lambda - 3 & 1 \\ 1 & \lambda - 3 \end{vmatrix} = 0$$

which reduces to

$$\begin{aligned}(\lambda - 3)^2 - 1 &= \lambda^2 - 6\lambda + 8\\ &= (\lambda - 2)(\lambda - 4)\\ &= 0\end{aligned}$$

Thus, the values of λ for which the given system has nontrivial solutions are $\lambda = 2$ and $\lambda = 4$.

2. Let a, b, c be scalars such that

$$ax^2 + b(x^2 - 1) + c(x + 2) = 0$$

or

$$(a + b)x^2 + cx + (-b + 2c) = 0$$

If this is valid for all values of x, we must have

$$a + b = 0, \qquad c = 0, \qquad -b + 2c = 0$$

which implies $a = b = c = 0$. Thus the given vectors x^2, $x^2 - 1$, $x + 2$ in the vector space P_2 are linearly independent. Since $\dim P_2 = 3$, the given vectors span the vector space P_2.

3. If $\mathbf{u}$ and $\mathbf{v}$ are orthogonal, then $\langle \mathbf{u}, \mathbf{v} \rangle = 0$. Thus

$$(1)(6) + (k)(5) + (1)(0) + (k)(k) = 0$$

or

$$k^2 + 5k + 6 = (k + 2)(k + 3) = 0$$

So, $k = -2$ and $k = -3$.

4. **(a)** From the formula for T we have

$$T(\mathbf{u}_1) = T(1) = x(1) = x$$

$$T(\mathbf{u}_2) = T(x) = x(x) = x^2$$

By inspection, the coordinate vectors for $T(\mathbf{u}_1)$ and $T(\mathbf{u}_2)$ relative to $\mathscr{B}'$ are

$$[T(\mathbf{u}_1)]_{\mathscr{B}'} = \begin{bmatrix} 0 \\ 1 \\ 0 \end{bmatrix}, \qquad [T(\mathbf{u}_2)]_{\mathscr{B}'} = \begin{bmatrix} 0 \\ 0 \\ 1 \end{bmatrix}$$

Thus, the matrix of T relative to $\mathscr{B}$ and $\mathscr{B}'$ is

$$A = \begin{bmatrix} 0 & 0 \\ 1 & 0 \\ 0 & 1 \end{bmatrix}$$

(b) By inspection, the coordinate vector of $\mathbf{p} = 2 - x$ relative to $\mathscr{B}$ is

$$[\mathbf{p}]_{\mathscr{B}} = \begin{bmatrix} 2 \\ -1 \end{bmatrix}$$

Therefore

$$[T(\mathbf{p})]_{\mathscr{B}'} = A[\mathbf{p}]_{\mathscr{B}} = \begin{bmatrix} 0 & 0 \\ 1 & 0 \\ 0 & 1 \end{bmatrix} \begin{bmatrix} 2 \\ -1 \end{bmatrix} = \begin{bmatrix} 0 \\ 2 \\ -1 \end{bmatrix}$$

Hence

$$\begin{aligned} T(\mathbf{p}) &= 0\mathbf{v}_1 + 2\mathbf{v}_2 + (-1)\mathbf{v}_3 \\ &= 0(1) + 2(x) - (x^2) \\ &= 2x - x^2 \end{aligned}$$

5. **(a)** Let $\mathscr{B} = \{\mathbf{e}_1, \mathbf{e}_2\}$ be the standard basis for R^2. Then, from the formula for T, we have

$$T(\mathbf{e}_1) = T\left(\begin{bmatrix} 1 \\ 0 \end{bmatrix}\right) = \begin{bmatrix} -2 \\ 1 \end{bmatrix}, \qquad T(\mathbf{e}_2) = T\left(\begin{bmatrix} 0 \\ 1 \end{bmatrix}\right) = \begin{bmatrix} 4 \\ 1 \end{bmatrix}$$

so that the standard matrix for T is

$$A = \begin{bmatrix} -2 & 4 \\ 1 & 1 \end{bmatrix}$$

(b) We know that $\mathbf{u}_1$ and $\mathbf{u}_2$ should be the eigenvectors of T, or equivalently, the eigenvectors of A obtained in **(a)**. The characteristic equation of A is

$$|\lambda I_2 - A| = \begin{vmatrix} \lambda + 2 & -4 \\ -1 & \lambda - 1 \end{vmatrix} = (\lambda + 2)(\lambda - 1) - 4 = (\lambda - 2)(\lambda + 3) = 0$$

Thus the eigenvalues of A are $\lambda = 2$ and $\lambda = -3$.

Let $\mathbf{x} = \begin{bmatrix} x_1 \\ x_2 \end{bmatrix}$ be an eigenvector of A corresponding to λ. Then $(\lambda I - A)\mathbf{x} = \mathbf{0}$, that is,

$$\begin{bmatrix} \lambda + 2 & -4 \\ -1 & \lambda - 1 \end{bmatrix} \begin{bmatrix} x_1 \\ x_2 \end{bmatrix} = \begin{bmatrix} 0 \\ 0 \end{bmatrix}$$

If $\lambda = 2$, we have

$$\begin{bmatrix} 4 & -4 \\ -1 & 1 \end{bmatrix} \begin{bmatrix} x_1 \\ x_2 \end{bmatrix} = \begin{bmatrix} 0 \\ 0 \end{bmatrix}$$

or

$$\begin{aligned} 4x_1 - 4x_2 &= 0 \\ -x_1 + x_2 &= 0 \end{aligned}$$

Thus, we get $x_1 = x_2$. So $\begin{bmatrix} 1 \\ 1 \end{bmatrix}$ is an eigenvector of A corresponding to $\lambda = 2$.

Similarly, if $\lambda = -3$, we have

$$\begin{bmatrix} -1 & -4 \\ -1 & -4 \end{bmatrix} \begin{bmatrix} x_1 \\ x_2 \end{bmatrix} = \begin{bmatrix} 0 \\ 0 \end{bmatrix}$$

or

$$-x_1 - 4x_2 = 0, \qquad \text{so } x_1 = -4x_2$$

Thus $\begin{bmatrix} -4 \\ 1 \end{bmatrix}$ is an eigenvector of A corresponding to $\lambda = -3$. Hence the desired basis is given by $\mathscr{B}' = \{\mathbf{u}_1, \mathbf{u}_2\}$ where

$$\mathbf{u}_1 = \begin{bmatrix} 1 \\ 1 \end{bmatrix}, \qquad \mathbf{u}_2 = \begin{bmatrix} -4 \\ 1 \end{bmatrix}$$

INDEX